T0198104

3G EVOLUTION: HSPA AND LTE FOR MOBILE BROADBAND

3G Evolution

HSPA and LTE for Mobile Broadband

Second edition

Erik Dahlman, Stefan Parkvall, Johan Sköld and Per Beming

AMSTERDAM • BOSTON • HEIDELBERG • LONDON • NEW YORK • OXFORD
PARIS • SAN DIEGO • SAN FRANCISCO • SINGAPORE • SYDNEY • TOKYO

Academic Press is an imprint of Elsevier

ELSEVIER

Academic Press is an imprint of Elsevier
Linacre House, Jordan Hill, Oxford, OX2 8DP
30 Corporate Drive, Burlington, MA 01803

First edition 2007
Second edition 2008

British Library Cataloguing in Publication Data
3G evolution : HSPA and LTE for mobile broadband. – 2nd ed.
  1. Broadband communication systems – Standards   2. Mobile communication
  systems – Standards   3. Cellular telephone systems – Standards
  I. Dahlman, Erik
  621.3'8546

Library of Congress Control Number: 2008931278

ISBN: 978-0-12-374538-5

For information on all Academic Press publications
visit our website at elsevierdirect.com

Typeset by Charon Tec Ltd., A Macmillan Company. (www.macmillansolutions.com)

Printed and bound in Great Britain by MPG Books Ltd, Bodmin, Cornwall

08 09 10 11   11 10 9 8 7 6 5 4 3 2 1

Contents

Part II: Technologies for 3G Evolution 27

List of Figures

List of Tables

Preface

During the past years, there has been a quickly rising interest in radio access technologies for providing mobile as well as nomadic and fixed services for voice, video, and data. The difference in design, implementation, and use between telecom and datacom technologies is also getting more blurred. One example is cellular technologies from the telecom world being used for broadband data and wireless LAN from the datacom world being used for voice over IP.

Today, the most widespread radio access technology for mobile communication is digital cellular, with the number of users passing 3 billion by 2007, which is almost half of the world's population. It has emerged from early deployments of an expensive voice service for a few car-borne users, to today's widespread use of third generation mobile-communication devices that provide a range of mobile services and often include camera, MP3 player, and PDA functions. With this widespread use and increasing interest in 3G, a continuing evolution ahead is foreseen.

This book describes the evolution of 3G digital cellular into an advanced broadband mobile access. The focus of this book is on the evolution of the 3G mobile communication as developed in the 3GPP (*Third Generation Partnership Project*) standardization, looking at the radio access and access network evolution.

This book is divided into five parts. Part I gives the background to 3G and its evolution, looking also at the different standards bodies and organizations involved in the process of defining 3G. It is followed by a discussion of the reasons and driving forces behind the 3G evolution. Part II gives a deeper insight into some of the technologies that are included, or are expected to be included as part of the 3G evolution. Because of its generic nature, Part II can be used as a background not only for the evolution steps taken in 3GPP as described in this book, but also for readers that want to understand the technology behind other systems, such as WiMAX and CDMA2000.

Part III describes the evolution of 3G WCDMA into *High Speed Packet Access* (HSPA). It gives an overview of the key features of HSPA and its continued evolution in the context of the technologies from Part II. Following this, the different uplink and downlink components are outlined and finally more detailed descriptions of how they work together are given.

Part IV introduces the *Long Term Evolution* (LTE) and *System Architecture Evolution* (SAE). As a start, the agreed requirements and objectives for LTE are described. This is followed by an introductory technical overview of LTE, where the most important technology components are introduced, also here, based on the generic technologies given in Part II. As a second step, a more detailed description of the protocol structure is given, with further details on the uplink and downlink transmission schemes and procedures, access procedures and flexible bandwidth operation. The system architecture evolution, applicable to both LTE and HSPA, is given with details of radio access network and core network. The ongoing work on LTE-Advanced is also presented.

Finally in Part V, an assessment is made on the 3G evolution. An evaluation of the performance puts the 3G evolution tracks in relation to the targets set in 3GPP. Through an overview of similar technologies developed in other standards bodies, it will be clear that the technologies adopted for the evolution in 3GPP are implemented in many other systems as well. Finally, looking into the future, it will be seen that the 3G evolution does not stop with the HSPA Evolution and LTE.

Acknowledgements

We thank all our colleagues at Ericsson for assisting in this project by helping with contributions to the book, giving suggestions and comments on the contents, and taking part in the huge team effort of developing HSPA and LTE.

The standardization process for 3G evolution involves people from all parts of the world, and we acknowledge the efforts of our colleagues in the wireless industry in general and in 3GPP RAN in particular. Without their work and contributions to the standardization, this book would not have been possible.

Finally, we are immensely grateful to our families for bearing with us and supporting us during the long process of writing this book.

List of Acronyms

3GPP	Third Generation Partnership Project
AAS	Adaptive Antenna System
ACK	Acknowledgement (in ARQ protocols)
ACK-CH	Acknowledgement Channel (for WiMAX)
ACLR	Adjacent Channel Leakage Ratio
ACS	Adjacent Channel Selectivity
ACIR	Adjacent Channel Interference Ratio
ACTS	Advanced Communications Technology and Services
AM	Acknowledged Mode (RLC configuration)
AMC	Adaptive Modulation and Coding
AMPR	Additional Maximum Power Reduction
AMPS	Advanced Mobile Phone System
AMR-WB	Adaptive MultiRate-WideBand
AP	Access Point
ARIB	Association of Radio Industries and Businesses
ARQ	Automatic Repeat-reQuest
ATDMA	Advanced Time Division Mobile Access
ATIS	Alliance for Telecommunications Industry Solutions
AWGN	Additive White Gaussian Noise
BCCH	Broadcast Control Channel
BCH	Broadcast Channel
BE	Best Effort Service
BER	Bit-Error Rate
BLER	Block-Error Rate
BM-SC	Broadcast Multicast Service Center
BPSK	Binary Phase-Shift Keying
BS	Base Station
BSC	Base Station Controller
BTC	Block Turbo Code
BTS	Base Transceiver Station
CC	Convolutional Code
CCCH	Common Control Channel
CCE	Control Channel Element
CCSA	China Communications Standards Association

CDD	Cyclic-Delay Diversity
CDF	Cumulative Density Function
CDM	Code-Division Multiplexing
CDMA	Code Division Multiple Access
CEPT	European Conference of Postal and Telecommunications Administrations
CN	Core Network
CODIT	Code-Division Test bed
CP	Cyclic Prefix
CPC	Continuous Packet Connectivity
CPICH	Common Pilot Channel
CQI	Channel-Quality Indicator
CQICH	Channel Quality Indication Channel (for WiMAX)
CRC	Cyclic Redundancy Check
C-RNTI	Cell Radio-Network Temporary Identifier
CS	Circuit Switched
CTC	Convolutional Turbo Code
CW	Continuous Wave
DCCH	Dedicated Control Channel
DCH	Dedicated Channel
DCI	Downlink Control Information
DFE	Decision-Feedback Equalization
DFT	Discrete Fourier Transform
DFTS-OFDM	DFT-spread OFDM, see also SC-FDMA
DL	Downlink
DL-SCH	Downlink Shared Channel
DPCCH	Dedicated Physical Control Channel
DPCH	Dedicated Physical Channel
DPDCH	Dedicated Physical Data Channel
DRS	Demodulation Reference Signal
DRX	Discontinuous Reception
DTCH	Dedicated Traffic Channel
DTX	Discontinuous Transmission
D-TxAA	Dual Transmit-Diversity Adaptive Array
DwPTS	The downlink part of the special subframe (for TDD operation).
E-AGCH	E-DCH Absolute Grant Channel
E-DCH	Enhanced Dedicated Channel
EDGE	Enhanced Data rates for GSM Evolution and Enhanced Data rates for Global Evolution

E-DPCCH	E-DCH Dedicated Physical Control Channel
E-DPDCH	E-DCH Dedicated Physical Data Channel
E-HICH	E-DCH Hybrid ARQ Indicator Channel
eNodeB	E-UTRAN NodeB
EPC	Evolved Packet Core
E-RGCH	E-DCH Relative Grant Channel
ErtPS	Extended Real-Time Polling Service
E-TFC	E-DCH Transport Format Combination
E-TFCI	E-DCH Transport Format Combination Index
ETSI	European Telecommunications Standards Institute
E-UTRA	Evolved UTRA
E-UTRAN	Evolved UTRAN
EV-DO	Evolution-Data Optimized (of CDMA2000 1x)
EV-DV	Evolution-Data and Voice (of CDMA2000 1x)
EVM	Error Vector Magnitude
FACH	Forward Access Channel
FBSS	Fast Base-Station Switching
FCC	Federal Communications Commission
FCH	Frame Control Header (for WiMAX)
FDD	Frequency Division Duplex
FDM	Frequency-Division Multiplex
FDMA	Frequency-Division Multiple Access
F-DPCH	Fractional DPCH
FEC	Forward Error Correction
FFT	Fast Fourier Transform
FIR	Finite Impulse Response
F-OSICH	Forward link Other Sector Indication Channel (for IEEE 802.20)
FPLMTS	Future Public Land Mobile Telecommunications Systems
FRAMES	Future Radio Wideband Multiple Access Systems
FTP	File Transfer Protocol
FUSC	Fully Used Subcarriers (for WiMAX)
FSTD	Frequency Shift Transmit Diversity
GERAN	GSM/EDGE Radio Access Network
GGSN	Gateway GPRS Support Node
GP	Guard Period (for TDD operation)
GPRS	General Packet Radio Services
GPS	Global Positioning System
G-RAKE	Generalized RAKE
GSM	Global System for Mobile communications

HARQ	Hybrid ARQ
HC-SDMA	High Capacity Spatial Division Multiple Access
H-FDD	Half-duplex FDD
HHO	Hard Handover
HLR	Home Location Register
HRPD	High Rate Packet Data
HSDPA	High-Speed Downlink Packet Access
HS-DPCCH	High-Speed Dedicated Physical Control Channel
HS-DSCH	High-Speed Downlink Shared Channel
HSPA	High-Speed Packet Access
HS-PDSCH	High-Speed Physical Downlink Shared Channel
HSS	Home Subscriber Server
HS-SCCH	High-Speed Shared Control Channel
HSUPA	High-Speed Uplink Packet Access
ICIC	Inter-Cell Interference Coordination
ICS	In-Channel Selectivity
IDFT	Inverse DFT
IEEE	Institute of Electrical and Electronics Engineers
IFDMA	Interleaved FDMA
IFFT	Inverse Fast Fourier Transform
IMS	IP Multimedia Subsystem
IMT-2000	International Mobile Telecommunications 2000
IP	Internet Protocol
IPsec	Internet Protocol security
IPv4	IP version 4
IPv6	IP version 6
IR	Incremental Redundancy
IRC	Interference Rejection Combining
ISDN	Integrated Services Digital Network
ITU	International Telecommunications Union
ITU-R	International Telecommunications Union-Radiocommunications Sector
Iu	The interface used for communication between the RNC and the core network.
Iu_cs	The interface used for communication between the RNC and the GSM/WCDMA circuit switched core network.
Iu_ps	The interface used for communication between the RNC and the GSM/WCDMA packet switched core network.
Iub	The interface used for communication between the NodeB and the RNC.
Iur	The interface used for communication between different RNCs.

J-TACS   Japanese Total Access Communication System

LAN   Local Area Network
LCID   Logical Channel Index
LDPC   Low-Density Parity Check Code
LMMSE   Linear Minimum Mean Square Error
LTE   Long-Term Evolution

MAC   Medium Access Control
MAN   Metropolitan Area Network
MAP   Map message (for WiMAX)
MBFDD   Mobile Broadband FDD (for IEEE 802.20)
MBMS   Multimedia Broadcast/Multicast Service
MBS   Multicast and Broadcast Service
MBSFN   Multicast-Broadcast Single Frequency Network
MBTDD   Mobile Broadband TDD (for IEEE 802.20)
MBWA   Mobile Broadband Wireless Access
MCCH   MBMS Control Channel
MC   Multi-Carrier
MCE   MBMS Coordination Entity
MCH   Multicast Channel
MCS   Modulation and Coding Scheme
MDHO   Macro-Diversity Handover
MIB   Master Information Block
MICH   MBMS Indicator Channel
MIMO   Multiple-Input Multiple-Output
ML   Maximum Likelihood
MLD   Maximum Likelihood Detection
MLSE   Maximum-Likelihood Sequence Estimation
MME   Mobility Management Entity
MMS   Multimedia Messaging Service
MMSE   Minimum Mean Square Error
MPR   Maximum Power Reduction
MRC   Maximum Ratio Combining
MSC   Mobile Switching Center
MSCH   MBMS Scheduling Channel
MTCH   MBMS Traffic Channel

NAK   Negative Acknowledgement (in ARQ protocols)
NAS   Non-Access Stratum (a functional layer between the core
                network and the terminal that supports signaling and user
                data transfer)

NMT	*Nordisk MobilTelefon (Nordic Mobile Telephony)*
NodeB	NodeB, a logical node handling transmission/reception in multiple cells. Commonly, but not necessarily, corresponding to a base station.
nrtPS	Non-Real-Time Polling Service
OFDM	Orthogonal Frequency-Division Multiplexing
OFDMA	Orthogonal Frequency-Division Multiple Access
OOB	Out-Of-Band (emissions)
OOK	On–Off Keying
OVSF	Orthogonal Variable Spreading Factor
PAN	Personal Area Network
PAPR	Peak-to-Average Power Ratio
PAR	Peak-to-Average Ratio (same as PAPR)
PARC	Per-Antenna Rate Control
PBCH	Physical Broadcast Channel
PCCH	Paging Control Channel
PCFICH	Physical Control Format Indicator Channel
PCG	Project Coordination Group (in 3GPP)
PCH	Paging Channel
PCI	Pre-coding Control Indication
PCS	Personal Communications Systems
PDC	Personal Digital Cellular
PDCCH	Physical Downlink Control Channel
PDCP	Packet Data Convergence Protocol
PDSCH	Physical Downlink Shared Channel
PDSN	Packet Data Serving Node
PDN	Packet Data Network
PDU	Protocol Data Unit
PF	Proportional Fair (a type of scheduler)
PHICH	Physical Hybrid-ARQ Indicator Channel
PHY	Physical layer
PHS	Personal Handy-phone System
PMCH	Physical Multicast Channel
PMI	Precoding-Matrix Indicator
PoC	Push to Talk over Cellular
PRACH	Physical Random Access Channel
PRB	Physical Resource Block
PS	Packet Switched
PSK	Phase Shift Keying

PSS   Primary Synchronization Signal
PSTN   Public Switched Telephone Networks
PUCCH   Physical Uplink Control Channel
PUSC   Partially Used Subcarriers (for WiMAX)
PUSCH   Physical Uplink Shared Channel

QAM   Quadrature Amplitude Modulation
QoS   Quality-of-Service
QPP   Quadrature Permutation Polynomial
QPSK   Quadrature Phase-Shift Keying

RAB   Radio Access Bearer
RACE   Research and development in Advanced Communications in Europe
RACH   Random Access Channel
RAN   Radio Access Network
RA-RNTI   Random Access RNTI
RAT   Radio Access Technology
RB   Resource Block
RBS   Radio Base Station
RF   Radio Frequency
RI   Rank Indicator
RIT   Radio Interface Technology
RLC   Radio Link Protocol
RNC   Radio Network Controller
RNTI   Radio-Network Temporary Identifier
ROHC   Robust Header Compression
RR   Round-Robin (a type of scheduler)
RRC   Radio Resource Control
RRM   Radio Resource Management
RS   Reference Symbol
RSN   Retransmission Sequence Number
RSPC   IMT-2000 radio interface specifications
RTP   Real Time Protocol
rtPS   Real-Time Polling Service
RTWP   Received Total Wideband Power
RV   Redundancy Version

S1   The interface between eNodeB and the Evolved Packet Core.
SA   System Aspects
SAE   System Architecture Evolution

S-CCPCH	Secondary Common Control Physical Channel
SC-FDMA	Single-Carrier FDMA
SDMA	Spatial Division Multiple Access
SDO	Standards Developing Organization
SDU	Service Data Unit
SEM	Spectrum Emissions Mask
SF	Spreading Factor
SFBC	Space-Frequency Block Coding
SFN	Single-Frequency Network or System Frame Number (in 3GPP)
SFTD	Space–Frequency Time Diversity
SGSN	Serving GPRS Support Node
SI	System Information message
SIB	System Information Block
SIC	Successive Interference Combining
SIM	Subscriber Identity Module
SINR	Signal-to-Interference-and-Noise Ratio
SIR	Signal-to-interference ratio
SMS	Short Message Service
SNR	Signal-to-noise ratio
SOHO	Soft Handover
SR	Scheduling Request
SRNS	Serving Radio Network Subsystem
SRS	Sounding Reference Signal
SSS	Secondary Synchronization Signal
STBC	Space–Time Block Coding
STC	Space–Time Coding
STTD	Space-Time Transmit Diversity
TACS	Total Access Communication System
TCP	Transmission Control Protocol
TC-RNTI	Temporary C-RNTI
TD-CDMA	Time Division-Code Division Multiple Access
TDD	Time Division Duplex
TDM	Time Division Multiplexing
TDMA	Time Division Multiple Access
TD-SCDMA	Time Division-Synchronous Code Division Multiple Access
TF	Transport Format
TFC	Transport Format Combination
TFCI	Transport Format Combination Index
TIA	Telecommunications Industry Association
TM	Transparent Mode (RLC configuration)

TR	Technical Report
TrCH	Transport Channel
TS	Technical Specification
TSG	Technical Specification Group
TSN	Transmission Sequence Number
TTA	Telecommunications Technology Association
TTC	Telecommunications Technology Committee
TTI	Transmission Time Interval
UCI	Uplink Control Information
UE	User Equipment, the 3GPP name for the mobile terminal
UGS	Unsolicited Grant Service
UL	Uplink
UL-SCH	Uplink Shared Channel
UM	Unacknowledged Mode (RLC configuration)
UMB	Ultra Mobile Broadband
UMTS	Universal Mobile Telecommunications System
UpPTS	The uplink part of the special subframe (for TDD operation).
USIM	UMTS SIM
US-TDMA	US Time Division Multiple Access standard
UTRA	Universal Terrestrial Radio Access
UTRAN	Universal Terrestrial Radio Access Network
VRB	Virtual Resource Block
WAN	Wide Area Network
WMAN	Wireless Metropolitan Area Network
WARC	World Administrative Radio Congress
WCDMA	Wideband Code Division Multiple Access
WG	Working Group
WiMAX	Worldwide Interoperability for Microwave Access
WLAN	Wireless Local Area Network
VoIP	Voice-over-IP
WP8F	Working Party 8F
WRC	World Radiocommunication Conference
X2	The interface between eNodeBs.
ZC	Zadoff-Chu
ZF	Zero Forcing
ZTCC	Zero Tailed Convolutional Code

Part I
Introduction

1

Background of 3G evolution

From the first experiments with radio communication by Guglielmo Marconi in the 1890s, the road to truly mobile radio communication has been quite long. To understand the complex 3G mobile-communication systems of today, it is also important to understand where they came from and how cellular systems have evolved from an expensive technology for a few selected individuals to today's global mobile-communication systems used by almost half of the world's population. Developing mobile technologies has also changed, from being a national or regional concern, to becoming a very complex task undertaken by global standards-developing organizations such as the *Third Generation Partnership Project* (3GPP) and involving thousands of people.

1.1 History and background of 3G

The cellular technologies specified by 3GPP are the most widely deployed in the world, with more than 2.6 billion users in 2008. The latest step being studied and developed in 3GPP is an evolution of 3G into an evolved radio access referred to as the *Long-Term Evolution* (LTE) and an evolved packet access core network in the *System Architecture Evolution* (SAE). By 2009–2010, LTE and SAE are expected to be first deployed.

Looking back to when it all it started, it begun several decades ago with early deployments of analog cellular services.

1.1.1 Before 3G

The US *Federal Communications Commission* (FCC) approved the first commercial car-borne telephony service in 1946, operated by AT&T. In 1947 AT&T also introduced the cellular concept of reusing radio frequencies, which became fundamental to all subsequent mobile-communication systems. Commercial mobile telephony continued to be car-borne for many years because of bulky

3

and power-hungry equipment. In spite of the limitations of the service, there were systems deployed in many countries during the 1950s and 1960s, but the users counted only in thousands at the most.

These first steps on the road of mobile communication were taken by the monopoly telephone administrations and wire-line operators. The big uptake of subscribers and usage came when mobile communication became an international concern and the industry was invited into the process. The first international mobile communication system was the analog NMT system (*Nordic Mobile Telephony*) which was introduced in the Nordic countries in 1981, at the same time as analog AMPS (*Advanced Mobile Phone Service*) was introduced in North America. Other analog cellular technologies deployed worldwide were TACS and J-TACS. They all had in common that equipment was still bulky, mainly car-borne, and voice quality was often inconsistent, with 'cross-talk' between users being a common problem.

With an international system such as NMT came the concept of 'roaming,' giving a service also for users traveling outside the area of their 'home' operator. This also gave a larger market for the mobile phones, attracting more companies into the mobile communication business.

The analog cellular systems supported 'plain old telephony services,' that is voice with some related supplementary services. With the advent of digital communication during the 1980s, the opportunity to develop a second generation of mobile-communication standards and systems, based on digital technology, surfaced. With digital technology came an opportunity to increase the capacity of the systems, to give a more consistent quality of the service, and to develop much more attractive truly mobile devices.

In Europe, the telecommunication administrations in CEPT[1] initiated the GSM project to develop a pan-European mobile-telephony system. The GSM activities were in 1989 continued within the newly formed *European Telecommunication Standards Institute* (ETSI). After evaluations of TDMA, CDMA, and FDMA-based proposals in the mid-1980s, the final GSM standard was built on TDMA. Development of a digital cellular standard was simultaneously done by TIA in the USA resulting in the TDMA-based IS-54 standard, later simply referred to as US-TDMA. A somewhat later development of a CDMA standard called IS-95 was completed by TIA in 1993. In Japan, a second-generation TDMA standard was also developed, usually referred to as PDC.

---

[1] The *European Conference of Postal and Telecommunications Administrations* (CEPT) consist of the telecom administrations from 48 countries.

All these standards were 'narrowband' in the sense that they targeted 'low-bandwidth' services such as voice. With the second-generation digital mobile communications came also the opportunity to provide data services over the mobile-communication networks. The primary data services introduced in 2G were text messaging (SMS) and circuit-switched data services enabling e-mail and other data applications. The peak data rates in 2G were initially 9.6 kbps. Higher data rates were introduced later in evolved 2G systems by assigning multiple time slots to a user and by modified coding schemes.

Packet data over cellular systems became a reality during the second half of the 1990s, with *General Packet Radio Services* (GPRS) introduced in GSM and packet data also added to other cellular technologies such as the Japanese PDC standard. These technologies are often referred to as 2.5G. The success of the wireless data service iMode in Japan gave a very clear indication of the potential for applications over packet data in mobile systems, in spite of the fairly low data rates supported at the time.

With the advent of 3G and the higher-bandwidth radio interface of UTRA (*Universal Terrestrial Radio Access*) came possibilities for a range of new services that were only hinted at with 2G and 2.5G. The 3G radio access development is today handled in 3GPP. However, the initial steps for 3G were taken in the early 1990s, long before 3GPP was formed.

What also set the stage for 3G was the internationalization of cellular standardization. GSM was a pan-European project, but quickly attracted worldwide interest when the GSM standard was deployed in a number of countries outside Europe. There are today only three countries worldwide where GSM is not deployed. A global standard gains in economy of scale, since the market for products becomes larger. This has driven a much tighter international cooperation around 3G cellular technologies than for the earlier generations.

1.1.2 Early 3G discussions

Work on a third-generation mobile communication started in ITU (*International Telecommunication Union*) in the 1980s. The radio communication sector ITU-R issued a first recommendation defining *Future Public Land Mobile Telecommunications Systems* (FPLMTS) in 1990, later revised in 1997 [48]. The name for 3G within ITU had by then changed from FPLMTS to IMT-2000. The World Administrative Radio Congress WARC-92 identified 230 MHz of spectrum for IMT-2000 on a worldwide basis. Of these 230 MHz, 2×60 MHz were identified as paired spectrum for FDD (*Frequency Division Duplex*) and 35 MHz as unpaired spectrum for TDD (*Time Division Duplex*), both for terrestrial use. Some spectrum was also set aside for satellite services. With that, the stage was set to specify IMT-2000.

Task Group 8/1 within ITU-R developed a range of recommendations for IMT-2000, defining a framework for services, network architectures, radio interface requirements, spectrum considerations, and evaluation methodology. Both a terrestrial and a satellite component were defined.

Task Group 8/1 defined the process for evaluating IMT-2000 technologies in ITU-R recommendation M.1225 [45]. The evaluation criteria set the target data rates for the 3G circuit-switched and packet-switched data services:

- Up to 2 Mbps in an indoor environment.
- Up to 144 kbps in a pedestrian environment.
- Up to 64 kbps in a vehicular environment.

These numbers became the benchmark that all 3G technologies were compared with. However, already today, data rates well beyond 2 Mbps can be seen in deployed 3G systems.

1.1.3   Research on 3G

In parallel with the widespread deployment and evolution of 2G mobile-communication systems during the 1990s, substantial efforts were put into 3G research activities. In Europe the partially EU-funded project *Research into Advanced Communications in Europe* (RACE) carried out initial 3G research in its first phase. 3G in Europe was named *Universal Mobile Telecommunications Services* (UMTS). In the second phase of RACE, the CODIT project (Code Division Test bed) and the ATDMA project (*Advanced TDMA Mobile Access*) further developed 3G concepts based on *Wideband CDMA* (WCDMA) and Wideband TDMA technologies. The next phase of related European research was *Advanced Communication Technologies and Services* (ACTS), which included the UMTS-related project *Future Radio Wideband Multiple Access System* (FRAMES). The FRAMES project resulted in a multiple access concept that included both Wideband CDMA and Wideband TDMA components.

At the same time parallel 3G activities were going on in other parts of the world. In Japan, the *Association of Radio Industries and Businesses* (ARIB) was in the process of defining a 3G wireless communication technology based on Wideband CDMA. Also in the US, a Wideband CDMA concept called WIMS was developed within the T1.P1[2] committee. Also Korea started work on Wideband CDMA at this time.

The FRAMES concept was submitted to the standardization activities for 3G in ETSI,[3] where other multiple access proposals were also introduced by the industry, including the Wideband CDMA concept from the ARIB standardization in Japan. The ETSI proposals were merged into five concept groups, which also meant that the Wideband CDMA proposals from Europe and Japan were merged.

1.1.4   3G standardization starts

The outcome of the ETSI process in early 1998 was the selection of *Wideband CDMA* (WCDMA) as the technology for UMTS in the paired spectrum (FDD) and TD-CDMA (*Time Division CDMA*) for the unpaired spectrum (TDD). There was also a decision to harmonize the parameters between the FDD and the TDD components.

The standardization of WCDMA went on in parallel in ETSI and ARIB until the end of 1998 when the *Third Generation Partnership Project* (3GPP) was formed by standards-developing organizations from all regions of the world. This solved the problem of trying to maintain parallel development of aligned specifications in multiple regions. The present organizational partners of 3GPP are ARIB (Japan), CCSA (China), ETSI (Europe), ATIS (USA), TTA (Korea) and TTC (Japan).

1.2   Standardization

1.2.1   The standardization process

Setting a standard for mobile communication is not a one-time job, it is an ongoing process. The standardization forums are constantly evolving their standards trying to meet new demands for services and features. The standardization process is different in the different forums, but typically includes the four phases illustrated in Figure 1.1:

1. *Requirements*, where it is decided what is to be achieved by the standard.
2. *Architecture*, where the main building blocks and interfaces are decided.

---

[2] The T1.P1 committee was part of T1 which presently has joined the ATIS standardization organization.
[3] The TDMA part of the FRAMES project was also fed into 2G standardization as the evolution of GSM into EDGE (Enhanced Data rates for GSM Evolution).

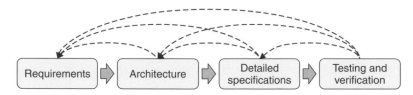

Figure 1.1   *The standardization phases and iterative process.*

3. *Detailed specifications*, where every interface is specified in detail.
4. *Testing and verification*, where the interface specifications are proven to
   work with real-life equipment.

These phases are overlapping and iterative. As an example, requirements can be
added, changed, or dropped during the later phases if the technical solutions call
for it. Likewise, the technical solution in the detailed specifications can change
due to problems found in the testing and verification phase.

Standardization starts with the *requirements* phase, where the standards body
decides what should be achieved with the standard. This phase is usually relatively short.

In the *architecture* phase, the standards body decides about the architecture, i.e.
the principles of how to meet the requirements. The architecture phase includes
decisions about reference points and interfaces to be standardized. This phase is
usually quite long and may change the requirements.

After the architecture phase, the *detailed specification* phase starts. It is in this
phase the details for each of the identified interfaces are specified. During the
detailed specification of the interfaces, the standards body may find that it has
to change decisions done either in the architecture or even in the requirements
phases.

Finally, the *testing and verification* phase starts. It is usually not a part of the
actual standardization in the standards bodies, but takes place in parallel through
testing by vendors and interoperability testing between vendors. This phase is
the final proof of the standard. During the testing and verification phase, errors
in the standard may still be found and those errors may change decisions in the
detailed standard. Albeit not common, changes may need to be done also to the
architecture or the requirements. To verify the standard, products are needed.
Hence, the implementation of the products starts after (or during) the detailed
specification phase. The testing and verification phase ends when there are

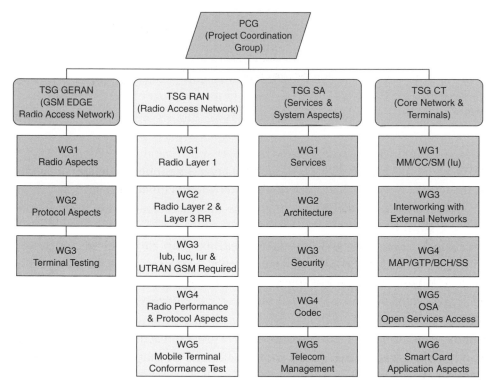

Figure 1.2 *3GPP organization.*

stable test specifications that can be used to verify that the equipment is fulfilling the standard.

Normally, it takes one to two years from the time when the standard is completed until commercial products are out on the market. However, if the standard is built from scratch, it may take longer time since there are no stable components to build from.

1.2.2 3GPP

The *Third-Generation Partnership Project* (3GPP) is the standards-developing body that specifies the 3G UTRA and GSM systems. 3GPP is a partnership project formed by the standards bodies ETSI, ARIB, TTC, TTA, CCSA and ATIS. 3GPP consists of several Technical Specifications Groups (TSGs), (see Figure 1.2).

A parallel partnership project called 3GPP2 was formed in 1999. It also develops 3G specifications, but for cdma2000, which is the 3G technology developed from the 2G CDMA-based standard IS-95. It is also a global project, and the organizational partners are ARIB, CCSA, TIA, TTA and TTC.

3GPP TSG RAN is the technical specification group that has developed WCDMA, its evolution HSPA, as well as LTE, and is in the forefront of the technology. TSG RAN consists of five working groups (WGs):

1. RAN WG1 dealing with the physical layer specifications.
2. RAN WG2 dealing with the layer 2 and layer 3 radio interface specifications.
3. RAN WG3 dealing with the fixed RAN interfaces, for example interfaces between nodes in the RAN, but also the interface between the RAN and the core network.
4. RAN WG4 dealing with the *radio frequency* (RF) and *radio resource management* (RRM) performance requirements.
5. RAN WG 5 dealing with the terminal conformance testing.

The scope of 3GPP when it was formed in 1998 was to produce global specifications for a 3G mobile system based on an evolved GSM core network, including the WCDMA-based radio access of the UTRA FDD and the TD-CDMA-based radio access of the UTRA TDD mode. The task to maintain and develop the GSM/EDGE specifications was added to 3GPP at a later stage. The UTRA (and GSM/EDGE) specifications are developed, maintained and approved in 3GPP. After approval, the organizational partners transpose them into appropriate deliverables as standards in each region.

In parallel with the initial 3GPP work, a 3G system based on TD-SCDMA was developed in China. TD-SCDMA was eventually merged into Release 4 of the 3GPP specifications as an additional TDD mode.

The work in 3GPP is carried out with relevant ITU recommendations in mind and the result of the work is also submitted to ITU. The organizational partners are obliged to identify regional requirements that may lead to options in the standard. Examples are regional frequency bands and special protection requirements local to a region. The specifications are developed with global roaming and circulation of terminals in mind. This implies that many regional requirements in essence will be global requirements for all terminals, since a roaming terminal has to meet the strictest of all regional requirements. Regional options in the specifications are thus more common for base stations than for terminals.

The specifications of all releases can be updated after each set of TSG meetings, which occur 4 times a year. The 3GPP documents are divided into releases, where

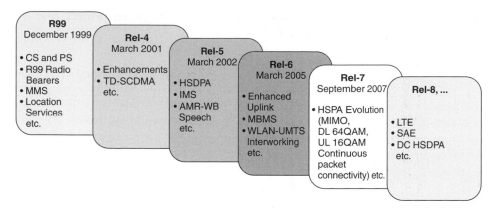

Figure 1.3 *Releases of 3GPP specifications for UTRA.*

each release has a set of features added compared to the previous release. The features are defined in Work Items agreed and undertaken by the TSGs. The releases up to Release 8 and some main features of those are shown in Figure 1.3. The date shown for each release is the day the content of the release was frozen. For historical reasons, the first release is numbered by the year it was frozen (1999), while the following releases are numbered 4, 5, etc.

For the WCDMA Radio Access developed in TSG RAN, Release 99 contains all features needed to meet the IMT-2000 requirements as defined by ITU. There are circuit-switched voice and video services, and data services over both packet-switched and circuit-switched bearers. The first major addition of radio access features to WCDMA is Release 5 with *High Speed Downlink Packet Access* (HSDPA) and Release 6 with *Enhanced Uplink*. These two are together referred to as HSPA and are described in more detail in Part III of this book. With HSPA, UTRA goes beyond the definition of a 3G mobile system and also encompasses broadband mobile data.

With the inclusion of an Evolved UTRAN (LTE) and the related *System Architecture Evolution* (SAE) in Release 8, further steps are taken in terms of broadband capabilities. The specific solutions chosen for LTE and SAE are described in Part IV of this book.

1.2.3 IMT-2000 activities in ITU

The present ITU work on 3G takes place in ITU-R *Working Party 5D*[4] (WP5D), where 3G systems are referred to as IMT-2000. WP5D does not write technical

---

[4] The work on IMT-2000 was moved from Working Party 8F to Working Party 5D in 2008.

specifications for IMT-2000, but has kept the role of defining IMT-2000, cooperating with the regional standardization bodies and to maintain a set of recommendations for IMT-2000.

The main IMT-2000 recommendation is ITU-R M.1457 [46], which identifies the IMT-2000 *radio interface specifications* (RSPC). The recommendation contains a 'family' of radio interfaces, all included on an equal basis. The family of six terrestrial radio interfaces is illustrated in Figure 1.4, which also shows what *Standards Developing Organizations* (SDO) or Partnership Projects produce the specifications. In addition, there are several IMT-2000 satellite radio interfaces defined, not illustrated in Figure 1.4.

For each radio interface, M.1457 contains an overview of the radio interface, followed by a list of references to the detailed specifications. The actual specifications are maintained by the individual SDOs and M.1457 provides URLs locating the specifications at each SDOs web archive.

With the continuing development of the IMT-2000 radio interfaces, including the evolution of UTRA to Evolved UTRA, the ITU recommendations also need to be updated. ITU-R WP5D continuously revises recommendation M.1457 and at the time of writing it is in its seventh version. Input to the updates is provided by the SDOs and Partnership Projects writing the standards. In the latest revision of ITU-R M.1457, LTE (or E-UTRA) is included in the family through the 3GPP family members for UTRA FDD and TDD, while UMB is included through CDMA2000, as shown in the figure. WiMAX is also included as the sixth family member for IMT-2000.

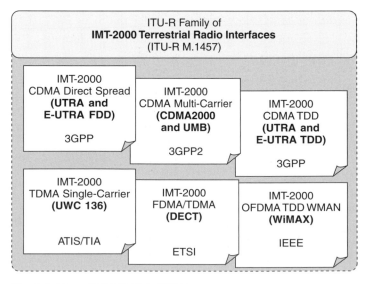

Figure 1.4   *The definition of IMT-2000 in ITU-R.*

In addition to maintaining the IMT-2000 specifications, the main activity in ITU-R WP5D is the work on systems beyond IMT-2000, named IMT-Advanced. ITU-R has concluded studies for IMT-Advanced of services and technologies, market forecasts, principles for standardization, estimation of spectrum needs, and identification of candidate frequency bands [47]. The spectrum work has involved sharing studies between IMT-Advanced and other technologies in those bands. In March 2008, ITU-R invited the submission of candidate *Radio Interface Technologies* (RIT) in a Circular letter [139]. Submission and evaluation of RITs will be ongoing through 2009 and 2010. The target date for the final ITU-R recommendation for the IMT-Advanced radio interface specifications is February 2011.

1.3 Spectrum for 3G and systems beyond 3G

Spectrum for 3G was first identified at the *World Administrative Radio Congress* WARC-92. Resolution 212 [60] identified the bands 1885–2025 and 2110–2200 MHz as intended for use by national administrations that want to implement IMT–2000. Of these 230 MHz of 3G spectrum, 2×30 MHz were intended for the satellite component of IMT–2000 and the rest for the terrestrial component. Parts of the bands were during the 1990s used for deployment of 2G cellular systems, especially in the Americas. The first deployment of 3G in 2001–2002 by Japan and Europe were done in this band allocation, and it is for that reason often referred to as the IMT-2000 'core band.'

Spectrum for IMT–2000 was also identified at the World Radiocommunication Conference WRC-2000 in Resolutions 223 and 224, where it was considered that an additional need for 160 MHz of spectrum for IMT-2000 was forecasted by ITU-R. The identification includes the bands used for 2G mobile systems in 806–960 MHz and 1710–1885 MHz, and 'new' 3G spectrum in the bands 2500–2690 MHz. The identification of bands assigned for 2G was also a recognition of the evolution of existing 2G mobile systems into 3G. Additional spectrum was identified at WRC'07 for IMT, encompassing both IMT-2000 and IMT-Advanced. The bands added are 450–470, 698–806, 2300–2400, and 3400–3600 MHz, but the applicability of the bands vary on a regional and national basis.

The somewhat diverging arrangement between regions of the frequency bands assigned to 3G means that there is not a single band that can be used for 3G roaming worldwide. Large efforts have however been put into defining a minimum set of bands that can be used to provide roaming. In this way, multi-band devices can provide efficient worldwide roaming for 3G. Release 8 of the 3GPP specifications includes 14 frequency bands for FDD and 8 for TDD. These are described in more detail in Chapter 20.

The worldwide frequency arrangements for IMT-2000 are outlined in ITU-R recommendation M.1036 [44]. The recommendation also identifies which parts of the spectrum that are paired and which are unpaired. For the paired spectrum, the bands for uplink (mobile transmit) and downlink (base-station transmit) are identified for *Frequency Division Duplex* (FDD) operation. The unpaired bands can for example be used for *Time Division Duplex* (TDD) operation. Note that the band that is most globally deployed for 3G is still 2 GHz.

3GPP first defined UTRA in Release 99 for the 2 GHz bands, with 2×60 MHz for UTRA FDD and $20 + 15$ MHz of unpaired spectrum for UTRA TDD. A separate definition was also made for the use of UTRA in the US PCS bands at 1900 MHz. The concept of frequency bands with separate and release-independent requirements were defined in Release 5 of the 3GPP specifications. The release-independence implies that a new frequency band added at a later release can be implemented also for earlier releases. All bands are also defined with consideration of what other bands may be deployed in the same region through special coexistence requirements for both base stations and terminals. These tailored requirements enable coexistence between 3G (and 2G) deployments in different bands in the same geographical area and even for co-location of base stations at the same sites using different bands.

2

The motives behind the 3G evolution

Before entering the detailed discussion on technologies being used or considered for the evolution of 3G mobile communication, it is important to understand the motivation for this evolution: that is, understanding the underlying driving forces. This chapter will try to highlight some of the driving forces giving the reader an understanding of where the technical requirements and solutions are coming from.

2.1 Driving forces

A key factor for success in any business is to understand the forces that will drive the business in the future. This is in particular true for the mobile-communication industry, where the rapid growth of the number of subscribers and the global presence of the technologies have attracted several new players that want to be successful. Both new operators and new vendors try to compete with the existing operators and vendors by adopting new technologies and standards to provide new and existing services better and at a lower cost than earlier systems. The existing operators and vendors will, of course, also follow or drive new technologies to stay ahead of competition. Thus, there is a key driving force in staying competitive or becoming competitive.

From the technical perspective, the development in areas like digital cameras and color displays enables new fancier services than the existing mobile-communication services. To be able to provide those services, the mobile-communication systems need to be upgraded or even replaced by new mobile-communication technologies. Similarly, the technical advancement in digital processors enables new and more powerful systems that not only can provide the new services, but also can provide the existing successful services better and to a lower cost than the dominant mobile-communication technologies of today. Thus, the key drivers are:

- staying competitive;
- services (better provisioning of old services as well as provisioning of new services);
- cost (more cost-efficient provisioning of old services as well as cost-effective provisioning of new services).

The technology advancement is necessary to provide new and more advanced services at a reasonable cost as well as to provide existing services in a better and more cost-efficient way.

2.1.1 Technology advancements

Technology advancements in many areas make it possible to build devices that were not possible 20, 10, or even 5 years ago. Even though Moore's law[1] is not a law of physics, it gives an indication of the rapid technology evolution for integrated circuits. This evolution enables faster processing/computing in smaller devices at lower cost. Similarly, the rapid development of color screens, small digital cameras, etc. makes it possible to envisage services to a device that were seen as utopia 10 years ago. For an example of the terminal development in the past 20 years, see Figure 2.1.

Figure 2.1 *The terminal development has been rapid the past 20 years.*

---

[1] Moore's law is an empirical observation, and states that with the present rate of technological development, the complexity of an integrated circuit, with respect to minimum component cost, will double in about 18 months.

The size and weight of the mobile terminals have been reduced dramatically during the past 20 years. The standby and talk times have also been extended dramatically and the end users do not need to re-charge their devices every day. Simple black-and-white (or brown-and-gray) numerical screens have evolved into color screens capable of showing digital photos at good quality. Mega-pixel-capable digital cameras have been added making the device more attractive to use. Thus, the mobile device has become a multi-purpose device, not only a mobile phone for voice communications.

In parallel to the technical development of the mobile devices, the mobile-communication technologies are developed to meet the demands of the new services enabled, and also to enable them wireless. The development of the digital signal processors enables more advanced receivers capable of processing mega-bits of data in a short time, and the introduction of the optical fibers enables high-speed network connections to the base stations. In sum, this enables a fast access to information on the Internet as well as a short roundtrip time for normal communications. Thus, new and fancier services are enabled by the technical development of the devices, and new and more efficient mobile-communication systems are enabled by a similar technical development.

2.1.2 Services

Delivering services to the end users is the fundamental goal of any mobile-communication system. Knowing them, understanding them, managing them, and charging them properly is the key for success. It is also the most difficult task being faced by the engineers developing the mobile-communication system of the future. It is very difficult to predict what service(s) will be popular in a 5- to 10-year perspective. In fact, the engineers have to design a system that can adapt to any service that might become popular and used in the future. Unfortunately, there are also technical limitations that need to be understood, and also the technical innovations that in the future enable new services.

2.1.2.1 Internet and IP technology

The success of the Internet and the IP-based services delivered over the Internet is more and more going wireless. This means that the mobile-communication systems are delivering more and more IP-based services, from the best effort-Internet data to voice-over-IP, for example in the shape of push-to-talk (PoC). Furthermore, in the wireless environment it is more natural to use, for example, location-based services and tracking services than in the fixed environment.

Thus, one can talk about mobile Internet services in addition to the traditional Internet services like browsing, e-mail, etc.[2]

Essentially, IP provides a transport mechanism that is service agnostic. Albeit there are several protocols on top of IP that are service-type specific (RTP, TCP, etc.), IP in itself is service agnostic. That enables service developers to develop services that only the imagination (and technology) sets the limit to. Thus, services will pop up, some will become popular for a while and then just fade away, whereas some others will never become a hit. Some services will become classics that will live and be used for a very long time.

2.1.2.2 *Traditional telephony services*
Going toward IP-based services obviously does not mean that traditional services that have been provided over the circuit-switched domain, in successful mobile-communication systems like GSM, will disappear. Rather, it means that the traditional circuit-switched services will be ported over the IP networks. One particular service is the circuit-switched telephony service that will be provided as VoIP service instead. Thus, both the new advanced services that are enabled by the technology advancement of the devices and the traditional circuit-switched services will be using IP as the transporting mechanism (and are therefore called IP-based services). Hence future mobile-communication networks, including the 3G evolution, need to be optimized for IP-based applications.

2.1.2.3 *Wide spectrum of service needs*
Trying to predict all the services that will be used over the mobile-communication systems 10 years from now is very difficult. The technology advancements in the various areas enable higher data rate connections, more memory on local devices, and more intelligent and easy to use man–machine interfaces. Furthermore, the human need of interaction and competition with other humans drives more intensive communication needs. All these combinations point toward applications and services that consume higher data rates and require lower delays compared to what today's mobile-communication systems can deliver.

However, the relative low-rate voice service will still be a very important component of the service portfolio that mobile-network operators wish to provide. In addition, services that have very relaxed delay requirements will also be there. Thus, not only high data rate services with a low-latency requirement, but also

---

[2] The common denominator between mobile Internet services and Internet services is the IP addressing technology with the IPv4 and IPv6 addresses identifying the end receiver. However, there is a need to handle the mobility provided by the cellular systems. Mobile IP is one possibility, but most of the cellular systems (if not all) have their own more efficient mobility mechanism.

low data rate best effort services will be provided. Furthermore, not only the data rate and delay are important to understand when talking about a service's need from a mobile-communication system, but also the setup time is very important, for example, a service can be totally useless if it takes too long to start it (for example making a phone call, downloading a web page). Thus the mobile-communication systems of the future, including the evolution of 3G mobile communication, need to be able to deliver short call setup times, low latency and a wide range of data rates.

2.1.2.4   Key design services

Since it is impossible to know what services that will be popular and since service possibilities and offers will differ with time and possible also with country, the future mobile-communication systems will need to be adaptive to the changing service environment. Luckily, there are a few known key services that span the technology space. Those are:

- *Real-time-gaming applications*: These have the characteristic to require small amount of data (game update information) relatively frequent with low delay requirement.[3] Only a limited delay jitter is tolerable. A first person shooter game like Counter Strike is an example of a game that has this characteristic.
- *Voice*: This has the characteristic to require small amounts of data (voice packets) frequently with no delay jitter. The end-to-end delay has to be small enough not to be noticeable.[4]
- *Interactive file download and upload applications*: These have the characteristics of requiring low delay and high data rates.
- *Background file download and upload applications*: These have the characteristics of accepting lower bit rates and longer delays. E-mail (mostly) is an example of background file download and upload.
- *Television*: This has the characteristics of streaming downlink to many users at the same time requiring low delay jitter. The service can tolerate delays, as long as it is approximately the same delay for all users in the neighborhood. The television service has moderate data rate requirements.

A mobile-communication system designed to handle these services and the services in between will be able to facilitate most services (see Figure 2.2). Unfortunately, the upper limit of the data rate demand and the lower limit of the delay requirement are difficult to provide in a cost-efficient manner. The designers

---

[3] The faster the data is delivered the better. Expert Counter Strike players look for game servers with a ping time of less than 50 ms.

[4] In 3G systems the end-to-end delay requirement for circuit-switched voice is approximately 400 ms. This delay is not disturbing humans in voice communications.

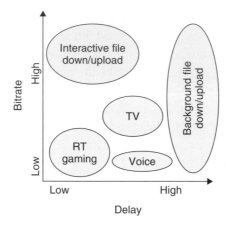

Figure 2.2   *The bit rate – delay service space that is important to cover when designing a new cellular system.*

of the mobile-communication systems need to stop at a reasonable level, a level that the technology available at the time of standardization can provide.

2.1.3   Cost and performance

There is another important driving factor for future mobile-communication systems and that is the cost of the service provisioning. The technology advancement that enables new services can also be utilized to provide better mobile-communication systems using more advanced technical features. Here IP technology is not only a key enabler to allow for new services to pop up, but also a way of reducing cost of new services. The reason is that IP as a bearer of new services can be used to introduce new services as they come, not requiring an extensive special design of the system. Of course, this requires that the devices used in the mobile-communication system can be programmed by third-party providers and that the operators allow third-party service providers to use their network for communication.

Another important factor is that operators need to provide the services to all the users. Not only one user needs to get the low delay, high data rate, etc. that its service needs, but all the users with their different service needs should be served efficiently. The processing capacity evolution and Moore's law help also for this problem. New techniques are enabled by the higher processing power in the devices – techniques that delivers more bits of data per hertz. Furthermore, the coverage is increased with more advanced antennas and receivers. This enables the operators to deliver the services to more users from one base station, thus requiring fewer sites. Fewer sites imply lower operational and capitalization costs. In essence, the operators need fewer base stations and sites to provide the service.

Obviously, all services would be 'happy' if they were provided with the highest data rate, lowest delay, and lowest jitter that the system can provide. Unfortunately, this is unattainable in practice and contradictory to the operator goal of an efficient system: in other words, the more delay a service can handle the more efficient the system can be. Thus, the cost of providing lowest possible delay, jitter and call setup time is somewhat in conflict with the need of the mobile network operator to provide it to all the users. Hence, there is a trade-off between user experience and system performance. The better the system performance is, the lower the cost of the network. However, the end users also need to get adequate performance which often is in conflict with the system performance, thus the operator cannot only optimize for system performance.

2.2 3G evolution: Two Radio Access Network approaches and an evolved core network

2.2.1 Radio Access Network evolution

TSG RAN organized a workshop on 3GPP long-term Evolution in the fall of 2004. The workshop was the starting point of the development of the Long-Term Evolution (LTE) radio interface. After the initial requirement phase in the spring of 2005, where the targets and objectives of LTE were settled, the technical specification group TSG SA launched a corresponding work on the System Architecture Evolution, since it was felt that the LTE radio interface needed a suitable evolved system architecture.

The result of the LTE workshop was that a study item in 3GPP TSG RAN was created in December 2004. The first 6 months were spent on defining the requirements, or design targets, for LTE. These were documented in a 3GPP technical report [86] and approved in June 2005. Chapter 13 will go through the requirements in more detail. Most notable are the requirements on high data rate at the cell edge and the importance of low delay, in addition to the normal capacity and peak data rate requirements. Furthermore, spectrum flexibility and maximum commonality between FDD and TDD solutions are pronounced.

During the fall 2005, 3GPP TSG RAN WG1 made extensive studies of different basic physical layer technologies and in December 2005 the TSG RAN plenary decided that the LTE radio access should be based on *OFDM* in the downlink and *single carrier FDMA* in the uplink.

TSG RAN and its working groups then worked on the LTE specifications and the specifications were approved in December 2007. However, 3GPP TSG RAN did not stop working on LTE when the first version of the specifications was completed. In fact, 3GPP will continue to evolve LTE towards LTE Advanced. Chapters 14–20 will go through the LTE radio interface in more detail.

At the same time as the LTE discussion was ongoing, 3GPP TSG RAN and its WGs continued to evolve the WCDMA system with more functionality, most notably MBMS and Enhanced Uplink. These additions were done in a backward compatible manner: that is, terminals of earlier releases can coexist on the same carrier in the same base station as terminals of the latest release. The main argument for the backward compatibility is that the installed base of equipment can be upgraded to handle the new features while still being capable of serving the old terminals. This is a cost-efficient addition of new features, albeit the new features are restricted by the solutions for the old terminals.

Naturally, HSPA[5] does not include all the technologies considered for LTE. Therefore, a study in 3GPP was initiated to see how far it is possible to take HSPA within the current spectrum allocation of 5 MHz and still respect the backward compatibility aspect. Essentially, the target with HSPA Evolution was, and still is, to reach near the characteristics of LTE when using a 5 MHz spectrum and at the same time being backward compatible. Chapters 8–12 will go through the HSPA and the HSPA Evolution in more detail.

Thus, the 3GPP 3G evolution standard has two parts: LTE and HSPA Evolution. Both parts have their merits. LTE can operate in new and more complex spectrum arrangements (although in the same spectrum bands as WCDMA and other 3G technologies) with the possibility for new designs that do not need to cater for terminals of earlier releases. HSPA Evolution can leverage on the installed base of equipment in the 5 MHz spectrum but needs to respect the backward compatibility of earlier terminals.

2.2.1.1 LTE drivers and philosophy

The 3GPP Long-Term Evolution is intended to be a mobile-communication system that can take the telecom industry into the 2020s. The philosophy behind LTE standardization is that the competence of 3GPP in specifying mobile-communication systems in general and radio interfaces in particular shall be used, but the result shall not be restricted by previous work in 3GPP. Thus, LTE does not need to be backward compatible with WCDMA and HSPA.

Leaving the legacy terminals behind, not being restricted by designs of the late 1990s, makes it possible to design a radio interface from scratch. In the LTE case,

---

[5] When operating with HSDPA and Enhanced UL, the system is known as HSPA.

the radio interface is purely optimized for IP transmissions not having to support ISDN traffic: that is, there is no requirement for support of GSM circuit-switch services, a requirement that WCDMA had. Furthermore, LTE also has a very large commonality of FDD and TDD operations, a situation that did not exist in 3GPP before LTE.

Instead new requirements have arisen, for example the requirement on spectrum flexibility, since the global spectrum situation becomes more and more complex. Operators get more and more scattered spectrum, spread over different bands with different contiguous bandwidths. LTE needs to be able to operate in all these bands and with the bandwidths that is available to the operator. However, in practice only a limited set of bandwidths can be used since otherwise the RF and filter design would be too costly. LTE is therefore targeted to operate in spectrum allocations from roughly 1 to 20 MHz. The spectrum flexibility support with the possibility to operate in other bandwidths than 5 MHz makes LTE very attractive for operators. The low-bandwidth operations are suitable for refarming of spectrum (for example GSM spectrum and CDMA2000 spectrum). The higher-bandwidth options are suitable for new deployments in unused spectrum, where it is more common to have larger chunks of contiguous spectrum.

Furthermore, when going to the data rates that LTE is targeting, achieving low delay and high data rates at the cell edges are more important requirements than the peak data rate. Thus, a more pronounced requirement for LTE is the low delay with high data rates at the cell edges than it was when WCDMA was designed in the late 1990s.

Although not backward compatible with WCDMA, LTE design is clearly influenced by the WCDMA and the HSPA work in 3GPP. It is the same body, the same people, and companies that are active and more importantly, WCDMA and HSPA protocols are a good foundation for the LTE design. The philosophy is to take what is good from WCDMA and HSPA, and redo those parts that have to be updated due to the new requirement situation: either there are new requirements such as the spectrum flexibility or there are requirements that no longer are valid such as the support of ISDN services. The technology advancement in the cellular area has, of course, also influenced the design choice of LTE.

2.2.1.2 HSPA evolution drivers and philosophy

WCDMA, HSDPA, and HSPA are in commercial operation throughout the world. This means that the infrastructure for HSPA is already in place with the network node sites, especially the base-station sites with their antenna arrangements and hardware. This equipment is serving millions of terminals with different characteristics and supported 3GPP releases. These terminals need to be supported by the WCDMA operators for many more years.

The philosophy of the HSPA Evolution work is to continue to add new and fancier technical features, and at the same time be able to serve the already existing terminals. This is the successful strategy of GSM that have added new features constantly since the introduction in the early 1990s. The success stems from the fact that there are millions of existing terminals at the launch time of the new features that can take the cost of the upgrade of the network for the initially few new terminals before the terminal fleet is upgraded. The time it takes to upgrade the terminal fleet is different from country to country, but a rule of thumb is that a terminal is used for 2 years before it is replaced. For HSPA Evolution that means that millions of HSPA-capable terminals need to be supported at launch. In other words, HSPA Evolution needs to be backward compatible with the previous releases in the sense that it is possible to serve terminals of earlier releases of WCDMA on the same carrier as HSPA-Evolution-capable terminals.

The backward compatibility requirement on the HSPA Evolution puts certain constraints on the technology that LTE does not need to consider, for example the physical layer fundamentals need to be the same as for WCDMA release 99. On the other hand, HSPA Evolution is built on the existing specifications and only those parts of the specifications that need to be upgraded are touched. Thus there is less standardization, implementation and testing work for HSPA Evolution than for LTE since the HSPA Evolution philosophy is to apply new more advanced technology on the existing HSPA standard. This will bring HSPA to a performance level comparable to LTE when compared on a 5 MHz spectrum allocation (see Chapter 23 for a performance comparison of HSPA Evolution and LTE).

2.2.2   An evolved core network: system architecture evolution

Roughly at the same time as LTE and HSPA Evolution was started, 3GPP decided to make sure that an operator can coexist easily between HSPA Evolution and LTE through an evolved core network, the *Evolved Packet Core*. This work was done under the umbrella *System Architecture Evolution* study item lead by TSG SA WG2.

The *System Architecture Evolution* study focused on how the 3GPP core network will evolve into the core network of the next decades. The existing core network was designed in the 1980s for GSM, extended during the 1990s' for GPRS and WCDMA. The philosophy of the SAE is to focus on the packet-switched domain,

Figure 2.3 *One HSPA and LTE deployment strategy: upgrade to HSPA Evolution, then deploy LTE as islands in the WCDMA/HSPA sea.*

and migrate away from the circuit-switched domain. This is done through the coming 3GPP releases ending up with the Evolved Packet Core.

Knowing that HSPA Evolution is backward compatible and knowing that the Evolved Packet Core will support both HSPA Evolution and LTE assures that LTE can be deployed in smaller islands and thus only where it is needed. A gradual deployment approach can be selected (see Figure 2.3). First the operator can upgrade its HSPA network to HSPA-Evolution-capable network, and then add LTE cells where capacity is lacking or where the operator wants to try out new services that cannot be delivered by HSPA Evolution. This approach reduces the cost of deployment since LTE do not need to be build for nationwide coverage from day one.

Part II
Technologies for 3G Evolution

3

High data rates in mobile communication

As discussed in Chapter 2, one main target for the evolution of 3G mobile communication is to provide the possibility for significantly higher end-user data rates compared to what is achievable with, for example, the first releases of the 3G standards. This includes the possibility for higher peak data rates but, as pointed out in the previous chapter, even more so the possibility for significantly higher data rates over the entire cell area, also including, for example, users at the cell edge. The initial part of this chapter will briefly discuss some of the more fundamental constraints that exist in terms of what data rates can actually be achieved in different scenarios. This will provide a background to subsequent discussions in the later part of the chapter, as well as in the following chapters, concerning different means to increase the achievable data rates in different mobile-communication scenarios.

3.1 High data rates: Fundamental constraints

In [70], Shannon provided the basic theoretical tools needed to determine the maximum rate, also known as the *channel capacity*, by which information can be transferred over a given communication channel. Although relatively complicated in the general case, for the special case of communication over a channel, e.g. a radio link, only impaired by additive white Gaussian noise, the channel capacity C is given by the relatively simple expression [50]

$$C = BW \cdot \log_2 \left(1 + \frac{S}{N} \right) \qquad (3.1)$$

where BW is the bandwidth available for the communication, S denotes the received signal power, and N denotes the power of the white noise impairing the received signal.

Already from (3.1) it should be clear that the two fundamental factors limiting the achievable data rate are the available received signal power, or more generally the available signal-power-to-noise-power ratio S/N, and the available bandwidth. To further clarify how and when these factors limit the achievable data rate, assume communication with a certain information rate R. The received signal power can then be expressed as $S = E_b \cdot R$ where E_b is the received energy per information bit. Furthermore, the noise power can be expressed as $N = N_0 \cdot BW$ where N_0 is the constant noise power spectral density measured in W/Hz.

Clearly, the information rate can never exceed the channel capacity. Together with the above expressions for the received signal power and noise power, this leads to the inequality

$$R \le C = BW \cdot \log_2 \left(1 + \frac{S}{N}\right) = BW \cdot \log_2 \left(1 + \frac{E_b \cdot R}{N_0 \cdot BW}\right) \tag{3.2}$$

or, by defining the radio-link *bandwidth utilization* $\gamma = R/BW$,

$$\gamma \le \log_2 \left(1 + \gamma \cdot \frac{E_b}{N_0}\right) \tag{3.3}$$

This inequality can be reformulated to provide a lower bound on the required received energy per information bit, normalized to the noise power density, for a given bandwidth utilization γ

$$\frac{E_b}{N_0} \ge \min\left\{\frac{E_b}{N_0}\right\} = \frac{2^\gamma - 1}{\gamma} \tag{3.4}$$

The rightmost expression, i.e. the minimum required E_b/N_0 at the receiver as a function of the bandwidth utilization is illustrated in Figure 3.1. As can be seen, for bandwidth utilizations significantly less than one, that is for information rates substantially smaller than the utilized bandwidth, the minimum required E_b/N_0 is relatively constant, regardless of γ. For a given noise power density, any increase of the information data rate then implies a similar relative increase in the minimum required signal power $S = E_b \cdot R$ at the receiver. On the other hand, for bandwidth utilizations larger than one the minimum required E_b/N_0 increases rapidly with γ. Thus, in case of data rates in the same order as or larger than the communication bandwidth, any further increase of the information data rate, without a corresponding increase in the available bandwidth, implies a larger, eventually much larger, relative increase in the minimum required received signal power.

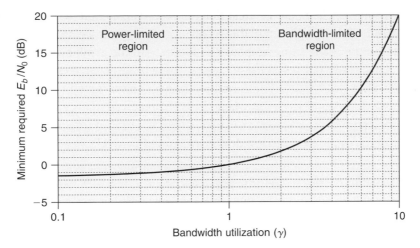

Figure 3.1 *Minimum required E_b/N_0 at the receiver as a function of bandwidth utilization.*

3.1.1 High data rates in noise-limited scenarios

From the discussion above, some basic conclusions can be drawn regarding the provisioning of higher data rates in a mobile-communication system when noise is the main source of radio-link impairment (a *noise-limited* scenario):

- The data rates that can be provided in such scenarios are always limited by the available received signal power or, in the general case, the received signal-power-to-noise-power ratio. Furthermore, any increase of the achievable data rate within a given bandwidth will require *at least* the same relative increase of the received signal power. At the same time, if sufficient received signal power can be made available, basically any data rate can, at least in theory, be provided within a given limited bandwidth.
- In case of low-bandwidth utilization, i.e. as long as the radio-link data rate is substantially lower than the available bandwidth, any further increase of the data rate requires *approximately the same* relative increase in the received signal power. This can be referred to as *power-limited* operation (in contrast to *bandwidth-limited* operation, see below) as, in this case, an increase in the available bandwidth does not substantially impact what received signal power is required for a certain data rate.
- On the other hand, in case of high-bandwidth utilization, i.e. in case of data rates in the same order as or exceeding the available bandwidth, any further increase in the data rate requires *a much larger* relative increase in the received signal power unless the bandwidth is increased in proportion to the increase in data rate. This can be referred to *bandwidth-limited operation* as, in this case, an increase in the bandwidth will reduce the received signal power required for a certain data rate.

Thus, to make efficient use of the available received signal power or, in the general case, the available signal-to-noise ratio, the transmission bandwidth should at least be of the same order as the data rates to be provided.

Assuming a constant transmit power, the received signal power can always be increased by reducing the distance between the transmitter and the receiver, thereby reducing the attenuation of the signal as it propagates from the transmitter to the receiver. Thus, in a noise-limited scenario it is at least in theory always possible to increase the achievable data rates, assuming that one is prepared to accept a reduction in the transmitter/receiver distance, that is a reduced range. In a mobile-communication system this would correspond to a reduced cell size and thus the need for more cell sites to cover the same overall area. Especially, pro-viding data rates in the same order as or larger than the available bandwidth, i.e. with a high-bandwidth utilization, would require a significant cell-size reduction. Alternatively, one has to accept that the high data rates are only available for mobile terminals in the center of the cell, i.e. not over the entire cell area.

Another means to increase the overall received signal power for a given transmit power is the use of additional antennas at the receiver side, also known as *receive-antenna diversity*. Multiple receive antennas can be applied at the base station (that is for the uplink) or at the mobile terminal (that is for the downlink). By proper combining of the signals received at the different antennas, the signal-to-noise ratio after the antenna combining can be increased in proportion to the number of receive antennas, thereby allowing for higher data rates for a given transmitter/receiver distance.

Multiple antennas can also be applied at the transmitter side, typically at the base station, and be used to focus a given total transmit power in the direction of the receiver, i.e. toward the target mobile terminal. This will increase the received signal power and thus, once again, allow for higher data rates for a given transmitter/receiver distance.

However, providing higher data rates by the use of multiple transmit or receive antennas is only efficient up to a certain level, i.e. as long as the data rates are power limited rather than bandwidth limited. Beyond this point, the achievable data rates start to saturate and any further increase in the number of transmit or receive antennas, although leading to a correspondingly improved signal-to-noise ratio at the receiver, will only provide a marginal increase in the achievable data rates. This saturation in achievable data rates can be avoided though, by the use of multiple antennas at both the transmitter *and* the receiver, enabling what can be referred to as *spatial multiplexing*, often also referred to as *MIMO* (Multiple-Input

Multiple-Output). Different types of multi-antenna techniques, including spatial multiplexing, will be discussed in more detail in Chapter 6. Multi-antenna techniques for the specific case of HSPA and LTE are discussed in Part III and IV of this book, respectively.

An alternative to increasing the received signal power is to reduce the noise power, or more exactly the noise power density, at the receiver. This can, at least to some extent, be achieved by more advanced receiver RF design, allowing for a reduced receiver noise figure.

3.1.2   Higher data rates in interference-limited scenarios

The discussion above assumed communication over a radio link only impaired by noise. However, in actual mobile-communication scenarios, interference from transmissions in neighbor cells, also referred to as *inter-cell interference*, is often the dominating source of radio-link impairment, more so than noise. This is especially the case in small-cell deployments with a high traffic load. Furthermore, in addition to inter-cell interference there may in some cases also be interference from other transmissions *within the current cell*, also referred to as *intra-cell interference*.

In many respects the impact of interference on a radio link is similar to that of noise. Especially, the basic principles discussed above apply also to a scenario where interference is the main radio-link impairment:

- The maximum data rate that can be achieved in a given bandwidth is limited by the available signal-power-to-interference-power ratio.
- Providing data rates larger than the available bandwidth (high-bandwidth utilization) is costly in the sense that it requires an un-proportionally high signal-to-interference ratio.

Also similar to a scenario where noise is the dominating radio-link impairment, reducing the cell size as well as the use of multi-antenna techniques are key means to increase the achievable data rates also in an interference-limited scenario:

- Reducing the cell size will obviously reduce the number of users, and thus also the overall traffic, per cell. This will reduce the relative interference level and thus allow for higher data rates.
- Similar to the increase in signal-to-noise ratio, proper combining of the signals received at multiple antennas will also increase the signal-to-interference ratio after the antenna combining.

- The use of beam-forming by means of multiple transmit antennas will focus the transmit power in the direction of the target receiver, leading to reduced interference to other radio links and thus improving the overall signal-to-interference ratio in the system.

One important difference between interference and noise is that interference, in contrast to noise, typically has a certain structure which makes it, at least to some extent, predictable and thus possible to further suppress or even remove completely. As an example, a dominant interfering signal may arrive from a certain direction in which case the corresponding interference can be further suppressed, or even completely removed, by means of *spatial processing* using multiple antennas at the receiver. This will be further discussed in Chapter 6. Also any differences in the spectrum properties between the target signal and an interfering signal can be used to suppress the interferer and thus reduce the overall interference level.

3.2   Higher data rates within a limited bandwidth: Higher-order modulation

As discussed in the previous section, providing data rates larger than the available bandwidth is fundamentally in-efficient in the sense that it requires un-proportionally high signal-to-noise and signal-to-interference ratios at the receiver. Still, bandwidth is often a scarce and expensive resource and, at least in some mobile-communication scenarios, high signal-to-noise and signal-to-interference ratios can be made available, e.g. in small-cell environments with a low traffic load or for mobile terminals close to the cell site. Future mobile-communication systems, including the evolution of 3G mobile communication, should be designed to be able to take advantage of such scenarios, that is should be able for offer very high data rates within a limited bandwidth when the radio conditions so allow.

A straightforward means to provide higher data rates within a given transmission bandwidth is the use of *higher-order modulation*, implying that the modulation alphabet is extended to include additional signaling alternatives and thus allowing for more bits of information to be communicated per modulation symbol.

In case of QPSK modulation, i.e. the modulation scheme used for the downlink in the first releases of the 3G mobile-communication standards (WCDMA and CDMA2000), the modulation alphabet consists of four different signaling alternatives. These four signaling alternatives can be illustrated as four different points in a two-dimensional plane (see Figure 3.2a). With four different signaling alternatives, QPSK allows for up to 2 bits of information to be communicated during each modulation-symbol interval. By extending to 16QAM modulation

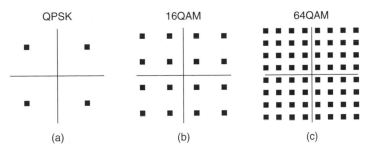

Figure 3.2 *Signal constellations for (a) QPSK, (b) 16QAM and (c) 64QAM.*

(Figure 3.2b), 16 different signaling alternatives are available. The use of 16QAM thus allows for up to 4 bits of information to be communicated per symbol interval. Further extension to 64QAM (Figure 3.2c), with 64 different signaling alternatives, allows for up to 6 bits of information to be communicated per symbol interval. At the same time, the bandwidth of the transmitted signal is, at least in principle, independent of the size of the modulation alphabet and mainly depends on the modulation rate, i.e. the number of modulation symbols per second. The maximum bandwidth utilization, expressed in bits/s/Hz, of 16QAM and 64QAM are thus, at least in principle, two and three times that of QPSK, respectively.

It should be pointed out that there are many other possible modulation schemes, in addition to those illustrated in Figure 3.2. One example is 8PSK consisting of eight signaling alternatives and thus providing up to 3 bits of information per modulation symbol. Readers are referred to [50] for a more thorough discussion on different modulation schemes.

The use of higher-order modulation provides the possibility for higher bandwidth utilization, that is the possibility to provide higher data rates within a given bandwidth. However, the higher bandwidth utilization comes at the cost of reduced robustness to noise and interference. Alternatively expressed, higher-order modulation schemes, such as 16QAM or 64QAM, require a higher E_b/N_0 at the receiver for a given bit-error probability, compared to QPSK. This is in line with the discussion in the previous section where it was concluded that high bandwidth utilization, i.e. a high information rate within a limited bandwidth, in general requires a higher receiver E_b/N_0.

3.2.1 Higher-order modulation in combination with channel coding

Higher-order modulation schemes such as 16QAM and 64QAM require, in themselves, a higher receiver E_b/N_0 for a given error rate, compared to QPSK.

However, in combination with channel coding the use of higher-order modulation will sometimes be more efficient, that is require a lower receiver E_b/N_0 for a given error rate, compared to the use of lower-order modulation such as QPSK. This may, for example, occur when the target bandwidth utilization implies that, with lower-order modulation, no or very little channel coding can be applied. In such a case, the additional channel coding that can be applied by using a higher-order modulation scheme such as 16QAM may lead to an overall gain in power efficiency compared to the use of QPSK.

As an example, if a bandwidth utilization of close to two information bits per modulation symbol is required, QPSK modulation would allow for very limited channel coding (channel-coding rate close to one). On the other hand, the use of 16QAM modulation would allow for a channel-coding rate in the order of one half. Similarly, if a bandwidth efficiency close to 4 information bits per modulation symbol is required, the use of 64QAM may be more efficient than 16QAM modulation, taking into account the possibility for lower-rate channel coding and corresponding additional coding gain in case of 64QAM. It should be noted that this does not speak against the general discussion in Section 3.1 where it was concluded that transmission with high-bandwidth utilization is inherently power in-efficient. The use of rate 1/2 channel coding for 16QAM obviously reduces the information data rate, and thus also the bandwidth utilization, to the same level as uncoded QPSK.

From the discussion above it can be concluded that, for a given signal-to-noise/ interference ratio, a certain combination of modulation scheme and channel-coding rate is optimal in the sense that it can deliver the highest-bandwidth utilization (the highest data rate within a given bandwidth) for that signal-to-noise/ interference ratio.

3.2.2   Variations in instantaneous transmit power

A general drawback of higher-order modulation schemes such as 16QAM and 64QAM, where information is encoded also in the instantaneous amplitude of the modulated signal, is that the modulated signal will have larger variations, and thus also larger peaks, in its instantaneous power. This can be seen from Figure 3.3 which illustrates the distribution of the instantaneous power, more specifically the probability that the instantaneous power is above a certain value, for QPSK, 16QAM, and 64QAM, respectively. Clearly, the probability for large peaks in the instantaneous power is higher in case of higher-order modulation.

Larger peaks in the instantaneous signal power imply that the transmitter power amplifier must be over-dimensioned to avoid that power-amplifier non-linearities,

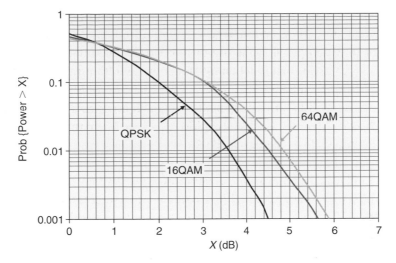

Figure 3.3 *Distribution of instantaneous power for different modulation schemes. Average power is same in all cases.*

occurring at high instantaneous power levels, cause corruption to the signal to be transmitted. As a consequence, the power-amplifier efficiency will be reduced, leading to increased power consumption. In addition, there will be a negative impact on the power-amplifier cost. Alternatively, the average transmit power must be reduced, implying a reduced range for a given data rate. High power-amplifier efficiency is especially important for the mobile terminal, i.e. in the uplink direction, due to the importance of low mobile-terminal power consumption and cost. For the base station, high power-amplifier efficiency, although far from irrelevant, is still somewhat less important. Thus, large peaks in the instantaneous signal power is less of an issue for the downlink compared to the uplink and, consequently, higher-order modulation is more suitable for the downlink compared to the uplink.

3.3 Wider bandwidth including multi-carrier transmission

As was shown in Section 3.1, transmission with a high-bandwidth utilization is fundamentally power in-efficient in the sense that it will require un-proportionally high signal-to-noise and signal-to-interference ratios for a given data rate. Providing very high data rates within a limited bandwidth, for example by means of higher-order modulation, is thus only possible in situations where relatively high signal-to-noise and signal-to-interference ratios can be made available, for example in small-cell environments with low traffic load or for mobile terminals close to the cell site.

Instead, to provide high data rates as efficiently as possible in terms of required signal-to-noise and signal-to-interference ratios, implying as good coverage as possible for high data rates, the transmission bandwidth should be at least of the same order as the data rates to be provided.

Having in mind that the provisioning of higher data rates with good coverage is one of the main targets for the evolution of 3G mobile communication, it can thus be concluded that support for even wider transmission bandwidth is an important part of this evolution.

However, there are several critical issues related to the use of wider transmission bandwidths in a mobile-communication system:

- Spectrum is, as already mentioned, often a scarce and expensive resource, and it may be difficult to find spectrum allocations of sufficient size to allow for very wideband transmission, especially at lower-frequency bands.
- The use of wider transmission and reception bandwidths has an impact on the complexity of the radio equipment, both at the base station and at the mobile terminal. As an example, a wider transmission bandwidth has a direct impact on the transmitter and the receiver sampling rates, and thus on the complexity and power consumption of digital-to-analog and analog-to-digital converters as well as front-end digital signal processing. RF components are also, in general, more complicated to design and more expensive to produce, the wider the bandwidth they are to handle.

The two issues above are mainly outside the scope of this book. However, a more specific technical issue related to wider-band transmission is the increased corruption of the transmitted signal due to time dispersion on the radio channel. Time dispersion occurs when the transmitted signal propagates to the receiver via multiple paths with different delays (see Figure 3.4a). In the frequency domain, a time-dispersive channel corresponds to a non-constant channel frequency response as illustrated in Figure 3.4b. This radio-channel *frequency selectivity* will corrupt the frequency-domain structure of the transmitted signal and lead to higher error rates for given signal-to-noise/interference ratios. Every radio channel is subject to frequency selectivity, at least to some extent. However, the extent to which the frequency selectivity impacts the radio communication depends on the bandwidth of the transmitted signal with, in general, larger impact for wider-band transmission. The amount of radio-channel frequency selectivity also depends on the environment with typically less frequency selectivity (less time dispersion) in case of small cells and in environments with few obstructions and potential reflectors, such as rural environments.

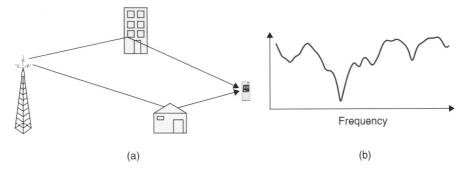

(a) (b)

Figure 3.4 *Multi-path propagation causing time dispersion and radio-channel frequency selectivity.*

It should be noted that Figure 3.4b illustrates a 'snapshot' of the channel frequency response. As a mobile terminal is moving through the environment, the detailed structure of the multi-path propagation, and thus also the detailed structure of the channel frequency response, may vary rapidly in time. The rate of the variations in the channel frequency response is related to the channel *Doppler spread*, f_D, defined as $f_D = v/c \cdot f_c$, where v is the speed of the mobile terminal, f_c is the carrier frequency (e.g. 2 GHz), and c is the speed of light.

Receiver-side *equalization* [50] has for many years been used to counteract signal corruption due to radio-channel frequency selectivity. Equalization has been shown to provide satisfactory performance with reasonable complexity at least up to bandwidths corresponding to the WCDMA bandwidth of 5 MHz (see, e.g. [29]). However, if the transmission bandwidth is further increased up to, for example 20 MHz, which is the target for the 3GPP *Long-Term Evolution*, the complexity of straightforward high-performance equalization starts to become a serious issue. One option is then to apply less optimal equalization, with a corresponding negative impact on the equalizer capability to counteract the signal corruption due to radio-channel frequency selectivity and thus a corresponding negative impact on the radio-link performance.

An alternative approach is to consider specific transmission schemes and signal designs that allow for good radio-link performance also in case of substantial radio-channel frequency selectivity without a prohibitively large receiver complexity. In the following, two such approaches to wider-band transmission will be discussed:

1. The use of different types of *multi-carrier transmission*, that is transmitting an overall wider-band signal as several more narrowband frequency-multiplexed signals, see below. One special case of multi-carrier transmission is *OFDM transmission* to be discussed in more detail in Chapter 4.

2. The use of specific *single-carrier* transmission schemes, especially designed to allow for efficient but still reasonably low-complexity equalization. This is further discussed in Chapter 5.

3.3.1 Multi-carrier transmission

One way to increase the overall transmission bandwidth, without suffering from increased signal corruption due to radio-channel frequency selectivity, is the use of so-called *multi-carrier transmission*. As illustrated in Figure 3.5, multi-carrier transmission implies that, instead of transmitting a single more wideband signal, multiple more narrowband signals, often referred to as *subcarriers*, are frequency multiplexed and jointly transmitted over the same radio link to the same receiver. By transmitting M signals in parallel over the same radio link, the overall data rate can be increased up to M times. At the same time, the impact in terms of signal corruption due to radio-channel frequency selectivity depends on the bandwidth of each subcarrier. Thus, the impact from a frequency-selective channel is essentially the same as for a more narrowband transmission scheme with a bandwidth that corresponds to the bandwidth of each subcarrier.

A drawback of the kind of multi-carrier evolution outlined in Figure 3.5, where an existing more narrowband radio-access technology is extended to a wider

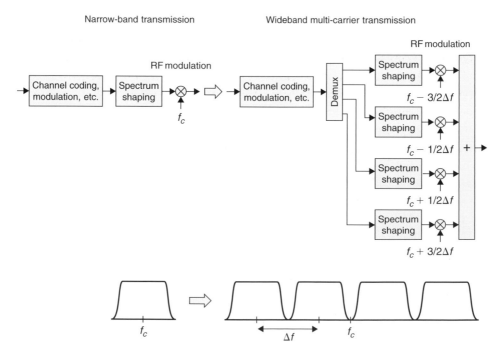

Figure 3.5   *Extension to wider transmission bandwidth by means of multi-carrier transmission.*

overall transmission bandwidth by the parallel transmission of M more narrow-band carriers, is that the spectrum of each subcarrier does typically not allow for very tight subcarrier 'packing.' This is illustrated by the 'valleys' in the overall multi-carrier spectrum outlined in the lower part of Figure 3.5. This has a some-what negative impact on the overall bandwidth efficiency of this kind of multi-carrier transmission.

As an example, consider a WCDMA multi-carrier evolution towards wider band-width. WCDMA has a modulation rate, also referred to as the *WCDMA chip rate*, of $f_{cr} = 3.84$ Mchips/s. However, due to spectrum shaping, even the theoretical WCDMA spectrum, not including spectrum widening due to transmitter imperfec-tions, has a bandwidth that significantly exceeds 3.84 MHz. More specifically, as can be seen in Figure 3.6, the theoretical WCDMA spectrum has a *raised-cosine* shape with roll-off $\alpha = 0.22$. As a consequence, the bandwidth outside of which the WCDMA theoretical spectrum equals zero is approximately 4.7 MHz (see right part of Figure 3.6).

For a straightforward multi-carrier extension of WCDMA, the subcarriers must thus be spaced approximately 4.7 MHz from each other to completely avoid inter-subcarrier interference. It should be noted though that a smaller subcarrier spacing can be used with only limited inter-subcarrier interference.

A second drawback of multi-carrier transmission is that, similar to the use of higher-order modulation, the parallel transmission of multiple carriers will lead to larger variations in the instantaneous transmit power. Thus, similar to the use of higher-order modulation, multi-carrier transmission will have a negative impact on the transmitter power-amplifier efficiency, implying increased transmitter power consumption and increased power-amplifier cost. Alternatively, the aver-age transmit power must be reduced, implying a reduced range for a given data rate. For this reason, similar to the use of higher-order modulation, multi-carrier

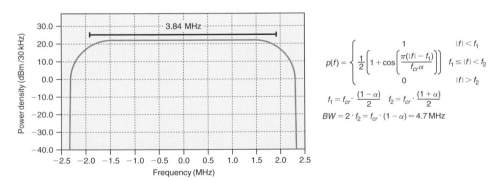

Figure 3.6 *Theoretical WCDMA spectrum. Raised-cosine shape with roll-off $\alpha = 0.22$.*

transmission is more suitable for the downlink (base-station transmission), com-pared to the uplink (mobile-terminal transmission), due to the higher importance of high power-amplifier efficiency at the mobile terminal.

The main advantage with the kind of multi-carrier extension outlined in Figure 3.5 is that it provides a very smooth evolution, in terms of both radio equipment and spectrum, of an already existing radio-access technology to wider transmission bandwidth and a corresponding possibility for higher data rates, especially for the downlink. In essence this kind of multi-carrier evolution to wider band-width can be designed so that, for legacy terminals not capable of multi-carrier reception, each downlink 'subcarrier' will appear as an original, more narrow-band carrier, while, for a multi-carrier-capable terminal, the network can make use of the full multi-carrier bandwidth to provide higher data rates.

The next chapter will discuss, in more detail, a different approach to multi-carrier transmission, based on so-called OFDM technique.

4

OFDM transmission

In this chapter, a more detailed overview of OFDM or *Orthogonal Frequency Division Multiplexing* will be given. OFDM has been adopted as the downlink transmission scheme for the 3GPP Long-Term Evolution (LTE) and is also used for several other radio technologies, e.g. WiMAX [116] and the DVB broadcast technologies [17].

4.1 Basic principles of OFDM

Transmission by means of OFDM can be seen as a kind of multi-carrier transmission. The basic characteristics of OFDM transmission, which distinguish it from a straightforward multi-carrier extension of a more narrowband transmission scheme as outlined in Figure 3.5 in the previous chapter, are:

- The use of a relatively large number of narrowband subcarriers. In contrast, a straightforward multi-carrier extension as outlined in Figure 3.5 would typically consist of only a few subcarriers, each with a relatively wide bandwidth. As an example, a WCDMA multi-carrier evolution to a 20 MHz overall transmission bandwidth could consist of four (sub)carriers, each with a bandwidth in the order of 5 MHz. In comparison, OFDM transmission may imply that several hundred subcarriers are transmitted over the same radio link to the same receiver.
- Simple rectangular pulse shaping as illustrated in Figure 4.1a. This corresponds to a sinc-square-shaped per-subcarrier spectrum, as illustrated in Figure 4.1b.
- Tight frequency-domain packing of the subcarriers with a subcarrier spacing $\Delta f = 1/T_u$, where T_u is the per-subcarrier modulation-symbol time (see Figure 4.2). The subcarrier spacing is thus equal to the per-subcarrier modulation rate $1/T_u$.

An illustrative description of a basic OFDM modulator is provided in Figure 4.3. It consists of a bank of N_c complex modulators, where each modulator corresponds to one OFDM subcarrier.

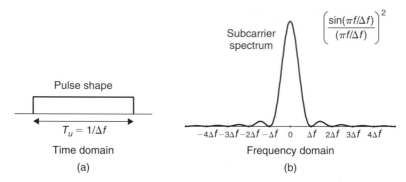

Figure 4.1   *(a) Per-subcarrier pulse shape and (b) spectrum for basic OFDM transmission.*

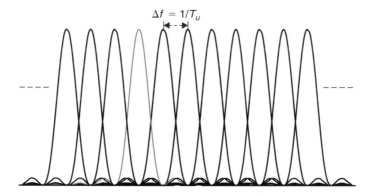

Figure 4.2   *OFDM subcarrier spacing.*

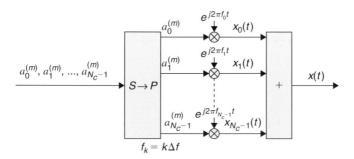

Figure 4.3   *OFDM modulation.*

In complex baseband notation, a basic OFDM signal $x(t)$ during the time interval $mT_u \leq t < (m + 1)T_u$ can thus be expressed as

$$x(t) = \sum_{k=0}^{N_c-1} x_k(t) = \sum_{k=0}^{N_c-1} a_k^{(m)} e^{j2\pi k \Delta f t} \tag{4.1}$$

where $x_k(t)$ is the kth modulated subcarrier with frequency $f_k = k \cdot \Delta f$ and $a_k^{(m)}$ is the, in general complex, modulation symbol applied to the kth subcarrier during the mth OFDM symbol interval, i.e. during the time interval $mT_u \leq t < (m + 1)T_u$. OFDM transmission is thus block based, implying that, during each OFDM symbol interval, N_c modulation symbols are transmitted in parallel. The modulation symbols can be from any modulation alphabet, such as QPSK, 16QAM, or 64QAM.

The number of OFDM subcarriers can range from less than one hundred to several thousand, with the subcarrier spacing ranging from several hundred kHz down to a few kHz. What subcarrier spacing to use depends on what types of environments the system is to operate in, including such aspects as the maximum expected radio-channel frequency selectivity (maximum expected time dispersion) and the maximum expected rate of channel variations (maximum expected Doppler spread). Once the subcarrier spacing has been selected, the number of subcarriers can be decided based on the assumed overall transmission bandwidth, taking into account acceptable out-of-band emission, etc. The selection of OFDM subcarrier spacing and number of subcarriers is discussed in somewhat more detail in Section 4.8.

As an example, for 3GPP LTE the basic subcarrier spacing equals 15 kHz. On the other hand, the number of subcarriers depends on the transmission bandwidth, with in the order of 600 subcarriers in case of operation in a 10 MHz spectrum allocation and correspondingly fewer/more subcarriers in case of smaller/larger overall transmission bandwidths.

The term *Orthogonal Frequency Division Multiplex* is due to the fact that two modulated OFDM subcarriers x_{k_1} and x_{k_2} are mutually *orthogonal* over the time interval $mT_u \leq t < (m + 1)T_u$, i.e.

$$\int_{mT_u}^{(m+1)T_u} x_{k_1}(t)x_{k_2}^*(t)\, dt = \int_{mT_u}^{(m+1)T_u} a_{k_1}a_{k_2}^* e^{j2\pi k_1 \Delta f t} e^{-j2\pi k_2 \Delta f t}\, dt = 0 \qquad \text{for } k_1 \neq k_2$$

$$(4.2)$$

Thus basic OFDM transmission can be seen as the modulation of a set of orthogonal functions $\varphi_k(t)$, where

$$\varphi_k(t) = \begin{cases} e^{j2\pi k \Delta f t} & 0 \leq t < T_u \\ 0 & \text{otherwise} \end{cases} \qquad (4.3)$$

The 'physical resource' in case of OFDM transmission is often illustrated as a time–frequency grid according to Figure 4.4 where each 'column'

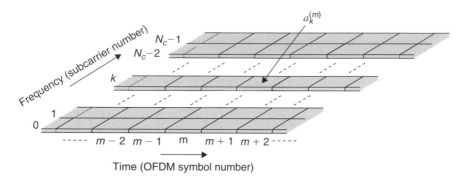

Figure 4.4   *OFDM time–frequency grid.*

corresponds to one OFDM symbol and each 'row' corresponds to one OFDM subcarrier.

4.2 OFDM demodulation

Figure 4.5 illustrates the basic principle of OFDM demodulation consisting of a bank of correlators, one for each subcarrier. Taking into account the orthogonality between subcarriers according to (4.2), it is clear that, in the ideal case, two OFDM subcarriers do not cause any interference to each other after demodulation. Note that this is the case despite the fact that the spectrum of neighbor subcarriers clearly overlap, as can be seen from Figure 4.2. Thus the avoidance of interference between OFDM subcarriers is not simply due to a subcarrier spectrum separation, which is, for example, the case for the kind of straightforward multi-carrier extension outlined in Figure 3.5 in the previous chapter. Rather, the subcarrier orthogonality is due to the *specific* frequency-domain structure of each subcarrier in combination with the *specific* choice of a subcarrier spacing Δf equal to the per-subcarrier symbol rate $1/T_u$. However, this also implies that, in contrast to the kind of multi-carrier transmission outlined in Section 3.3.1 of the previous chapter, any corruption of the frequency-domain structure of the OFDM subcarriers, e.g. due to a frequency-selective radio channel, may lead to a loss of inter-subcarrier orthogonality and thus to interference between subcarriers. To handle this and to make an OFDM signal truly robust to radio-channel frequency selectivity, *cyclic-prefix insertion* is typically used, as will be further discussed in Section 4.4.

4.3 OFDM implementation using IFFT/FFT processing

Although a bank of modulators/correlators according to Figures 4.3 and 4.5 can be used to illustrate the basic principles of OFDM modulation and demodulation, respectively, these are not the most appropriate modulator/demodulator structures for actual implementation. Actually, due to its specific structure and the selection

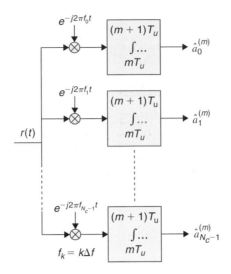

Figure 4.5 *Basic principle of OFDM demodulation.*

of a subcarrier spacing Δf equal to the per-subcarrier symbol rate $1/T_u$, OFDM allows for low-complexity implementation by means of computationally efficient *Fast Fourier Transform* (FFT) processing.

To confirm this, consider a time-discrete (sampled) OFDM signal where it is assumed that the sampling rate f_s is a multiple of the subcarrier spacing Δf, i.e. $f_s = 1/T_s = N \cdot \Delta f$. The parameter N should be chosen so that the sampling theorem [50] is sufficiently fulfilled.[1] As $N_c \cdot \Delta f$ can be seen as the nominal bandwidth of the OFDM signal, this implies that N should exceed N_c with a sufficient margin.

With these assumptions, the time-discrete OFDM signal can be expressed as[2]

$$x_n = x(nT_s) = \sum_{k=0}^{N_c-1} a_k e^{j2\pi k \Delta f n T_s} = \sum_{k=0}^{N_c-1} a_k e^{j2\pi kn/N} = \sum_{k=0}^{N-1} a'_k e^{j2\pi kn/N} \quad (4.4)$$

where

$$a'_k = \begin{cases} a_k & 0 \le k < N_c \\ 0 & N_c \le k < N \end{cases} \quad (4.5)$$

Thus, the sequence x_n, i.e. the sampled OFDM signal, is the size-N Inverse Discrete Fourier Transform (IDFT) of the block of modulation symbols

---

[1] An OFDM signal defined according to (4.1) in theory has an infinite bandwidth and thus the sampling theorem can never be fulfilled completely.

[2] From now on the index m on the modulation symbols, indicating the OFDM-symbol number, will be ignored unless especially needed.

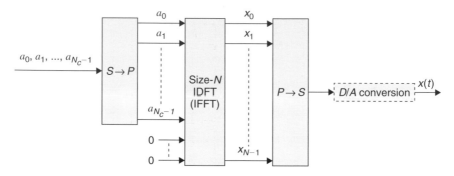

Figure 4.6   *OFDM modulation by means of IFFT processing.*

a_0, a_1, ..., a_{N_c-1} extended with zeros to length N. OFDM modulation can thus be implemented by means of IDFT processing followed by digital-to-analog conversion, as illustrated in Figure 4.6. Especially, by selecting the IDFT size N equal to 2^m for some integer m, the OFDM modulation can be implemented by means of implementation-efficient radix-2 Inverse Fast Fourier Transform (IFFT) processing. It should be noted that the ratio N/N_c, which could be seen as the over-sampling of the time-discrete OFDM signal, could very well be, and typically is, a non-integer number. As an example and as already mentioned, for 3GPP LTE the number of subcarriers N_c is approximately 600 in case of a 10 MHz spectrum allocation. The IFFT size can then, for example, be selected as $N = 1024$. This corresponds to a sampling rate $f_s = N \cdot \Delta f = 15.36$ MHz, where $\Delta f = 15$ kHz is the LTE subcarrier spacing.

It is important to understand that IDFT/IFFT-based implementation of an OFDM modulator, and even more so the exact IDFT/IFFT size, are just transmitter-implementation choices and not something that would be mandated by any radio-access specification. As an example, nothing forbids the implementation of an OFDM modulator as a set of parallel modulators as illustrated in Figure 4.3. Also nothing prevents the use of a larger IFFT size, e.g. a size-2048 IFFT size, even in case of a smaller number of OFDM subcarriers.

Similar to OFDM modulation, efficient FFT processing can be used for OFDM demodulation, replacing the bank of N_c parallel demodulators of Figure 4.5 with sampling with some sampling rate $f_s = 1/T_s$, followed by a size-N DFT/FFT, as illustrated in Figure 4.7.

4.4   Cyclic-prefix insertion

As described in Section 4.2, an uncorrupted OFDM signal can be demodulated without any interference between subcarriers. One way to understand this *subcarrier orthogonality* is to recognize that a modulated subcarrier $x_k(t)$ in

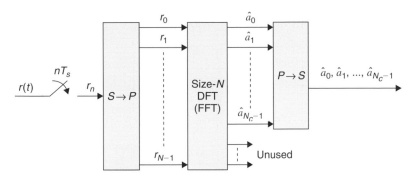

Figure 4.7 *OFDM demodulation by means of FFT processing.*

(4.1) consists of an integer number of periods of complex exponentials during the demodulator integration interval $T_u = 1/\Delta f$.

However, in case of a time-dispersive channel the orthogonality between the sub-carriers will, at least partly, be lost. The reason for this loss of subcarrier orthogonality in case of a time-dispersive channel is that, in this case, the demodulator correlation interval for one path will overlap with the symbol boundary of a different path, as illustrated in Figure 4.8. Thus, the integration interval will not necessarily correspond to an integer number of periods of complex exponentials of that path as the modulation symbols a_k may differ between consecutive symbol intervals. As a consequence, in case of a time-dispersive channel there will not only be inter-symbol interference within a subcarrier but also interference between subcarriers.

Another way to explain the interference between subcarriers in case of a time-dispersive channel is to have in mind that time dispersion on the radio channel is equivalent to a frequency-selective channel frequency response. As clarified in Section 4.2, orthogonality between OFDM subcarriers is not simply due to frequency-domain separation but due to the *specific* frequency-domain structure of each subcarrier. Even if the frequency-domain channel is constant over a bandwidth corresponding to the main lobe of an OFDM subcarrier and only the subcarrier side lobes are corrupted due to the radio-channel frequency selectivity, the orthogonality between subcarriers will be lost with inter-subcarrier interference as a consequence. Due to the relatively large side lobes of each OFDM subcarrier, already a relatively limited amount of time dispersion or, equivalently, a relatively modest radio-channel frequency selectivity may cause non-negligible interference between subcarriers.

To deal with this problem and to make an OFDM signal truly insensitive to time dispersion on the radio channel, so-called *cyclic-prefix insertion* is typically used in case of OFDM transmission. As illustrated in Figure 4.9, cyclic-prefix

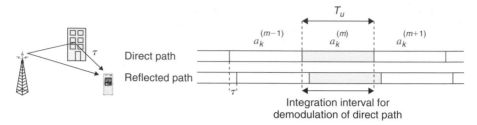

Figure 4.8 *Time dispersion and corresponding received-signal timing.*

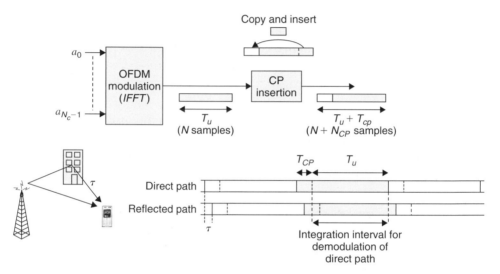

Figure 4.9 *Cyclic-prefix insertion.*

insertion implies that the last part of the OFDM symbol is copied and inserted at the beginning of the OFDM symbol. Cyclic-prefix insertion thus increases the length of the OFDM symbol from T_u to $T_u + T_{CP}$, where T_{CP} is the length of the cyclic prefix, with a corresponding reduction in the OFDM symbol rate as a consequence. As illustrated in the lower part of Figure 4.9, if the correlation at the receiver side is still only carried out over a time interval $T_u = 1/\Delta f$, subcarrier orthogonality will then be preserved also in case of a time-dispersive channel, as long as the span of the time dispersion is shorter than the cyclic-prefix length.

In practice, cyclic-prefix insertion is carried out on the time-discrete output of the transmitter IFFT. Cyclic-prefix insertion then implies that the last N_{CP} samples of the IFFT output block of length N is copied and inserted at the beginning of the block, increasing the block length from N to $N + N_{CP}$. At the receiver side, the corresponding samples are discarded before OFDM demodulation by means of, for example, DFT/FFT processing.

Cyclic-prefix insertion is beneficial in the sense that it makes an OFDM signal insensitive to time dispersion as long as the span of the time dispersion does not exceed the length of the cyclic prefix. The drawback of cyclic-prefix insertion is that only a fraction $T_u/(T_u + T_{CP})$ of the received signal power is actually utilized by the OFDM demodulator, implying a corresponding power loss in the demodulation. In addition to this power loss, cyclic-prefix insertion also implies a corresponding loss in terms of bandwidth as the OFDM symbol rate is reduced without a corresponding reduction in the overall signal bandwidth.

One way to reduce the relative overhead due to cyclic-prefix insertion is to reduce the subcarrier spacing Δf, with a corresponding increase in the symbol time T_u as a consequence. However, this will increase the sensitivity of the OFDM transmission to fast channel variations, that is high Doppler spread, as well as different types of frequency errors, see further Section 4.8.

It is also important to understand that the cyclic prefix does not necessarily have to cover the entire length of the channel time dispersion. In general, there is a trade-off between the power loss due to the cyclic prefix and the signal corruption (inter-symbol and inter-subcarrier interference) due to residual time dispersion not covered by the cyclic prefix and, at a certain point, further reduction of the signal corruption due to further increase of the cyclic-prefix length will not justify the corresponding additional power loss. This also means that, although the amount of time dispersion typically increases with the cell size, beyond a certain cell size there is often no reason to increase the cyclic prefix further as the corresponding power loss due to a further increase of the cyclic prefix would have a larger negative impact, compared to the signal corruption due to the residual time dispersion not covered by the cyclic prefix [15].

4.5 Frequency-domain model of OFDM transmission

Assuming a sufficiently large cyclic prefix, the linear convolution of a time-dispersive radio channel will appear as a circular convolution during the demodulator integration interval T_u. The combination of OFDM modulation (IFFT processing), a time-dispersive radio channel, and OFDM demodulation (*FFT* processing) can then be seen as a *frequency-domain* channel as illustrated in Figure 4.10, where the frequency-domain channel taps $H_0, \ldots, H_{N_c-1}$ can be directly derived from the channel impulse response.

The demodulator output b_k in Figure 4.10 is the transmitted modulation symbol a_k scaled and phase rotated by the complex frequency-domain channel tap H_k and impaired by noise n_k. To properly recover the transmitted symbol for

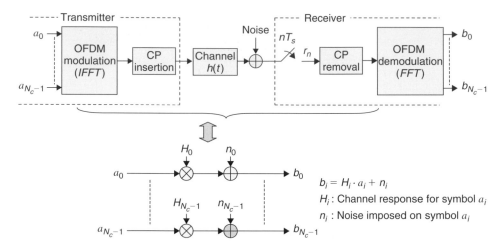

Figure 4.10   *Frequency-domain model of OFDM transmission/reception.*

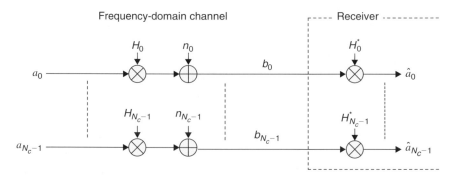

Figure 4.11   *Frequency-domain model of OFDM transmission/reception with 'one-tap equalization' at the receiver.*

further processing, for example data demodulation and channel decoding, the receiver should multiply b_k with the complex conjugate of H_k as illustrated in Figure 4.11. This is often expressed as a *one-tap equalizer* being applied to each received subcarrier.

4.6   Channel estimation and reference symbols

As described above, to demodulate the transmitted modulation symbol a_k and allow for proper decoding of the transmitted information at the receiver side, scaling with the complex conjugate of the frequency-domain channel tap H_k should be applied after OFDM demodulation (*FFT* processing) (see Figure 4.11). To be able to do this, the receiver obviously needs an estimate of the frequency-domain channel taps $H_0, \ldots, H_{N_c-1}$.

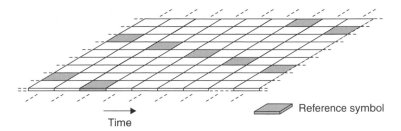

Time

Reference symbol

Figure 4.12   *Time-frequency grid with known reference symbols.*

The frequency-domain channel taps can be estimated indirectly by first estimating the channel impulse response and, from that, calculate an estimate of H_k. However, a more straightforward approach is to estimate the frequency-domain channel taps directly. This can be done by inserting known *reference symbols*, sometimes also referred to as *pilot symbols*, at regular intervals within the OFDM time-frequency grid, as illustrated in Figure 4.12. Using knowledge about the reference symbols, the receiver can estimate the frequency-domain channel around the location of the reference symbol. The reference symbols should have a sufficiently high density in both the time and the frequency domain to be able to provide estimates for the entire time/frequency grid also in case of radio channels subject to high frequency and/or time selectivity.

Different more or less advanced algorithms can be used for the channel estimation, ranging from simple averaging in combination with linear interpolation to Minimum-Mean-Square-Error (MMSE) estimation relying on more detailed knowledge of the channel time/frequency-domain characteristics. Readers are referred to, for example, [31] for a more in-depth discussion on channel estimation for OFDM.

4.7   Frequency diversity with OFDM: Importance of channel coding

As discussed in Section 3.3 in the previous chapter, a radio channel is always subject to some degree of frequency selectivity, implying that the channel quality will vary in the frequency domain. In case of a single wideband carrier, such as a WCDMA carrier, each modulation symbol is transmitted over the entire signal bandwidth. Thus, in case of the transmission of a single wideband carrier over a highly frequency-selective channel (see Figure 4.13a), each modulation symbol will be transmitted both over frequency bands with relatively good quality (relatively high signal strength) and frequency bands with low quality (low signal strength). Such transmission of information over multiple frequency bands with different instantaneous channel quality is also referred to as *frequency diversity*.

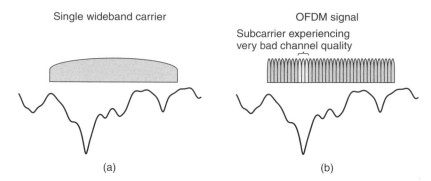

Figure 4.13   *(a) Transmission of single wideband carrier and (b) OFDM transmission over a frequency-selective channel.*

On the other hand, in case of OFDM transmission each modulation symbol is mainly confined to a relatively narrow bandwidth. Thus, in case of OFDM transmission over a frequency-selective channel, certain modulation symbols may be fully confined to a frequency band with very low instantaneous signal strength as illustrated in Figure 4.13b. Thus, the individual modulation symbols will typically not experience any substantial frequency diversity even if the channel is highly frequency selective over the overall OFDM transmission bandwidth. As a consequence, the basic error-rate performance of OFDM transmission over a frequency-selective channel is relatively poor and especially much worse than the basic error rate in case of a single wideband carrier.

However, in practice channel coding is used in most cases of digital communication and especially in case of mobile communication. Channel coding implies that each bit of information to be transmitted is spread over several, often very many, code bits. If these coded bits are then, via modulation symbols, mapped to a set of OFDM subcarriers that are well distributed over the overall transmission bandwidth of the OFDM signal, as illustrated in Figure 4.14, each information bit will experience frequency diversity in case of transmission over a radio channel that is frequency selective over the transmission bandwidth, despite the fact that the subcarriers, and thus also the code bits, will not experience any frequency diversity. Distributing the code bits in the frequency domain, as illustrated in Figure 4.14, is sometimes referred to as *frequency interleaving*. This is similar to the use of time-domain interleaving to benefit from channel coding in case of fading that varies in time.

Thus, in contrast to the transmission of a single wideband carrier, channel coding (combined with frequency interleaving) is an essential component in order for OFDM transmission to be able to benefit from frequency diversity on a frequency-selective channel. As channel coding is typically anyway used

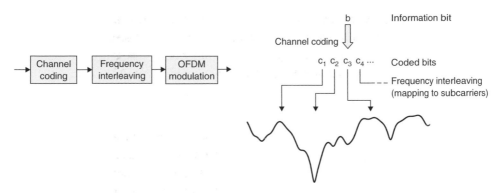

Figure 4.14 *Channel coding in combination with frequency-domain interleaving to provide frequency diversity in case of OFDM transmission.*

in most cases of mobile communication this is not a very serious drawback, especially taking into account that a significant part of the available frequency diversity can be captured already with a relatively high code rate.

4.8 Selection of basic OFDM parameters

If OFDM is to be used as the transmission scheme in a mobile-communication system, the following basic OFDM parameters need to be decided on:

- The subcarrier spacing Δf.
- The number of subcarriers N_c, which, together with the subcarrier spacing, determines the overall transmission bandwidth of the OFDM signal.
- The cyclic-prefix length T_{CP}. Together with the subcarrier spacing $\Delta f = 1/T_u$, the cyclic-prefix length determines the overall OFDM symbol time $T = T_{CP} + T_u$ or, equivalently, the OFDM symbol rate.

4.8.1 OFDM subcarrier spacing

There are two factors that constrain the selection of the OFDM subcarrier spacing:

- The OFDM subcarrier spacing should be as small as possible (T_u as large as possible) to minimize the relative cyclic-prefix overhead $T_{CP}/(T_u + T_{CP})$, see further Section 4.8.3.
- A too small subcarrier spacing increases the sensitivity of the OFDM transmission to Doppler spread and different kinds of frequency inaccuracies.

A requirement for the OFDM subcarrier orthogonality (4.2) to hold at the receiver side, i.e. after the transmitted signal has propagated over the radio channel, is that the instantaneous channel does not vary noticeably during the demodulator correlation interval T_u (see Figure 4.5). In case of such channel variations, e.g. due to very high Doppler spread, the orthogonality between subcarriers will be lost with inter-subcarrier interference as a consequence. Figure 4.15 illustrates the subcarrier signal-to-interference ratio due to inter-subcarrier interference between two neighbor subcarriers, as a function of the normalized Doppler spread. When considering Figure 4.15, it should be had in mind that a subcarrier will be subject to interference from multiple subcarriers on both sides,[3] that is the overall inter-subcarrier interference from all subcarriers will be higher than what is illustrated in Figure 4.15.

In practice, the amount of inter-subcarrier interference that can be accepted very much depends on the service to be provided and to what extent the received signal is anyway corrupted due to noise and other impairments. As an example, on the cell border of large cells the signal-to-noise/interference ratio will anyway be relatively low, with relatively low achievable data rates as a consequence. A small amount of additional inter-subcarrier interference, for example, due to Doppler spread, may then be more or less negligible. At the same time, in high signal-to-noise/interference scenarios, for example, in small cells with low traffic or close to the base station, where high data rates are to be provided, the same amount of inter-subcarrier interference may have a much more negative impact.

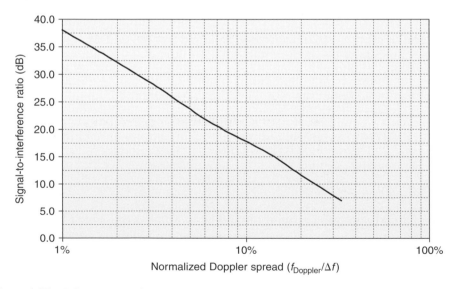

Figure 4.15 *Subcarrier interference as a function of the normalized Doppler spread $f_{Doppler}/\Delta f$.*

---

[3] Except for the subcarrier at the edge of the spectrum.

It should also be noted that, in addition to Doppler spread, inter-subcarrier interference will also be due to different transmitter and receiver inaccuracies, such as frequency errors and phase noise.

4.8.2 Number of subcarriers

Once the subcarrier spacing has been selected based on environment, expected Doppler spread and time dispersion, etc., the number of subcarriers can be determined based on the amount of spectrum available and the acceptable out-of-band emissions.

The basic bandwidth of an OFDM signal equals $N_c \cdot \Delta f$, i.e. the number of subcarriers multiplied by the subcarrier spacing. However, as can be seen in Figure 4.16, the spectrum of a basic OFDM signal falls off very slowly outside the basic OFDM bandwidth and especially much slower than for a WCDMA signal. The reason for the large *out-of-band emission* of a basic OFDM signal is the use of rectangular pulse shaping (Figure 4.1), leading to per-subcarrier side lobes that fall off relatively slowly. However, in practice, straightforward filtering or *time-domain windowing* [117] will be used to suppress a main part of the OFDM out-of-band emissions. Thus, in practice, typically in the order of 10% guard-band is needed for an OFDM signal implying, as an example, that in a spectrum allocation of 5 MHz, the basic OFDM bandwidth $N_c \cdot \Delta f$ could be in the order of 4.5 MHz. Assuming, for example, a subcarrier spacing of 15 kHz as selected for LTE, this corresponds to approximately 300 subcarriers in 5 MHz.

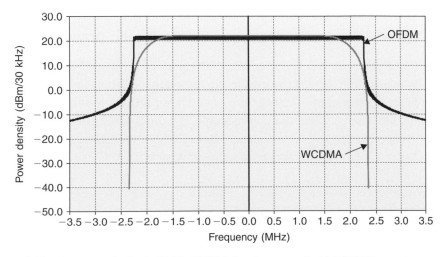

Figure 4.16 *Spectrum of a basic 5 MHz OFDM signal compared with WCDMA spectrum.*

4.8.3   Cyclic-prefix length

In principle, the cyclic-prefix length T_{CP} should cover the maximum length of the time-dispersion expected to be experienced. However, as already discussed, increasing the length of the cyclic prefix, without a corresponding reduction in the subcarrier spacing Δf, implies an additional overhead in terms of power as well as bandwidth. Especially the power loss implies that, as the cell size grows and the system performance becomes more power limited, there is a trade-off between the loss in power due to the cyclic prefix and the signal corruption due to time dispersion not covered by the cyclic prefix. As already mentioned, this implies that, although the amount of time dispersion typically increases with the cell size, beyond a certain cell size there is often no reason to increase the cyclic prefix further as the corresponding power loss would have a larger negative impact, compared to the signal corruption due to the residual time dispersion not covered by the cyclic prefix [15].

One situation where a longer cyclic prefix may be needed is in the case of multi-cell transmission using SFN (Single-Frequency Network), as further discussed in Section 4.11.

Thus, to be able to optimize performance to different environments, some OFDM-based systems support multiple cyclic-prefix lengths. The different cyclic-prefix lengths can then be used in different transmission scenarios:

- Shorter cyclic prefix in small-cell environments to minimize the cyclic-prefix overhead.
- Longer cyclic prefix in environments with extreme time dispersion and especially in case of SFN operation.

4.9   Variations in instantaneous transmission power

According to Section 3.3.1, one of the drawbacks of multi-carrier transmission is the corresponding large variations in the instantaneous transmit power, implying a reduced power-amplifier efficiency and higher mobile-terminal power consumption, alternatively that the power-amplifier output power has to be reduced with a reduced range as a consequence. Being a kind of multi-carrier transmission scheme, OFDM is subject to the same drawback.

However, a large number of different methods have been proposed to reduce the large power peaks of an OFDM signal:

- In case of *tone reservation* [78], a subset of the OFDM subcarriers are not used for data transmission. Instead, these subcarriers are modulated in such a way that the largest peaks of the overall OFDM signal are suppressed, allowing for a reduced power-amplifier back-off. One drawback of tone reservation is the bandwidth loss due to the fact that a number of subcarriers are not available for actual data transmission. The calculation of what modulation to apply to the reserved tones can also be of relatively high complexity.

- In case of *pre-filtering* or *pre-coding*, linear processing is applied to the sequence of modulation symbols before OFDM modulation. DFT-spread-OFDM, to be described in the next chapter, can be seen as one kind of pre-filtering.
- In case of *selective scrambling* [23], the coded-bit sequence to be transmitted is scrambled with a number of different scrambling codes. Each scrambled sequence is then OFDM modulated and the signal with the lowest peak power is selected for transmission. After OFDM demodulation at the receiver side, descrambling and subsequent decoding is carried out for all the possible scrambling sequences. Only the decoding carried out for the scrambling code actually used for the transmission will provide a correct decoding result. A drawback of selective scrambling is an increased receiver complexity as multiple decodings need to be carried out in parallel.

Readers are referred to the references above for a more in-depth discussion on different peak-reduction schemes.

4.10 OFDM as a user-multiplexing and multiple-access scheme

The discussion has, until now, implicitly assumed that all OFDM subcarriers are transmitted from the same transmitter to a certain receiver, i.e.:

- downlink transmission of all subcarriers to a *single* mobile terminal;
- uplink transmission of all subcarriers from a *single* mobile terminal.

However, OFDM can also be used as a *user-multiplexing* or *multiple-access scheme*, allowing for simultaneous frequency-separated transmissions to/from multiple mobile terminals (see Figure 4.17).

In the downlink direction, OFDM as a user-multiplexing scheme implies that, in each OFDM symbol interval, different subsets of the overall set of available subcarriers are used for transmission to *different* mobile terminals (see Figure 4.17a).

Similarly, in the uplink direction, OFDM as a *user-multiplexing* or *multiple-access scheme* implies that, in each OFDM symbol interval, different subsets of

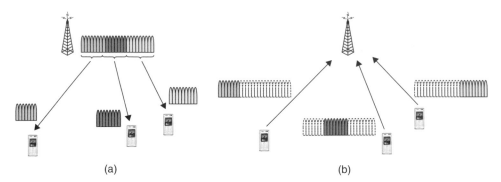

Figure 4.17   *OFDM as a user-multiplexing/multiple-access scheme: (a) downlink and (b) uplink.*

the overall set of subcarriers are used for data transmission from *different* mobile terminals (see Figure 4.17b). In this case, the term Orthogonal Frequency Division Multiple Access or OFDMA is also often used.[4]

Figure 4.17 assumes that *consecutive* subcarriers are used for transmission to/from the same mobile terminal. However, distributing the subcarriers to/from a mobile terminal in the frequency domain is also possible as illustrated in Figure 4.18. The benefit of such *distributed* user multiplexing or *distributed* multiple access is a possibility for additional frequency diversity as each transmission is spread over a wider bandwidth.

In the case when OFDMA is used as an uplink multiple-access scheme, i.e. in case of frequency multiplexing of OFDM signals from multiple mobile terminals, it is critical that the transmissions from the different mobile terminals arrive approximately time aligned at the base station. More specifically, the transmissions from the different mobile terminals should arrive at the base station with a timing misalignment less than the length of the cyclic prefix to preserve orthogonality between subcarriers received from different mobile terminals and thus avoid inter-user interference.

Due to the differences in distance to the base station for different mobile terminals and the corresponding differences in the propagation time (which may far exceed the length of the cyclic prefix), it is therefore necessary to control the uplink transmission timing of each mobile terminal (see Figure 4.19). Such *transmit-timing control* should adjust the transmit timing of each mobile terminal to ensure that uplink transmissions arrive approximately time aligned at the base station. As the propagation time changes as the mobile terminal is moving

---

[4]The term OFDMA is sometimes also used to denote the use of OFDM to multiplex multiple users in the downlink as illustrated in Figure 4.17a.

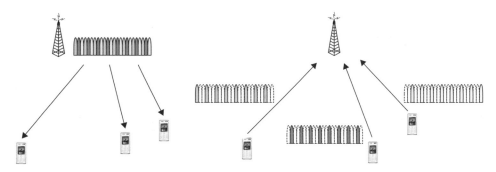

Figure 4.18 *Distributed user multiplexing.*

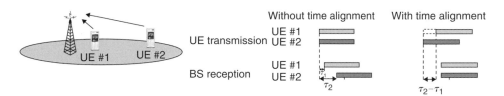

Figure 4.19 *Uplink transmission-timing control.*

within the cell, the transmit-timing control should be an active process, continuously adjusting the exact transmit timing of each mobile terminal.

Furthermore, even in case of perfect transmit-timing control, there will always be some interference between subcarriers e.g. due to frequency errors. Typically this interference is relatively low in case of reasonable frequency errors, Doppler spread, etc. (see Section 4.8). However, this assumes that the different subcarriers are received with at least approximately the same power. In the uplink, the propagation distance and thus the path loss of the different mobile-terminal transmissions may differ significantly. If two terminals are transmitting with the same power, the received-signal strengths may thus differ significantly, implying a potentially significant interference from the stronger signal to the weaker signal unless the subcarrier orthogonality is perfectly retained. To avoid this, at least some degree of *uplink transmit-power control* may need to be applied in case of uplink OFDMA, reducing the transmit power of user terminals close to the base station and ensuring that all received signals will be of approximately the same power.

4.11 Multi-cell broadcast/multicast transmission and OFDM

The provisioning of broadcast/multicast services in a mobile-communication system implies that the same information is to be *simultaneously* provided to multiple mobile terminals, often dispersed over a large area corresponding to a large number of cells as shown in Figure 4.20. The broadcast/multicast information may be a TV news clip, information about the local weather conditions, stock-market information, or any other kind of information that, at a given time instant, may be of interest to a large number of people.

When the same information is to be provided to multiple mobile terminals within a cell it is often beneficial to provide this information as a single 'broadcast' radio transmission covering the entire cell and simultaneously being received by all relevant mobile terminals (Figure 4.21a), rather than providing the information by means of individual transmissions to each mobile terminal (unicast transmission, see Figure 4.21b).

As a broadcast transmission according to Figure 4.21a has to be dimensioned to reach also the worst-case mobile terminals, including mobile terminals at the cell border, it will be relative costly in terms of the recourses (base-station transmit power) needed to provide a certain broadcast-service data rate. Alternatively, taking into account the limited signal-to-noise ratio that can be achieved at, for

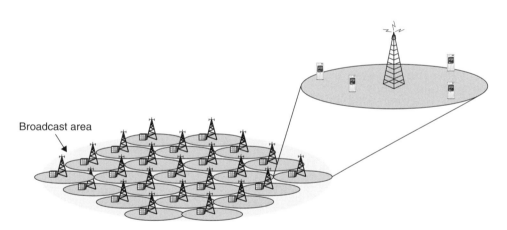

Figure 4.20   *Broadcast scenario.*

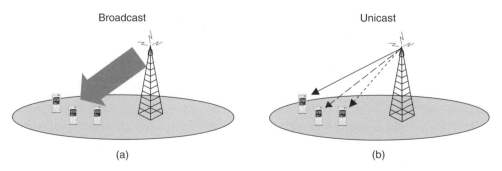

Figure 4.21   *Broadcast vs. Unicast transmission. (a) Broadcast and (b) Unicast.*

example, the cell edge, the achievable broadcast data rates may be relatively limited, especially in case of large cells. One way to increase the broadcast data rates would then be to reduce the cell size, thereby increasing the cell-edge receive power. However, this will increase the number of cells to cover a certain area and is thus obviously negative from a cost-of-deployment point-of-view.

However, as discussed above, the provisioning of broadcast/multicast services in a mobile-communication network typically implies that identical information is to be provided over a large number of cells. In such a case, the resources (downlink transmit power) needed to provide a certain broadcast data rate can be considerably reduced if mobile terminals at the cell edge can utilize the received power from broadcast transmissions from multiple cells when detecting/decoding the broadcast data.

Especially large gains can be achieved if the mobile terminal can simultaneously receive and combine the broadcast transmissions from multiple cells *before* decoding. Such *soft combining* of broadcast/multicast transmissions from multiple cells has already been adopted for WCDMA Multimedia Broadcast/Multicast Service (MBMS) [102], as discussed in Chapter 11.

In case of WCDMA, each cell is transmitting on the downlink using different cell-specific scrambling codes (see Chapter 8). This is true also in case of MBMS. Thus the broadcast transmissions received from different cells are not identical but just provide the same broadcast/multicast information. As a consequence, the mobile terminal must still explicitly identify what cells to include in the soft combining. Furthermore, although the soft combining significantly increases the overall received power for mobile terminals at the cell border and thus significantly improve the overall broadcast efficiency, the broadcast transmissions from the different cells will still interfere with each other. This will limit the achievable broadcast-transmission signal-to-interference ratio and thus limit the achievable broadcast data rates.

One way to mitigate this and further improve the provisioning of broadcast/multicast services in a mobile-communication network is to ensure that the broadcast transmissions from different cells *are truly identical* and *transmitted mutually time aligned*. In this case, the transmissions received from multiple cells will, as seen from the mobile terminal, appear as a single transmission subject to severe multi-path propagation as illustrated in Figure 4.22. The transmission of identical time-aligned signals from multiple cells, especially in the case of provisioning of broadcast/multicast services, is sometimes referred to as Single-Frequency Network (SFN) operation [124].

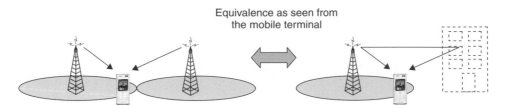

Figure 4.22 *Equivalence between simulcast transmission and multi-path propagation.*

In case of such identical time-aligned transmissions from multiple cells, the 'inter-cell interference' due to transmissions in neighboring cells will, from a terminal point of view, be replaced by signal corruption due to time dispersion. If the broadcast transmission is based on OFDM with a cyclic prefix that covers the main part of this 'time dispersion,' the achievable broadcast data rates are thus only limited by noise, implying that, especially in smaller cells, very high broadcast data rates can be achieved. Furthermore, in contrast to the explicit (in the receiver) multi-cell soft combining of WCDMA MBMS, the OFDM receiver does not need to explicitly identify the cells to be soft combined. Rather, all transmissions that fall within the cyclic prefix will 'automatically' be captured by the receiver.

5

Wider-band 'single-carrier' transmission

The previous chapters discussed multi-carrier transmission in general (Section 3.3.1) and OFDM transmission in particular (Chapter 4) as means to allow for very high overall transmission bandwidth while still being robust to signal corruption due to radio-channel frequency selectivity.

However, as discussed in the previous chapter, a drawback of OFDM modulation, as well as any kind of multi-carrier transmission, is the large variations in the instantaneous power of the transmitted signal. Such power variations imply a reduced power-amplifier efficiency and higher power-amplifier cost. This is especially critical for the uplink, due to the high importance of low mobile-terminal power consumption and cost.

As discussed in Chapter 4, several methods have been proposed on how to reduce the large power variations of an OFDM signal. However, most of these methods have limitations in terms of to what extent the power variations can be reduced. Furthermore, most of the methods also imply a significant computational complexity and/or a reduced link performance.

Thus, there is an interest to consider also wider-band *single-carrier* transmission as an alternative to multi-carrier transmission, especially for the uplink, i.e. for mobile-terminal transmission. It is then necessary to consider what can be done to handle the corruption to the signal waveform that will occur in most mobile-communication environments due to radio-channel frequency selectivity.

5.1 Equalization against radio-channel frequency selectivity

Historically, the main method to handle signal corruption due to radio-channel frequency selectivity has been to apply different forms of *equalization* [50] at the receiver side. The aim of equalization is to, by different means, compensate for the channel frequency selectivity and thus, at least to some extent, restore the original signal shape.

5.1.1 Time-domain linear equalization

The most basic approach to equalization is the time-domain linear equalizer, consisting of a linear filter with an impulse response $w(\tau)$ applied to the received signal (see Figure 5.1).

By selecting different filter impulse responses, different receiver/equalizer strategies can be implemented. As an example, in DS-CDMA-based systems a so-called RAKE receiver structure has historically often been used. The RAKE receiver is simply the receiver structure of Figure 5.1 where the filter impulse response has been selected to provide *channel-matched filtering*

$$w(\tau) = h^*(-\tau) \tag{5.1}$$

that is the filter response has been selected as the complex conjugate of the time-reversed channel impulse response. This is also often referred to as a *Maximum-Ratio Combining* (MRC) filter setting [50].

Selecting the receiver filter according to the MRC criterion, that is as a channel-matched filter, maximizes the post-filter signal-to-noise ratio (thus the term maximum-ratio combining). However, MRC-based filtering does not provide any compensation for any radio-channel frequency selectivity, that is no equalization. Thus, MRC-based receiver filtering is appropriate when the received signal is mainly impaired by noise or interference from other transmissions but not when a main part of the overall signal corruption is due to the radio-channel frequency selectivity.

Another alternative is to select the receiver filter to fully compensate for the radio-channel frequency selectivity. This can be achieved by selecting the receiver-filter impulse response to fulfill the relation

$$h(\tau) \otimes w(\tau) = 1 \tag{5.2}$$

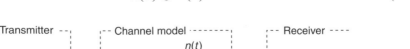

Figure 5.1   *General time-domain linear equalization.*

where '$\otimes$' denotes linear convolution. This selecting of the filter setting, also known as *Zero-Forcing* (ZF) equalization [50], provides full compensation for any radio-channel frequency selectivity (complete equalization) and thus full suppression of any related signal corruption. However, zero-forcing equalization may lead to a large, potentially very large, increase in the noise level after equalization and thus to an overall degradation in the link performance. This will especially be the case when the channel has large variations in its frequency response.

A third and, in most cases, better alternative is to select a filter setting that provides a trade-off between signal corruption due to radio-channel frequency selectivity, and noise/interference. This can for example be done by selecting the filter to minimize the mean-square error between the equalizer output and the transmitted signal, i.e. to minimize

$$\varepsilon = E\{|\hat{s}(t) - s(t)|^2\} \tag{5.3}$$

This is also referred to as a *Minimum Mean-Square Error* (MMSE) equalizer setting [50].

In practice, the linear equalizer has most often been implemented as a *time-discrete FIR filter* [52] with L filter taps applied to the sampled received signal, as illustrated in Figure 5.2. In general, the complexity of such a time-discrete equalizer grows relatively rapidly with the bandwidth of the signal to be equalized:

- A more wideband signal is subject to relatively more radio-channel frequency selectivity or, equivalently, relatively more time dispersion. This implies that the equalizer needs to have a larger span (larger length L, that is more filter taps) to be able to properly compensate for the channel frequency selectivity.

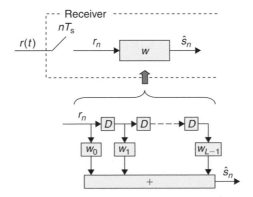

Figure 5.2 *Linear equalization implemented as a time-discrete FIR filter.*

- A more wideband signal leads to a correspondingly higher sampling rate for the received signal. Thus also the receiver-filter processing needs to be carried out with a correspondingly higher rate.

It can be shown [32] that the time-discrete MMSE equalizer setting $\bar{w} = [w_0$ $w_1, w_{L-1}]^H$ is given by the expression

$$\bar{w} = R^{-1}\bar{p} \tag{5.4}$$

In this expression, R is the *channel-output auto-correlation matrix* of size $L \times L$, which depends on the channel impulse response as well as on the noise level, and $\bar{p}$ is the *channel-output/channel-input cross-correlation vector* of size $L \times 1$ that depends on the channel impulse response.

Especially in case of a large equalizer span (large L, the time-domain MMSE equalizer may be of relatively high complexity:

- The equalization itself (the actual filtering) may be of relatively high complexity according to above.
- Calculation of the MMSE equalizer setting, especially the calculation of the inverse of the size $L \times L$ correlation matrix R, may be of relatively high complexity.

5.1.2   Frequency-domain equalization

A possible way to reduce the complexity of linear equalization is to carry out the equalization in the frequency domain [24], as illustrated in Figure 5.3. In case of such *frequency-domain linear equalization*, the equalization is carried out block-wise with block size N. The sampled received signal is first transformed into the frequency domain by means of a size-N DFT. The equalization is then carried out as frequency-domain filtering, with the frequency-domain filter taps $W_0, ..., W_{N-1}$, for example being the DFT of the corresponding time-domain filter taps $w_0, ..., w_{L-1}$ of Figure 5.2. Finally, the equalized frequency-domain signal is transformed back to the time domain by means of a size-N inverse DFT. The block size-N should preferably be selected as $N = 2^n$ for some integer n to allow for computational-efficient radix-2 FFT/IFFT implementation of the DFT/IDFT processing.

For each processing block of size N, the frequency-domain equalization basically consists of:

- A size-N DFT/FFT.
- N complex multiplications (the frequency-domain filter).
- A size-N inverse DFT/FFT.

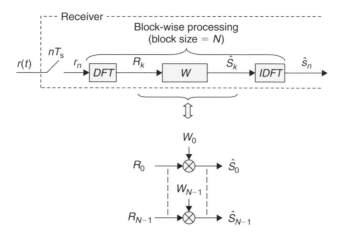

Figure 5.3 *Frequency-domain linear equalization.*

Especially in case of channels with extensive frequency selectivity, implying the need for a large span of a time-domain equalizer (large equalizer length L, equalization in the frequency domain according to Figure 5.3 can be of significantly less complexity, compared to time-domain equalization illustrated in Figure 5.2.

However, there are two issues with frequency-domain equalization:

- The time-domain filtering of Figure 5.2 implements a time-discrete *linear convolution*. In contrast, frequency-domain filtering according to Figure 5.3 corresponds to *circular convolution* in the time domain. Assuming a time-domain equalizer of length L, this implies that the first $L - 1$ samples at the output of the frequency-domain equalizer *will not* be identical to the corresponding output of the time-domain equalizer.
- The frequency-domain filter taps W_0, ..., W_{N-1} can be determined by first determining the pulse response of the corresponding time-domain filter and then transforming this filter into the frequency domain by means of a DFT. However, as mentioned above, determining e.g. the MMSE time-domain filter may be relative complex in case of a large equalizer length L.

One way to address the first issue is to apply an *overlap* in the block-wise processing of the frequency-domain equalizer as outlined in Figure 5.4, where the overlap should be at least $L - 1$ samples. With such an overlap, the first $L - 1$ ('incorrect') samples at the output of the frequency-domain equalizer can be *discarded* as the corresponding samples are also (correctly) provided as the last part of the previously received/equalized block. The drawback with this kind of 'overlap-and-discard' processing is a computational overhead, that is somewhat higher receiver complexity.

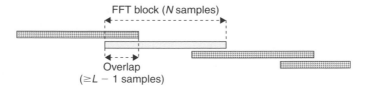

Figure 5.4   *Overlap-and-discard processing.*

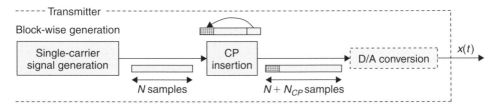

Figure 5.5   *Cyclic-prefix insertion in case of single-carrier transmission.*

An alternative approach that addresses both of the above issues is to apply *cyclic-prefix insertion* at the transmitter side (see Figure 5.5). Similar to OFDM, cyclic-prefix insertion in case of single-carrier transmission implies that a cyclic prefix of length N_{CP} samples is inserted block-wise at the transmitter side. The transmitter-side block size should be the same as the block size N used for the receiver-side frequency-domain equalization.

With the introduction of a cyclic prefix, the channel will, from a receiver point of view, appear as a circular convolution over a receiver processing block of size N. Thus there is no need for any receiver overlap-and-discard processing. Furthermore, the frequency-domain filter taps can now be calculated directly from an estimate of the sampled channel frequency response without first determining the time-domain equalizer setting. As an example, in case of an MMSE equalizer the frequency-domain filter taps can be calculated according to

$$W_k = \frac{H_k^*}{|H_k|^2 + N_0} \tag{5.5}$$

where N_0 is the noise power and H_k is the sampled channel frequency response. For large equalizer lengths, this calculation is of much lower complexity compared to the time-domain calculation discussed in the previous section.

The drawback of cyclic-prefix insertion in case of single-carrier transmission is the same as for OFDM, that is it implies an overhead in terms of both power and bandwidth. One method to reduce the relative cyclic-prefix overhead is to increase the block size N of the frequency-domain equalizer. However, for the block-wise equalization to be accurate, the channel needs to be approximately constant over a time span corresponding to the size of the processing block. This constraint provides an upper limit on the block size N that depends on the rate of the channel variations. Note that this is similar to the constraint on the OFDM subcarrier spacing $\Delta f = 1/T_u$ depending on the rate of the channel variations, as discussed in Chapter 4.

5.1.3   Other equalizer strategies

The previous sections discussed different approaches to linear equalization as a means to counter-act signal corruption of a wideband signal due to radio-channel frequency selectivity. However, there are also other approaches to equalization:

- *Decision-Feedback Equalization* (DFE) [50] implies that previously detected symbols are fed back and used to cancel the contribution of the corresponding transmitted symbols to the overall signal corruption. Such decision feedback is typically used in combination with time-domain linear filtering, where the linear filter transforms the channel response to a shape that is more suitable for the decision-feedback stage. Decision feedback can also very well be used in combination with frequency-domain linear equalization [24].
- *Maximum-Likelihood* (ML) detection, also known as *Maximum Likelihood Sequence Estimation* (MLSE), [25] is strictly speaking not an equalization scheme but rather a receiver approach where the impact of radio-channel time dispersion is explicitly taken into account in the receiver-side detection process. Fundamentally, an ML detector uses the entire received signal to decide on the most likely transmitted sequence, taking into account the impact of the time dispersion on the received signal. To implement maximum-likelihood detection the *Viterbi algorithm* [26] is often used.[1] However, although maximum-likelihood detection based on the Viterbi algorithm has been extensively used for 2G mobile communication such as GSM, it is foreseen to be too complex to be applied for the 3G evolution due to the much wider transmission bandwidth leading to both much more extensive channel frequency selectivity and much higher sampling rates.

5.2   Uplink FDMA with flexible bandwidth assignment

In practice, there are obviously often multiple mobile terminals within a cell and thus multiple mobile terminals that should share the overall uplink radio resource within the cell by means of the *uplink intra-cell multiple-access scheme*.

---

[1] Thus the term *Viterbi equalizer* is sometimes used for the ML detector.

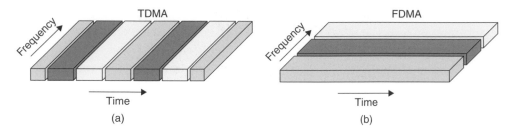

Figure 5.6   *Orthogonal multiple access: (a) TDMA and (b) FDMA.*

In case of mobile communication based on WCDMA and cdma2000, uplink transmissions within a cell are mutually *non-orthogonal* and the base-station receiver relies on the processing gain due to channel coding and additional direct-sequence spreading to suppress the intra-cell interference. Although non-orthogonal multiple access can, fundamentally, provide higher capacity compared to orthogonal multiple access, in practice the possibility for mutually orthogonal uplink transmissions within a cell is often beneficial from a system-performance point-of-view.

Mutually orthogonal uplink transmissions within a cell can be achieved in the time domain (*Time Division Multiple Access*, TDMA), as illustrated in Figure 5.6a. TDMA implies that different mobile terminals within in a cell transmit in different non-overlapping time intervals. In each such time interval, the full system bandwidth is then assigned for uplink transmission from a single mobile terminal.

Alternatively, mutually orthogonal uplink transmissions within a cell can be achieved in the frequency domain (*Frequency Division Multiple Access*, FDMA) as illustrated in Figure 5.6b, that is by having mobile terminals transmit in different frequency bands.

To be able to provide high-rate packet-data transmission, it should be possible to assign the entire system bandwidth for transmission from a single mobile terminal. At the same time, due to the burstiness of most packet-data services, in many cases mobile terminals will have no uplink data to transmit. Thus, for efficient packet-data access, a TDMA component should always be part of the uplink multiple-access scheme.

However, relying on only TDMA to provide orthogonality between uplink transmissions within a cell could be bandwidth inefficient, especially in case of a very wide overall system bandwidth. As discussed in Chapter 3, a wide bandwidth is needed to support high data rates in a power-efficient way. However, the data rates that can be achieved over a radio link are, in many cases, limited by the available signal power (power-limited operation) rather than by the available

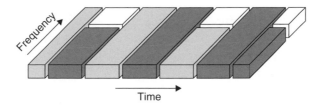

Figure 5.7 *FDMA with flexible bandwidth assignment.*

bandwidth. This is especially the case for the uplink, due to the, in general, more limited mobile-terminal transmit power. Allocating the entire system bandwidth to a single mobile terminal could, in such cases, be highly inefficient in terms of bandwidth utilization. As an example, allocating 20 MHz of transmission bandwidth to a mobile terminal in a scenario where the achievable uplink data rate, due to mobile-terminal transmit-power limitations, is anyway limited to, for example, a few 100 kbps would obviously imply a very inefficient usage of the overall available bandwidth. In such cases, a smaller transmission bandwidth should be assigned to the mobile terminal and the remaining part of the overall system bandwidth should be used for uplink transmissions from other mobile terminals. Thus, in addition to TDMA, an uplink transmission scheme should preferably allow for orthogonal user multiplexing also in the frequency domain, i.e. FDMA.

At the same time, it should be possible to allocate the entire overall transmission bandwidth to a single mobile terminal when the channel conditions are such that the wide bandwidth can be efficiently utilized, that is when the achievable data rates are not power limited. Thus an orthogonal uplink transmission scheme should allow for *FDMA with flexible bandwidth assignment* as illustrated in Figure 5.7.

Flexible bandwidth assignment is straightforward to achieve with an OFDM-based uplink transmission scheme by dynamically allocating different number of subcarriers to different mobile terminals depending on their instantaneous channel conditions. In the next section, it will be discussed how this can also be achieved in case of low-PAR 'single-carrier' transmission, more specifically by means of so-called *DFT-spread OFDM*.

5.3 DFT-spread OFDM

DFT-spread OFDM (DFTS-OFDM) is a transmission scheme that can combine the desired properties discussed in the previous sections, i.e.:

- Small variations in the instantaneous power of the transmitted signal ('single-carrier' property).

- Possibility for low-complexity high-quality equalization in the frequency domain.
- Possibility for FDMA with flexible bandwidth assignment.

Due to these properties, DFTS-OFDM has been selected as the uplink transmission scheme for LTE, that is the long-term 3G evolution, see further Part IV of this book.

5.3.1 Basic principles

The basic principle of DFTS-OFDM transmission is illustrated in Figure 5.8. One way to interpret DFTS-OFDM is as normal OFDM with a *DFT-based precoding*. Similar to OFDM modulation, DFTS-OFDM relies on block-based signal generation. In case of DFTS-OFDM, a block of M modulation symbols from some modulation alphabet, e.g. QPSK or 16QAM, is first applied to a size-M DFT. The output of the DFT is then applied to consecutive inputs (subcarriers) of an OFDM modulator where, in practice, the OFDM modulator will be implemented as a size-N *inverse* DFT (IDFT) with $N > M$ and where the unused inputs of the IDFT are set to zero. Typically, the IDFT size N is selected as $N = 2^n$ for some integer n to allow for the IDFT to be implemented by means of computationally efficient radix-2 IFFT processing. Also similar to normal OFDM, a cyclic prefix is preferable inserted for each transmitted block. As discussed in Section 5.1.2, the presence of a cyclic prefix allows for straightforward low-complexity frequency-domain equalization at the receiver side.

If the DFT size M would equal the IDFT size N, the cascaded DFT/IDFT processing would obviously completely cancel out each other. However, if M is smaller than N and the remaining inputs to the IDFT are set to zero, the output of the IDFT will be a signal with 'single-carrier' properties, i.e. a signal with low power variations, and with a bandwidth that depends on M. More specifically,

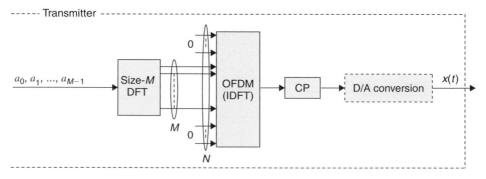

Figure 5.8   *DFTS-OFDM signal generation.*

assuming a sampling rate f_s at the output of the IDFT, the nominal bandwidth of the transmitted signal will be $BW = M/N \cdot f_s$. Thus, by varying the block size M the instantaneous bandwidth of the transmitted signal can be varied, allowing for flexible-bandwidth assignment. Furthermore, by shifting the IDFT inputs to which the DFT outputs are mapped, the transmitted signal can be shifted in the frequency domain, as will be further discussed in Section 5.3.3.

To have a high degree of flexibility in the instantaneous bandwidth, given by the DFT size M, it is typically not possible to ensure that M can be expressed as 2^m for some integer m. However, as long as M can be expressed as a product of relatively small prime numbers, the DFT can still be implemented as relatively low-complexity non-radix-2 FFT processing. As an example, a DFT size $M = 144$ can be implemented by means of a combination of radix-2 and radix-3 FFT processing ($144 = 3^2 \cdot 2^4$).

The main benefit of DFTS-OFDM, compared to normal OFDM, is reduced variations in the instantaneous transmit power, implying the possibility for increased power-amplifier efficiency. This benefit of DFTS-OFDM is illustrated in Figure 5.9, which illustrates the distribution of the *Peak-to-Average-power Ratio* (PAR) for DFTS-OFDM and conventional OFDM. The PAR is defined as the peak power within one DFT block (one OFDM symbol) normalized by the average signal power. It should be noted that the PAR distribution is *not* the same as the distribution of the instantaneous transmit power as illustrated in Figure 3.3. Historically, PAR distributions have often been used to illustrate the power variations of OFDM.

As can be seen in Figure 5.9, the PAR is significantly lower for DFTS-OFDM, compared to OFDM. In case of 16QAM modulation, the PAR of DFTS-OFDM

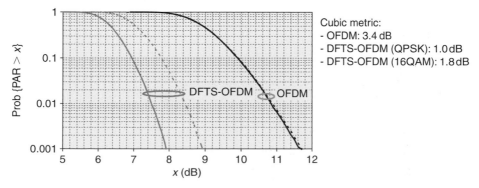

Figure 5.9 *PAR distribution for OFDM and DFTS-OFDM, respectively. Solid curve: QPSK. Dashed curve: 16QAM.*

increases somewhat as expected (compare Figure 3.3). On the other hand, in case of OFDM the PAR distribution is more or less independent of the modulation scheme. The reason is that, as the transmitted OFDM signal is the sum of a large number of independently modulated subcarriers, the instantaneous power has an approximately exponential distribution, regardless of the modulation scheme applied to the different subcarriers.

Although the PAR distribution can be used to qualitatively illustrate the difference in power variations between different transmission schemes, it is not a very good measure to more accurately quantify the impact of the power variations on e.g. the required power-amplifier back-off.

A better measure of the impact on the required power-amplifier back-off and the corresponding impact on the power-amplifier efficiency is given by the so-called *cubic metric* [49]. The cubic metric is a measure of the amount of additional back-off needed for a certain signal wave form, relative to the back-off needed for some reference wave form.

As can be seen from Figure 5.9, the cubic metric (given to the right of the graph) follows the same trend as the PAR. However, the differences in cubic metric are somewhat smaller than the corresponding differences in PAR.

5.3.2 DFTS-OFDM receiver

The basic principle of demodulation of a DFTS-OFDM signal is illustrated in Figure 5.10. The operations are basically the reverse of those for the DFTS-OFDM signal generation of Figure 5.8, i.e. size-N DFT (FFT) processing, removal of the frequency samples not corresponding to the signal to be received, and size-*M* inverse DFT processing.

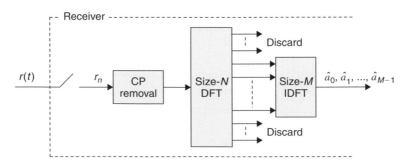

Figure 5.10   *Basic principle of DFTS-OFDM demodulation.*

In the ideal case with no signal corruption on the radio channel, DFTS-OFDM demodulation according to Figure 5.10 will perfectly restore the block of transmitted symbols. However, in case of a time dispersive or, equivalently, a frequency-selective radio channel, the DFTS-OFDM signal will be corrupted, with 'self-interference' as a consequence. This can be understood in two ways:

1. Being a wideband single-carrier signal, the DFTS-OFDM spread signal is, obviously, corrupted in case of a time-dispersive channel.
2. If the channel is frequency selective over the span of the DFT, the inverse DFT at the receiver will not be able to correctly reconstruct the original block of transmitted symbols.

Thus, in case of DFTS-OFDM, an equalizer is needed to compensate for the radio-channel frequency selectivity. Assuming the basic DFTS-OFDM demodulator structure according to Figure 5.10, frequency-domain equalization as discussed in Section 5.1.2 is especially applicable to DFTS-OFDM transmission (see Figure 5.11).

5.3.3 User multiplexing with DFTS-OFDM

As mentioned above, by dynamically adjusting the transmitter DFT size and, consequently, also the size of the block of modulation symbols $a_0, a_1. \ldots, a_{M-1}$, the nominal bandwidth of the DFTS-OFDM signal can be dynamically adjusted. Furthermore, by shifting the IDFT inputs to which the DFT outputs are mapped, the exact frequency-domain 'position' of the signal to be transmitted can be adjusted. By these means, DFTS-OFDM allows for uplink FDMA with flexible bandwidth assignment as illustrated in Figure 5.12.

Figure 5.12a illustrates the case of multiplexing the transmissions from two mobile terminals with equal bandwidth assignments, that is equal DFT sizes M, while Figure 5.12b illustrates the case of differently sized bandwidth assignments.

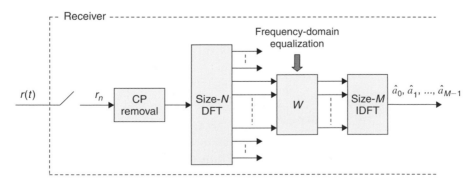

Figure 5.11 *DFTS-OFDM demodulator with frequency-domain equalization.*

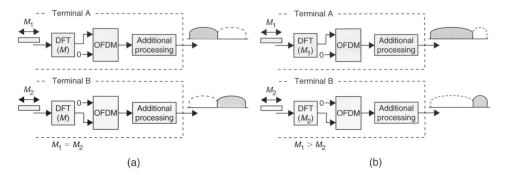

Figure 5.12   *Uplink user multiplexing in case of DFTS-OFDM. (a) Equal-bandwidth assignment and (b) unequal-bandwidth assignment.*

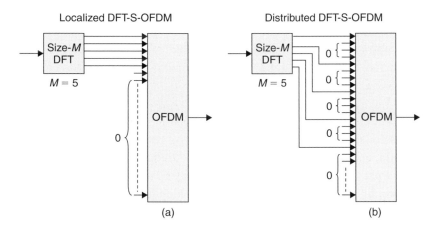

Figure 5.13   *Localized DFTS-OFDM vs. Distributed DFTS-OFDM.*

5.3.4   *Distributed DFTS-OFDM*

What has been illustrated in Figure 5.8 can more specifically be referred to as *Localized DFTS-OFDM*, referring to the fact that the output of the DFT is mapped to *consecutive* inputs of the OFDM modulator. An alternative is to map the output of the DFT to *equidistant* inputs of the OFDM modulator with zeros inserted in between, as illustrated in Figure 5.13. This is can also be referred to as *Distributed DFTS-OFDM*.

Figure 5.14 illustrates the basic structure of the transmitted spectrum in case of localized and distributed DFTS-OFDM, respectively. Although the spectrum of the localized DFTS-OFDM signal clearly indicates a single-carrier transmission this is not as clearly seen from the spectrum of the distributed DFTS-OFDM signal. However, it can be shown that a distributed DFTS-OFDM signal has similar power variations as localized DFTS-OFDM. Actually, it can be shown

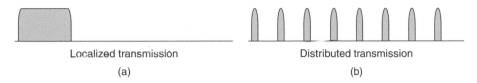

Localized transmission
(a)

Distributed transmission
(b)

Figure 5.14 *Spectrum of localized and distributed DFTS-OFDM signals.*

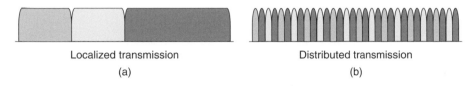

Localized transmission
(a)

Distributed transmission
(b)

Figure 5.15 *User multiplexing in case of localized and distributed DFTS-OFDM.*

that a distributed DFTS-OFDM signal is equivalent to so-called *Interleaved FDMA* (IFDMA) [67]. The benefit of distributed DFTS-OFDM, compared to localized DFTS-OFDM, is the possibility for additional frequency diversity as even a low-rate distributed DFTS-OFDM signal (small DFT size M can be spread over a potentially very large overall transmission bandwidth.

User multiplexing in the frequency domain as well as flexible bandwidth allocation is possible also in case of distributed DFTS-OFDM. However, in this case, the different users are 'interleaved in the frequency domain, as illustrated in Figure 5.15b (thus the alternative term *"Interleaved FDMA"*).' As a consequence, distributed DFTS-OFDM is more sensitivity to frequency errors and has higher requirements on power control, compared to localized DFTS-OFDM. This is similar to the case of localized OFDM vs. distributed OFDM as discussed in Section 4.10.

6

Multi-antenna techniques

Multi-antenna techniques can be seen as a joint name for a set of techniques with the common theme that they rely on the use of multiple antennas at the receiver and/or the transmitter, in combination with more or less advanced signal processing. Multi-antenna techniques can be used to achieve improved system performance, including improved system capacity (more users per cell) and improved coverage (possibility for larger cells), as well as improved service provisioning, for example, higher per-user data rates. This chapter will provide a general overview of different multi-antenna techniques applicable to the 3G evolution. How multi-antenna techniques are specifically applied to HSPA and its evolution and to LTE is discussed in somewhat more details in Part III and IV, respectively.

6.1   Multi-antenna configurations

An important characteristic of any multi-antenna configuration is the distance between the different antenna elements, to a large extent due to the relation between the antenna distance and the mutual correlation between the radio-channel fading experienced by the signals at the different antennas.

The antennas in a multi-antenna configuration can be located relatively far from each other, typically implying a relatively low mutual correlation. Alternatively, the antennas can be located relatively close to each other, typically implying a high mutual fading correlation, that is in essence that the different antennas experience the same, or at least very similar, instantaneous fading. Whether high or low correlation is desirable depends on what is to be achieved with the multi-antenna configuration (diversity, beam-forming, or spatial multiplexing) as discussed further below.

What actual antenna distance is needed for low, alternatively high, fading correlation depends on the wavelength or, equivalently, the carrier frequency used for the radio communication. However, it also depends on the deployment scenario.

In case of base-station antennas in typical macro-cell environments (relatively large cells, relatively high base-station antenna positions, etc.), an antenna distance in the order of ten wavelengths is typically needed to ensure a low mutual fading correlation. At the same time, for a mobile terminal in the same kind of environment, an antenna distance in the order of only half a wavelength (0.5λ) is often sufficient to achieve relatively low mutual correlation [41]. The reason for the difference between the base station and the mobile terminal in this respect is that, in the macro-cell scenario, the multi-path reflections that cause the fading mainly occur in the near-zone around the mobile terminal. Thus, as seen from the mobile terminal, the different paths will typically arrive from a wide angle, implying a low fading correlation already with a relatively small antenna distance. At the same time, as seen from the (macro-cell) base station the different paths will typically arrive within a much smaller angle, implying the need for significantly larger antenna distance to achieve low fading correlation.[1]

On the other hand, in other deployment scenarios, such as micro-cell deployments with base-station antennas below roof-top level and indoor deployments, the environment as seen from the base station is more similar to the environment as seen from the mobile terminal. In such scenarios, a smaller base-station antenna distance is typically sufficient to ensure relatively low mutual correlation between the fading experienced by the different antennas.

The above discussion assumed antennas with the same polarization direction. Another means to achieve low mutual fading correlation is to apply different polarization directions for the different antennas [41]. The antennas can then be located relatively close to each other, implying a compact antenna arrangement, while still experiencing low mutual fading correlation.

6.2 Benefits of multi-antenna techniques

The availability of multiple antennas at the transmitter and/or the receiver can be utilized in different ways to achieve different aims:

- Multiple antennas at the transmitter and/or the receiver can be used to provide additional diversity against fading on the radio channel. In this case, the channels experienced by the different antennas should have low mutual correlation, implying the need for a sufficiently large inter-antenna distance (*spatial diversity*), alternatively the use of different antenna polarization directions (*polarization diversity*).

---

[1] Although the term 'arrive' is used above, the situation is exactly the same in case of multiple *transmit* antennas. Thus, in case of multiple base-station transmit antennas in a macro-cell scenario, the antenna distance typically needs to be a number of wavelengths to ensure low fading correlation while, in case of multiple transmit antennas at the mobile terminal, an antenna distance of a fraction of a wavelength is sufficient.

- Multiple antennas at the transmitter and/or the receiver can be used to 'shape' the overall antenna beam (transmit beam and receive beam, respectively) in a certain way, for example, to maximize the overall antenna gain in the direction of the target receiver/transmitter or to suppress specific dominant interfering signals. Such *beam-forming* can be based either on high or low fading correlation between the antennas, as is further discussed in Section 6.4.2.
- The simultaneous availability of multiple antennas at the transmitter and the receiver can be used to create what can be seen as multiple parallel communication 'channels' over the radio interface. This provides the possibility for very high bandwidth utilization without a corresponding reduction in power efficiency or, in other words, the possibility for very high data rates within a limited bandwidth without an un-proportionally large degradation in terms of coverage. Herein we will refer to this as *spatial multiplexing*. It is often also referred to as MIMO (Multi-Input Multi-Output) antenna processing.

6.3 Multiple receive antennas

Perhaps the most straightforward and historically the most commonly used multi-antenna configuration is the use of multiple antennas at the receiver side. This is often referred to as *receive diversity* or *RX diversity* even if the aim of the multiple receive antennas is not always to achieve additional diversity against radio-channel fading.

Figure 6.1 illustrates the basic principle of linear combining of signals $r_1, \ldots, r_{N_R}$ received at N_R different antennas, with the received signals being multiplied by complex weight factors $w_1^* \ldots w_{N_R}^*$ before being added together. In vector notation this *linear receive-antenna combining* can be expressed as[2]

$$\hat{s} = [w_1^* \ldots w_{N_R}^*] \cdot \begin{bmatrix} r_1 \\ \vdots \\ r_{N_R} \end{bmatrix} = \bar{w}^H \cdot \bar{r} \tag{6.1}$$

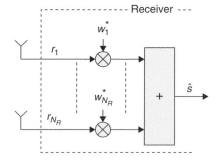

Figure 6.1 *Linear receive-antenna combining.*

---

[2] Note that the weight factors are expressed as complex conjugates of $w_1, \ldots, w_{N_R}$.

What is outlined in (6.1) and Figure 6.1 is linear receive-antenna combining in general. Different specific antenna-combining approaches then differ in the exact choice of the weight vector $\bar{w}$.

Assuming that the transmitted signal is only subject to non-frequency-selective fading and (white) noise, i.e., there is no radio-channel time dispersion, the signals received at the different antennas in Figure 6.1 can be expressed as

$$\bar{r} = \begin{pmatrix} r_1 \\ \vdots \\ r_{N_R} \end{pmatrix} = \begin{pmatrix} h_1 \\ \vdots \\ h_{N_R} \end{pmatrix} \cdot s + \begin{pmatrix} n_1 \\ \vdots \\ n_{N_R} \end{pmatrix} = \bar{h} \cdot s + \bar{n} \qquad (6.2)$$

where s is the transmitted signal, the vector $\bar{h}$ consists of the N_R complex channel gains, and the vector $\bar{n}$ consists of the noise impairing the signals received at the different antennas (see also Figure 6.2).

One can easily show that, to maximize the signal-to-noise ratio after linear combining, the weight vector $\bar{w}$ should be selected as [50]

$$\bar{w}_{MRC} = \bar{h} \qquad (6.3)$$

This is also known as *Maximum-Ratio Combining* (MRC). The MRC weights fulfill two purposes:

- Phase rotate the signals received at the different antennas to compensate for the corresponding channel phases and ensure that the signals are phase aligned when added together (coherent combining).
- Weight the signals in proportion to their corresponding channel gains, that is apply higher weights for stronger received signals.

In case of mutually uncorrelated antennas, that is sufficiently large antenna distances or different polarization directions, the channel gains h_1, ..., h_{N_R} are

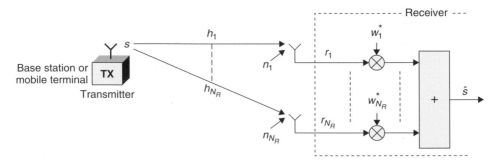

Figure 6.2 *Linear receive-antenna combining.*

uncorrelated and the linear antenna combining provides diversity of order N_R. In terms of receiver-side beam-forming, selecting the antenna weights according to (6.3) corresponds to a receiver beam with maximum gain N_R in the direction of the target signal. Thus, the use of multiple receive antennas may increase the post-combiner signal-to-noise ratio in proportion to the number of receive antennas.

MRC is an appropriate antenna-combining strategy when the received signal is mainly impaired by noise. However, in many cases of mobile communication the received signal is mainly impaired by interference from other transmitters within the system, rather than by noise. In a situation with a relatively large number of interfering signals of approximately equal strength, maximum-ratio combining is typically still a good choice as, in this case, the overall interference will appear relatively 'noise-like' with no specific direction-of-arrival. However, in situations where there is a single dominating interferer (or, in the general case, a limited number of dominating interferers), as illustrated in Figure 6.3, improved perform-ance can be achieved if, instead of selecting the antenna weights to maximize the received signal-to-noise ratio after antenna combining (MRC), the antenna weights are selected so that the interferer is suppressed. In terms of receiver-side beam-forming this corresponds to a receiver beam with high attenuation in the direction of the interferer, rather than focusing the receiver beam in the direction of the tar-get signal. Applying receive-antenna combining with a target to suppress specific interferers is often referred to as *Interference Rejection Combining* (IRC) [38].

In the case of a single dominating interferer as outlined in Figure 6.3, expression (6.2) can be extended according to

$$\bar{r} = \begin{pmatrix} r_1 \\ \vdots \\ r_{N_R} \end{pmatrix} = \begin{pmatrix} h_1 \\ \vdots \\ h_{N_R} \end{pmatrix} \cdot s + \begin{pmatrix} h_{I,1} \\ \vdots \\ h_{I,N_R} \end{pmatrix} \cdot s_I + \begin{pmatrix} n_1 \\ \vdots \\ n_{N_R} \end{pmatrix} = \bar{h} \cdot s + \bar{h}_I \cdot s_I + \bar{n} \qquad (6.4)$$

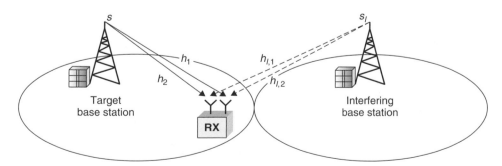

Figure 6.3 *Downlink scenario with a single dominating interferer (special case of only two receive antennas).*

where s_I is the transmitted interferer signal and the vector $\bar{h}_I$ consists of the complex channel gains from the interferer to the N_R receive antennas. By applying expression (6.1) to (6.4) it is clear that the interfering signal will be completely suppressed if the weight vector $\bar{w}$ is selected to fulfill the expression

$$\bar{w}^H \cdot \bar{h}_I = 0 \qquad (6.5)$$

In the general case, (6.5) has $N_R - 1$ non-trivial solutions, indicating flexibility in the weight-vector selection. This flexibility can be used to suppress additional dominating interferers. More specifically, in the general case of N_R receive antennas there is a possibility to, at least in theory, completely suppress up to $N_R - 1$ separate interferers. However, such a choice of antenna weights, providing complete suppression of a number of dominating interferers, may lead to a large, potentially very large, increase in the noise level after the antenna combining. This is similar to the potentially large increase in the noise level in case of a Zero-Forcing equalizer as discussed in Chapter 5.

Thus, similar to the case of linear equalization, a better approach is to select the antenna weight vector $\bar{w}$ to minimize the mean square error

$$\varepsilon = E\{|s - s|^2\} \qquad (6.6)$$

also known as the *Minimum Mean Square Error* (MMSE) combining [41, 52].

Although Figure 6.3 illustrates a downlink scenario with a dominating interfering base station, IRC can also be applied to the uplink to suppress interference from specific mobile terminals. In this case, the interfering mobile terminal may either be in the same cell as the target mobile terminal (*intra-cell-interference*) or in a neighbor cell (*inter-cell-interference*) (see Figure 6.4a and b, respectively). Suppression of intra-cell interference is relevant in case of a non-orthogonal uplink, that is when multiple mobile terminals are transmitting simultaneously using the same time-frequency resource. Uplink intra-cell-interference

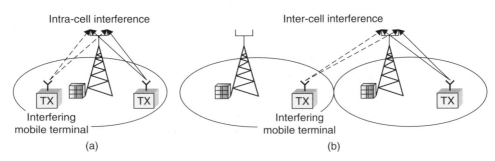

Figure 6.4 *Receiver scenario with one strong interfering mobile terminal: (a) Intra-cell interference and (b) Inter-cell interference.*

suppression by means of IRC is sometimes also referred to as *Spatial Division Multiple Access* (SDMA) [68, 69].

As discussed in Chapter 3, in practice a radio channel is always subject to at least some degree of time dispersion or, equivalently, frequency selectivity, causing corruption to a wide-band signal. As discussed in Chapter 5, one method to counteract such signal corruption is to apply linear equalization, either in the time or frequency domain.

It should be clear from the discussion above that linear receive-antenna combining has many similarities to linear equalization:

- Linear time-domain (frequency-domain) filtering/equalization as described in Chapter 5 implies that linear processing is applied to signals received at different time instances (different frequencies) with a target to maximize the post-equalizer SNR (MRC-based equalization), alternatively to suppress signal corruption due to radio-channel frequency selectivity (zero-forcing equalization, MMSE equalization, etc.).
- Linear receive-antenna combining implies that linear processing is applied to signals received at different antennas, i.e. processing in the spatial domain, with a target to maximize the post-combiner SNR (MRC-based combining), alternatively to suppress specific interferers (IRC based on e.g. MMSE).

Thus, in the general case of frequency-selective channel and multiple receive antennas, two-dimensional time/space linear processing/filtering can be applied as illustrated in Figure 6.5 where the linear filtering can be seen as a generalization of the antenna weights of Figure 6.1. The filters should be jointly selected to minimize the overall impact of noise, interference and signal corruption due to radio-channel frequency selectivity.

Alternatively, especially in the case when cyclic-prefix insertion has been applied at the transmitter side, two-dimensional frequency/space linear processing can be applied as illustrated in Figure 6.6. The frequency-domain weights

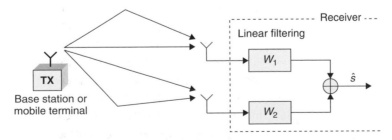

Figure 6.5 *Two-dimensional space/time linear processing (two receive antennas).*

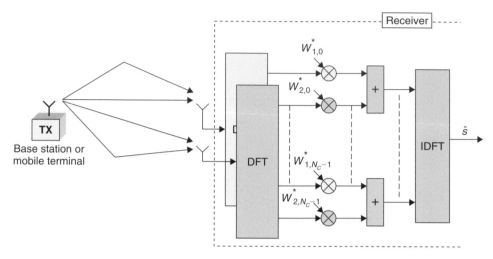

Figure 6.6   *Two-dimensional space/frequency linear processing (two receive antennas).*

should then be jointly selected to minimize the overall impact of noise, interference, and signal corruption due to radio-channel frequency selectivity.

The frequency/space processing outlined in Figure 6.6, without the IDFT, is also applicable if receive diversity is to be applied to OFDM transmission. In the case of OFDM transmission, there is no signal corruption due to radio-channel frequency selectivity. Thus, the frequency-domain coefficients of Figure 6.6 can be selected taking into account only noise and interference. In principle, this means the antenna-combining schemes discussed above (MRC and IRC) are applied on a *per-subcarrier* basis.

Note that, although Figure 6.5 and Figure 6.6 assume two receive antennas, the corresponding receiver structures can straightforwardly be extended to more than two antennas.

6.4   Multiple transmit antennas

As an alternative or complement to multiple receive antennas, diversity and beam-forming can also be achieved by applying multiple antennas at the transmitter side. The use of multiple transmit antennas is primarily of interest for the downlink, i.e. at the base station. In this case, the use of multiple transmit antennas provides an opportunity for diversity and beam-forming without the need for additional receive antennas and corresponding additional receiver chains at the mobile terminal. On the other hand, due to complexity reasons the use of multiple transmit antennas for the uplink, i.e. at the mobile terminal, is less attractive.

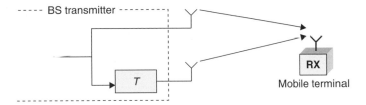

Figure 6.7 *Two-antenna delay diversity.*

In this case, it is typically preferred to apply additional receive antennas and corresponding receiver chains at the base station.

6.4.1 Transmit-antenna diversity

If no knowledge of the downlink channels of the different transmit antennas is available at the transmitter, multiple transmit antennas cannot provide beam-forming but only diversity. For this to be possible, there should be low mutual correlation between the channels of the different antennas. As discussed in Section 6.1, this can be achieved by means of sufficiently large distance between the antennas, alternatively by the use of different antenna polarization directions. Assuming such antenna configurations, different approaches can be taken to realize the diversity offered by the multiple transmit antennas.

6.4.1.1 Delay diversity

As discussed in Chapter 4, a radio channel subject to time dispersion, with the transmitted signal propagating to the receiver via multiple, independently fading paths with different delays, provides the possibility for multi-path diversity or, equivalently, frequency diversity. Thus multi-path propagation is actually beneficial in terms of radio-link performance, assuming that the amount of multi-path propagation is not too extensive and that the transmission scheme includes tools to counteract signal corruption due to the radio-channel frequency selectivity, for example, by means of OFDM transmission or the use of advanced receiver-side equalization.

If the channel in itself is not time dispersive, the availability of multiple transmit antennas can be used to create *artificial time dispersion* or, equivalently, *artificial frequency selectivity* by transmitting identical signals with different relative delays from the different antennas. In this way, the antenna diversity, i.e. the fact that the fading experienced by the different antennas have low mutual correlation, can be transformed into frequency diversity. This kind of *delay diversity* is illustrated in Figure 6.7 for the special case of two transmit antennas. The relative delay T should be selected to ensure a suitable amount of frequency selectivity over the bandwidth of the signal to be transmitted. It should be noted that, although Figure 6.7 assumes two transmit antennas, delay diversity can straightforwardly be extended to more than two transmit antennas with different relative delays for each antenna.

Delay diversity is in essence invisible to the mobile terminal, which will simply see a single radio-channel subject to additional time dispersion. Delay diversity can thus straightforwardly be introduced in an existing mobile-communication system without requiring any specific support in a corresponding radio-interface standard. Delay diversity is also applicable to basically any kind of transmission scheme that is designed to handle and benefit from frequency-selective fading, including, for example, WCDMA and CDMA2000.

6.4.1.2 Cyclic-delay diversity

Cyclic-Delay Diversity (CDD) [6] is similar to delay diversity with the main difference that cyclic-delay diversity operates block-wise and applies *cyclic shifts*, rather than linear delays, to the different antennas (see Figure 6.8). Thus cyclic-delay diversity is applicable to block-based transmission schemes such as OFDM and DFTS-OFDM.

In case of OFDM transmission, a cyclic shift of the time-domain signal corresponds to a frequency-dependent phase shift before OFDM modulation, as illustrated in Figure 6.8b. Similar to delay diversity, this will create artificial frequency selectivity as seen by the receiver.

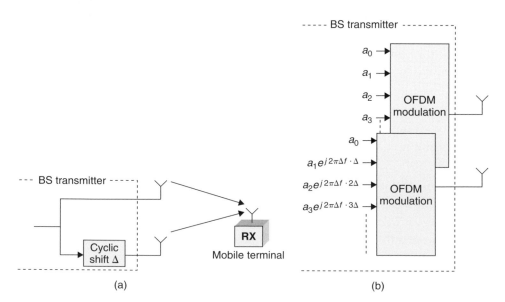

Figure 6.8   *Two-antenna Cyclic-Delay Diversity (CDD).*

Also similar to delay diversity, CDD can straightforwardly be extended to more than two transmit antennas with different cyclic shifts for each antenna.

6.4.1.3 Diversity by means of space–time coding

Space-time coding is a general term used to indicate multi-antenna transmission schemes where modulation symbols are mapped in the time and spatial (transmit-antenna) domain to capture the diversity offered by the multiple transmit antennas. Two-antenna space–time block coding (STBC), more specifically a scheme referred to as *Space–Time Transmit Diversity* (STTD), has been part of the 3G WCDMA standard already from its first release [94].

As shown in Figure 6.9, STTD operates on pairs of modulation symbols. The modulation symbols are directly transmitted on the first antenna. However, on the second antenna the order of the modulation symbols within a pair is reversed. Furthermore, the modulation symbols are sign-reversed and complex-conjugated as illustrated in Figure 6.9.

In vector notation STTD transmission can be expressed as

$$\overline{r} = \begin{pmatrix} r_{2n} \\ r_{2n+1}^* \end{pmatrix} = \begin{pmatrix} h_1 & -h_2 \\ h_2^* & h_1^* \end{pmatrix} \cdot \begin{pmatrix} s_{2n} \\ s_{2n+1}^* \end{pmatrix} = H \cdot \overline{s} \tag{6.7}$$

where r_{2n} and r_{2n+1} are the received symbols during the symbol intervals $2n$ and $2n + 1$, respectively.[3] It should be noted that this expression assumes that the channel coefficients h_1 and h_2 are constant over the time corresponding to two consecutive symbol intervals, an assumption that is typically valid. As the matrix H is a scaled unitary matrix, the sent symbols s_{2n} and s_{2n+1} can be recovered from the received symbols r_{2n} and r_{2n+1}, without any interference between the symbols, by applying the matrix $W = H^{-1}$ to the vector $\overline{r}$.

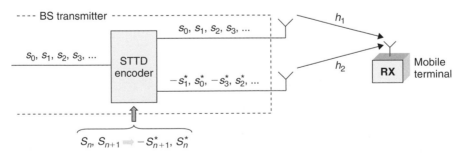

Figure 6.9 *WCDMA Space–Time Transmit Diversity (STTD).*

---

[3] Note that, for convenience, complex-conjugates have been applied for the second elements of $\overline{r}$ and $\overline{s}$.

The two-antenna space–time coding of Figure 6.9 can be said to be of rate one, implying that the input symbol rate is the same as the symbol rate at each antenna, corresponding to a bandwidth utilization of one. Space–time coding can also be extended to more than two antennas. However, in case of complex-valued modulation, such as QPSK or 16/64QAM, space–time codes of rate one without any inter-symbol interference (*orthogonal space–time codes*) only exist for two antennas [74]. If inter-symbol interference is to be avoided in case of more than two antennas, space–time codes with rate less than one must be used, corresponding to reduced bandwidth utilization.

6.4.1.4   *Diversity by means of space–frequency coding*

Space–frequency block coding (SFBC) is similar to space–time block coding, with the difference that the encoding is carried out in the antenna/frequency domains rather than in the antenna/time domains. Thus, space-frequency coding is applicable to OFDM and other 'frequency-domain' transmission schemes. The space–frequency equivalence to STTD (which could be referred to as Space–Frequency Transmit Diversity, SFTD) is illustrated in Figure 6.10. As can be seen, the block of (frequency-domain) modulation symbols a_0, a_1, a_2, a_3, ... is directly mapped to OFDM carriers of the first antenna, while the block of symbols $-a_1^*, a_0^*, -a_3^*, a_2^*, ...$ is mapped to the corresponding subcarriers of the second antenna.

Similar to space–time coding, the drawback of space–frequency coding is that there is no straightforward extension to more than two antennas unless a rate reduction is acceptable.

Comparing Figure 6.10 with the right-hand side of Figure 6.8, it can be noted that the difference between SFBC and two-antenna cyclic-delay diversity in essence

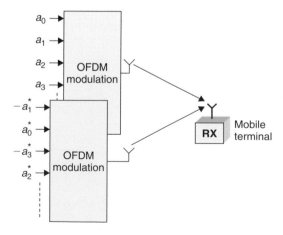

Figure 6.10   *Space–Frequency Transmit Diversity assuming two transmit antennas.*

lies in how the block of frequency-domain modulation symbols are mapped to the second antenna. The benefits of SFBC compared to CDD is that SFBC provides diversity on modulation-symbol level while CDD, in case of OFDM, must rely on channel coding in combination with frequency-domain interleaving to provide diversity.

6.4.2 Transmitter-side beam-forming

If some knowledge of the downlink channels of the different transmit antennas, more specifically some knowledge of the relative channel phases, is available at the transmitter side, multiple transmit antennas can, in addition to diversity, also provide beam-forming, that is the shaping of the overall antenna beam in the direction of a target receiver. In general, such beam-forming can increase the signal-strength at the receiver with up to a factor N_T, i.e. in proportion to the number of transmit antennas. When discussing transmission schemes relying on multiple transmit antennas to provide beam-forming one can distinguish between the cases of *high* and *low mutual antenna correlation*, respectively.

High mutual antenna correlation typically implies an antenna configuration with a small inter-antenna distance as illustrated in Figure 6.11a. In this case, the channels between the different antennas and a specific receiver are essentially the same, including the same radio-channel fading, except for a direction-dependent phase difference. The overall transmission beam can then be steered in different directions by applying *different phase shifts* to the signals to be transmitted on the different antennas, as illustrated in Figure 6.11b.

This approach to transmitter side beam-forming, with different phase shifts applied to highly correlated antennas, is sometimes referred to as 'classical' beam-forming. Due to the small antenna distance, the overall transmission beam will be relatively wide and any adjustments of the beam direction, in practice adjustments

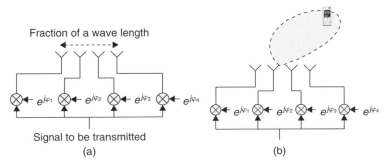

(a) (b)

Figure 6.11 *Classical beam-forming with high mutual antennas correlation: (a) antenna configuration and (b) beam-structure.*

of the antenna phase shifts, will typically be carried out on a relatively slow basis. The adjustments could, for example, be based on estimates of the direction to the target mobile terminal derived from uplink measurements. Furthermore, due to the assumption of high correlation between the different transmit antennas, classical beam-forming cannot provide any diversity against radio-channel fading but only an increase of the received signal strength.

Low mutual antenna correlation typically implies either a sufficiently large antenna distance, as illustrated in Figure 6.12, or different antenna polarization directions. With low mutual antenna correlation, the basic beam-forming principle is similar to that of Figure 6.11, that is the signals to be transmitted on the different antennas are multiplied by different complex weights. However, in contrast to classical beam-forming, the antenna weights should now take general complex values, i.e. both the phase and the amplitude of the signals to be transmitted on the different antennas can be adjusted. This reflects the fact that, due to the low mutual antenna correlation, both the phase and the instantaneous gain of the channels of each antenna may differ.

Applying different complex weights to the signals to be transmitted on the different antennas can be expressed, in vector notation, as applying a *pre-coding vector* $\bar{v}$ to the signal to be transmitted according to

$$\bar{s} = \begin{pmatrix} s_1 \\ \vdots \\ s_{N_T} \end{pmatrix} = \begin{pmatrix} v_1 \\ \vdots \\ v_{N_T} \end{pmatrix} \cdot s = \bar{v} \cdot s \tag{6.8}$$

It should be noted that classical beam-forming according to Figure 6.11 can also be described according to (6.8), that is as *transmit-antenna pre-coding*, with the constraint that the antenna weights are limited to unit-gain and only provide phase shifts to the different transmit antennas.

Assuming that the signals transmitted from the different antennas are only subject to non-frequency-selective fading and white noise, that is there is no radio-channel

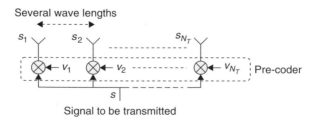

Figure 6.12 *Pre-coder-based beam-forming in case of low mutual antenna correlation.*

time dispersion, it can be shown [37] that, in order to maximize the received signal power, the pre-coding weights should be selected according to

$$v_i = \frac{h_i^*}{\sqrt{\sum_{k=1}^{N_T} |h_k|^2}} \tag{6.9}$$

that is as the complex conjugate of the corresponding channel coefficient h_i and with a normalization to ensure a fixed overall transmit power. The pre-coding vector thus:

- Phase rotates the transmitted signals to compensate for the instantaneous channel phase and ensure that the received signals are received phase aligned.
- Allocates power to the different antennas with, in general, more power being allocated to antennas with good instantaneous channel conditions (high channel gain $|h_i|$).
- Ensures an overall unit (or any other constant) transmit power.

A key difference between classical beam-forming according to Figure 6.11, assuming high mutual antenna correlation and beam-forming according to Figure 6.12, assuming low mutual antenna correlation, is that, in the later case, there is a need for more detailed channel knowledge, including estimates of the instantaneous channel fading. Updates to the pre-coding vector are thus typically done on a relatively short time scale to capture the fading variations. As the adjustment of the pre-coder weights takes into account also the instantaneous fading, including the instantaneous channel gain, fast beam-forming according to Figure 6.12 also provides diversity against radio-channel fading.

Furthermore, at least in case of communication based on *Frequency Division Duplex* (FDD), with uplink and downlink communication taking place in different frequency bands, the fading is typically uncorrelated between the downlink and uplink. Thus, in case of FDD, only the mobile terminal can determine the downlink fading. The mobile terminal would then report an estimate of the downlink channel to the base station by means of uplink signaling. Alternatively, the mobile terminal may, in itself, select a suitable pre-coding vector from a limited set of possible pre-coding vectors, the so-called pre-coder code-book, and report this to the base station.

On the other hand, in case of *Time Division Duplex* (TDD), with uplink and downlink communication taking place in the same frequency band but in separate non-overlapping time slots, there is typically a high fading correlation between the downlink and uplink. In this case, the base station could, at least in

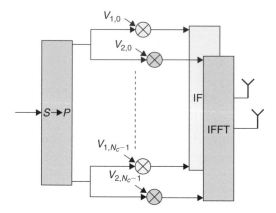

Figure 6.13   *Per-subcarrier pre-coding in case of OFDM (two transmit antennas).*

theory, determine the instantaneous downlink fading from measurements on the uplink, thus avoiding the need for any feedback. Note however that this assumes that the mobile terminal is continuously transmitting on the uplink.

The above discussion assumed that the channel gain was constant in the frequency-domain, that is there was no radio-channel frequency selectivity. In case of a frequency-selective channel there is obviously not a single channel coefficient per antennas based on which the antennas weights can be selected according to (6.9). However, in case of OFDM transmission, each subcarrier will typically experience a frequency-non-selective channel. Thus, in case of OFDM transmission, the pre-coding of Figure 6.12 can be carried out on a per-subcarrier basis as outlined in Figure 6.13, where the pre-coding weights of each subcarrier should be selected according to (6.9).

It should be pointed out that in case of single-carrier transmission, such as WCDMA, the one-weight-per-antenna approach outlined in Figure 6.12 can be extended to take into account also a time-dispersive/frequency-selective channel [77].

6.5 Spatial multiplexing

The use of multiple antennas at both the transmitter and the receiver can simply be seen as a tool to further improve the signal-to-noise/interference ratio and/or achieve additional diversity against fading, compared to the use of only multiple receive antennas or multiple transmit antennas. However, in case of multiple antennas at both the transmitter and the receiver there is also the possibility for so-called *spatial multiplexing*, allowing for more efficient utilization of high signal-to-noise/interference ratios and significantly higher data rates over the radio interface.

6.5.1 Basic principles

It should be clear from the previous sections that multiple antennas at the receiver and the transmitter can be used to improve the receiver signal-to-noise ratio in proportion to the number of antennas by applying beam-forming at the receiver and the transmitter. In the general case of N_T transmit antennas and N_R receive antennas, the receiver signal-to-noise ratio can be made to increase in proportion to the product $N_T \times N_R$. As discussed in Chapter 3, such an increase in the receiver signal-to-noise ratio allows for a corresponding increase in the achievable data rates, assuming that the data rates are power limited rather than bandwidth limited. However, once the bandwidth-limited range-of-operation is reached, the achievable data rates start to saturate unless the bandwidth is also allowed to increase.

One way to understand this saturation in achievable data rates is to consider the basic expression for the normalized channel capacity

$$\frac{C}{BW} = \log_2\left(1 + \frac{S}{N}\right) \tag{6.10}$$

where, by means of beam-forming, the signal-to-noise ratio S/N can be made to grow proportionally to $N_T \times N_R$. In general, $\log_2(1 + x)$ is proportional to x for small x, implying that, for low signal-to-noise ratios, the capacity grows approximately proportionally to the signal-to-noise ratio. However, for larger x, $\log_2(1 + x) \approx \log_2(x)$, implying that, for larger signal-to-noise ratios, capacity grows only logarithmically with the signal-to-noise ratio.

However, in the case of multiple antennas at the transmitter *and* the receiver it is, *under certain conditions*, possible to create up to $N_L = \min\{N_T, N_R\}$ parallel 'channels' each with N_L times lower signal-to-noise ratio (the signal power is 'split' between the channels), i.e. with a channel capacity

$$\frac{C}{BW} = \log_2\left(1 + \frac{N_R}{N_L} \cdot \frac{S}{N}\right) \tag{6.11}$$

As there are now N_L parallel channels, each with a channel capacity given by (6.11), the overall channel capacity for such a multi-antenna configuration is thus given by

$$\begin{aligned}
\frac{C}{BW} &= N_L \cdot \log_2\left(1 + \frac{N_R}{N_L} \cdot \frac{S}{N}\right) \\
&= \min\{N_T, N_R\} \cdot \log_2\left(1 + \frac{N_R}{\min\{N_T, N_R\}} \cdot \frac{S}{N}\right)
\end{aligned} \tag{6.12}$$

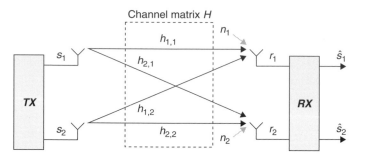

Figure 6.14   *2 × 2-antenna configuration.*

Thus, *under certain conditions*, the channel capacity can be made to grow essentially linearly with the number of antennas, avoiding the saturation in the data rates. We will refer to this as *Spatial Multiplexing*. The term MIMO (Multiple-Input/Multiple-Output) antenna processing is also very often used, although the term strictly speaking refers to all cases of multiple transmit antennas and multiple receive antennas, including also the case of combined transmit and receive diversity.[4]

To understand the basic principles how multiple parallel channels can be created in case of multiple antennas at the transmitter and the receiver, consider a 2 × 2 antenna configuration, that is two transmit antennas and two receive antennas, as outlined in Figure 6.14. Furthermore, assume that the transmitted signals are only subject to non-frequency-selective fading and white noise, i.e. there is no radio-channel time dispersion.

Based on Figure 6.14, the received signals can be expressed as

$$\bar{r} = \begin{pmatrix} r_1 \\ r_2 \end{pmatrix} = \begin{pmatrix} h_{1,1} & h_{1,2} \\ h_{2,1} & h_{2,2} \end{pmatrix} \cdot \begin{pmatrix} s_1 \\ s_2 \end{pmatrix} + \begin{pmatrix} n_1 \\ n_2 \end{pmatrix} = H \cdot \bar{s} + \bar{n} \tag{6.13}$$

where H is the 2 × 2 *channel matrix*. This expression can be seen as a generalization of (6.2) in Section 6.3 to multiple transmit antennas with different signals being transmitted from the different antennas.

Assuming no noise and that the channel matrix H is invertible, the vector $\bar{s}$, and thus both signals s_1 and s_2, can be perfectly recovered at the receiver, with no residual interference between the signals, by multiplying the received vector $\bar{r}$ with a matrix $W = H^{-1}$ (see also Figure 6.15).

---

[4] The case of a single transmit antenna and multiple receive antennas is, consequently, often referred to as SIMO (Single-Input/Multiple-Output). Similarly, the case of multiple transmit antennas and a single receiver antenna can be referred to as MISO (Multiple-Input/Single-Output).

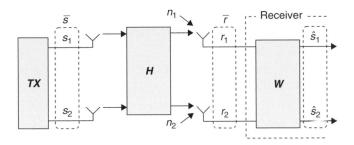

Figure 6.15 *Linear reception/demodulation of spatially multiplexed signals.*

$$\begin{pmatrix} \hat{s}_1 \\ \hat{s}_2 \end{pmatrix} = W \cdot \bar{r} = \begin{pmatrix} s_1 \\ s_2 \end{pmatrix} + H^{-1} \cdot \bar{n} \tag{6.14}$$

This is illustrated in Figure 6.15.

Although the vector $\bar{s}$ can be perfectly recovered in case of no noise, as long as the channel matrix H is invertible, (6.14) also indicates that the properties of H will determine to what extent the joint demodulation of the two signals will increase the noise level. More specifically, the closer the channel matrix is to being a singular matrix the larger the increase in the noise level.

One way to interpret the matrix W is to realize that the signals transmitted from the two transmit antennas are two signals causing interference to each other. The two receive antennas can then be used to carry out IRC, in essence completely suppressing the interference from the signal transmitted on the second antenna when detecting the signal transmitted at the first antenna and vice versa. The rows of the receiver matrix W simply implement such IRC.

In the general case, a multiple-antenna configuration will consist of N_T transmit antennas and N_R receive antennas. As discussed above, in such a case, the number of parallel signals that can be spatially multiplexed is, at least in practice, upper limited by $N_L = \min\{N_T, N_R\}$. This can intuitively be understood from the fact that:

- Obviously no more than N_T different signals can be transmitted from N_T transmit antennas, implying a maximum of N_T spatially multiplexed signals.
- With N_R receive antennas, a maximum of $N_R - 1$ interfering signals can be suppressed, implying a maximum of N_R spatially multiplexed signals.

However, in many cases, the number of spatially multiplexed signals, or the *order of the spatial multiplexing*, will be less than N_L given above:

- In case of very bad channel conditions (low signal-to-noise ratio) there is no gain of spatial multiplexing as the channel capacity is anyway a linear function of the signal-to-noise ratio. In such a case, the multiple transmit and receive antennas should be used for beam-forming to improve the signal-to-noise ratio, rather than for spatial multiplexing.

- In more general case, the spatial-multiplexing order should be determined based on the properties of the size $N_R \times N_T$ channel matrix. Any excess antennas should then be used to provide beam-forming. Such combined beam-forming and spatial multiplexing can be achieved by means of *pre-coder-based* spatial multiplexing, as discussed below.

6.5.2   Pre-coder-based spatial multiplexing

Linear pre-coding in case of spatial multiplexing implies that linear processing by means of a size $N_T \times N_L$ pre-coding matrix is applied at the transmitter side as illustrated in Figure 6.16. In line with the discussion above, in the general case N_L is equal or smaller than N_T, implying that N_L signal are spatially multiplexed and transmitted using N_T transmit antennas.

It should be noted that pre-coder-based spatial multiplexing can be seen as a generalization of pre-coder-based beam-forming as described in Section 6.4.2 with the pre-coding vector of size $N_T \times 1$ replaced by a pre-coding matrix of size $N_T \times N_L$.

The pre-coding of Figure 6.16 can serve two purposes:

- In the case when the number of signals to be spatially multiplexed equals the number of transmit antennas ($N_L = N_T$), the pre-coding can be used to 'orthogonalize' the parallel transmissions, allowing for improved signal isolation at the receiver side.
- In the case when the number signals to be spatially multiplexed is less than the number of transmit antennas ($N_{Lg} < N_T$), the pre-coding in addition provides

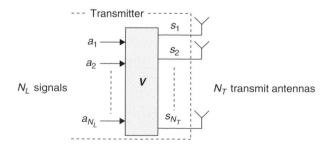

Figure 6.16   *Pre-coder-based spatial multiplexing.*

the mapping of the N_L spatially multiplexed signals to the N_T transmit antennas including the combination of spatially multiplexing and beam-forming.

To confirm that pre-coding can improve the isolation between the spatially multiplexed signals, express the channel matrix H as its singular-value decomposition [36]

$$H = W \cdot \Sigma \cdot V^H \tag{6.15}$$

where the columns of V and W each form an orthonormal set and Σ is an $N_L \times N_L$ diagonal matrix with the N_L strongest eigenvalues of $H^H H$ as its diagonal elements. By applying the matrix V as pre-coding matrix at the transmitter side and the matrix W^H at the receiver side, one arrives at an equivalent channel matrix $H' = \Sigma$ (see Figure 6.17). As H' is a diagonal matrix there is thus no interference between the spatially multiplexed signals at the receiver. At the same time, as both V and W have orthonormal columns, the transmit power as well as the demodulator noise level (assuming spatially white noise) are unchanged.

Clearly, in case of pre-coding each received signal will have a certain 'quality,' depending on the eigenvalues of the channel matrix (see right part of Figure 6.17). This indicates potential benefits of applying dynamic link adaptation in the *spatial domain*, that is the adaptive selection of the coding rates and/or modulation schemes for each signal to be transmitted.

As the pre-coding matrix will never perfectly match the channel matrix in practice, there will always be some residual interference between the spatially multiplexed signals. This interference can be taken care of by means of additional receiver-size linear processing according to Figure 6.15 or non-linear processing as discussed in Section 6.5.3 below.

To determine the pre-coding matrix V, knowledge about the channel matrix H is obviously needed. Similar to pre-coder-based beam-forming, a common approach is to have the receiver estimate the channel and decide on a suitable

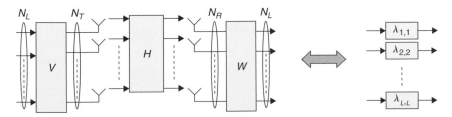

Figure 6.17 *Orthogonalization of spatially multiplexed signals by means of pre-coding.* $\lambda_{i,i}$ *is the ith eigenvalue of the matrix* $H^H H$.

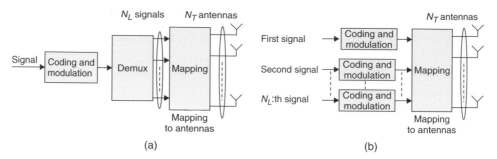

Figure 6.18   *Single-codeword transmission (a) vs. multi-codeword transmission (b).*

pre-coding matrix from a set of available pre-coding matrices (the pre-coder *code-book*). The receiver then feedback information about the selected pre-coding matrix to the transmitter.

6.5.3   Non-linear receiver processing

The previous sections discussed the use of linear receiver processing to jointly recover spatially multiplexed signals. However, improved demodulation performance can be achieved if non-linear receiver processing can be applied in case of spatial multiplexing.

The 'optimal' receiver approach for spatially multiplexed signals is to apply *Maximum-Likelihood* (ML) detection [25]. However, in many cases ML detection is too complex to use. Thus, several different proposals have been made for reduced complexity almost ML schemes (see, e.g. [39]).

Another non-linear approach to the demodulation of spatially multiplexed signals is to apply *Successive Interference Cancellation* (SIC) [71]. Successive Interference Cancellation is based on an assumption that the spatially multiplexed signals are separately coded before the spatial multiplexing. This is often referred to as *Multi-Codeword* transmission, in contrast to *Single-Codeword* transmission where the spatially multiplexed signals are assumed to be jointly coded (Figure 6.18). It should be understood that, also in the case of multi-codeword transmission, the data may originate from the same source but then de-multiplexed into different signals to be spatially multiplexed before channel coding.

As shown in Figure 6.19, in case of successive interference cancellation the receiver first demodulates and decodes one of the spatially multiplexed signals. The corresponding decoded data is then, if correctly decoded, re-encoded and subtracted from the received signals. A second spatially multiplexed signal can then be demodulated and decoded without, at least in the ideal case, any interference

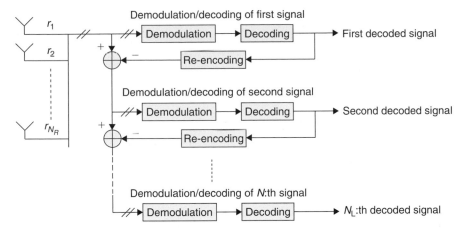

Figure 6.19 *Demodulation/decoding of spatially multiplexed signals based on Successive Interference Cancellation.*

from the first signal, that is with an improved signal-to-interference ratio. The decoded data of the second signal is then, if correctly decoded, re-encoded and subtracted from the received signal before decoding of a third signal. These iterations continue until all spatially multiplexed signals have been demodulated and decoded.

Clearly, in case of Successive Interference Cancellation, the first signals to be decoded are subject to higher interference level, compared to later decoded signals. To work properly, there should thus be a differentiation in the robustness of the different signals with, at least in principle, the first signal to be decoded being more robust than the second signal, the second signal being more robust than the third signal, etc. Assuming multi-codeword transmission according to Figure 6.18b, this can be achieved by applying different modulation schemes and coding rates to the different signals with, typically, lower-order modulation and lower coding rate, implying a lower data rate, for the first signals to be decoded. This is often referred to as *Per-Antenna Rate Control* (PARC) [53].

7

Scheduling, link adaptation and hybrid ARQ

One key characteristic of mobile radio communication is the typically rapid and significant variations in the instantaneous channel conditions. There are several reasons for these variations. Frequency-selective fading will result in rapid and random variations in the channel attenuation. Shadow fading and distance-dependent path loss will also affect the average received signal strength significantly. Finally, the interference at the receiver due to transmissions in other cells and by other terminals will also impact the interference level. Hence, to summarize, there will be rapid, and to some extent random, variations in the experienced quality of each radio link in a cell, variations that must be taken into account and preferably exploited.

In this chapter, some of the techniques for handling variations in the instantaneous radio-link quality will be discussed. *Channel-dependent scheduling* in a mobile-communication system deals with the question of how to share, between different users (different mobile terminals), the radio resource(s) available in the system to achieve as efficient resource utilization as possible. Typically, this implies to minimize the amount of resources needed per user and thus allow for as many users as possible in the system, while still satisfying whatever quality-of-service requirements that may exist. Closely related to scheduling is *link adaptation*, which deals with how to set the transmission parameters of a radio link to handle variations of the radio-link quality.

Both channel-dependent scheduling and link adaptation tries to exploit the channel variations through appropriate processing *prior* to transmission of the data. However, due to the random nature of the variations in the radio-link quality, perfect adaptation to the instantaneous radio-link quality is never possible. *Hybrid ARQ,* which requests retransmission of erroneously received data packets, is therefore useful. This can be seen as a mechanism for handling variations in the instantaneous radio-link quality *after* transmission and nicely complements channel-dependent scheduling and link adaptation. Hybrid ARQ also serves the purpose of handling random errors due to, for example, noise in the receiver.

7.1   Link adaptation: Power and rate control

Historically, *dynamic transmit-power control* has been used in CDMA-based mobile-communication systems such as WCDMA and cdma2000 to compensate for variations in the instantaneous channel conditions. As the name suggests, dynamic power control dynamically adjusts the radio-link transmit power to compensate for variations and differences in the instantaneous channel conditions. The aim of these adjustments is to maintain a (near) constant E_b/N_0 at the receiver to successfully transmit data without a too high error probability. In principle, transmit-power control increases the power at the transmitter when the radio link experiences poor radio conditions (and vice versa). Thus, the transmit power is in essence inversely proportional to the channel quality as illustrated in Figure 7.1(a). This results in a basically constant data rate, regardless of the

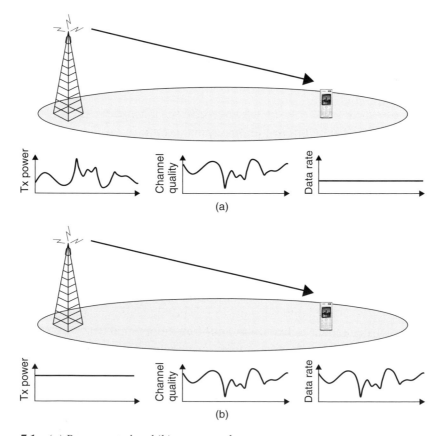

Figure 7.1   *(a) Power control and (b) rate control.*

channel variations. For services such as circuit-switched voice, this is a desirable property. Transmit-power control can be seen as one type of link adaptation, that is the adjustment of transmission parameters, in this case the transmit power, to adapt to differences and variations in the instantaneous channel conditions to maintain the received E_b/N_0 at a desired level.

However, in many cases of mobile communication, especially in case of packet-data traffic, there is not a strong need to provide a certain constant data rate over a radio link. Rather, from a user perspective, the data rate provided over the radio interface should simply be as 'high as possible.' Actually, even in case of typical 'constant-rate' services such as voice and video, (short-term) variations in the data rate are often not an issue, as long as the average data rate remains constant, assuming averaging over some relatively short time interval. In such cases, that is when a constant data rate is not required, an alternative to transmit-power control is link adaptation by means of *dynamic rate control*. Rate control does not aim at keeping the instantaneous radio-link data rate constant, regardless of the instantaneous channel conditions. Instead, with rate control, the data rate is dynamically adjusted to compensate for the varying channel conditions. In situations with advantageous channel conditions, the data rate is increased and vice versa. Thus, rate control maintains the $E_b/N_0 \sim P/R$ at the desired level, not by adjusting the transmission power P, but rather by adjusting the data rate R. This is illustrated in Figure 7.1(b).

It can be shown that rate control is more efficient than power control [11, 28]. Rate control in principle implies that the power amplifier is always transmitting at full power and therefore efficiently utilized. Power control, on the other hand, results in the power amplifier in most situations not being efficiently utilized as the transmission power is less than its maximum.

In practice, the radio-link data rate is controlled by adjusting the modulation scheme and/or the channel coding rate. In case of advantageous radio-link conditions, the E_b/N_0 at the receiver is high and the main limitation of the data rate is the bandwidth of the radio link. Hence, in such situations higher-order modulation, for example 16QAM or 64QAM, together with a high code rate is appropriate as discussed in Chapter 3. Similarly, in case of poor radio-link conditions, QPSK and low-rate coding is used. For this reason, link adaptation by means of rate control is sometimes also referred to as *Adaptive Modulation and Coding* (AMC).

7.2 Channel-dependent scheduling

Scheduling controls the allocation of the shared resources among the users at each time instant. It is closely related to link adaptation and often scheduling and link adaptation is seen as one joint function. The scheduling principles, as well as which resources that are shared between users, differ depending on the radio-interface characteristics, for example, whether uplink or downlink is considered and whether different users' transmissions are mutually orthogonal or not.

7.2.1   Downlink scheduling

In the downlink, transmissions to different mobile terminals within a cell are typically mutually orthogonal, implying that, at least in theory, there is no interference between the transmissions (no intra-cell interference). Downlink intra-cell orthogonality can be achieved in time domain, *Time Division Multiplexing* (TDM); in the frequency domain, *Frequency-Domain Multiplexing* (FDM); or in the code domain, *Code Domain Multiplexing* (CDM). In addition, the spatial domain can also be used to separate users, at least in a quasi-orthogonal way, through different antenna arrangements. This is sometimes referred to as *Spatial Division Multiplexing* (SDM), although it in most cases is used in combination with one or several of the above multiplexing strategies and not discussed further in this chapter.

For packet data, where the traffic often is very bursty, it can be shown that TDM is preferable from a theoretical point-of-view [40, 111] and is therefore typically the main component in the downlink [35, 62]. However, as discussed in Chapter 5, the TDM component is often combined with sharing of the radio resource also in the frequency domain (FDM) or in the code domain (CDM). For example, in case of HSDPA (see Chapter 9), downlink multiplexing is a combination of TDM and CDM. On the other hand, in case of LTE (see Part IV), downlink multiplexing is a combination of TDM and FDM. The reasons for sharing the resources not only in the time domain will be elaborated further upon later in this section.

When transmissions to multiple users occur in parallel, either by using FDM or CDM, there is also an instantaneous sharing of the total available cell transmit power. In other words, not only the time/frequency/code resources are shared resources, but also the power resource in the base station. In contrast, in case of sharing only in the time domain there is, per definition, only a single transmission at a time and thus no instantaneous sharing of the total available cell transmit power.

For the purpose of discussion, assume initially a TDM-based downlink with a single user being scheduled at a time. In this case, the utilization of the radio resources is maximized if, at each time instant, all resources are assigned to the user with the best instantaneous channel condition:

- In case of link adaptation based on power control, this implies that the lowest possible transmit power can be used for a given data rate and thus minimizes the interference to transmissions in other cells for a given link utilization.

- In case of link adaptation based on rate control, this implies that the highest data rate is achieved for a given transmit power, or, in other words, for a given interference to other cells, the highest link utilization is achieved.

However, if applied to the downlink, transmit-power control in combination with TDM scheduling implies that the total available cell transmit power will, in most cases, not be fully utilized. Thus, rate control is generally preferred [11, 40, 51, 62].

The strategy outlined above is an example of channel-dependent scheduling, where the scheduler takes the instantaneous radio-link conditions into account. Scheduling the user with the instantaneously best radio link conditions is often referred to as *max-C/I* (or *maximum rate*) scheduling. Since the radio conditions for the different radio links within a cell typically vary independently, at each point in time there is almost always a radio link whose channel quality is near its peak (see Figure 7.2). Thus, the channel eventually used for transmission will typically have a high quality and, with rate control, a correspondingly high data rate can be used. This translates into a high system capacity. The gain obtained by transmitting to users with favorable radio-link conditions is commonly known as multi-user diversity; the gains are larger, the larger the channel variations and the larger the number of users in a cell. Hence, in contrast to the traditional view that fast fading, that is rapid variations in the radio-link quality, is an undesirable effect that has to be combated, with the possibility for channel-dependent scheduling *fading is in fact potentially beneficial and should be exploited.*

Mathematically, the max-C/I (maximum rate) scheduler can be expressed as scheduling user k given by

$$k = \arg \max_i R_i$$

where R_i is the instantaneous data rate for user i. Although, from a system capacity perspective, max-C/I scheduling is beneficial, this scheduling principle

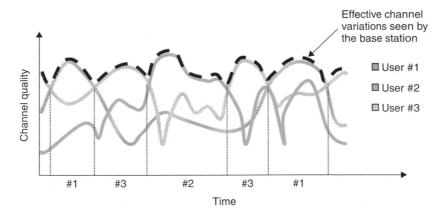

Figure 7.2 *Channel-dependent scheduling.*

will not be fair in all situations. If all mobile terminals are, on average, experiencing similar channel conditions and large variations in the instantaneous channel conditions are only due to, for example, fast multi-path fading, all users will experience the same average data rate. Any variations in the instantaneous data rate are rapid and often not even noticeable by the user. However, in practice different mobile terminals will experience also differences in the (short-term) average channel conditions, for example, due to differences in the distance and shadow fading between the base station and the mobile terminal. In this case, the channel conditions experienced by one mobile terminal may, for a relatively long time, be worse than the channel conditions experienced by other mobile terminals. A pure max-C/I-scheduling strategy may then, in essence, 'starve' the mobile terminal with the bad channel conditions, and the mobile terminal with bad channel conditions will never be scheduled. This is illustrated in Figure 7.3a where a max-C/I scheduler is used to schedule between two different users with different average channel quality. Virtually all the time the same user is scheduled. Although resulting in the highest system capacity, this situation is often not acceptable from a quality-of-service point-of-view.

An alternative to the max-C/I scheduling strategy is so-called *round-robin* scheduling, illustrated in Figure 7.3b. This scheduling strategy let the users take turns in using the shared resources, without taking the instantaneous channel conditions into account. Round-robin scheduling can be seen as fair scheduling in the sense that the same amount of radio resources (the same amount of time) is given to each communication link. However, round-robin scheduling is not fair in the

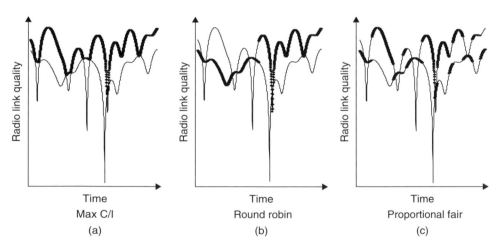

Figure 7.3 *Example of three different scheduling behaviors for two users with different average channel quality: (a) max C/I, (b) round robin, and (c) proportional fair. The selected user is shown with bold lines.*

sense of providing the same service quality to all communication links. In that case more radio resources (more time) must be given to communication links with bad channel conditions. Furthermore, as round-robin scheduling does not take the instantaneous channel conditions into account in the scheduling process it will lead to lower overall system performance but more equal service quality between different communication links, compared to max-C/I scheduling.

Thus what is needed is a scheduling strategy that is able to utilize the fast channel variations to improve the overall cell throughput while still ensuring the same average user throughput for all users or at least a certain minimum user throughput for all users. When discussing and comparing different scheduling algorithms it is important to distinguish between different types of variations in the service quality:

- Fast variations in the service quality corresponding to, for example, fast multi-path fading and fast variations in the interference level. For many packet-data applications, relatively large short-term variations in service quality are often acceptable or not even noticeable to the user.
- More long-term differences in the service quality between different communication links corresponding to, for example, differences in the distance to the cell site and shadow fading. In many cases there is a need to limit such long-term differences in service quality.

A practical scheduler should thus operate somewhere in-between the max-C/I scheduler and the round-robin scheduler, that is try to utilize fast variations in channel conditions as much as possible while still satisfying some degree of fairness between users.

One example of such a scheduler is the proportional-fair scheduler [33, 34, 119], illustrated in Figure 7.3c. In this strategy, the shared resources are assigned to the user with the *relatively* best radio-link conditions, that is, at each time instant user k is selected for transmission according to

$$k = \arg \max_i \frac{R_i}{\bar{R}_i}$$

where R_i is the instantaneous data rate for user i and $\bar{R}_i$ is the average data rate for user i. The average is calculated over a certain averaging period T_{PF}. To ensure efficient usage of the short-term channel variations and, at the same time, limit the long-term differences in service quality to an acceptable level, the time constant T_{PF} should be set longer than the time constant for the short-term variations. At the same time T_{PF} should be sufficiently short so that quality variations within the interval T_{PF} are not strongly noticed by a user. Typically, T_{PF} can be set to be in the order of one second.

In the above discussion, it was assumed that all the radio resources in the downlink were assigned to a single user at a time, that is scheduling were done purely in the time domain using TDM between users. However, in several situations, TDM is complemented by CDM or FDM. In principle, there are two reasons for not relying solely on TDM in the downlink :

- In case of insufficient payload, that is the amount of data to transfer to a user is not sufficiently large to utilize the full channel capacity, and a fraction of resources could be assigned to another user, either through FDM or CDM.
- In case channel variations in the frequency domain are exploited through FDM as discussed further below.

The scheduling strategies in these cases can be seen as generalizations of the schemes discussed for the TDM-only cases above. For example, to handle small payloads, a greedy filling approach can be used, where the scheduled user is selected according to max-C/I (or any other scheduling scheme). Once this user has been assigned resources matching the amount of data awaiting transmission, the second best user according to the scheduling strategy is selected and assigned (a fraction of) the residual resources and so on.

Finally, it should also be noted that the scheduling algorithm typically is a base-station-implementation issue and nothing that is normally specified in any standard. What need to be specified in a standard to support channel-dependent scheduling are channel-quality measurements/reports and the signaling needed for dynamic resource allocation.

7.2.2   Uplink scheduling

The previous section discussed scheduling from a downlink perspective. However, scheduling is equally applicable to uplink transmissions and to a large extent the same principles can be reused although there are some differences between the two.

Fundamentally, the uplink power resource is *distributed* among the users, while in the downlink the power resource is *centralized* within the base station. Furthermore, the maximum uplink transmission power of a single terminal is typically significantly lower than the output power of a base station. This has a significant impact on the scheduling strategy. Unlike the downlink, where pure TDMA often can be used, uplink scheduling typically has to rely on sharing in the frequency and/or code domain in addition to the time domain as a single terminal may not have sufficient power for efficiently utilizing the link capacity.

Channel-dependent scheduling is, similar to the downlink case, beneficial also in the uplink case. However, the characteristics of the underlying radio interface, most notably whether the uplink relies on orthogonal or non-orthogonal multiple access and the type of link adaptation scheme used, also have a significant impact on the uplink scheduling strategy.

In case of a non-orthogonal multiple-access scheme such as CDMA, power control is typically essential for proper operation. As discussed earlier in this chapter, the purpose of power control is to control the received E_b/N_0 such that the received information can be recovered. However, in a non-orthogonal multiple-access setting, power control also serves the purpose of controlling the amount of interference affecting *other* users. This can be expressed as the *maximum tolerable interference level* at the base station is a shared resource. Even if it, from a single user's perspective, would be beneficial to transmit at full power to maximize the data rate, this may not be acceptable from an interference perspective as other terminals in this case may not be able to successfully transfer any data. Thus, with non-orthogonal multiple access, scheduling a terminal when the channel conditions are favorable may not directly translate into a higher data rate as the interference generated to other simultaneously transmitting terminals in the cell must be taken into account. Stated differently, the *received* power (and thus the data rate) is, thanks to power control, in principle constant, regardless of the channel conditions at the time of transmission, while the *transmitted* power depends on the channel conditions at the time of transmission. Hence, even though channel-dependent scheduling in this example does not give a direct gain in terms of a higher data rate from the terminal, channel-dependent scheduling will still provide a gain for the system in terms of lower intra-cell interference.

The above discussion on non-orthogonal multiple access was simplified in the sense that no bounds on the terminals transmission power were assumed. In practice, the transmission power of a terminal is upper-bounded, both due to implementation and regulatory reasons, and scheduling a terminal for transmission in favorable channel conditions decreases the probability that the terminal has insufficient power to utilize the channel capacity.

In case of orthogonal multiple-access scheme, intra-cell power control is fundamentally not necessary and the benefits with channel-dependent scheduling become more similar to the downlink case. In principle, from an intra-cell perspective, a terminal can transmit at full power and the scheduler assigns a suitable part of the orthogonal resources (in practice a suitable part of the overall bandwidth) to the terminal for transmission. The remaining orthogonal resources can be assigned to other users. However, implementation constraints, for example leakage between the received signals or limited dynamic range in the receiver circuitry, may pose restrictions on the maximum tolerable power difference between the signals from simultaneously transmitting terminals. As a consequence, a certain degree of power control may be necessary, making the situation somewhat similar to the non-orthogonal case.

The discussion on non-orthogonal and orthogonal multiple access mainly considered intra-cell multiple access. However, in many practical systems universal frequency reuse between cells is used. In this case, the inter-cell multiple access is non-orthogonal, regardless of the intra-cell multiple access, which sets limits on the allowable transmission power from a terminal.

Regardless of whether orthogonal or non-orthogonal multiple access is used, the same basic scheduling principles as for the downlink can be used. A max-C/I scheduler would assign all the uplink resources to the terminal with the best uplink channel conditions. Neglecting any power limitations in the terminal, this would result in the highest capacity (in an isolated cell) [40].

In case of a non-orthogonal multiple-access scheme, *greedy filling* is one possible scheduling strategy [51]. With greedy filling, the terminal with the best radio conditions is assigned as high data rate as possible. If the interference level at the receiver is smaller than the maximum tolerable level, the terminal with the second best channel conditions is allowed to transmit as well, continuing with more and more terminals until the maximum tolerable interference level at the receiver is reached. This strategy maximizes the air interface utilization but is achieved at the cost of potentially large differences in data rates between users. In the extreme case, a user at the cell border with poor channel conditions may not be allowed to transmit at all.

Strategies between greedy filling and max-C/I can also be envisioned, for example different proportional-fair strategies. This can be achieved by including a weighting factor for each user, proportional to the ratio between the instantaneous and average data rates, into the greedy filling algorithm.

The schedulers above all assume knowledge of the instantaneous radio-link conditions; knowledge that can be hard to obtain in the uplink scenario as discussed in Section 7.2.4 below. In situations when no information about the uplink radio-link quality is available at the scheduler, *round-robin* scheduling can be used. Similar to the downlink, round-robin implies terminals taking turns in transmitting, thus creating a TDMA-like operation with inter-user orthogonality in the time domain. Although the round-robin scheduler is simple, it is far from the optimal scheduling strategy.

However, as already discussed in Chapter 5, the transmission power in a terminal is limited and therefore additional sharing of the uplink resources in frequency and/or code domain is required. This also impacts the scheduling decisions. For example, terminals far from the base station typically operate in the power-limited region, in contrast to terminals close to the base stations which often are in the bandwidth-limited region (for a discussion on power-limited vs. band-width-limited operation, see Chapter 3). Thus, for a terminal far from the base station, increasing the bandwidth will not result in an increased data rate and it is better to only assign a small amount of the bandwidth to this terminal and assign the remaining bandwidth to other terminals. On the other hand, for terminals close to the base station, an increase in the assigned bandwidth will provide a higher data rate.

7.2.3 Link adaptation and channel-dependent scheduling in the frequency domain

In the previous section, TDM-based scheduling was assumed and it was explained how, in this case, channel variations in the time domain could be utilized to improve system performance by applying channel-dependent scheduling, especially in combination with dynamic rate control. However, if the scheduler has access to the frequency domain, for example through the use of OFDM transmission, scheduling and link adaptation can also take place in the frequency domain.

Link adaptation in the frequency domain implies that, based on knowledge about the instantaneous channel conditions also in the frequency domain, that is knowledge about the attenuation as well as the noise/interference level of, in the extreme case, every OFDM subcarrier, the power and/or the data rate of each OFDM carrier can be individually adjusted for optimal utilization.

Similarly, channel-dependent scheduling in the frequency domain implies that, based on knowledge about the instantaneous channel conditions also in the frequency domain, different subcarriers are used for transmission to or from different mobile terminals. The scheduling gains from exploiting variations in the frequency-domain are similar to those obtained from time-domain variations.

Obviously, in situations where the channel quality varies significantly with the frequency while the channel quality only varies slowly with time, channel-dependent scheduling in the frequency domain can enhance system capacity. An example of such a situation is a wideband indoor system with low mobility, where the quality only varies slowly with time.

7.2.4 Acquiring on channel-state information

To select a suitable data rate, in practice a suitable modulation scheme and channel-coding rate, the transmitter needs information about the radio-link channel conditions. Such information is also required for the purpose of channel-dependent scheduling. In case of a system based on frequency-division duplex (FDD), only the receiver can accurately estimate the radio-link channel conditions.

For the downlink, most systems provide a downlink signal of a predetermined structure, known as the downlink pilot or the downlink reference signal. This reference signal is transmitted from the base station with a constant power and can be used by the mobile terminal to estimate the instantaneous downlink channel conditions. Information about the instantaneous downlink conditions can then be reported to the base station.

Basically, what is relevant for the transmitter is an estimate reflecting the channel conditions *at the time of transmission*. Hence, in principle, the terminal could apply a predictor, trying to predict the future channel conditions and report this predicted value to the base station. However, as this would require specification of prediction algorithms and how they would operate when the terminal is moving at different speeds, most practical systems simply report the measured channel conditions to the base station. This can be seen as a very simple predictor, basically assuming the conditions in the near future will be similar to the current conditions. Thus, the more rapid the time-domain channel variations are, the less efficient link adaptation is.

As there inevitably will be a delay between the point in time when the terminal measured the channel conditions and the application of the reported value in the transmitter, channel-dependent scheduling and link adaptation typically operates at its best at low terminal mobility. If the terminal starts to move at a high speed, the measurement reports will be outdated once reported to the base station. In such cases, it is often preferable to perform link adaptation on the long-term average channel quality and rely on hybrid ARQ with soft combining for the rapid adaptation.

For the uplink, estimation of the uplink channel conditions is not as straightforward as there is typically not any reference signal transmitted with constant power from each mobile terminal. Discussions on how to estimate uplink channel conditions is provided in Chapter 10 for HSPA and Chapter 17 for LTE.

In case of a system with time-division duplex (TDD) where uplink and downlink communication are time multiplexed within the same frequency band, the uplink signal attenuation could be estimated from downlink measurements of the mobile terminal, due to the reciprocity of also the multi-path fading in case of TDD. However, it should then be noted that this may not provide full knowledge of the downlink channel conditions. As an example, the interference situations at the mobile terminal and the base station are different also in case of TDD.

7.2.5 Traffic behavior and scheduling

It should be noted that there is little difference between different scheduling algorithms at low system load, that is when only one or, in some cases, a few users have data waiting for transmission at the base station at each scheduling instant. The differences between different scheduling algorithms are primarily visible at high load. However, not only the load, but also the traffic behavior affects the overall scheduling performance.

As discussed above, channel-dependent scheduling tries to exploit short-term variations in radio quality. Generally speaking, a certain degree of long-term fairness in service quality is desirable, which should be accounted for in the scheduler design. However, since system throughput decreases the more fairness is enforced, a trade-off between fairness and system throughput is necessary. In this trade-off, it is important to take traffic characteristics into account as they have a significant influence on the trade-off between system throughput and service quality.

To illustrate this, consider three different downlink schedulers :

1. *Round-robin (RR) scheduler,* where channel conditions are not taken into account.
2. *Proportional-fair (PF) scheduler,* where short-term channel variations are exploited while maintaining the long-term average user data rate.
3. *Max-C/I scheduler,* where the user with the best instantaneous channel quality in absolute terms is scheduled.

For a full buffer scenario when there is always data available at the base station for all terminals in the cell, a max-C/I scheduler will result in no, or a very low, user throughput for users at the cell edge with a low average channel quality. The reason is the fundamental strategy of the max-C/I scheduler – all resources are allocated for transmission to the terminal whose channel conditions support the highest data rate. Only in the rare, not to say unlikely, case of a cell-edge user having better conditions than a cell-center user, for example due to a deep fading dip for the cell-center user, will the cell-edge user be scheduled. A proportional-fair scheduler, on the other hand, will ensure some degree of fairness by selecting the user supporting the highest data rate relative to its average data rate. Hence, users tend to be scheduled on their fading peaks, regardless of the absolute quality. Thus, also users on the cell edge will be scheduled, thereby resulting in some degree of fairness between users.

For a scenario with bursty packet data, the situation is different. In this case, the users' buffers will be finite and in many cases also empty. For example, a web page has a certain size and after transmitting the page, there is no more data to be transmitted to the terminal in question until the users requests a new page by clicking on a link. In this case, a max-C/I scheduler can still provide a certain degree of fairness. Once the buffer for the user with the highest C/I has been emptied, another user with non-empty buffers will have the highest C/I and be scheduled and so on. This is the reason for the difference between full buffer and web-browsing traffic illustrated in Figure 7.4. The proportional-fair scheduler has similar performance in both scenarios.

Clearly, the degree of fairness introduced by the traffic properties depends heavily on the actual traffic; a design made with certain assumptions may be less desirable in an actual network where the traffic pattern may be different from the assumptions made during the design. Therefore, relying solely on the traffic properties for fairness is not a good strategy, but the discussion above emphasizes the need to not only design the scheduler for the full buffer case.

7.3   Advanced retransmission schemes

Transmissions over wireless channels are subject to errors, for example due to variations in the received signal quality. To some degree, such variations can be counteracted through link adaptation as discussed above. However, receiver noise and unpredictable interference variations cannot be counteracted. Therefore, virtually all wireless communications systems employ some form of *Forward Error Correction* (FEC), tracing its roots to the pioneering work by Claude Shannon in 1948 [70]. There is a rich literature in the area of error-correcting coding, see for example [42, 66] and the references therein, and a detailed description is beyond the scope of this book. In short, the basic principle

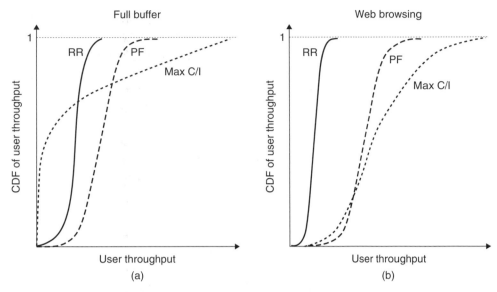

Figure 7.4 *Illustration of the principle behavior of different scheduling strategies: (a) for full buffers and (b) for web browsing traffic model.*

beyond forward error-correcting coding is to introduce redundancy in the transmitted signal. This is achieved by adding *parity bits* to the information bits prior to transmission (alternatively, the transmission could consists of parity bits alone, depending on the coding scheme used). The parity bits are computed from the information bits using a method given by the coding structure used. Thus, the number of bits transmitted over the channel is larger then the number of original information bits and a certain amount of *redundancy* has been introduced in the transmitted signal.

Another approach to handle transmissions errors is to use *Automatic Repeat Request* (ARQ). In an ARQ scheme, the receiver uses an error-detecting code, typically a Cyclic Redundancy Check (CRC), to detect if the received packet is in error or not. If no error is detected in the received data packet, the received data is declared error-free and the transmitter is notified by sending a positive acknowledgment (ACK). On the other hand, if an error is detected, the receiver discards the received data and notifies the transmitter via a return channel by sending a negative acknowledgment (NAK). In response to a NAK, the transmitter retransmits the same information.

Virtually all modern communication systems, including WCDMA and cdma2000, employ a combination of forward error-correcting coding and ARQ, a combination known as *hybrid ARQ*. Hybrid ARQ uses forward error correcting codes to correct a subset of all errors and rely on error detection to detect uncorrectable errors. Erroneously received packets are discarded and the receiver requests retransmissions of corrupted packets. Thus, it is a combination of FEC and ARQ as described above. Hybrid ARQ was first proposed in [120] and numerous publications on hybrid ARQ have appeared since, see [42] and references therein. Most practical hybrid ARQ schemes are built around a CRC code for error detection and convolutional or Turbo codes for error correction, but in principle any error-detecting and error-correcting code can be used.

7.4   Hybrid ARQ with soft combining

The hybrid ARQ operation described above discards erroneously received packets and requests retransmission. However, despite that the packet was not possible to decode, the received signal still contains information, which is lost by discarding erroneously received packets. This shortcoming is addressed by *hybrid ARQ with soft combining*. In hybrid ARQ with soft combining, the erroneously received packet is stored in a buffer memory and later combined with the retransmission to obtain a single, combined packet which is more reliable than its constituents. Decoding of the error-correcting code operates on the combined signal. If the decoding fails (typically a CRC code is used to detect this event), a retransmission is requested.

Retransmission in any hybrid ARQ scheme must, by definition, represent the same set of information bits as the original transmission. However, the set of coded bits transmitted in each retransmission may be selected differently as long as they represent the same set of information bits. Hybrid ARQ with soft combining is therefore usually categorized into *Chase combining* and *Incremental Redundancy*, depending on whether the retransmitted bits are required to be identical to the original transmission or not.

Chase combining, where the retransmissions consist of the same set of coded bits as the original transmission, was first proposed in [9]. After each retransmission, the receiver uses maximum-ratio combining to combine each received channel bit with any previous transmissions of the same bit and the combined signal is fed to the decoder. As each retransmission is an identical copy of the original transmission, retransmissions with Chase combining can be seen as additional repetition coding. Therefore, as no new redundancy is transmitted, Chase combining does not give any additional coding gain but only increases the accumulated received E_b/N_0 for each retransmission (Figure 7.5).

Several variants of Chase combining exist. For example, only a subset of the bits transmitted in the original transmission might be retransmitted, so-called partial Chase combining. Furthermore, although combining is often done after

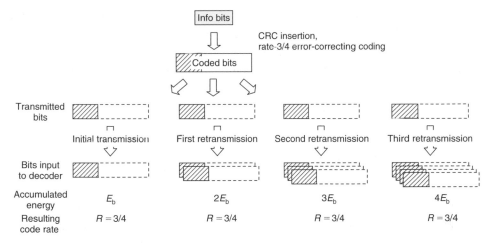

Figure 7.5   *Example of Chase combining.*

demodulation but before channel decoding, combining can also be done at the modulation symbol level before demodulation, as long as the modulation scheme is unchanged between transmission and retransmission.

With *Incremental Redundancy* (IR), each retransmission does not have to be identical to the original transmission. Instead, *multiple sets* of coded bits are generated, each of them representing the same set of information bits [56, 114]. Whenever a retransmission is required, the retransmission typically uses a different set of coded bits than the previous transmission. The receiver combines the retransmission with previous transmission attempts of the same packet. As the retransmission may contain additional parity bits, not included in the previous transmission attempts, the resulting code rate is generally lowered by a retransmission. Furthermore, each retransmission does not necessarily have to consist of the same number of coded bits as the original and, in general, also the modulation scheme can be different for different retransmissions. Hence, incremental redundancy can be seen as a generalization of Chase combining or, stated differently, Chase combining is a special case of incremental redundancy.

Typically, incremental redundancy is based on a low-rate code and the different redundancy versions are generated by puncturing the output of the encoder. In the first transmission only a limited number of the coded bits are transmitted, effectively leading to a high-rate code. In the retransmissions, additional coded bits are transmitted. As an example, assume a basic rate-1/4 code. In the first transmission, only every third coded bit is transmitted, effectively giving a rate-3/4 code as illustrated in Figure 7.6. In case of a decoding error and a subsequent request for a retransmission, additional bits are transmitted, effectively

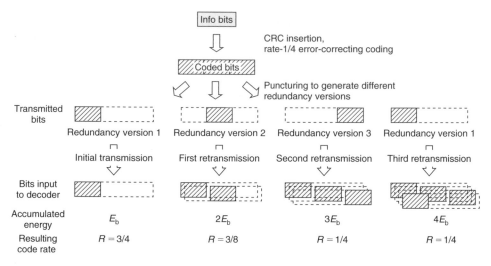

Figure 7.6   *Example of incremental redundancy.*

leading to a rate-3/8 code. After a second retransmission the code rate is 1/4. In case of more than two retransmissions, already transmitted coded bits would be repeated. In addition to a gain in accumulated received E_b/N_0, incremental redundancy also results in a coding gain for each retransmission. The gain with IR compared to Chase is larger for high initial code rates while at lower initial coding rates, Chase combining is almost as good as IR [10]. Furthermore, as shown in [27], the performance gain of incremental redundancy compared to Chase combining can also depend on the relative power difference between the transmission attempts.

With incremental redundancy, the code used for the first transmission should provide good performance not only when used alone, but also when used in combination with the code for the second transmission. The same holds for subsequent retransmissions. Thus, as the different redundancy versions typically are generated through puncturing of a low-rate mother code, the puncturing patterns should be defined such that all the code bits used by a high-rate code should also be part of any lower-rate codes. In other words, the resulting code rate R_i after transmission attempt i, consisting of the coded bits from redundancy versions RV_k, $k = 1,\ldots, i$, should have similar performance as a good code designed directly for rate R_i. Examples of this for convolutional codes are the so-called rate-compatible convolutional codes [30].

In the discussion so far, it has been assumed that the receiver has received all the previously transmitted redundancy versions. If all redundancy versions provide the same amount of information about the data packet, the order of the redundancy versions are not critical. However, for some code structures, not all redundancy versions are of equal importance. One example hereof is Turbo codes, where the systematic bits are of higher importance than the parity bits. Hence, the initial transmission should at least include all the systematic bits and some parity bits. In the retransmission (s), parity bits not part of the initial transmission can be included. However, if the initial transmission was received with poor quality or not at all, a retransmission with only parity bits is not appropriate as a retransmission of (some of) the systematic bits provides better performance. Incremental redundancy with Turbo codes can therefore benefit from multiple levels of feedback, for example by using two different negative acknowledgments – NAK to request additional parity bits and LOST to request a retransmission of the systematic bits. In general, the problem of determining the amount of systematic and parity bits in a retransmission based on the signal quality of previous transmission attempts is non-trivial.

Hybrid ARQ with soft combining, regardless of whether Chase or incremental redundancy is used, leads to an implicit reduction of the data rate by means of retransmissions and can thus be seen as *implicit* link adaptation. However, in contrast to link adaptation based on explicit estimates of the instantaneous channel conditions, hybrid ARQ with soft combining implicitly adjusts the coding rate based on the result of the decoding. In terms of overall throughput this kind of implicit link adaptation can be superior to explicit link adaptation as additional redundancy is only added *when needed*, that is when previous higher-rate transmissions were not possible to decode correctly. Furthermore, as it does not try to predict any channel variations, it works equally well, regardless of the speed of which the terminal is moving. Since implicit link adaptation can provide a gain in system throughput, a valid question is why explicit link adaptation at all is necessary. One major reason for having explicit link adaptation is the reduced delay. Although relying on implicit link adaptation alone is sufficient from a system throughput perspective, the end-user service quality may not be acceptable from a delay perspective.

Part III
HSPA

8

WCDMA evolution: HSPA and MBMS

The first step in the evolution of WCDMA radio access is the introduction of *High-Speed Downlink Packet Access* (HSDPA) in Release 5 of the 3GPP/ WCDMA specifications. Although packet-data communication is supported already in the first release of the WCDMA standard, HSDPA brings further enhancements to the provisioning of packet-data services in WCDMA, both in terms of system and end-user performance. The downlink packet-data enhancements of HSDPA are complemented by *Enhanced Uplink*, introduced in Release 6 of the 3GPP/WCDMA specifications. HSDPA and Enhanced Uplink are often jointly referred to as *High-Speed Packet Access* (HSPA).

As discussed in Chapter 2, important requirements for cellular systems providing packet-data services are high data rates and low delays while, as at the same time, maintaining good coverage and providing high capacity. To achieve this, HSPA introduces several of the basic techniques described in Part II into WCDMA, such as higher order modulation, fast (channel-dependent) scheduling and rate control, and fast hybrid ARQ with soft combining. Altogether, HSPA provides downlink and uplink data rates up to approximately 14 and 5.7 Mbit/s, respectively, and significantly reduced round trip times and improved capacity, compared to Release 99.

3GPP Release 6 also brings efficient support for broadcast services into WCDMA through the introduction of *Multimedia Broadcast Multicast Services* (MBMS), suitable for applications like mobile TV. With MBMS, multiple terminals may receive the same broadcast transmission instead of the network transmitting the same information individually to each of the users. Naturally, this will lead to an improvement in the resource utilization when several users in a cell receive the same content. As a consequence of the broadcast transmission, user-specific adaptation of the transmission parameters cannot be used and diversity as a mean to maintain good coverage is important. For MBMS, macro-diversity through multi-cell reception is employed for this purpose.

The evolution of the WCDMA radio access continues and will continue also in the future. For example, 3GPP Release 7 introduces several new features. MIMO is a tool to further improve capacity and especially the HSPA peak data rates. *Continuous Packet Connectivity* aims at providing an 'always-on' service perception for the end-user.

In parallel to enhancements of the radio-access specifications, there is an ongoing effort in *advanced* receiver structures. These improvements, which can provide a significant gain in both system and end-user performance, are to a large extent implementation specific, although the receiver-performance requirements as such are subject to standardization.

As a summary, the evolution of WCDMA is illustrated in Figure 8.1.

In the following chapters, the evolution of WCDMA is discussed. Chapter 9 provides a description of HSDPA, including how several of the basic technologies from Part II have been incorporated in the WCDMA radio access. A similar description for Enhanced Uplink is provided in Chapter 10. MBMS is described in Chapter 11. Finally, Chapter 12 describes Release 7 features such as MIMO, Continuous Packet Connectivity, and advanced receivers, representing the evolution of HSPA. The chapters have a similar structure. The first part of each chapter provides a high-level overview of the respective feature. This can be used to get a quick understanding of how HSPA and MBMS work and can to a large extent be read on its own. The overview is followed by a description on how the

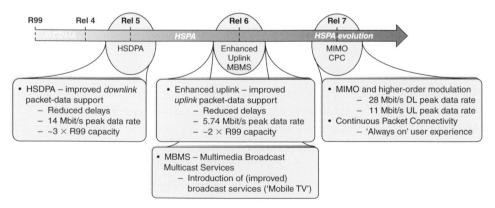

Figure 8.1   *WCDMA evolution.*

key technologies have been integrated into WCDMA and why certain solutions were chosen. Finally, at the end of the chapter, additional details are provided.

The remainder of this chapter contains an overview of the first release of WCDMA. The intention is to provide the reader with an understanding of the foundation upon which HSPA and MBMS are built. The reader already familiar with WCDMA may want to skip the remainder of this chapter.

8.1 WCDMA: Brief overview

This section provides a brief overview of WCDMA Release 99 to serve as a background to the subsequent sections. As a thorough walk through of WCDMA is a topic of its own and beyond the scope of this book, the reader is referred to other books and articles [4, 14, 121] for a detailed description. WCDMA is a versatile and highly flexible radio interface that can be configured to meet the requirements from a large number of services, but the focus for the description below is the functionality commonly used to support packet-data transmissions. The main purpose is to provide a brief background to WCDMA to put the enhancements described later on into a perspective.

8.1.1 Overall architecture

WCDMA is based on a hierarchical architecture [103] with the different nodes and interfaces as illustrated in Figure 8.2. A terminal, also referred to as *User Equipment* (UE) in 3GPP terminology, communicates with one (or several) NodeBs. In the WCDMA architecture, the term *NodeB* refers to a logical node, responsible for physical-layer processing such as error-correcting coding, modulation and spreading, as well as conversion from baseband to the radio-frequency signal transmitted from the antenna. A NodeB is handling transmission and reception in one or several cells. Three-sector sites are common, where each NodeB is handling transmissions in three cells, although other arrangements of the cells belonging to one NodeB can be thought of, for example, a large number of indoor cell or several cells along a highway belonging to the same NodeB. Thus, a base station is a possible implementation of a NodeB.

The *Radio Network Controller* (RNC) controls multiple NodeBs. The number of NodeBs connected to one RNC varies depending on the implementation and deployment, but up to a few hundred NodeBs per RNC is not uncommon. The RNC is, among other things, in charge of call setup, quality-of-service handling, and management of the radio resources in the cells for which it is responsible. The ARQ protocol, handling retransmissions of erroneous data, is also located

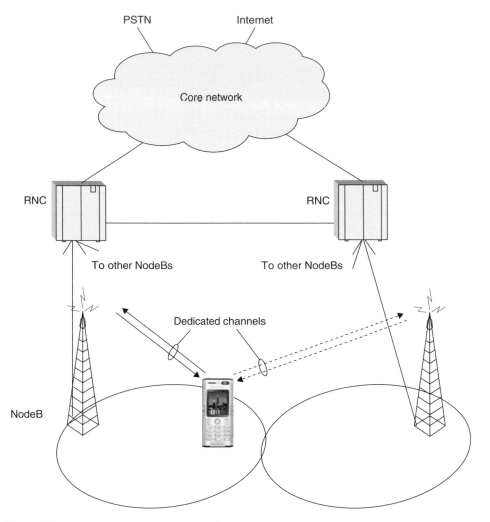

Figure 8.2   *WCDMA radio-access network architecture.*

in the RNC. Thus, in Release 99, most of the 'intelligence' in the radio-access network resides in the RNC, while the NodeBs mainly acts as modems.

Finally, the RNCs are connected to the Internet and the traditional wired telephony network through the core network.

Most modern communication systems structure the processing into different *layers* and WCDMA is no exception. The layered approach is beneficial as it provides a certain structure to the overall processing where each layer is responsible for a specific part of the radio-access functionality. The different protocol layers used in WCDMA are illustrated in Figure 8.3 and briefly described [98].

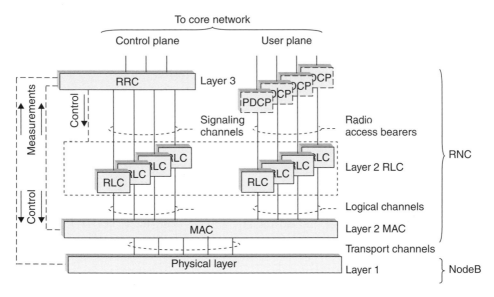

Figure 8.3   *WCDMA protocol architecture.*

User data from the core network, for example in the form of IP packets, are first processed by the *Packet Data Convergence Protocol* (PDCP) which performs (optional) header compression. IP packets have a relatively large header, 40 bytes for IPv4 and 60 bytes for IPv6, and to save radio-interface resources, header compression is beneficial.

The *Radio Link Protocol* (RLC) is, among other things, responsible for segmentation of the IP packets into smaller units known as RLC *Protocol Data Units* (RLC PDUs). At the receiving end, the RLC performs the corresponding reassembly of the received segments. The RLC also handles the ARQ protocol. For packet-data services, error-free delivery of data is often essential and the RLC can therefore be configured to request retransmissions of erroneous RLC PDUs in this case. For each incorrectly received PDU, the RLC requests a retransmission. The need for a retransmission is indicated by the RLC entity at the receiving end to its peer RLC entity at the transmitting end by means of status reports.

The *Medium Access Control* (MAC) layer offers services to the RLC in the form of so-called *Logical Channels*. The MAC layer can multiplex data from multiple logical channels. It is also responsible for determining the *Transport Format* of the data sent to the next layer, the *physical layer*. In essence, the transport format is the instantaneous data rate used over the radio link and is better understood once the interface between the MAC and physical layer has been described. This interface is specified through so-called *Transport Channels* over which data in the form of *transport blocks* are transferred. There are several transport channels defined in WCDMA; for an overview of those and the allowed mappings of logical channels to transport channels, the reader is referred to [98].

In each *Transmission Time Interval* (TTI) one or several transport blocks are fed from the MAC layer to the physical layer, which performs coding, interleaving, multiplexing, spreading, etc., prior to data transmission. Thus, for WCDMA, the TTI is the time which the interleaver spans and the time it takes to transmit the transport block over the radio interface. A larger TTI implies better time diversity, but also a longer delay. In the first release, WCDMA relies on TTI lengths of 10, 20, 40, and 80 ms. As will be seen later, HSPA introduces additional 2 ms TTI to reduce the latency.

To support different data rates, the MAC can vary the transport format between consecutive TTIs. The transport format consists of several parameters describing how the data shall be transmitted in a TTI. By varying the transport-block size and/or the number of transport blocks, different data rates can be realized.

At the bottom of the protocol stack is the *physical layer*. The physical layer is responsible for operations such as coding, spreading and data modulation, as well as modulation of the radio-frequency carrier.

PDCP, RLC, MAC, and physical layer, are configured by the *Radio Resource Control* (RRC) protocol. RRC performs admission control, handover decisions, and active set management for soft handover. By setting the parameters of the RLC, MAC, and physical layers properly, the RRC can provide the necessary quality-of-service requested by the core network for a certain service.

On the network side, the MAC, RLC and RRC entities in Release 99 are all located in the RNC while the physical layer is mainly located in the NodeB. The same entities also exist in the UE. For example, the MAC in the UE is responsible for selecting the transport format for uplink transmissions from a set of formats configured by the network. However, the handling of the radio resources in the cell is controlled by the RRC entity in the network and the UE obeys the RRC decisions taken in the network.

8.1.2 Physical layer

The basis for the physical layer of WCDMA is spreading of the data to be transmitted to the chip rate, which equals 3.84 Mchip/s. It is also responsible for coding, transport-channel multiplexing, and modulation of the radio-frequency carrier. A simplified illustration of the physical layer processing is provided in Figure 8.4.

To each transport block to be transmitted, the transmitter physical layer adds a Cyclic Redundancy Check (CRC) for the purpose of error detection. If the receiver detects an error, the RLC protocol in the receiver is informed and requests a retransmission. After CRC attachment, the data is encoded using a rate-1/3 Turbo coder (convolutional coding is also supported by WCDMA, but typically not used for packet-data services). Rate matching by means of puncturing or repetition of the coded bits is used to fine-tune the code rate and multiple coded and interleaved transport channels can be multiplexed together, forming a single stream of bits to be spread to the chip rate and subsequently modulated. For the downlink, QPSK modulation is used, while the uplink uses BPSK. The resulting stream of modulation symbols is mapped to a *physical*

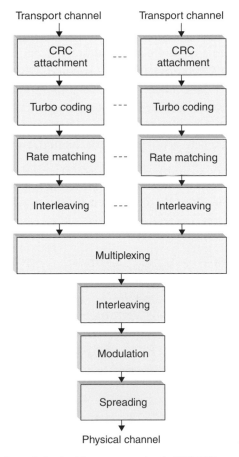

Figure 8.4 *Simplified view of physical layer processing in WCDMA.*

channel and subsequently digital-to-analog converted and modulated onto a radio-frequency carrier. In principle, each physical channel corresponds to a unique spreading code, used to separate transmissions to/from different users.

The spreading operation actually consists of two steps: spreading to the chip rate by using orthogonal *channelization codes* with a length equal to the symbol time followed by scrambling using non-orthogonal scrambling sequences with a length equal to the 10 ms radio frame.

The channelization codes are so-called *Orthogonal Variable Spreading Factor* (OVSF) codes, defined by the tree structure in Figure 8.5. A key property of the OVSF codes is the mutual orthogonality between data streams spread with different OVSF codes, even if the data rate, and therefore the spreading factors, are different. This holds as long as the different OVSF codes are selected from different branches of the channelization code tree and transmitted with the same timing. By selecting different spreading factors, physical channels with different data rates can be provided.

In the downlink, data to a certain user, including the necessary control information, is carried on one QPSK-modulated *Dedicated Physical Channel* (DPCH) corresponding to one OVSF code. By varying the spreading factor, different DPCH symbol rates can be provided. The spreading factor is determined by the transport format selected by the MAC layer. There is also a possibility to use multiple physical channels to a single user for the highest data rates.

Some downlink channelization codes are pre-allocated for specific purposes. One of the most important is the code used for reference signal transmission,

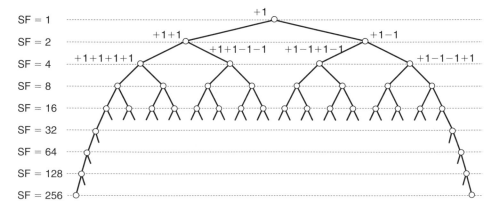

Figure 8.5 *Channelization codes.*

in WCDMA known as the *Common Pilot Channel* (CPICH). The CPICH contains known data and is used as a reference for downlink channel estimation by all UEs in the cell. There is also a pre-allocated control channel carrying cell-specific control information.

Since physical channels are separated by OVSF codes, transmissions on different physical channels are orthogonal and will not interfere with each other. The WCDMA downlink is therefore often said to be *orthogonal*. However, at the receiver side, orthogonality will partly be lost in case of a frequency-selective channel. As discussed in Chapter 5, this will cause signal corruption to the received signal and result in interference between different codes used for downlink transmission. Such signal corruption can be counteracted by an equalizer.

WCDMA supports asynchronous operation, where transmissions from different cells are not time synchronized. To separate different cells, the scrambling is cell specific in the downlink. A terminal receiving transmissions from one cell will therefore be interfered by transmissions in neighboring cells as the scrambling sequences are non-orthogonal. This interference will be suppressed by the terminal receiver with a factor proportional to the processing gain (the processing gain is given by the spreading factor divided by the code rate after rate matching).

In the uplink, data is carried on a BPSK-modulated *Dedicated Physical Data Channel* (DPDCH). Similar to the downlink, different data rates are realized by using different spreading factors for the DPDCH. Coherent demodulation is used also in the uplink, which requires a channel estimate. Unlike the downlink, where a common pilot signal is used, uplink transmissions originate from different locations. Therefore, a common pilot cannot be used and each user must have a separate pilot signal. This is carried on the *Dedicated Physical Control Channel* (DPCCH). The DPCCH also carries information about the transport format of the data transmitted on the DPDCH. This information is required by the physical layer in the NodeB for proper demodulation.

User-specific scrambling is used in the WCDMA uplink and the channelization codes are only used to separate different physical channels from the same terminal. The same set of channelization codes can therefore be used by multiple UEs. As the transmissions from different terminals are not time synchronized, separation of different terminals by the use of OVSF codes is not possible. Hence, the uplink is said to be non-orthogonal and different users' transmissions will interfere with each other.

The fact that the uplink is non-orthogonal implies that fast closed-loop power control is an essential feature of WCDMA. With fast closed-loop power control, the NodeB measures the received signal-to-interference ratio (SIR) on the DPCCH received from each terminal and 1500 times per second commands the terminals to adjust its transmission power accordingly. The target of the power control is to ensure that the received SIR of the DPCCH is at an appropriate level for each user. Note that the required SIR depends on the data rate. This SIR target may very well be different for different users. If the SIR is below the target, that is too low for proper demodulation, the NodeB commands the UE to increase its transmission power. Similarly, if the received SIR is above the target and unnecessarily high, the UE is instructed to lower its transmission power. If power control would not be present, the inter-user interference could cause the transmissions from some users to be non-decodable. This is often referred to as the *near-far problem* – transmissions from users near the base station will be received at a significantly higher power level than transmissions from users far from the base station, making demodulation of far users impossible unless (closed-loop) power control is used.

Fast closed-loop power control is used in the downlink as well, although not primarily motivated by the near-far problem as the downlink is orthogonal. However, it is still useful to combat the fast fading by varying the transmission power; when the channel conditions are favorable, less transmission power is used and vice versa. This typically results in a lower average transmitted power than for the corresponding non-power-controlled case, thereby reducing the average (inter-cell) interference and improving the system capacity.

Soft handover, or *macro diversity* as it is also called, is a key feature of WCDMA. It implies that a terminal is communicating with multiple cells, in the general case with multiple NodeBs, simultaneously, and is mainly used for terminals close to the cell border to improve performance. The set of cells the UE is communicating with is known as the *active set*. The RNC determines, based on measurements from the UE, the set of cells forming the active set.

In the downlink, soft handover implies that data to a UE is transmitted simultaneously from multiple cells. This provides diversity against fast fading – the likelihood that the instantaneous channel conditions from all cells simultaneously are disadvantageous is lower the larger the number of cells in the active set.

Uplink soft handover implies the transmission from the UE is received in multiple cells. The cells can often belong to different NodeBs. Reception of data at multiple locations is fundamentally beneficial as it provides diversity against fast fading. In case the cells receiving the uplink transmission are located in the same NodeB, combining of the received signals occurs in the physical-layer receiver. Typically, a RAKE receiver is used for this purpose, although other receiver structures are also possible. However, if the cells belong to different NodeBs, combining cannot take place in the RAKE receiver. Instead, each NodeB tries to decode the received signal and forwards correctly received data units to the RNC. As long as at least one NodeB received the data correctly, the transmission is successful and the RNC can discard any duplicates in the case multiple NodeBs correctly received the transmission and only forward one copy of each correctly received data unit.[1] Missing data units in case none of the NodeBs received a correct copy of the data unit can be detected by the RLC protocol and a retransmission can be requested. Soft handover is one of the main reasons why the RLC is located in the RNC and not in the NodeB.

In addition to reception at multiple cells, uplink soft handover also implies power control from multiple cells. As it is sufficient if the transmission is received in one cell, the UE lowers its transmission power if the power control mechanism in at least one of the cells commands the UE to do so. Only if all the cells request the terminal to increase its transmission power is the power increased. This mechanism, known as or-of-the-downs, ensures the average transmission power to be kept as low as possible. Due to the non-orthogonal property of the uplink, any reduction in average interference directly translates into an increased capacity.

8.1.3 Resource handling and packet-data session

With the above discussion in mind, a typical packet-data session in the first release of WCDMA can be briefly outlined. At call setup, when a connection is established between a UE and the radio-access network, the RNC checks the amount of resources the UE needs during the session.

In the downlink, the resources in principle consist of channelization codes and transmission power. Since the RNC is responsible for allocation of resources, it knows the fraction of the channelization code tree not reserved for any user. Measurements in the NodeB provide the RNC with information about the average amount of available transmission power.

In the uplink, thanks to its non-orthogonal nature, there is no channelization code limitation. Instead, the resource consists of the amount of additional interference the cell can tolerate. To quantify this, the term *noise rise* or *rise-over-thermal* is

---

[1] Formally, the selection and duplicate removal mechanisms in case of inter-NodeB macro diversity are part of the physical layer. Hence, the physical layer spans both the NodeB and the RNC.

often used when discussing uplink operation. Noise rise, defined as $(I_0 + N_0)/N_0$, where I_0 and N_0 are the power spectral densities due to uplink transmissions and background noise, respectively, is a measure of the increase in interference in the cell due to the transmission activity. For example, 0 dB noise rise indicates an unloaded system (no interference from other users) and 3 dB noise rise implies a power spectral density due to uplink transmission equal to the noise spectral density. Although noise rise as such is not of major interest, it has a close relation to coverage and uplink load. A large noise rise would result in loss of coverage of some channels – a terminal may not have sufficient transmission power available to reach the required E_b/N_0 at the base station. Therefore, the radio resource control in the RNC must keep the noise rise within certain limits. The NodeB provides uplink measurements which enables the RNC to estimate the uplink load.

Provided there are sufficient resources in both uplink and downlink, the RNC admits the UE into the cell and configures a dedicated (physical) channel in each direction. Several parameters are configured, one of them being the set of transport formats the MAC layer in the UE and the RNC are allowed to select from for uplink and downlink transmissions, respectively. Hence, the RNC reserves resources corresponding to the highest data rate the UE may transmit with during the call. In the downlink, a fraction of the code tree need to be reserved corresponding to the smallest spreading factor that may be required during the session. Similarly, in the uplink, the RNC must ensure that the maximum interference in the cell is not exceeded even if the UE transmits at its highest data rate.

During the packet-data call, the data rate of the transmission may vary, depending on the traffic pattern. As a consequence, the transmission power will vary depending on the instantaneous data rate. However, the amount of the code tree allocated for a certain user in the downlink remains constant throughout the packet call (unless the parameters are reconfigured). Furthermore, as downlink transport format selection is located in the RNC, it is not aware of the instantaneous power consumption in the NodeB. The RRC must therefore use a certain margin when admitting users into the system to ensure that sufficient transmission power is available for any transport format the MAC may select. Similarly, in the uplink, the RRC can only observe the average uplink interference and must therefore have sufficient margins to handle the case of all UEs suddenly selecting the transport format corresponding to the highest data rate and therefore the highest transmission power. In other words, the amount of resources reserved for a user is fairly static during the period of the packet call. This is suitable for services with a relatively constant data rate, for example voice services or interactive video transmission, but, as will be seen in the subsequent chapters significant enhancements are possible for packet-data services with rapidly varying resource requirements.

9

High-Speed Downlink Packet Access

The introduction of *High-Speed Downlink Packet Access*, (HSDPA), implies a major extension of the WCDMA radio interface, enhancing the WCDMA downlink packet-data performance and capabilities in terms of higher peak data rate, reduced latency and increased capacity. This is achieved through the introduction of several of the techniques described in Part II, including *higher-order modulation*, *rate control*, *channel-dependent scheduling*, and *hybrid ARQ with soft combining*. The HSDPA specifications are found in [100] and the references therein.

9.1 Overview

9.1.1 *Shared-channel transmission*

A key characteristic of HSDPA is the use of *Shared-Channel Transmission*. Shared-channel transmission implies that a certain fraction of the total downlink radio resources available within a cell, channelization codes and transmission power in case of WCDMA, is seen as a common resource that is dynamically shared between users, primarily in the time domain. The use of shared-channel transmission, in WCDMA implemented through the *High-Speed Downlink Shared Channel* (HS-DSCH) as described below, enables the possibility to rapidly allocate a large fraction of the downlink resources for transmission of data to a specific user. This is suitable for packet-data applications which typically have bursty characteristics and thus rapidly varying resource requirements.

The basic HS-DSCH code- and time- structure is illustrated in Figure 9.1. The HS-DSCH code resource consists of a set of channelization codes of spreading factor 16, see upper part of Figure 9.1 where the number of codes available for HS-DSCH transmission is configurable between 1 and 15. Codes not reserved

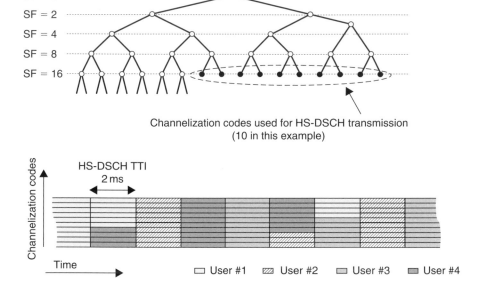

Channelization codes used for HS-DSCH transmission
(10 in this example)

Figure 9.1   *Time- and code-domain structure for HS-DSCH.*

for HS-DSCH transmission are used for other purposes, for example related control signaling, MBMS services, or circuit-switched services.

The dynamic allocation of the HS-DSCH code resource for transmission to a specific user is done on 2 ms TTI basis, see lower part of Figure 9.1. The use of such a short TTI for HSDPA reduces the overall delay and improves the tracking of fast channel variations exploited by the rate control and the channel-dependent scheduling as discussed below.

In addition to being allocated a part of the overall code resource, a certain part of the total available cell power should also be allocated for HS-DSCH transmission. Note that the HS-DSCH is not power controlled but rate controlled as discussed below. This allows the remaining power, after serving other channels, to be used for HS-DSCH transmission and enables efficient exploitation of the overall available power resource.

9.1.2   Channel-dependent scheduling

Scheduling controls to which user the shared-channel transmission is directed at a given time instant. The scheduler is a key element and to a large extent determines the overall system performance, especially in a highly loaded network. In each TTI, the scheduler decides to which user(s) the HS-DSCH should be transmitted and, in close cooperation with the rate-control mechanism, at what data rate.

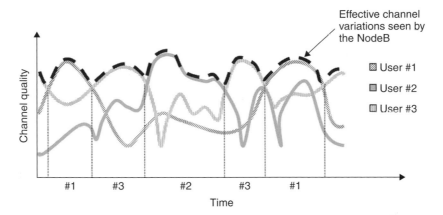

Figure 9.2 *Channel-dependent scheduling for HSDPA.*

In Chapter 7, it was discussed how a significant increase in capacity can be obtained if the radio-channel conditions are taken into account in the scheduling decision, so-called *channel-dependent scheduling*. Since the radio conditions for the radio links to different UEs within a cell typically vary independently, at each point in time there is almost always a radio link whose channel quality is near its peak, see Figure 9.2. As this radio link is likely to have good channel quality, a high data rate can be used for this radio link. This translates into a high system capacity. The gain obtained by transmitting to users with favorable radio-link conditions is commonly known as *multi-user diversity* and the gains are larger, the larger the channel variations and the larger the number of users in a cell. Thus, in contrast to the traditional view that fast fading is an undesirable effect that has to be combated, with the possibility for channel-dependent scheduling fading is potentially beneficial and should be exploited.

Several different scheduling strategies were discussed in Chapter 7. As discussed, a practical scheduler strategy exploits the short-term variations, for example due to multi-path fading and fast interference variations, while maintaining some degree of long-term fairness between the users. In principle, the larger the long-term unfairness, the higher the cell capacity. A trade-off between fairness and capacity is therefore required.

In addition to the channel conditions, traffic conditions are also taken into account by the scheduler. For example, there is obviously no purpose in scheduling a user with no data awaiting transmission, regardless of whether the channel conditions are beneficial or not. Furthermore, some services should preferably be given higher priority. As an example, streaming services should be ensured a relatively constant long-term data rate while background services such as file download have less stringent requirements on a constant long-term data rate.

9.1.3   Rate control and higher-order modulation

In Chapter 7, rate control was discussed and, for packet-data services, shown to be a more efficient tool for link adaptation, compared to the fast power control typically used in CDMA-based systems, especially when used together with channel-dependent scheduling.

For HSDPA, rate control is implemented by dynamically adjusting the channel-coding rate as well as dynamically selecting between QPSK and 16QAM modulation. *Higher-order modulation* such as 16QAM allows for higher bandwidth utilization than QPSK, but requires higher received E_b/N_0 as described in Chapter 3. Consequently, 16QAM is mainly useful in advantageous channel conditions. The data rate is selected independently for each 2 ms TTI by the NodeB and the rate control mechanism can therefore track rapid channel variations.

9.1.4   Hybrid ARQ with soft combining

Fast *hybrid ARQ with soft combining* allows the terminal to request retransmission of erroneously received transport blocks, effectively fine-tuning the effective code rate and compensating for errors made by the link-adaptation mechanism. The terminal attempts to decode each transport block it receives and reports to the NodeB its success or failure 5 ms after the reception of the transport block. This allows for rapid retransmissions of unsuccessfully received data and significantly reduces the delays associated with retransmissions compared to Release 99.

Soft combining implies that the terminal does not discard soft information in case it cannot decode a transport block as in traditional hybrid-ARQ protocols, but combines soft information from previous transmission attempts with the current retransmission to increase the probability of successful decoding. Incremental redundancy, IR, is used as the basis for soft combining in HSDPA, that is the retransmissions may contain parity bits not included in the original transmission. From Chapter 7, it is known that IR can provide significant gains when the code rate for the initial transmission attempts is high as the additional parity bits in the retransmission results in a lower overall code rate. Thus, IR is mainly useful in bandwidth-limited situations, for example, when the terminal is close to the base station and the amount of channelization codes, and not the transmission power, limits the achievable data rate. The set of coded bits to use for the retransmission is controlled by the NodeB, taking the available UE memory into account.

9.1.5 Architecture

From the previous discussion it is clear that the basic HSDPA techniques rely on fast adaptation to rapid variations in the radio conditions. Therefore, these techniques need to be placed close to the radio interface on the network side, that is in the NodeB. At the same time, an important design objective of HSDPA was to retain the Release 99 functional split between layers and nodes as far as possible. Minimization of the architectural changes is desirable as it simplifies introduction of HSDPA in already deployed networks and also secures operation in environments where not all cells have been upgraded with HSDPA functionality. Therefore, HSDPA introduces a new MAC sub-layer in the NodeB, the *MAC-hs,* responsible for scheduling, rate control and hybrid-ARQ protocol operation. Hence, apart from the necessary enhancements to the RNC such as admission control of HSDPA users, the introduction of HSDPA mainly affects the NodeB (Figure 9.3).

Each UE using HSDPA will receive HS-DSCH transmission from one cell, the *serving cell.* The serving cell is responsible for scheduling, rate control, hybrid ARQ, and all other MAC-hs functions used by HSDPA. Uplink soft handover

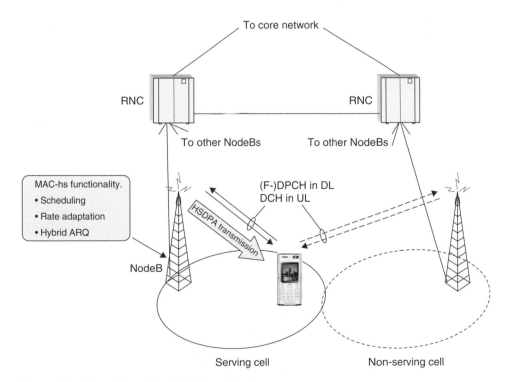

Figure 9.3   *Illustration of the HSDPA architecture.*

is supported, in which case the uplink data transmission will be received in multiple cells and the UE will receive power control commands from multiple cells.

Mobility from a cell supporting HSDPA to a cell that is not supporting HSDPA is easily handled. Uninterrupted service to the user can be provided, albeit at a lower data rate, by using channel switching in the RNC and switch the user to a dedicated channel in the non-HSDPA cell. Similarly, a user equipped with an HSDPA-capable terminal may be switched from a dedicated channel to HSDPA when the user enters a cell with HSDPA support.

9.2   Details of HSDPA

9.2.1   HS-DSCH: Inclusion of features in WCDMA Release 5

The High-Speed Downlink Shared Channel (HS-DSCH), is the transport channel used to support shared-channel transmission and the other basic technologies in HSDPA, namely channel-dependent scheduling, rate control (including higher-order modulation), and hybrid ARQ with soft combining. As discussed in the introduction and illustrated in Figure 9.1, the HS-DSCH corresponds to a set of channelization codes, each with spreading factor 16. Each such channelization code is also known as an HS-PDSCH – *High-Speed Physical Downlink Shared Channel*.

In addition to HS-DSCH, there is a need for other channels as well, for example for circuit-switched services and for control signaling. To allow for a trade-off between the amount of code resources set aside for HS-DSCH and the amount of code resource used for other purposes, the number of channelization codes available for HS-DSCH can be configured, ranging from 1 to 15 codes. Codes not reserved for HS-DSCH transmission are used for other purposes, for example related control signaling and circuit-switched services. The first node in the code tree can never be used for HS-DSCH transmission as this node includes mandatory physical channels such as the common pilot.

Sharing of the HS-DSCH code resource should primarily take place in the time domain. The reason is to fully exploit the advantages of channel-dependent scheduling and rate control, since the quality at the terminal varies in the time domain, but is (almost) independent of the set of codes (physical channels) used for transmission. However, sharing of the HS-DSCH code resource in the code domain is also supported as illustrated in Figure 9.1. With code-domain sharing, two or more UEs are scheduled simultaneously by using different parts of the common code resource (different sets of physical channels). The reasons for code-domain sharing are twofold: support of terminals that arc, for complexity reasons, not able to despread the full set of codes, and efficient support of small payloads when the transmitted data does not require the full set of allocated HS-DSCH codes. In either of these cases, it is obviously a waste of resources to assign the full code resource to a single terminal.

In addition to being allocated a part of the overall code resource, a certain part of the total available cell power should also be used for HS-DSCH transmission. To maximize the utilization of the power resource in the base station, the remaining power after serving other, power-controlled channels, should preferably be used for HS-DSCH transmission as illustrated in Figure 9.4. In principle, this results in a (more or less) constant transmission power in a cell. Since the HS-DSCH is rate controlled as discussed below, the HS-DSCH data rate can be selected to match the radio conditions and the amount of power instantaneously available for HS-DSCH transmission.

To obtain rapid allocation of the shared resources, and to obtain a small end-user delay, the TTI should be selected as small as possible. At the same time, a too small TTI would result in excessive overhead as control signaling is required for each transmission. For HSDPA, this trade-off resulted in the selection of a 2 ms TTI.

Downlink control signaling is necessary for the operation of HS-DSCH in each TTI. Obviously, the identity of the UE(s) currently being scheduled must be signaled as well as the physical resource (the channelization codes) used for transmission to this UE. The UE also needs to be informed about the transport format used for the transmission as well as hybrid-ARQ-related information. The resource and transport-format information consists of the part of the code

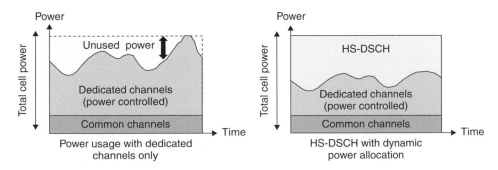

Figure 9.4 *Dynamic power usage with HS-DSCH.*

tree used for data transmission, the modulation scheme used, and the transport-block size. The downlink control signaling is carried on the *High-Speed Shared Control Channel* (HS-SCCH), transmitted in parallel to the HS-DSCH using a separate channelization code. The HS-SCCH is a shared channel, received by all UEs for which an HS-DSCH is configured to find out whether the UE has been scheduled or not.

Several HS-SCCHs can be configured in a cell, but as the HS-DSCH is shared mainly in the time domain and only the currently scheduled terminal needs to receive the HS-SCCH, there is typically only one or, if code-domain sharing is supported in the cell, a few HS-SCCHs configured in each cell. However, each HS-DSCH-capable terminal is required to be able to monitor up to four HS-SCCHs. Four HS-SCCH has been found to provide sufficient flexibility in the scheduling of multiple UEs; if the number was significantly smaller the scheduler would have been restricted in which UEs to schedule simultaneously in case of code-domain sharing.

HSDPA transmission also requires uplink control signaling as the hybrid-ARQ mechanism must be able to inform the NodeB whether the downlink transmission was successfully received or not. For each downlink TTI in which the UE has been scheduled, an ACK or NAK will be sent on the uplink to indicate the result of the HS-DSCH decoding. This information is carried on the uplink *High-Speed Dedicated Physical Control Channel* (HS-DPCCH). One HS-DPCCH is set up for each UE with an HS-DSCH configured. In addition, the NodeB needs information about the instantaneous downlink channel conditions at the UE for the purpose of channel-dependent scheduling and rate control. Therefore, each UE also measures the instantaneous downlink channel conditions and transmits a *Channel-Quality Indicator* (CQI), on the HS-DPCCH.

In addition to HS-DSCH and HS-SCCH, an HSDPA terminal need to receive power control commands for support of fast closed-loop power control of the uplink in the same way as any WCDMA terminal. This can be achieved by a downlink dedicated physical channel, DPCH, for each UE. In addition to power control commands, this channel can also be used for user data not carried on the HS-DSCH, for example circuit-switched services.

In Release 6, support for *fractional DPCH,* F-DPCH, is added to reduce the consumption of downlink channelization codes. In principle, the only use for a dedicated channel in the downlink is to carry power control commands to the UE in order to adjust the uplink transmission. If all data transmissions, including higher-layer signaling radio bearers, are mapped to the HS-DSCH, it is a waste

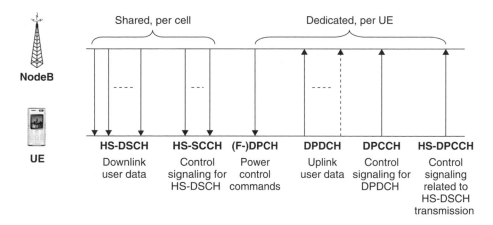

Figure 9.5 *Channel structure with HSDPA.*

of scarce code resources to use a dedicated channel with spreading factor 256 per UE for power control only. The F-DPCH resolves this by allowing multiple UEs to share a single downlink channelization code.

To summarize, the overall channel structure with HSDPA is illustrated in Figure 9.5.

Neither the HS-PDSCH, nor the HS-SCCH, is subject to downlink macro-diversity or soft handover. The basic reason is the location of the HS-DSCH scheduling in the NodeB. Hence, it is not possible to simultaneously transmit the HS-DSCH to a single UE from multiple NodeBs, which prohibits the use of inter-NodeB soft handover. Furthermore, it should be noted that within each cell, multi-user diversity is exploited by the channel-dependent scheduler. Basically, the scheduler only transmits to a user when the instantaneous radio conditions are favorable and thus the additional gain from macro-diversity is reduced.

However, the uplink channels, as well as any dedicated downlink channels, can be in soft handover. As these channels are not subject to channel-dependent scheduling, macro-diversity provides a direct coverage benefit.

9.2.2 MAC-hs and physical-layer processing

As mentioned in the introduction, the MAC-hs is a new sub-layer located in the NodeB and responsible for the HS-DSCH scheduling, rate control and hybrid-ARQ protocol operation. To support these features, the physical layer has also been enhanced with the appropriate functionality, for example support for soft

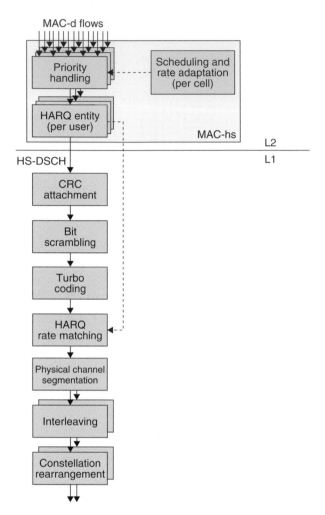

Figure 9.6   *MAC-hs and physical-layer processing.*

combining in the hybrid ARQ. In Figure 9.6, the MAC-hs and physical-layer processing is illustrated.

The MAC-hs consists of scheduling, priority handling, transport-format selection (rate control), and the protocol parts of the hybrid-ARQ mechanism. Data, in the form of a single transport block with dynamic size, passes from the MAC-hs via the HS-DSCH transport channel to the HS-DSCH physical-layer processing.

The HS-DSCH physical-layer processing is straightforward. A 24-bit CRC is attached to each transport block. The CRC is used by the UE to detect errors in the received transport block.

Demodulation of 16QAM, which is one of the modulation schemes supported by the HS-DSCH, requires amplitude knowledge at the receiver in order to correctly form the soft values prior to Turbo decoding. This is different from QPSK, where no such knowledge is required as all information is contained in the phase of the received signal. To ease the estimation of the amplitude reference in the receiver, the bits after CRC attachment are scrambled. This results in a sufficiently random sequence out from the Turbo coder to cause both inner and outer signal points in the 16QAM constellation to be used, thereby aiding the UE in the estimation of the amplitude reference. Note that bit scrambling is done regardless of the modulation scheme used, even if it is strictly speaking only needed in case of 16QAM modulation.

The fundamental coding scheme in HSDPA is rate-1/3 Turbo coding. To obtain the code rate selected by the rate-control mechanism in the MAC-hs, rate matching, that is, puncturing or repetition, is used to match the number of coded bits to the number of physical-channel bits available. The rate-matching mechanism is also part of the physical-layer hybrid-ARQ and is used to generate different redundancy versions for incremental redundancy. This is done through the use of different puncturing patterns; different bits are punctured for initial transmissions and retransmission.

Physical-channel segmentation distributes the bits to the channelization codes used for transmission, followed by channel interleaving.

Constellation rearrangement is used only for 16QAM. If Chase combining is used in combination with 16QAM, a gain in performance can be obtained if the signal constellation is changed between retransmissions. This is further elaborated upon below.

9.2.3 Scheduling

One of the basic principles for HSDPA is the use of channel-dependent scheduling. The scheduler in the MAC-hs controls what part of the shared code and power resource is assigned to which user in a certain TTI. It is a key component and to a large extent determines the overall HSDPA system performance, especially in a loaded network. At lower loads, only one or a few users are available for scheduling and the differences between different scheduling strategies are less pronounced.

Although the scheduler is implementation specific and not specified by 3GPP, the overall goal of most schedulers is to take advantage of the channel variations between users and preferably schedule transmissions to a user when the channel conditions are advantageous. As discussed in Chapter 7, several scheduling strategies are possible. However, efficient scheduling strategies require at least:

- information about the instantaneous channel conditions at the UE,
- information about the buffer status and priorities of the data flows.

Information about the instantaneous channel quality at the UE is typically obtained through a 5-bit Channel-Quality Indicator (CQI), which each UE feed back to the NodeB at regular intervals. The CQI is calculated at the UE based on the signal-to-noise ratio of the received common pilot. Instead of expressing the CQI as a received signal quality, the CQI is expressed as a recommended transport-block size, taking into account also the receiver performance. This is appropriate as the quantity of relevance is the instantaneous data rate a terminal can support rather than the channel quality alone. Hence, a terminal with a more advanced receiver, being able to receive data at a higher rate at the same channel quality, will report a larger CQI than a terminal with a less advanced receiver, all other conditions being identical.

In addition to the instantaneous channel quality, the scheduler should typically also take buffer status and priority levels into account. Obviously UEs for which there is no data awaiting transmission should not be scheduled. There could also be data that is important to transmit within a certain maximum delay, regardless of the channel conditions. One important example hereof is RRC signaling, for example, related to cell change in order to support mobility, which should be delivered to the UE as soon as possible. Another example, although not as time critical as RRC signaling, is streaming services, which has an upper limit on the acceptable delay of a packet to ensure a constant average data rate. To support priority handling in the scheduling decision, a set of priority queues is defined into which the data is inserted according to the priority of the data as illustrated in Figure 9.7. The scheduler selects data from these priority queues for transmission based on the channel conditions, the priority of the queue, and any other relevant information. To efficiently support streaming applications, which require a minimum average data rate, there is a possibility for the RNC to 'guarantee' this data rate by providing information about the average data rate to the scheduler in the NodeB. The scheduler can take this constraint into account in the scheduling process.

9.2.4   Rate control

As described in Chapter 7, rate control denotes the process of adjusting the data rate to match the instantaneous radio conditions. The data rate is adjusted by

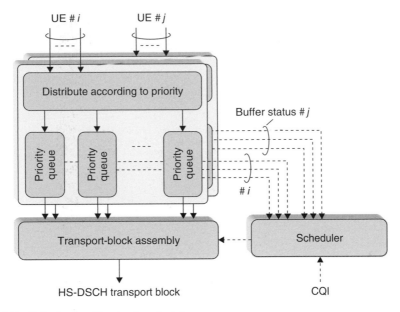

Figure 9.7 *Priority handling in the scheduler.*

changing the modulation scheme and the channel-coding rate. For each TTI, the rate-control mechanism in the scheduler selects, for the scheduled user(s), the transport format(s) and channelization-code resources to use. The transport format consists of the modulation scheme (QPSK or 16QAM) and the transport-block size.

The resulting code rate after Turbo coding and rate matching is given implicitly by the modulation scheme, the transport-block size, and the channelization-code set allocated to the UE for the given TTI. The number of coded bits after rate matching is given by the modulation scheme and the number of channelization codes, while the number of information bits prior coding is given by the transport-block size. Hence, by adjusting some or all of these parameters, the overall code rate can be adjusted.

Rate control is implemented by allowing the MAC-hs to set the transport format independently for each 2 ms HS-DSCH TTI. Hence, both the modulation scheme and the instantaneous code rate can be adjusted to obtain a data rate suitable for the current radio conditions. The relatively short TTI of 2 ms allows the rate control to track reasonable rapid variations in the instantaneous channel quality.

The HS-DSCH transport-block size can take one of 254 values. These values, illustrated in Figure 9.8, are listed in the specifications and therefore known

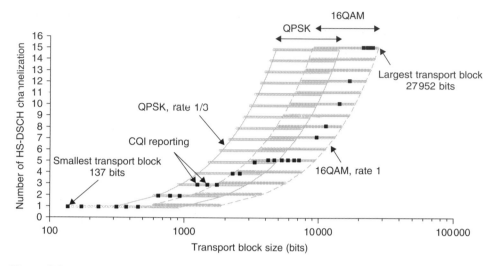

Figure 9.8 *Transport-block sizes vs. the number of channelization codes for QPSK and 16QAM modulation. The transport-block sizes used for CQI reporting are also illustrated.*

to both the UE and the NodeB. Thus, there is no need for (re)configuration of transport-block sizes at channel setup or when switching serving cell, which reduces the amount of overhead associated with mobility. Each combination of HS-DSCH channelization codes and modulation scheme defines a subset containing 63 out of the 254 different transport-block sizes and the 6-bit 'HS-DSCH transport-block size information' indicates one out of the 63 transport-block sizes possible for this subset. With this scheme, transport-block sizes in the range of 137–27 952 bits can be signaled, with channel-coding rates ranging from 1/3 up to 1.

For retransmissions, there is a possibility for a code rate >1. This is achieved by exploiting the fact that the transport-block size cannot change between transmission and retransmission. Hence, instead of signaling the transport-block size for the retransmission, a reserved value can be used, indicating that no transport-block-size information is provided by the HS-SCCH and the value from the original transmission should be used. This is useful for additional scheduling flexibility, for example to retransmit only a small amount of parity bits in case the latest CQI report indicates that the UE was 'almost' able to decode the original transmission.

As stated in the introduction, the primary way of adapting to rapid variations in the instantaneous channel quality is rate control as no fast closed-loop power control is specified for HS-DSCH. This does not imply that the HS-DSCH transmission power cannot change for other reasons, for example due to variations in the power required by other downlink channels. In Figure 9.4 on page 147, an example of a dynamic HS-DSCH power allocation scheme is illustrated, where the HS-DSCH uses all power not used by other channels. Of course, the overall interference created in the cell must be taken into account when allocating the amount of HS-DSCH power. This is the responsibility of the radio-resource control in the RNC, which can set an upper limit on the power used by the NodeB for the HS-DSCHs and all HS-SCCHs.[1] As long as the NodeB stays within this limit, the power allocation for HSDPA is up to the NodeB implementation. Corresponding measurements, used by the NodeB to report the current power usage to the RNC are also defined. Knowledge about the amount of power used for non-HSDPA channels is useful to the admission control functionality in the RNC. Without this knowledge, the RNC would not be able to determine whether there are resources left for non-HSDPA users trying to enter the cell.

Unlike QPSK, the demodulation of 16QAM requires an amplitude reference at the UE. How this is achieved is implementation specific. One possibility is to use a channel estimate formed from the common pilot and obtain the ratio between the HS-DSCH and common pilot received powers through averaging over 2 ms. The instantaneous amplitude estimate necessary for 16QAM demodulation can then be obtained from the common pilot and the estimated offset. This is the reason for the bit level scrambling prior Turbo coding in Figure 9.6; with scrambling both the inner and outer signal points in the 16QAM constellation will be used with a high probability and an accurate estimate of the received HS-DSCH power can be formed.

What criteria to use for the rate control, that is, the transport-format selection process in the MAC-hs, are implementation specific and not defined in the standard. Principally, the target for the rate control is to select a transport format resulting in transmitting an as large transport block as possible with a reasonable error probability, given the instantaneous channel conditions. Naturally, selecting a transport-block size larger than the amount of data to be transmitted in a TTI is not useful, regardless of whether the instantaneous radio conditions allows for a larger transport block to be transmitted. Hence, the transport-format selection does not only depend on the instantaneous radio conditions, but also on the instantaneous source traffic situation.

Since the rate control typically depends on the instantaneous channel conditions, rate control relies on the same estimate of the instantaneous radio quality

---

[1] If the cell is configured to support E-DCH as well, this limit also covers the power used for the related E-DCH downlink control signaling. See Chapter 10.

at the UE as the scheduler. As discussed above, this knowledge is typically obtained from the CQI although other quantities may also be useful. This is further elaborated upon in Section 9.3.6.

9.2.5 Hybrid ARQ with soft combining

The hybrid-ARQ functionality spans both the MAC-hs and the physical layer. As the MAC-hs is located in the NodeB, erroneous transport blocks can be rapidly retransmitted. Hybrid-ARQ retransmissions are therefore significantly less costly in terms of delay compared to RLC-based retransmissions. There are two fundamental reasons for this difference:

- There is no need for signaling between the NodeB and the RNC for the hybrid-ARQ retransmission. Consequently, any Iub/Iur delays are avoided for retransmissions. Handling retransmission in the NodeB is also beneficial from a pure Iub/Iur capacity perspective; hybrid-ARQ retransmissions come at no cost in terms of transport-network capacity.
- The RLC protocol is typically configured with relatively infrequent status reports of erroneous data blocks (once per several TTIs) to reduce the signaling load, while the HSDPA hybrid-ARQ protocol allows for frequent status reports (once per TTI), thus reducing the roundtrip time.

In HSDPA, the hybrid ARQ operates per transport block or, equivalently, per TTI that is, whenever the HS-DSCH CRC indicates an error, a retransmission representing the same information as the original transport block is requested. As there is a single transport block per TTI, the content of the whole TTI is retransmitted in case of an error. This reduces the amount of uplink signaling as a single ACK/NAK bit per TTI is sufficient. Furthermore, studies during the HSDPA design phase indicated that the benefits of having multiple transport blocks per TTI with the possibility for individual retransmissions were quite small. A major source of transmission errors are sudden interference variations on the channel and errors in the link-adaptation mechanism. Thanks to the short TTI, the channel is relatively static during the transmission of a transport block and in most cases errors are evenly distributed over the TTI. This limits the potential benefits of individual retransmissions.

Incremental redundancy is the basic scheme for soft combining, that is, retransmissions may consist of a different set of coded bits than the original transmission. Different redundancy versions, that is, different sets of coded bits, are generated as part of the rate-matching mechanism. The rate matcher uses puncturing (or repetition) to match the number of code bits to the number

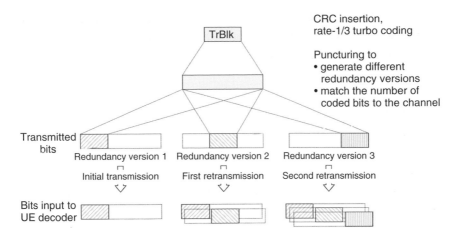

Figure 9.9 *Generation of redundancy versions.*

of physical channel bits available. By using different puncturing patterns, different sets of coded bits, that is different redundancy versions, result. This is illustrated in Figure 9.9. Note that Chase combining is a special case of incremental redundancy; the NodeB decides whether to use incremental redundancy or Chase combining by selecting the appropriate puncturing pattern for the retransmission.

The UE receives the coded bits and attempts to decode them. In case the decoding attempts fails, the UE buffers the received soft bits and requests a retransmission by sending a NAK. Once the retransmission occurs, the UE combines the buffered soft bits with the received soft bits from the retransmission and tries to decode the combination.

For soft combining to operate properly, the UE need to know whether the transmission is a retransmission of previously transmitted data or whether it is transmission of new data. For this purpose, the downlink control signaling includes a new-data indicator, used by the UE to control whether the soft buffer should be cleared (the current transmission is new data) or whether soft combining of the soft buffer and the received soft bits should take place (retransmission).

To minimize the delay associated with a retransmission, the outcome of the decoding in the UE should be reported to the NodeB as soon as possible. At the same time, the amount of overhead from the feedback signaling should be minimized. This lead to the choice of a stop-and-wait structure for HSDPA, where a single bit is transmitted from the UE to the NodeB a well-specified time,

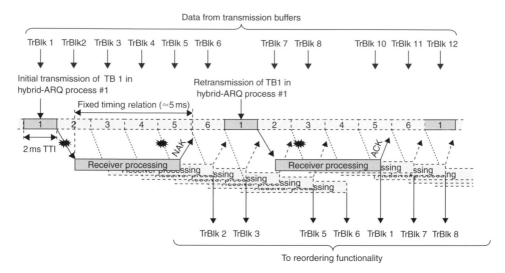

Figure 9.10   *Multiple hybrid-ARQ process (six in this example).*

approximately 5 ms, after the reception of a transport block. To allow for continuous transmission to a single UE, multiple stop-and-wait structures, or *hybrid-ARQ processes*, are operated in parallel as illustrated in Figure 9.10. Hence, for each user there is one *hybrid-ARQ entity*, each consisting of *multiple* hybrid-ARQ processes.

The number of hybrid-ARQ processes should match the roundtrip time between the UE and NodeB, including their respective processing time, to allow for continuous transmission to a UE. Using a larger number of processes than motivated by the roundtrip time does not provide any gains but introduces unnecessary delays between retransmissions.

Since the NodeB processing time may differ between different implementations, the number of hybrid-ARQ processes is configurable. Up to eight processes can be set up for a user, although a typical number of processes is six. This provides approximately 2.8 ms of processing time in the NodeB from the reception of the ACK/NAK until the NodeB can schedule a (re)transmission to the UE in the same hybrid-ARQ process.

Downlink control signaling is used to inform the UE which of the hybrid-ARQ processes that is used for the current TTI. This is important information to the UE as it is needed to do soft combining with the correct soft buffer; each hybrid-ARQ process has its own soft buffer.

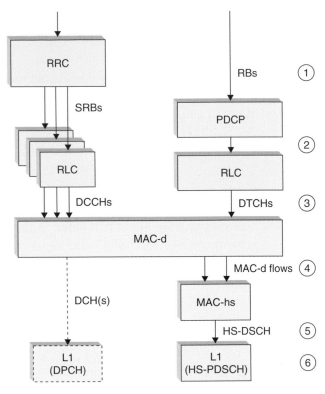

Figure 9.11 *Protocol configuration when HS-DSCH is assigned. The numbers in the rightmost part of the figure corresponds to the numbers to the right in Figure 9.12.*

One result of having multiple independent hybrid-ARQ processes operated in parallel is that decoded transport blocks may appear out-of-sequence. For example, a retransmission may be needed in hybrid-ARQ process number one, while process number two did successfully receive the data after the first transmission attempts. Therefore, the transport block transmitted in process number two will be available for forwarding to higher layers at the receiver side before the transport block transmitted in process number one, although the transport blocks were originally transmitted in a different order. This is illustrated in Figure 9.10. As the RLC protocol assumes data to appear in the correct order, a reordering mechanism is used between the outputs from the multiple hybrid-ARQ processes and the RLC protocol. The reordering mechanism is described in more detail in Section 9.3.4.

9.2.6 Data flow

To illustrate the flow of user data through the different layers, an example radio-interface protocol configuration is shown in Figure 9.11. For the UE in this example, an IP-based service is assumed, where the user data is mapped to the HS-DSCH.

For signaling purposes in the radio network, several signaling radio bearers are configured in the control plane. In Release 5, signaling radio bearers cannot be mapped to the HS-DSCH, and consequently dedicated transport channels must be used, while this restriction is removed in Release 6 to allow for operation completely without dedicated transport channels in the downlink.

Figure 9.12 illustrates the data flow at the reference points shown in Figure 9.11. In this example an IP-based service is assumed. The PDCP performs (optional) IP header compression. The output from the PDCP is fed to the RLC protocol entity. After possible concatenation, the RLC SDUs are segmented into smaller blocks of typically 40 bytes. An RLC PDU is comprised of a data segment and the RLC header. If logical-channel multiplexing is performed in MAC-d, a 4-bit header is added to form a MAC-d PDU. In MAC-hs, a number of MAC-d PDUs, possibly of variable size, are assembled and a MAC-hs header is attached

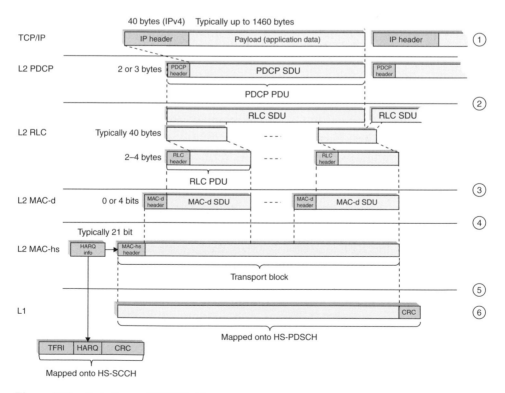

Figure 9.12   *Data flow at UTRAN side.*

to form one transport block, subsequently coded and transmitted by the physical layer.

9.2.7 Resource control for HS-DSCH

With the introduction of HSDPA, parts of the radio resource management are handled by the NodeB instead of the RNC. This is a result of introducing channel-dependent scheduling and rate control in the NodeB in order to exploit rapid channel variations. However, the RNC still has the overall responsibility for radio-resource management, including admission control and handling of inter-cell interference. Therefore, new measurement reports from the NodeB to the RNC have been introduced, as well as mechanisms for the RNC to set the limits within which the NodeB are allowed to handle the HSDPA resources[2] in the cell.

To limit the transmission power used for HSDPA, the RNC can set the maximum amount of power the NodeB is allowed to use for HSDPA-related downlink transmissions. This ensures that the RNC has control of the maximum amount of interference a cell may generate to neighboring cells. Within the limitation set by the RNC, the NodeB is free to manage the power spent on the HSDPA downlink channels. If the quantity is absent (or larger than the tot al NodeB power), the NodeB may use all available power for downlink transmissions on the HS-DSCH and HS-SCCH.

Admission control in the RNC needs to take into account the amount of power available in the NodeB. Only if there is a sufficient amount of transmission power available in the NodeB can a new user be admitted into the cell. The *Transmitted carrier power* measurement is available for this purpose. However, with the introduction of HSDPA, the NodeB can transmit at full power, even with a single user in the cell, to maximize the data rates. To the admission control in the RNC, it would appear as the cell is fully loaded and no more users would be admitted. Therefore, a new measurement, *Transmitted carrier power of all codes not used for HS-PDSCH or HS-SCCH,* is introduced, which can be used in admission control to determine whether new users can be admitted into the cell or not (Figure 9.13).

In addition to the power-related signaling discussed above, there is also signaling useful to support streaming services. To efficiently support streaming, where

---

[2]Note that many of these measurements were extended in Rel6 to include Enhanced Uplink Downlink control channels in addition to the HSDPA-related channels.

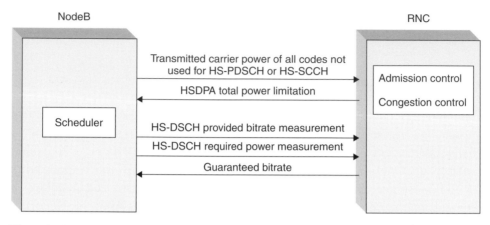

Figure 9.13   *Measurements and resource limitations for HSDPA.*

a certain minimum data rate needs to be provided on average, the RNC can signal the *MAC-hs Guaranteed Bit Rate*. The scheduler can use this information to
ensure that, averaged over a longer period of time, a sufficiently high data rate is
provided for a certain MAC-d priority queue. To monitor the fulfillment of this,
and to be able to observe the load in the cell due to these restrictions, the NodeB
can report the required transmission power for each priority class configured by
the RNC in order to identify 'costly' UEs. The NodeB can also report the data
rate, averaged over 100 ms, it actually provides for each priority class.

9.2.8   Mobility

Mobility for HSDPA, that is, change of serving cell, is handled through RRC
signaling using similar procedures as for dedicated channels. The basics for
mobility are network-controlled handover and UE measurement reporting.
Measurements are reported from the UE to the RNC, which, based on the measurements, reconfigures the UE and involved NodeBs, resulting in a change of
serving cell.

Several measurement mechanisms are specified already in the first release of
WCDMA and used for, for example, active set update, hard handover, and intra-
frequency measurements. One example is Measurement Event 1D, 'change of
best cell,' which is reported by the UE whenever the common-pilot strength
from a neighboring cell (taking any measurement offsets into account) becomes
stronger than for the current best cell. This can be used to determine when to
switch the HS-DSCH serving cell as illustrated in Figure 9.14. Updates of the
active set are not included in this example; it is assumed that both the source
serving cell and the target serving cell are part of the active set.

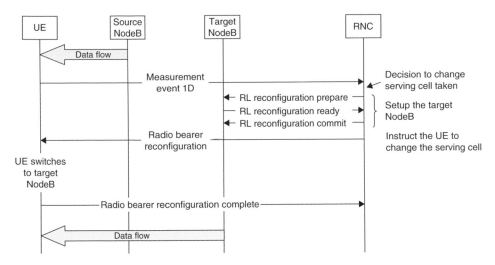

Figure 9.14 *Change of serving cell for HSPA. It is assumed that both the source and target NodeB are part of the active set.*

The reconfiguration of the UE and involved NodeBs can be either synchronous or asynchronous. With synchronous reconfiguration, an activation time is defined in the reconfiguration message, ensuring that all involved parties change their reconfiguration at the same time. Due to unknown delays between the NodeB and the RNC, as well as processing and protocol delays, a suitable margin may need to be taken into account in the choice of activation time. Asynchronous reconfiguration implies that the involved nodes obey the reconfiguration message as soon it is received. However, in this case, data transmission from the new cell may start before the UE has been switched from the old cell, which would result in some data loss that has to be retransmitted by the RLC protocol. Hence, synchronous reconfigurations are typically used for HS-DSCH serving cell change. The MAC-hs protocol is reset when moving from one NodeB to another. Thus the hybrid-ARQ protocol state is not transferred between the two NodeBs. Any packet losses at the time of cell change are instead handled by the RLC protocol.

Related to mobility is the flow control between the NodeB and the RNC, used to control the amount of data buffered in the MAC-hs in the NodeB and avoid overflow in the buffers. The requirements on the flow control are, to some extent, conflicting as it shall ensure that MAC-hs buffers should be large enough to contain a sufficient amount of data to fully utilize the physical channel resources (in case of advantageous channel conditions), while at the same time MAC-hs buffers should be kept as small as possible to minimize the amounts of packets that need to be resent to a new NodeB in case of inter-NodeB handover.

9.2.9   UE categories

To allow for a range of UE implementations, different UE capabilities are specified. The UE capabilities are divided into a number of parameters, which are sent from the UE at the establishment of a connection and if/when the UE capabilities are changed during an ongoing connection. The UE capabilities may then be used by the network to select a configuration that is supported by the UE. Several of the UE capabilities applicable to other channels are valid for HS-DSCH as well, but there are also some HS-DSCH-specific capabilities.

Basically, the physical-layer UE capabilities are used to limit the requirements for three different UE resources: the despreading resource, the soft-buffer memory used by the hybrid-ARQ functionality, and the Turbo decoder. The despreading resource is limited in terms of the maximum number of HS-PDSCH codes the UE simultaneously needs to despread. Three different capabilities exist in terms of de-spreading resources, corresponding to the capability to despread a maximum of 5, 10, or 15 physical channels (HS-PDSCH).

The amount of soft-buffer memory is in the range of 14 400–172 800 soft bits, depending on the UE category. Note that this is the total amount of buffer memory for all hybrid-ARQ processes, not the value per process. The memory is divided among the multiple hybrid-ARQ processes, typically with an equal amount of memory per process although non-equal allocation is also possible.

The requirements on the Turbo-decoding resource are defined through two parameters: the maximum number of transport-channel bits that can be received within an HS-DSCH TTI and the minimum inter-TTI interval, that is the distance in time between subsequent transport blocks. The decoding time in a Turbo decoder is roughly proportional to the number of information bits which thus provides a limit on the required processing speed. In addition, for low-end UEs, there is a possibility to avoid continuous data transmission by specifying an inter-TTI interval larger than one.

In order to limit the number of possible combinations of UE capabilities and to avoid parameter combinations that do not make sense, the UE capability parameters relevant for the physical layer are lumped into 12 different *categories* as illustrated in Table 9.1.

9.3   Finer details of HSDPA

9.3.1   Hybrid ARQ revisited: Physical-layer processing

Hybrid ARQ with soft combining has been described above, although some details of the physical-layer and protocol operation were omitted in order to

Table 9.1 *HSDPA UE categories [99].*

HS-DSCH category	Maximum number of HS-DSCH codes received	Minimum inter-TTI interval	Maximum transport-block size soft bits		Maximum number of schemes	Supported modulation
1	5	3	7298	(3.6 Mbit/s)	19200	16QAM, QPSK
2	5	3	7298	(3.6 Mbit/s)	28800	16QAM, QPSK
3	5	2	7298	(3.6 Mbit/s)	28800	16QAM, QPSK
4	5	2	7298	(3.6 Mbit/s)	38400	16QAM, QPSK
5	5	1	7298	(3.6 Mbit/s)	57600	16QAM, QPSK
6	5	1	7298	(3.6 Mbit/s)	67200	16QAM, QPSK
7	10	1	14411	(7.2 Mbit/s)	115200	16QAM, QPSK
8	10	1	14411	(7.2 Mbit/s)	134400	16QAM, QPSK
9	15	1	20251	(10.1 Mbit/s)	172800	16QAM, QPSK
10	15	1	27952	(14 Mbit/s)	172800	16QAM, QPSK
11	5	2	3630	(1.8 Mbit/s)	14400	QPSK
12	5	1	3630	(1.8 Mbit/s)	28800	QPSK

simplify the description. This section provides a more detailed description of the processing, aiming at filling the missing gaps.

As already mentioned, the hybrid ARQ operates on a single transport block, that is, whenever the HS-DSCH CRC indicates an error, a retransmission representing the same information as the original transport block is requested. Since there is a single transport block per TTI, this implies that it is not possible to mix transmissions and retransmissions within the same TTI.

Since incremental redundancy is the basic hybrid-ARQ soft-combining scheme, retransmissions generally consist of a different set of coded bits. Furthermore, the modulation scheme, the channelization-code set, and the transmission power can be different compared to the original transmission. Incremental redundancy generally has better performance, especially for high initial code rates, but poses higher requirements on the soft buffering in the UE since soft bits from all transmission attempts must be buffered prior to decoding. Therefore, the NodeB needs to have knowledge about the soft-buffer size in the UE (for each active hybrid-ARQ process). Coded bits that do not fit within the soft buffer shall not be transmitted. For HSDPA, this problem is solved through the use of *two-stage rate matching*. The first rate-matching stage limits the number of coded bits to what is possible to fit in the soft buffer, while the second rate-matching stage generates the different redundancy versions.

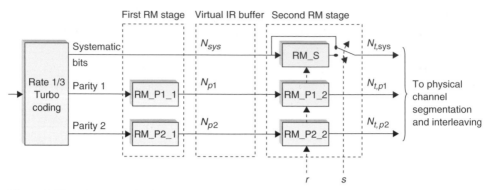

Figure 9.15   *The principle of two-stage rate matching.*

Each rate-matching stage uses several identical rate-matching blocks, denoted RM in Figure 9.15. An RM block can be configured to puncture or repeat every nth bit.

The first rate-matching stage is used to limit the number of coded bits to the available UE soft buffer for the hybrid-ARQ process currently being addressed. A sufficient number of coded bits are punctured to ensure that all coded bits at the output of the first rate-matching stage will fit in the soft buffer (known as *virtual IR buffer* at the transmitter side). Hence, depending on the soft-buffer size in the UE, the lowest code rate may be higher than the rate-1/3 mother code rate in the Turbo coder. Note that, if the number of bits from the channel coding does not exceed the UE soft-buffering capability, the first rate-matching stage is transparent and no puncturing is performed.

The second rate-matching stage serves two purposes:

- Matching the number of bits in the virtual IR buffer to the number of available channel bits. The number of available channel bits is given by the size of the channelization-code set and the modulation scheme selected for the TTI.
- Generating different sets of coded bits as controlled by the two redundancy-version parameters r and s, described below.

Equal repetition for all three streams is applied if the number of available channel bits is larger than the number of bits in the virtual IR buffer, otherwise puncturing is applied.

To support full incremental redundancy, that is, to have the possibility to transmit only/mainly parity bits in a retransmission, puncturing of systematic bits is possible as controlled by the parameter s. Setting $s = 1$ implies that the systematic bits are prioritized and puncturing is primarily applied with an equal amount to the two parity-bit streams. On the other hand, for a transmission prioritizing the parity bits, $s = 0$ and primarily the systematic bits are punctured. If, for a transmission prioritizing the systematic bits, the number of coded bits is larger than the number of physical channel bits, despite all the parity bits have been punctured, further puncturing is applied to the systematic bits. Similarly, if puncturing the systematic bits is not sufficient for a transmission prioritizing the parity bits, puncturing is applied to the parity bits as well.

For good performance, all systematic bits should be transmitted in the initial transmission, corresponding to $s = 1$, and the code rate should be set to less than one. For the retransmission (assuming the initial transmission did not succeed), different strategies can be applied. If the NodeB received neither ACK, nor NAK, in response to the initial transmission attempt, the UE may have missed the initial transmission. Setting $s = 1$ also for the retransmission is therefore appropriate. This is also the case if NAK is received and Chase combining is used for retransmissions. However, if a NAK is received and incremental redundancy is used, that is, the parity bits should be prioritized, setting $s = 0$ is appropriate.

The parameter r controls the puncturing pattern in each rate-matching block in Figure 9.15 and determines which bits to puncture. Typically, $r = 0$ is used for the initial transmission attempt. For retransmissions, the value of r is typically increased, effectively leading to a different puncturing pattern. Thus, by varying r, multiple, possibly partially overlapping, sets of coded bits representing the same set of information bits can be generated. It should be noted that changing the number of channel bits by changing the modulation scheme or the number of channelization codes also affects which coded bits that are transmitted even if the r and s parameters are unchanged between the transmission attempts.

With the two-stage rate-matching scheme, both incremental redundancy and Chase combining can easily be supported. By setting $s = 1$ and $r = 0$ for all transmission attempts, the same set of bits is used for the retransmissions as for the original transmission, that is Chase combining. Incremental redundancy is easily achieved by setting $s = 1$ and using $r = 0$ for the initial transmission, while retransmissions use $s = 0$ and $r > 0$. Partial IR, that is, incremental redundancy with the systematic bits included in each transmission, results if $s = 1$ for all the retransmissions as well as the initial transmission.

In Figure 9.16, a simple numerical example is shown to further illustrate the operation of the physical-layer hybrid-ARQ processing of data. Assume that,

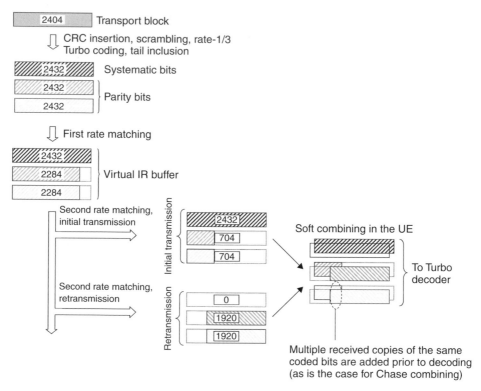

Figure 9.16   *An example of the generation of different redundancy versions in the case of IR. The numbers indicate the number of bits after the different stages using the example case in the text.*

as an example, a transport block of 2404 bits is to be transmitted using one of the hybrid-ARQ processes. Furthermore, assume the hybrid-ARQ process in question is capable of buffering at most 7000 soft values due to memory limitations in the UE and the soft memory configuration set by higher layers. Finally, the channel can carry 3840 coded bits in this example (QPSK modulation, 4 channelization codes).

A 24-bit CRC is appended to the transport block, rate-1/3 Turbo coding is applied and a 12-bit tail appended, resulting in 7296 coded bits. The coded bits are fed to the first stage rate matching, which punctures parity bits such that 2432 systematic bits and 2×2284 parity bits, in total 7000 bits, are fed to the second-stage rate-matching block. Since at most 7000 coded bits can be transmitted, the lowest possible code rate is $2432/7000 = 0.35$, which is slightly higher than the mother code's rate of 1/3 due to the limited soft buffer in the UE.

For the initial transmission, the second-stage rate matching matches the 7000 coded bits to the 3840 channel bits by puncturing the parity bits only. This is achieved by using $r = 0$ and $s = 1$, that is, a self-decodable transmission, and the resulting code rate is $2432/3840 = 0.63$.

For retransmissions, either Chase combining or incremental redundancy can be used, as chosen by the NodeB. If Chase combining is used by setting $s = 1$ and $r = 0$, the same 3840 bits as used for the initial transmission are retransmitted (assuming unchanged modulation scheme and channelization-code set). The resulting effective code rate remains 0.63 as no additional parity has been transmitted, but an energy gain has been obtained as, in total, twice the amount of energy has been transmitted for each bit. Note that this example assumed identical transport formats for the initial transmission and the retransmission.

If incremental redundancy is used for the retransmission, for example, by using $s = 0$ and $r = 1$, the systematic bits are punctured and only parity bits are retransmitted, of which 3840 (out of 4568 parity bits available after the first stage rate matching) fit into the physical channel. Note that some of these parity bits were included already in the original transmission as the number of unique parity bits is not large enough to fill both the original transmission and the retransmissions. After the retransmission, the resulting code rate is $2432/7000 = 0.35$. Hence, contrary to Chase combining, there is a coding gain in addition to the energy gain.

9.3.2 Interleaving and constellation rearrangement

For 16QAM, two of the four bits carried by each modulation symbol will be more reliable at the receiver due to the difference in the number of nearest neighbors in the constellation. This is in contrast to QPSK, where both bits are of equal reliability. Furthermore, for Turbo codes, systematic bits are of greater importance in the decoding process, compared to parity bits. Hence, it is desirable to map as many of the systematic bits as possible to the more reliable positions in a 16QAM symbol. A dual interleaver scheme, illustrated in Figure 9.17, has been adopted for HS-DSCH in order to control the mapping of systematic and parity bits onto the 16QAM modulation symbols.

For QPSK, only the upper interleaver in Figure 9.17 is used, while for 16QAM, two identical interleavers are used in parallel. Systematic bits are primarily fed into the upper interleaver, whereas parity bits are primarily fed into the lower interleaver. The 16QAM constellation is defined such that the output from the upper interleaver is mapped onto the reliable bit positions and the output from the lower interleaver onto the less reliable positions.

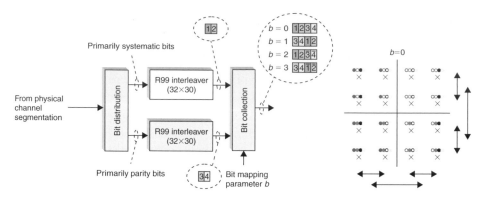

Figure 9.17 *The channel interleaver for the HS-DSCH. The shaded parts are only used for 16QAM. Colors illustrate the mapping order for a sequence of 4 bits, where a bar on top of the figure denotes bit inversion.*

If 16QAM is used in conjunction with hybrid ARQ using Chase combining, there is a performance gain by rearranging the 16QAM symbol constellations between multiple transmission attempts as this provides an averaging effect among the reliability of the bits. However, note that this gain is only available for retransmissions and not for the initial transmission. Furthermore, the gains with constellation rearrangement in combination with incremental redundancy are minor. Hence, its use is mainly applicable when Chase combining is used.

Constellation rearrangement is obtained through bit manipulations in the bit collector block and is controlled by a four-state bit mapping parameter, controlling two independent operations. First, the output of the two interleavers can be swapped. Second, the output of the lower interleaver (or the upper interleaver if swapping is used) can be inverted. In essence, this results in the selection of one out of four different signal constellations for 16QAM.

9.3.3 Hybrid ARQ revisited: Protocol operation

As stated earlier, each hybrid-ARQ entity is capable of supporting multiple (up to eight) stop-and-wait hybrid-ARQ processes. The motivation behind this is to allow for continuous transmission to a single UE, which cannot be achieved by a single stop-and-wait scheme. The number of hybrid-ARQ processes is configurable by higher-layer signaling. Preferably, the number of hybrid-ARQ processes is chosen to match the roundtrip time, consisting of the TTI itself, any radio-interface delay in downlink and uplink, the processing time in the UE, and the processing time in the NodeB.

The protocol design assumes a well-defined time between the end of the received transport block and the transmission of the ACK/NAK as discussed in

Section 9.2.5. In essence, this time is the time the UE has available for decoding of the received data. From a delay perspective, this time should be as small as possible, but a too small value would put unrealistic requirements on the UE processing speed. Although in principle the time could be made a UE capability, this was not felt necessary and a value of 5 ms was agreed as a good trade-off between performance and complexity. This value affects the number of hybrid-ARQ processes necessary. Typically, a total of six processes are configured, which leaves around 2.8 ms for processing of retransmissions in the NodeB.

Which of the hybrid-ARQ processes that is used for the current transmission is controlled by the scheduler and explicitly signaled to the UE. Note that the hybrid-ARQ processes can be addressed in any order. The amount of soft-buffering memory available in the UE is semi-statically split between the different hybrid-ARQ processes. Thus the larger the number of hybrid-ARQ processes is, the smaller the amount of soft-buffer memory available to a hybrid-ARQ process for incremental redundancy. The split of the total soft-buffer memory between the hybrid-ARQ processes is controlled by the RNC and does not necessarily have to be such that the soft-buffer memory per hybrid-ARQ process is the same. Some hybrid-ARQ processes can be configured to use more soft-buffer memory than others, although the typical case is to split the available memory equally among the processes.

Whenever the current transmission is not a retransmission, the NodeB MAC-hs increments the single-bit new-data indicator. Hence, for each new transport block, the bit is toggled. The indicator is used by the UE to clear the soft buffer for initial transmissions since, by definition, no soft combining should be done for an initial transmission. The indicator is also used to detect error cases in the status signaling, for example, if the 'new-data' indicator is not toggled despite the fact that the previous data for the hybrid-ARQ process in question was correctly decoded and acknowledged, an error in the uplink signaling has most likely occurred. Similarly, if the indicator is toggled but the previous data for the hybrid-ARQ process was not correctly decoded, the UE will replace the data previously in the soft buffers with the new received data.

Errors in the status (ACK/NAK) signaling will impact the overall performance. If an ACK is misinterpreted as a NAK, an unnecessary hybrid-ARQ retransmission will take place, leading to a (small) reduction in the throughput. On the other hand, misinterpreting a NAK as an ACK will lead to loss of data as the NodeB will not perform a hybrid-ARQ retransmission despite the UE was not able to successfully decode the data. Instead, the missing data has to be retransmitted by the RLC protocol, a more time-consuming procedure than hybrid-ARQ retransmissions. Therefore, the requirements on the ACK/NAK errors are typically asymmetric with $\Pr\{NAK|ACK\} = 10^{-2}$ and $\Pr\{ACK|NAK\} = 10^{-3}$ (or 10^{-4}) as typical values. With these error probabilities, the impact on the end-user TCP performance due to hybrid-ARQ signaling errors is small [75].

9.3.4   In-sequence delivery

The multiple hybrid-ARQ processes cannot themselves ensure in-sequence delivery as there is no interaction between the processes. Hence, in-sequence delivery must be implemented on top of the hybrid-ARQ processes and a reordering queue in the UE MAC-hs is used for this purpose. Related to the reordering queues in the UE are the priority queues in the NodeB, used for handling priorities in the scheduling process.

The NodeB MAC-hs receives MAC-d PDUs in one or several MAC-d flows. Each such MAC-d PDU has a priority assigned to it and MAC-d PDUs with different priorities can be mixed in the same MAC-d flow. The MAC-d flows are split if necessary and the MAC-d PDUs are sorted into priority queues as illustrated in Figure 9.18. Each priority queue corresponds to a certain MAC-d flow and a certain MAC-d priority, where RRC signaling is used to set up the mapping between the priority queues and the MAC-d flows. Hence, the scheduler in the MAC-hs can, if desired, take the priorities into account when making

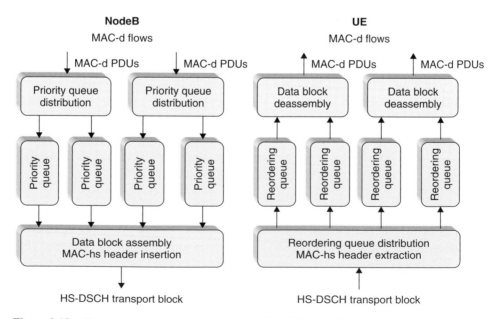

Figure 9.18   *The priority queues in the NodeB MAC-hs (left) and the reordering queues in the UE MAC-hs (right).*

the scheduling decision. One or several MAC-d PDUs from one of the priority queues are assembled into a data block, where the number of MAC-d PDUs and the priority queue selection is controlled by the scheduler. A MAC-hs header containing, among others, queue identity and a transmission sequence number, is added to form a transport block. The transport block is forwarded to the physical layer for further processing. As there is only a single transmission sequence number and queue identity in the transport block, all MAC-d PDUs within the same transport block come from the same priority queue. Thus, mixing MAC-d PDUs from different priority queues within the same TTI is not possible.

In the UE, the reordering-queue identity is used to place the received data block, containing received MAC-d PDUs, into the correct reordering queue as illustrated in Figure 9.18. Each reordering queue corresponds to a priority queue in the NodeB, although the priority queues buffer MAC-d PDUs, while the reordering queues buffer data blocks. Within each reordering queue, the transmission sequence number, sent in the MAC-hs header is used to ensure in-sequence delivery of the MAC-d PDUs. The transmission sequence number is unique within the reordering queue, but not between different reordering queues.

The basic idea behind reordering, illustrated in Figure 9.19, is to store data blocks in the reordering queue until all data blocks with lower sequence numbers have been delivered. As an example, at time t_0 in Figure 9.19, the NodeB has transmitted data blocks with sequence numbers 0 through 3. However, the data block with sequence number 1 has not yet reached the MAC-hs reordering queue in the UE, possibly due to hybrid-ARQ retransmissions or errors in the hybrid-ARQ uplink signaling. Data block 0 has been disassembled into MAC-d

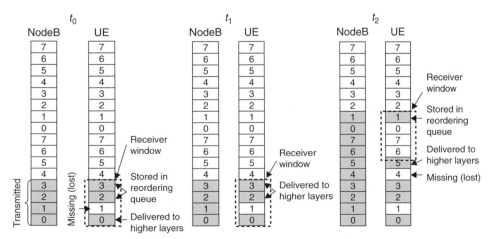

Figure 9.19 *Illustration of the principles behind reordering queues.*

PDUs and delivered to upper layers by the UE MAC-hs, while data blocks 2 and 3 are buffered in the reordering queue since data block 1 is missing.

Evidently, there is a risk of stalling the reordering queue if missing data blocks (data block 1 in this example) are not successfully received within a finite time. Therefore, a timer-based stall avoidance mechanism is defined for the MAC-hs. Whenever a data block is successfully received but cannot be delivered to higher layers, a timer is started. In Figure 9.19, this occurs when data block 2 is received since data block 1 is missing in the reordering buffer. Note that there is at maximum one stall avoidance timer active. Therefore no timer is started upon reception of data block 3 as there is already one active timer started for data block 2. Upon expiration of the timer, which occurs at time t_1 in Figure 9.19, data block 1 is considered to be lost. Any subsequent data blocks up to the first missing data block are to be disassembled into MAC-d PDUs and delivered to higher layers. In Figure 9.19, data blocks 2 and 3 are delivered to higher layers.

Relying on the timer-based mechanism alone would limit the possible values of the timer and limit the performance if the sequence numbers are to be kept unique. Hence, a window-based stall avoidance mechanism is defined in addition to the timer-based mechanism to ensure a consistent UE behavior. If a data block with a sequence number higher than the end of the window is received by the reordering function, the data block is inserted into the reordering buffer at the position indicated by the sequence number. The receiver window is advanced such that the received data block forms the last data block within the window. Any data blocks not within the window after the window advancement are delivered to higher layers. In the example in Figure 9.19, the window size of 4 is used, but the MAC-hs window size is configurable by RRC. In Figure 9.19, a data block with sequence number 1 is received at time t_2, which causes the receiver window to be advanced to cover sequence numbers 6 through 1. Data block 4 is considered to be lost, since it is now outside the window whereas data block 5 is disassembled and delivered to higher layers. In order for the reordering functionality in the UE to operate properly, the NodeB should not retransmit MAC-hs PDUs with sequence numbers lower than the highest transmitted sequence number minus the UE receiver window size.

9.3.5 MAC-hs header

To support reordering and de-multiplexing of MAC-d PDUs in the UE as discussed above, the necessary information needs to be signaled to the UE. As this information is required only after successful decoding of a transport block, in-band signaling in the form of a MAC-hs header can be used.

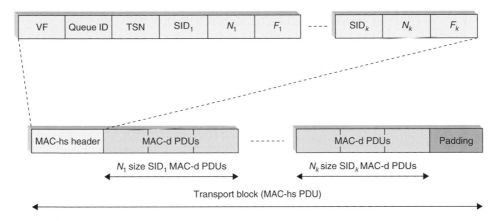

Figure 9.20 *The structure of the MAC-hs header.*

The MAC-hs header contains

- reordering-queue identity,
- Transmission Sequence Number (TSN),
- number and size of the MAC-d PDUs.

The structure of the MAC-hs header is illustrated in Figure 9.20. The Version Flag (VF) is identical to zero and reserved for future extensions of the MAC-hs header. The 3-bit Queue ID identifies the reordering queue to be used in the receiver. All MAC-d PDUs in one MAC-hs PDU belong to the same reordering queue. The 6-bit TSN field identifies the transmission sequence number of the MAC-hs data block. The TSN is unique within a reordering buffer but not between different reordering buffers. Together with the Queue ID, the TSN provides support for in-sequence delivery as described in the previous section.

The MAC-hs payload consists of one or several MAC-d PDUs. The 3-bit *SID*, size index identifier, provides the MAC-d PDU size and the 7-bit *N* field identifies the number of MAC-d PDUs. The flag *F* is used to indicate the end of the MAC-hs header. One set of *SID*, *N*, and *F* is used for each set of consecutive MAC-d PDUs and multiple MAC-d PDU sizes are supported by forming groups of MAC-d PDUs of equal size. Note that all the MAC-d PDUs within a data block must be in consecutive order since the sequence numbering is per data block. Hence, if a sequence of MAC-d PDUs with sizes given by SID_1, SID_2, SID_1 is to be transmitted, three groups has to be formed despite that there are only two MAC-d PDU sizes. Finally, the MAC-hs PDU is padded (if necessary) such that the MAC-hs PDU size equals a suitable block size. It should be noted that, in most cases, there is only a single MAC-d PDU size and, consequently, only a single set of *SID, N*, and *F*.

9.3.6   CQI and other means to assess the downlink quality

Obviously, some of the key HSDPA functions, primarily scheduling and rate control, rely on rapid adaptation of the transmission parameters to the instantaneous channel conditions as experienced by the UE. The NodeB is free to form an estimate of the channel conditions using any available information, but, as already discussed, uplink control signaling from the UEs in the form of a Channel-Quality Indicator (CQI), is typically used.

The CQI does not explicitly indicate the channel quality, but rather the data rate supported by the UE given the current channel conditions. More specifically, the CQI is a recommended transport-block size (which is equivalent to a recommended data rate).

The reason for not reporting an explicit channel-quality measure is that different UEs might support different data rates in identical environments, depending on the exact receiver implementation. By reporting the data rate rather than an explicit channel-quality measure, the fact that a UE has a relatively better receiver can be utilized to provide better service (higher data rates) to such a UE. It is interesting to note that this provides a benefit with advanced receiver structures for the end user. For a power-controlled channel, the gain from an advanced receiver is seen as a lower transmit power at the NodeB, thus providing a benefit for the network but not the end user. This is in contrast to the HS-DSCH using rate control, where a UE with an advanced receiver can receive the HS-DSCH with higher data rate compared to a standard receiver.

Each 5-bit CQI value corresponds to a given transport-block size, modulation scheme, and number of channelization codes. These values are shown in Figure 9.8 on page 153 (assuming a high-end terminal, capable of receiving 15 codes). Different tables are used for different UE categories as a UE shall not report a CQI exceeding its capabilities. For example, a UE only supporting 5 codes shall not report a CQI corresponding to 15 codes, while a 15-code UE may do so. Therefore, power offsets are used for channel qualities exceeding the UE capabilities. A power offset of x dB indicates that the UE can receive a certain transport-block size, but at x dB lower transmission power than the CQI report was based upon. This is illustrated in Table 9.2 for some different UE categories. UEs belonging to category 1–6 can only receive up to 5 HS-DSCH channelization codes and therefore must use a power offset for the highest CQI values, while category 10 UEs are able to receive up to 15 codes.

The CQI values listed are sorted in ascending order and the UE shall report the highest CQI for which transmission with parameters corresponding to the

Table 9.2 *Example of CQI reporting for two different UE categories [97].*

CQI value	Transport-block size		Modulation scheme	Number of HS-DSCH channelization codes		Power offset (dB)	
	Category 1–6	Category 10		Category 1–6	Category 10	Category 1–6	Category 10
0		N/A			Out of range		
1		137	QPSK	1		0	
2		173	QPSK	1		0	
3		233	QPSK	1		0	
4		317	QPSK	1		0	
5		377	QPSK	1		0	
6		461	QPSK	1		0	
7		650	QPSK	2		0	
8		792	QPSK	2		0	
9		931	QPSK	2		0	
10		1262	QPSK	3		0	
11		1483	QPSK	3		0	
12		1742	QPSK	3		0	
13		2279	QPSK	4		0	
14		2583	QPSK	4		0	
15		3319	QPSK	5		0	
16		3565	16QAM	5		0	
17		4189	16QAM	5		0	
18		4664	16QAM	5		0	
19		5287	16QAM	5		0	
20		5887	16QAM	5		0	
21		6554	16QAM	5		0	
22		7168	16QAM	5		0	
23	7168	9719	16QAM	5	7	−1	0
24	7168	11418	16QAM	5	8	−2	0
25	7168	14411	16QAM	5	10	−3	0
26	7168	17237	16QAM	5	12	−4	0
27	7168	21754	16QAM	5	15	−5	0
28	7168	23370	16QAM	5	15	−6	0
29	7168	24222	16QAM	5	15	−7	0
30	7168	25558	16QAM	5	15	−8	0

CQI result in a block error probability not exceeding 10%. The CQI values are chosen such that an increase in CQI by one step corresponds to approximately 1 dB increase in the instantaneous carrier-to-interference ratio on an AWGN channel. Measurements on the common pilot form the basis for the CQI. The CQI represents the instantaneous channel conditions in a predefined 3-slot interval ending one slot prior to the CQI transmission. Specifying which

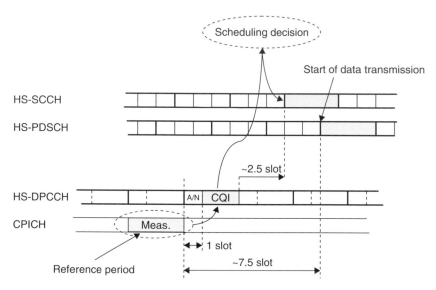

Figure 9.21   *Timing relation for the CQI reports.*

interval the CQI relates to allows the NodeB to track changes in the channel quality between the CQI reports by using the power control commands for the associated downlink (F-) DPCH as described below. The timing of the CQI reports and the earliest possible time the report can be used for scheduling purposes is illustrated in Figure 9.21.

The rate of the channel-quality reporting is configurable in the range of one report per 2–160 ms. The CQI reporting can also be switched off completely.

As the scheduling and rate-adaptation algorithms are vendor specific, it is possible to perform rate control based on other criteria than the UE reports as well, either alone or in combination. Using the transmit power level of the associated DPCH is one such possibility, where a high transmit power indicates unfavorable channel conditions and a low DPCH transmit power indicates favorable conditions. Since the power level is a relative measure of the channel quality and not reflects an absolute subjective channel quality, this technique is advantageously combined with infrequent UE quality reports. The UE reports provide an absolute quality and the transmission power of the power-controlled DPCH can be used to update this quality report between the reporting instances. This combined scheme works quite well and can significantly reduce the frequency of the UE CQI reports as long as the DPCH is not in soft handover. In soft handover the transmit power of the different radio links involved in the soft handover are power controlled such that the combined received signal is of sufficient quality. Consequently, the DPCH transmit power at the serving

HS-DSCH cell does not necessarily reflect the perceived UE channel quality. Hence, more frequent UE quality reports are typically required in soft handover scenarios.

9.3.7 Downlink control signaling: HS-SCCH

The HS-SCCH, sometimes referred to as the *shared control channel*, is a shared downlink physical channel that carries control signaling information needed for a UE to be able to properly despread, demodulate and decode the HS-DSCH.

In each 2 ms interval corresponding to one HS-DSCH TTI, one HS-SCCH carries physical-layer signaling to a single UE. As HSDPA supports HS-DSCH transmission to multiple users in parallel by means of code multiplexing, see Section 9.1.1, multiple HS-SCCH may be needed in a cell. According to the specification, a UE should be able to decode four HS-SCCHs in parallel. However, more than four HS-SCCHs can be configured within a cell, although the need for this is rare.

HS-SCCH uses a spreading factor of 128 and has a time structure based on a subframe of length 2 ms which is the same length as the HS-DSCH TTI. The following information is carried on the HS-SCCH:

- The HS-DSCH transport format, consisting of:
  - HS-DSCH channelization-code set [7 bits]
  - HS-DSCH modulation scheme, QPSK /16QAM [1 bit]
  - HS-DSCH transport-block size information [6 bits].
- Hybrid-ARQ-related information, consisting of:
  - hybrid-ARQ process number [3 bits]
  - redundancy version [3 bits]
  - new-data indicator [1 bit].
- A UE ID that identifies the UE for which the HS-SCCH information is intended [16 bits]. As will be described below, the UE ID is not explicitly transmitted but implicitly included in the CRC calculation and HS-SCCH channel coding.

As described in Section 9.2.4, the HS-DSCH transport block can take 1 out of 254 different sizes. Each combination of channelization-code-set size and modulation scheme corresponds to a subset of these transport-block sizes, where each subset consists of 63 possible transport-block sizes. The 6-bit 'HS-DSCH transport-block size information' indicates which out of the 63 possible transport-block sizes is actually used for the HS-DSCH transmission in the corresponding TTI. The transport-block sizes have been defined to make full use of code rates ranging from 1/3 to 1 for initial transmissions. For retransmissions, instantaneous code rates larger than one can be achieved by indicating 'the transport-block size is identical to the previous transmission in this hybrid-ARQ process.' This is indicated setting the 'HS-DSCH transport-block size information' field to 111111. This is useful for additional scheduling flexibility, for example to retransmit only a small amount of parity bit in case the latest CQI report indicates the UE 'almost' was able to decode the original transmission.

Requirements on when different parts of the HS-SCCH information need to be available to the UE has affected the detailed structure of the HS-SCCH channel coding and physical-channel mapping. For UE complexity reasons, it is beneficial if the channelization-code set is known to the UE prior to the start of the HS-DSCH transmission. Otherwise, the UE would have to buffer the received signal on a sub-chip level prior to despreading or, alternatively, despread all potential HS-DSCH codes up to the maximum of 15 codes. Knowing the modulation scheme prior to the HS-DSCH subframe is also preferred as it allows for 'on-the-fly' demodulation. On the other hand, the transport-block size and the hybrid-ARQ-related information are only needed at HS-DSCH decoding/soft combining, which can anyway not start until the end of the HS-DSCH TTI. Thus, the HS-SCCH information is split into two parts:

- Part 1 consisting of channelization-code set and modulation scheme [total of 8 bits].
- Part 2 consisting of transport-block size and hybrid-ARQ-related parameters [total of 13 bits].

The HS-SCCH coding, physical-channel mapping and timing relation to the HS-DSCH transmission is illustrated in Figure 9.22. The HS-DSCH channel coding is based on rate-1/3 convolutional coding, carried out separately for part 1 and part 2. Part 1 is coded and rate matched to 40 bits to fit into the first slot of the HS-SCCH subframe. Before mapping to the physical channel, the coded part 1 is scrambled by a 40 bits UE-specific bit sequence. The sequence is derived from the 16 bits UE ID using rate-1/2 convolutional coding followed by puncturing. With the scheme of Figure 9.22, the part 1 information can be decoded after one slot of the HS-SCCH subframe. Furthermore, in case of more than one HS-SCCH, the UE can find the correct HS-SCCH from the soft metric of the channel decoder already after the first slot. One possible way for the UE to utilize the soft metric for determining which (if any) of the multiple HS-SCCHs that carries control information for the UE is to form the log-likelihood ratio between the most likely code word and the second most likely code word for each HS-SCCH. The HS-SCCH with the largest ratio is likely to be intended for the UE and can be selected for further decoding of the part-2 information.

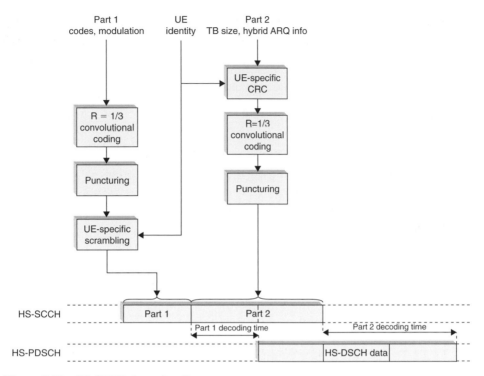

Figure 9.22 *HS-SCCH channel coding.*

Part 2 is coded and rate matched to 80 bits to fit into the second and third slot of the HS-SCCH. Part 2 includes a UE-specific CRC for error detection. The CRC is calculated over all the information bits, both part 1 and part 2, as well as the UE identity. The identity is not explicitly transmitted, but by including its ID when calculating the CRC at the receiver, the UE can decide whether it was the intended recipient or not. If the transmission is intended for another UE, the CRC will not check.

In case of HS-DSCH transmission to a single UE in consecutive TTIs, the UE must despread the HS-SCCH in parallel to the HS-DSCH channelization codes. To reduce the number of required despreaders, the same HS-SCCH shall be used when HS-DSCH transmission is carried out in consecutive TTI. This implies that, when simultaneously receiving HS-DSCH, the UE only needs to despread a single HS-SCCH.

In order to avoid waste of capacity, the HS-SCCH transmit power should be adjusted to what is needed to reach the intended UE. Similar information used for rate control of the HS-DSCH, for example the CQI reports, can be used to power control the HS-SCCH.

9.3.8   Downlink control signaling: F-DPCH

As described in Section 9.2.1, for each UE for which HS-DSCH can be transmitted, there is also an associated downlink DPCH. In principle, if all data transmission, including RRC signaling, is mapped to the HS-DSCH, there is no need to carry any user data on the DPCH. Consequently, there is no need for downlink Transport-Format Combination Indicator (TFCI) or dedicated pilots on such a DPCH. In this case, the only use for the downlink DPCH in case of HS-DSCH transmission is to carry power control commands to the UE in order to adjust the uplink transmission power. This fact is exploited by the F-DPCH or *fractional DPCH,* introduced in Release 6 as a means to reduce the amount of downlink channelization codes used for dedicated channels. Instead of allocating one DPCH with spreading factor 256 for the sole purpose of transmitting one power control command per slot, the F-DPCH allows up to ten UEs to share a single channelization code for this purpose. In essence, the F-DPCH is a slot format supporting TPC bits only. Two TPC bits (one QPSK symbol) is transmitted in one tenth of a slot, using a spreading factor 256, and the rest of the slot remains unused. By setting the downlink timing of multiple UEs appropriately, as illustrated in Figure 9.23, up to ten UEs can then share a single channelization code. This can also be seen as time-multiplexing power control commands to several users on one channelization code.

9.3.9   Uplink control signaling: HS-DPCCH

For operation of the hybrid-ARQ protocol and to provide the NodeB with knowledge about the instantaneous downlink channel conditions, uplink control signaling is required. This signaling is carried on an additional new uplink physical channel, the HS-DPCCH, using a channelization code separate from the conventional uplink DPCCH. The use of a separate channelization code for the HS-DPCCH makes the HS-DPCCH 'invisible' to non-HSDPA-capable base stations and allows for the uplink being in soft handover even if not all NodeBs in the active set support HSDPA.

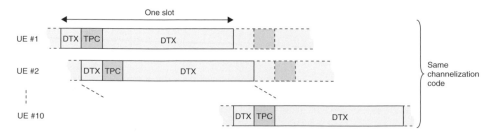

Figure 9.23   *Fractional DPCH (F-DPCH), introduced in Release 6.*

The HS-DPCCH uses a spreading factor of 256 and is transmitted in parallel with the other uplink channels as illustrated Figure 9.24. To reduce the uplink pcak-to-average ratio, the channelization code used for HS-DPCCH and if the HS-DPCCH is mapped to the I or Q branch of this code depends on the maximum number of DPDCHs used by the transport-format combination set configured for the UE.

As the HS-DPCCH spreading factor is 256, the HS-DPCCH allows for a total of 30 channel bits per 2 ms subframe (3 slots). The HS-DPCCH information is divided in such a way that the hybrid-ARQ acknowledgement is transmitted in the first slot of the subframe while the channel-quality indication is transmitted in the second and third slot, see Figure 9.24.

In order to minimize the hybrid-ARQ roundtrip time, the HS-DPCCH trans-mission timing is not slot aligned to the other uplink channels. Instead, the HS-DPCCH timing is defined relative to the end of the subframe carrying the corresponding HS-DSCH data as illustrated in Figure 9.24. The timing is such that there are approximately 7.5 slots (19 200 chips) of UE processing time avail-able, from the end of the HS-DSCH TTI to the transmission of the corresponding uplink hybrid-ARQ acknowledgement. If the HS-DPCCH had been slot aligned to the uplink DPCCH, there would have been an uncertainty of one slot in the HS-DSCH /HS-DPCCH timing. This uncertainty would have reduced the processing time available for the UE /NodeB by one slot.

Due to the alignment between the uplink HS-DPCCH and the downlink HS-DSCH, the HS-DPCCH will not necessarily be slot aligned with the uplink DPDCH/DPCCH. However, note that the HS-DPCCH is always aligned to the uplink DPCCH/DPDCH on a 256-chip basis in order to keep uplink orthogonality. As a consequence, the HS-DPCCH cannot have a completely fixed transmit tim-ing relative to the received HS-DSCH. Instead the HS-DPCCH transmit timing

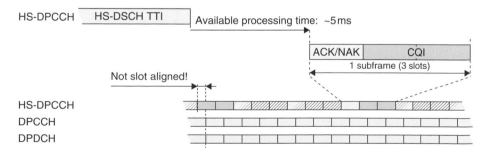

Figure 9.24 *Basic structure of uplink signaling with IQ/code-multiplexed HS-DPCCH.*

varies in an interval 19 200 chips to 19 200 + 255 chips. Note that CQI and ACK/ NAK are transmitted independently of each other. In subframes where no ACK/ NAK or CQI is to be transmitted, nothing is transmitted in the corresponding HS-DPCCH field.

The hybrid-ARQ acknowledgement consists of a single information bit, ACK or NAK, indicating whether the HS-DSCH was correctly decoded (the CRC checked) or not. ACK or NAK is only transmitted in case the UE correctly received the HS-SCCH control signaling. If no HS-SCCH control signaling intended for the UE was detected, nothing is transmitted in the ACK/NAK field (DTX). This reduces the uplink load as only the UEs to which HS-DSCH data was actually sent in a TTI transmit an ACK/NAK on the uplink. The single-bit ACK is repetition coded into 10 bits to fit into the first slot of a HS-DPCCH subframe.

Reliable reception of the uplink ACK/NAK requires a sufficient amount of energy. In some situations where the UE is power limited, it may not be possible to collect enough energy by transmitting the ACK/NAK over a single slot. Therefore, there is a possibility to configure the UE to repeat the ACK/NAK in N subsequent ACK/NAK slots. Naturally, when the UE is configured to transmit repeated acknowledgements, it cannot receive HS-DSCH data in consecutive TTIs, as the UE would then not be able to acknowledge all HS-DSCH data. Instead there must be at least $N - 1$ idle 2 ms subframes between each HS-DSCH TTI in which data is to be received. Examples when repetition of the acknowledgements can be useful are very large cells, or in some soft handover situations. In soft handover, the uplink can be power controlled by multiple NodeBs. If any of the non-serving NodeBs has the best uplink, the received HS-DPCCH quality at the serving NodeB may not be sufficient and repetition may therefore be necessary.

As mentioned earlier, the impact of ACK-to-NAK and NAK-to-ACK errors are different, leading to different requirements. In addition, the DTX-to-ACK error case also has to be handled. If the UE misses the scheduling information and the NodeB misinterprets the DTX as ACK, data loss in the hybrid ARQ will occur. An asymmetric decision threshold in the ACK/NAK detector should therefore preferably be used as illustrated in Figure 9.25. Based on the noise variance at the ACK/NAK detector, the threshold can be computed to meet a certain DTX-to-ACK error probability, for example, 10^{-2}, after which the transmission power of the ACK and NAK can be set to meet the remaining error requirements (ACK-to-NAK and NAK-to-ACK).

In Release 6 of the WCDMA specifications, an enhancement to the ACK/NAK signaling has been introduced. In addition to the ACK and NAK, the UE may also

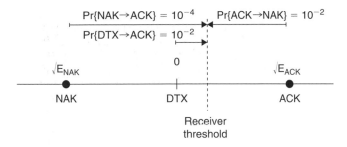

Figure 9.25 *Detection threshold for the ACK/NAK field of HS-DPCCH.*

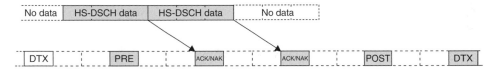

Figure 9.26 *Enhanced ACK/NAK using PRE and POST.*

transmit two additional code words, PRE and POST, on the HS-DPCCH. A UE configured to use the enhancement will transmit PRE and POST in the subframes preceding and succeeding, respectively, the ACK/NAK (unless these subframes were used by the ACK/NAK for other transport blocks). Thus, an ACK will cause a transmission spanning multiple subframes and the power can therefore be reduced while maintaining the same ACK-to-NAK error rate (Figure 9.26).

The CQI consists of five information bits. A (20,5) block code is used to code this information to 20 bits, which corresponds to two slots on the HS-DPCCH. Similarly to the ACK/NAK, repetition of the CQI field over multiple 2 ms subframes is possible and can be used to provide improved coverage.

10
Enhanced Uplink

Enhanced Uplink, also known as *High-Speed Uplink Packet Access* (HSUPA), has been introduced in WCDMA Release 6. It provides improvements in WCDMA uplink capabilities and performance in terms of higher data rates, reduced latency, and improved system capacity, and is therefore a natural complement to HSDPA. Together, the two are commonly referred to as High-Speed Packet Access (HSPA). The specifications of Enhanced Uplink can be found in [101] and the references therein.

10.1 Overview

At the core of Enhanced Uplink are two basic technologies used also for HSDPA – fast scheduling and fast hybrid ARQ with soft combining. For similar reasons as for HSDPA, Enhanced Uplink also introduces a short 2 ms uplink TTI. These enhancements are implemented in WCDMA through a new transport channel, the *Enhanced Dedicated Channel* (E-DCH).

Although the same technologies are used both for HSDPA and Enhanced Uplink, there are fundamental differences between them, which have affected the detailed implementation of the features:

- In the downlink, the shared resource is transmission power and the code space, both of which are located in *one* central node, the NodeB. In the uplink, the shared resource is the amount of allowed uplink interference, which depends on the transmission power of *multiple distributed* nodes, the UEs.
- The scheduler and the transmission buffers are located in the same node in the downlink, while in the uplink the scheduler is located in the NodeB while the data buffers are distributed in the UEs. Hence, the UEs need to signal buffer status information to the scheduler.
- The WCDMA uplink, also with Enhanced Uplink, is inherently non-orthogonal, and subject to interference between uplink transmissions within the same

cell. This is in contrast to the downlink, where different transmitted channels are *orthogonal*. Fast power control is therefore essential for the uplink to handle the near-far problem.[1] The E-DCH is transmitted with a power offset relative to the power-controlled uplink control channel and by adjusting the maximum allowed power offset, the scheduler can control the E-DCH data rate. This is in contrast to HSDPA, where a (more or less) constant transmission power with rate adaptation is used.

- Soft handover is supported by the E-DCH. *Receiving* data from a terminal in multiple cells is fundamentally beneficial as it provides diversity, while transmission from multiple cells in case of HSDPA is cumbersome and with questionable benefits as discussed in the previous chapter. Soft handover also implies *power control by multiple cells*, which is necessary to limit the amount of interference generated in neighboring cells and to maintain backward compatibility and coexistence with UE not using the E-DCH for data transmission.

- In the downlink, higher-order modulation, which trades power efficiency for bandwidth efficiency, is useful to provide high data rates in some situations, for example when the scheduler has assigned a small number of channelization codes for a transmission but the amount of available transmission power is relatively high. The situation in the uplink is different; there is no need to share channelization codes between users and the channel coding rates are therefore typically lower than for the downlink. Hence, unlike the downlink, higher-order modulation is less useful in the uplink macro-cells and therefore not part of the first release of enhanced uplink.[2]

With these differences in mind, the basic principles behind Enhanced Uplink can be discussed.

10.1.1 Scheduling

For Enhanced Uplink, the scheduler is a key element, controlling *when* and *at what data rate* the UE is allowed to transmit. The higher the data rate a terminal is using, the higher the terminal's received power at the NodeB must be to maintain the E_b/N_0 required for successful demodulation. By increasing the transmission power, the UE can transmit at a higher data rate. However, due to the non-orthogonal uplink, the received power from one UE represents interference for other terminals. Hence, the shared resource for Enhanced Uplink is the

---

[1] The near-far problem describes the problem of detecting a weak user, located far from the transmitter, when a user close to the transmitter is active. Power control ensured the signals are received at a similar strength, therefore, enabling detection of both users' transmissions.

[2] Uplink higher-order modulation is introduced in Release 7; see Chapter 12 for further details.

amount of tolerable interference in the cell. If the interference level is too high, some transmissions in the cell, control channels and non-scheduled uplink transmissions, may not be received properly. On the other hand, a too low interference level may indicate that UEs are artificially throttled and the full system capacity not exploited. Therefore, Enhanced Uplink relies on a scheduler to give users with data to transmit permission to use an as high data rate as possible without exceeding the maximum tolerable interference level in the cell.

Unlike HSDPA, where the scheduler and the transmission buffers both are located in the NodeB, the data to be transmitted resides in the UEs for the uplink case. At the same time, the scheduler is located in the NodeB to coordinate different UEs transmission activities in the cell. Hence, a mechanism for communicating the scheduling decisions to the UEs and to provide buffer information from the UEs to the scheduler is required. The scheduling framework for Enhanced Uplink is based on *scheduling grants* sent by the NodeB scheduler to control the UE transmission activity and *scheduling requests* sent by the UEs to request resources. The scheduling grants control the maximum allowed E-DCH-to-pilot power ratio the terminal may use; a larger grant implies the terminal may use a higher data rate but also contributes more to the interference level in the cell. Based on measurements of the (instantaneous) interference level, the scheduler controls the scheduling grant in each terminal to maintain the interference level in the cell at a desired target (Figure 10.1).

In HSDPA, typically a single user is addressed in each TTI. For Enhanced Uplink, the implementation-specific uplink scheduling strategy in most cases schedules multiple users in parallel. The reason is the significantly smaller transmit power of a terminal compared to a NodeB: a single terminal typically cannot utilize the full cell capacity on its own.

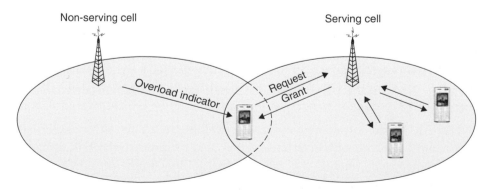

Figure 10.1 *Enhanced Uplink scheduling framework.*

Inter-cell interference also needs to be controlled. Even if the scheduler has allowed a UE to transmit at a high data rate based on an acceptable intra-cell interference level, this may cause non-acceptable interference to neighboring cells. Therefore, in soft handover, the *serving cell* has the main responsibility for the scheduling operation, but the UE monitors scheduling information from all cells with which the UE is in soft handover. The non-serving cells can request all its non-served users to lower their E-DCH data rate by transmitting an *overload indicator* in the downlink. This mechanism ensures a stable network operation.

Fast scheduling allows for a more relaxed connection admission strategy. A larger number of bursty high-rate packet-data users can be admitted to the system as the scheduling mechanism can handle the situation when multiple users need to transmit in parallel. If this creates an unacceptably high interference level in the system, the scheduler can rapidly react and restrict the data rates they may use. Without fast scheduling, the admission control would have to be more conservative and reserve a margin in the system in case of multiple users transmitting simultaneously.

10.1.2   Hybrid ARQ with soft combining

Fast hybrid ARQ with soft combining is used by Enhanced Uplink for basically the same reason as for HSDPA – to provide robustness against occasional transmission errors. A similar scheme as for HSDPA is used. For each transport block received in the uplink, a single bit is transmitted from the NodeB to the UE to indicate successful decoding (ACK) or to request a retransmission of the erroneously received transport block (NAK).

One main difference compared to HSDPA stems from the use of soft handover in the uplink. When the UE is in soft handover, this implies that the hybrid ARQ protocol is *terminated in multiple cells*. Consequently, in many cases, the transmitted data may be successfully received in some NodeBs but not in others. From a UE perspective, it is sufficient if at least one NodeB successfully receives the data. Therefore, in soft handover, all involved NodeBs attempt to decode the data and transmits an ACK or a NAK. If the UE receives an ACK from at least one of the NodeBs, the UE considers the data to be successfully received.

Hybrid ARQ with soft combining can be exploited not only to provide robustness against unpredictable interference, but also to improve the link efficiency to increase capacity and/or coverage. One possibility to provide a data rate of x Mbit/s is to transmit at x Mbit/s and set the transmission power to target a low error probability (in the order of a few percent) in the first transmission attempt.

Alternatively, the same resulting data rate can be provided by transmitting using n times higher data rate at an unchanged transmission power and use multiple hybrid ARQ retransmissions. From the discussion in Chapter 7, this approach on average results in a lower cost per bit (a lower E_b/N_0) than the first approach. The reason is that, on average, less than n transmissions will be used. This is sometimes known as *early termination gain* and can be seen as implicit rate adaptation. Additional coded bits are only transmitted when necessary. Thus, the code rate after retransmissions is determined by what was needed by the instantaneous channel conditions. This is exactly what rate adaptation also tries to achieve, the main difference being that rate adaptation tries to find the correct code rate prior to transmission. The same principle of implicit rate adaptation can also be used for HS-DSCH in the downlink to improve the link efficiency.

10.1.3 Architecture

For efficient operation, the scheduler should be able to exploit rapid variations in the interference level and the channel conditions. Hybrid ARQ with soft combining also benefits from rapid retransmissions as this reduces the cost of retransmissions. These two functions should therefore reside close to the radio-interface. As a result, and for similar reasons as for HSDPA, the scheduling and hybrid ARQ functionalities of Enhanced Uplink are located in the NodeB. Furthermore, also similar to the HSDPA design, it is preferable to keep all radio-interface layers above MAC intact. Hence, ciphering, admission control, etc., is still under the control of the RNC. This also allows for a smooth introduction of Enhanced Uplink in selected areas; in cells not supporting E-DCH transmissions, channel switching can be used to map the user's data flow onto the DCH instead.

Following the HSDPA design philosophy, a new MAC entity, the *MAC-e*, is introduced in the UE and NodeB. In the NodeB, the MAC-e is responsible for support of fast hybrid ARQ retransmissions and scheduling, while in the UE, the MAC-e is responsible for selecting the data rate within the limits set by the scheduler in the NodeB MAC-e.

When the UE is in soft handover with multiple NodeBs, different transport blocks may be successfully decoded in different NodeBs. Consequently, one transport block may be successfully received in one NodeB while another NodeB is still involved in retransmissions of an earlier transport block. Therefore, to ensure in-sequence delivery of data blocks to the RLC protocol, a reordering functionality is required in the RNC in the form of a new MAC entity, the MAC-es. In soft handover, multiple MAC-e entities are used per UE as the data is received in multiple cells. However, the MAC-e in the serving cell

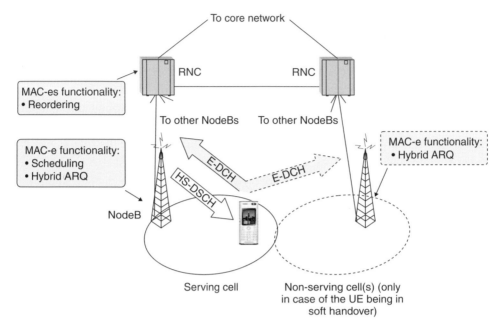

Figure 10.2   *The architecture with E-DCH (and HS-DSCH) configured.*

has the main responsibility for the scheduling; the MAC-e in a non-serving cell is mainly handling the hybrid ARQ protocol (Figure 10.2).

10.2   Details of Enhanced Uplink

To support uplink scheduling and hybrid ARQ with soft combining in WCDMA, a new transport-channel type, the Enhanced Dedicated Channel (E-DCH) has been introduced in Release 6. The E-DCH can be configured simultaneously with one or several DCHs. Thus, high-speed packet-data transmission on the E-DCH can occur at the same time as services using the DCH from the same UE.

A low delay is one of the key characteristics of Enhanced Uplink and required for efficient packet-data support. Therefore, a short TTI of 2 ms is supported by the E-DCH to allow for rapid adaptation of transmission parameters and reduction of the end-user delays associated with packet-data transmission. Not only does this reduce the cost of a retransmission, the transmission time for the initial transmission is also reduced. Physical-layer processing delay is typically proportional to the amount of data to process and the shorter the TTI, the smaller the amount of data to process in each TTI for a given data rate. At the same time, in deployments with relatively modest data rates, for example in large cells, a longer TTI may be beneficial as the payload in a 2 ms TTI can become unnecessarily small

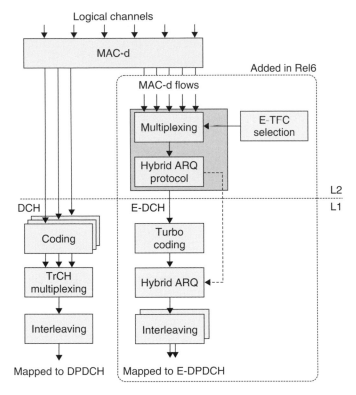

Figure 10.3 *Separate processing of E-DCH and DCH.*

and the associated relative overhead too large. Hence, the E-DCH supports two TTI lengths, 2 and 10 ms, and the network can configure the appropriate value. In principle, different UEs can be configured with different TTIs.

The E-DCH is mapped to a set of uplink channelization codes known as *E-DCH Dedicated Physical Data Channels* (E-DPDCHs). Depending on the instantaneous data rate, the number of E-DPDCHs and their spreading factors are both varied.

Simultaneous transmission of E-DCH and DCH is possible as discussed above. Backward compatibility requires the E-DCH processing to be invisible to a NodeB not supporting Enhanced Uplink. This has been solved by separate processing of the DCH and E-DCH and mapping to different channelization code sets as illustrated in Figure 10.3. If the UE is in soft handover with multiple cells, of which some does not support Enhanced Uplink, the E-DCH transmission is invisible to these cells. This allows for a gradual upgrade of an existing network. An additional benefit with the structure is that it simplifies the introduction of the 2 ms TTI and also provides greater freedom in the selection of hybrid ARQ processing.

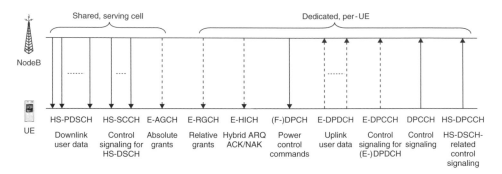

Figure 10.4 *Overall channel structure with HSDPA and Enhanced Uplink. The new channels introduced as part of Enhanced Uplink are shown with dashed lines.*

Downlink control signaling is necessary for the operation of the E-DCH. The downlink, as well as uplink, control channels used for E-DCH support are illustrated in Figure 10.4, together with the channels used for HSDPA.

Obviously, the NodeB needs to be able to request retransmissions from the UE as part of the hybrid ARQ mechanism. This information, the ACK/NAK, is sent on a new downlink dedicated physical channel, the *E-DCH Hybrid ARQ Indicator Channel* (E-HICH). Each UE with E-DCH configured receives one E-HICH of its own from each of the cells which the UE is in soft handover with.

Scheduling grants, sent from the scheduler to the UE to control when and at what data rate the UE is transmitting, can be sent to the UE using the shared *E-DCH Absolute Grant Channel* (E-AGCH). The E-AGCH is sent from the serving cell only as this is the cell having the main responsibility for the scheduling operation and is received by all UEs with an E-DCH configured. In addition, scheduling grant information can also be conveyed to the UE through an *E-DCH Relative Grant Channel* (E-RGCH). The E-AGCH is typically used for large changes in the data rate, while the E-RGCH is used for smaller adjustments during an ongoing data transmission. This is further elaborated upon in the discussion on scheduling operation below.

Since the uplink by design is non-orthogonal, fast closed-loop power control is necessary to address the near-far problem. The E-DCH is no different from any other uplink channel and is therefore power controlled in the same way as other uplink channels. The NodeB measures the received signal-to-interference ratio and sends power control commands in the downlink to the UE to adjust the transmission power. Power control commands can be transmitted using DPCH or, to save channelization codes, the fractional DPCH, F-DPCH.

In the uplink, control signaling is required to provide the NodeB with the necessary information to be able to demodulate and decode the data transmission. Even though, in principle, the serving cell could have this knowledge as it has issued the scheduling grants, the non-serving cells in soft handover clearly do not have this information. Furthermore, as discussed below, the E-DCH also supports non-scheduled transmissions. Hence, there is a need for out-band control signaling in the uplink, and the *E-DCH Dedicated Physical Control Channel* (E-DPCCH) is used for this purpose.

10.2.1 MAC-e and physical layer processing

Similar to HSDPA, short delays and rapid adaptation are important aspects of the Enhanced Uplink. This is implemented by introducing the MAC-e, a new entity in the NodeB responsible for scheduling and hybrid ARQ protocol operation. The physical layer is also enhanced to provide the necessary support for a short TTI and for soft combining in the hybrid ARQ mechanism.

In soft handover, uplink data can be received in multiple NodeBs. Consequently, there is a need for a MAC-e entity in each of the involved NodeBs to handle the hybrid ARQ protocol. The MAC-e in the serving cell is, in addition, responsible for handling the scheduling operation.

To handle the Enhanced Uplink processing in the terminal, there is also a MAC-e entity in the UE. This can be seen in Figure 10.5, where the Enhanced Uplink processing in the UE is illustrated. The MAC-e in the UE consists of MAC-e multiplexing, transport format selection, and the protocol parts of the hybrid ARQ mechanism.

Mixed services, for example simultaneous file upload and VoIP, are supported. Hence, as there is only a single E-DCH transport channel, data from multiple MAC-d flows can be multiplexed through MAC-e multiplexing. The different services are in this case typically transmitted on different MAC-d flows as they may have different quality-of-service requirements.

Only the UE has accurate knowledge about the buffer situation and power situation in the UE at the time of transmission of a transport block in the uplink. Hence, the UE is allowed to autonomously select the data rate or, strictly speaking, the *E-DCH Transport Format Combination* (E-TFC). Naturally, the UE needs to take the scheduling decisions into account in the transport format selection; the scheduling decision represents an upper limit of the data rate the UE is not allowed to exceed. However, it may well use a lower data rate, for example if the

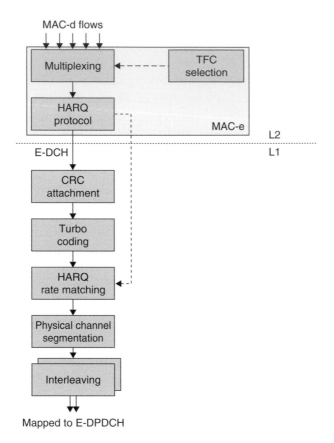

Figure 10.5 *MAC-e and physical-layer processing.*

transmit power does not support the scheduled data rate. E-TFC selection, including MAC-e multiplexing, is discussed further in conjunction with scheduling.

The hybrid ARQ protocol is similar to the one used for HSDPA, that is multiple stop-and-wait hybrid ARQ processes operated in parallel. There is one major difference though – when the terminal is in soft handover with several NodeBs, the hybrid ARQ protocol is terminated in multiple nodes.

Physical layer processing is straightforward and has several similarities with the HS-DSCH physical layer processing. From the MAC-e in the UE, data is passed to the physical layer in the form of one transport block per TTI on the E-DCH. Compared to the DCH coding and multiplexing chain, the overall structure of the E-DCH physical layer processing is simpler as there is only a single E-DCH and hence no transport channel multiplexing.

A 24-bit CRC is attached to the single E-DCH transport block to allow the hybrid ARQ mechanism in the NodeB to detect any errors in the received transport block. Coding is done using the same rate 1/3 Turbo coder as used for HSDPA.

The physical layer hybrid ARQ functionality is implemented in a similar way as for HSDPA. Repetition or puncturing of the bits from the Turbo coder is used to adjust the number of coded bits to the number of channel bits. By adjusting the puncturing pattern, different redundancy versions can be generated.

Physical channel segmentation distributes the coded bits to the different channelization codes used, followed by interleaving and modulation.

10.2.2 Scheduling

Scheduling is one of the fundamental technologies behind Enhanced Uplink. In principle, scheduling is possible already in the first version of WCDMA, but Enhanced Uplink supports a significantly faster scheduling operation thanks to the location of the scheduler in the NodeB.

The responsibility of the scheduler is to control *when* and *at what data rate* a UE is allowed to transmit, thereby controlling the amount of interference affecting other users at the NodeB. This can be seen as controlling each UE's consumption of common resources, which in case of Enhanced Uplink is the amount of tolerable interference, that is the total received power at the base station. The amount of common uplink resources a terminal is using depends on the data rate used. Generally, the higher the data rate, the larger the required transmission power and thus the higher the resource consumption.

The term *noise rise* or *rise-over-thermal* is often used when discussing uplink operation. Noise rise, defined as $(I_0 + N_0)/N_0$ where N_0 and I_0 are the noise and interference power spectral densities, respectively, is a measure of the increase in interference in the cell due to the transmission activity. For example, 0 dB noise rise indicates an unloaded system and 3 dB noise rise implies a power spectral density due to uplink transmission equal to the noise spectral density. Although noise rise as such is not of major interest, it has a close relation to coverage and uplink load. A too large noise rise would result in loss of coverage for some channels – a terminal may not have sufficient transmission power available to reach the required E_b/N_0 at the base station. Hence, the uplink scheduler must keep the noise rise within acceptable limits.

Channel-dependent scheduling, which typically is used in HSDPA, is possible for the uplink as well, but it should be noted that the benefits are different. As fast power control is used for the uplink, a terminal transmitting when the channel conditions are favorable will generate the same amount of interference in the cell as a terminal transmitting in unfavorable channel conditions, given the same data rate for the two. This is in contrast to HSDPA, where in principle a constant transmission power is used and the data rates are adapted to the channel conditions, resulting in a higher data rate for users with favorable radio conditions. However, for the uplink the transmission power will be different for the two terminals. Hence, the amount of interference generated in *neighboring* cells will differ. Channel-dependent scheduling will therefore result in a lower noise rise in the system, thereby improving capacity and/or coverage.

In practical cases, the transmission power a UE is limited by several factors, both regulatory restrictions and power amplifier implementation restrictions. For WCDMA, different power classes are specified limiting the maximum power the UE can use to, where 21 dBm is a common value of the maximum power. This affects the discussion on uplink scheduling, making channel-dependent scheduling beneficial also from an intra-cell perspective. A UE scheduled when the channel conditions are beneficial encounters a reduced risk of hitting its transmission power limitation. This implies that the UE is likely to be able to transmit at a higher data rate if scheduled to transmit at favorable channel conditions. Therefore, taking channel conditions into account in the uplink scheduling decisions will improve the capacity, although the difference between non-channel-dependent and channel-dependent scheduling in most cases not are as large as in the downlink case.

Round-robin scheduling is one simple example of an uplink scheduling strategy, where terminals take turn in transmitting in the uplink. Similar to round-robin scheduling in HSDPA, this results in TDMA-like operation and avoids intra-cell interference due to the non-orthogonal uplink. However, as the maximum transmission power of the terminals is limited, a single terminal may not be able to fully utilize the uplink capacity when transmitting and thus reducing the uplink capacity in the cell. The larger the cells, the higher the probability that the UE does not have sufficient transmit power available.

To overcome this, an alternative is to assign the same data rate to all users having data to transmit and to select this data rate such that the maximum cell load is respected. This results in maximum fairness in terms of the same data rate for all users, but does not maximize the capacity of the cell. One of the benefits though is the simple scheduling operation – there is no need to estimate the uplink channel quality and the transmission power status for each UE. Only the buffer status of each UE and the total interference level in the cell is required.

With *greedy filling*, the terminal with the best radio conditions is assigned as high data rate as possible. If the interference level at the receiver is smaller than the maximum tolerable level, the terminal with the second best channel conditions is allowed to transmit as well, continuing with more and more terminals until the maximum tolerable interference level at the receiver is reached. This strategy maximizes the radio-interface utilization but is achieved at the cost of potentially large differences in data rates between users. In the extreme case, a user at the cell border with poor channel conditions may not be allowed to transmit at all.

Strategies between these two can also be considered such as different proportional fair strategies. This can be achieved by including a weighting factor for each user, proportional to the ratio between the instantaneous and average data rates, into the greedy filling algorithm. In a practical scenario, it is also necessary to take the transport network capacity and the processing resources in the base station into account in the scheduling decision, as well as the priorities for different data flows.

The above discussion of different scheduling strategies assumed all UEs having an infinite amount of data to transmit (full buffers). Similarly as the discussion for HSDPA, the traffic behavior is important to take into account when comparing different scheduling strategies. Packet-data applications are typically bursty in nature with large and rapid variations in their resource requirements. Hence, the overall target of the scheduler is to allocate a large fraction of the shared resource to users momentarily requiring high data rates, while at the same time ensuring stable system operation by keeping the noise rise within limits.

A particular benefit of fast scheduling is the fact that it allows for a more relaxed connection admission strategy. For the DCH, admission control typically has to reserve resources relative to the peak data rate as there are limited means to recover from an event when many or all users transmit simultaneously with their maximum rate. Admission relative to the peak rate results in a rather conservative admission strategy for bursty packet-data applications. With fast scheduling, a larger number of packet-data users can be admitted since fast scheduling provides means to control the load in case many users request for transmission simultaneously.

10.2.2.1 *Scheduling framework for Enhanced Uplink*
The scheduling framework for Enhanced Uplink is generic in the sense the control signaling allows for several different scheduling implementations. One major difference between uplink and downlink scheduling is the location of the scheduler and the information necessary for the scheduling decisions.

In HSDPA, the scheduler and the buffer status are located at the same node, the NodeB. Hence, the scheduling strategy is completely implementation dependent and there is no need to standardize any buffer status signaling to support the scheduling decisions.

In Enhanced Uplink, the scheduler is still located in the NodeB to control the transmission activity of different UEs, while the buffer status information is distributed among the UEs. In addition to the buffer status, the scheduler also needs information about the available transmission power in the UE: if the UE is close to its maximum transmission power there is no use in scheduling a (significantly) higher data rate. Hence, there is a need to specify signaling to convey buffer status and power availability information from the UE to the NodeB.

The basis for the scheduling framework is *scheduling grants* sent by the NodeB to the UE and limiting the E-DCH data rate and *scheduling requests* sent from the UE to the NodeB to request permission to transmit (at a higher rate than currently allowed). Scheduling decisions are taken by the *serving cell*, which has the main responsibility for scheduling as illustrated in Figure 10.6 (in case of simultaneous HSDPA and Enhanced Uplink, the same cell is the serving cell for both). However, when in soft handover, the non-serving cells have a possibility to influence the UE behavior to control the inter-cell interference.

Providing the scheduler with the necessary information about the UE situation, taking the scheduling decision based on this information, and communicating

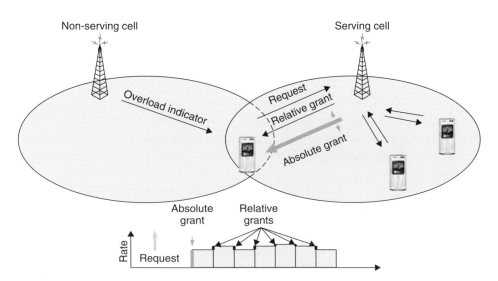

Figure 10.6   *Overview of the scheduling operation.*

the decision back to the UE takes a non-zero amount of time. The situation at the UE in terms of buffer status and power availability may therefore be different at the time of transmission compared to the time of providing the information to the NodeB UE buffer situation. For example, the UE may have less data to transmit than assumed by the scheduler, high-priority data may have entered the transmission buffer or the channel conditions may have worsened such that the UE has less power available for data transmission. To handle such situations and to exploit any interference reductions due to a lower data rate, the scheduling grant does not set the E-DCH data rate, but rather an *upper limit* of the resource usage. The UE select the data rate or, more precisely, the *E-DCH Transport Format Combination* (E-TFC) within the restrictions set by the scheduler.

The *serving grant* is an internal variable in each UE, used to track the maximum amount of resource the UE is allowed to use. It is expressed as a maximum E-DPDCH-to-DPCCH power ratio and the UE is allowed to transmit from any MAC-d flow and using any transport block size as long as it does not exceed the serving grant. Hence, the scheduler is responsible for scheduling between UEs, while the UEs themselves are responsible to schedule between MAC-d flows according to rules in the specifications. Basically, a high-priority flow should be served before a low-priority flow.

Expressing the serving grant as a maximum power ratio is motivated by the fact that the fundamental quantity the scheduler is trying to control is uplink interference, which is directly proportional to transmission power. The E-DPDCH transmission power is defined relative to the DPCCH to ensure the E-DPDCH is affected by the power control commands. As the E-DPDCH transmission power typically is significantly larger than the DPCCH transmission power, the E-DPDCH-to-DPCCH power ratio is roughly proportional to the total transmission power, $(P_{E\text{-}2DPDCH} + P_{DPCCH})/P_{DPCCH} \approx P_{E\text{-}DPDCH}/P_{DPCCH}$, and thus setting a limit on the maximum E-DPCCH-to-DPCCH power ration corresponds to control of the maximum transmission power of the UE.

The NodeB can update the serving grant in the UE by sending an *absolute grant* or a *relative grant* to the UE (Figure 10.7). Absolute grants are transmitted on the shared E-AGCH and are used for absolute changes of the serving grant. Typically, these changes are relatively large, for example to assign the UE a high data rate for an upcoming packet transmission.

Relative grants are transmitted on the E-RGCH and are used for relative changes of the serving grant. Unlike the absolute grants, these changes are small; the change in transmission power due to a relative grant is typically in the order of

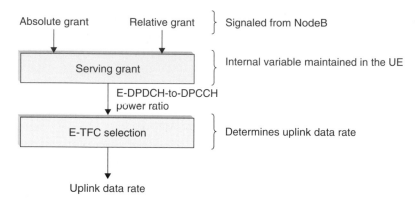

Figure 10.7 *The relation between absolute grant, relative grant and serving grant.*

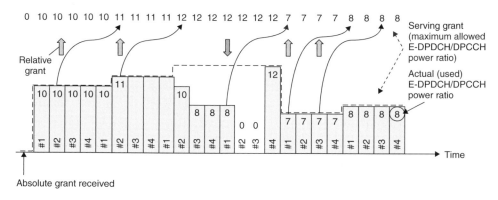

Figure 10.8 *Illustration of relative grant usage.*

1 dB. Relative grants can be sent from both serving and, in case of the UE being in soft handover, also from the non-serving cells. However, there is a significant difference between the two and the two cases deserve to be treated separately.

Relative grants from the serving cell are dedicated for a single UE, that is each UE receives its own relative grant to allow for individual adjustments of the serving grants in different UEs. The relative grant is typically used for small, possibly frequent, updates of the data rate during an ongoing packet transmission. A relative grant from the serving cell can take one of the three values: 'UP,' 'HOLD,' or 'DOWN.' The 'up' ('down') command instructs the UE to increase (decrease) the serving grant, that is to increase (decrease) the maximum allowed E-DPDCH-to-DPCCH power ratio compared to the last used power ratio, where the last used power ratio refers to the previous TTI in the same hybrid ARQ process. The 'hold' command instructs the UE not to change the serving grant. An illustration of the operation is found in Figure 10.8.

Relative grants from non-serving cells are used to control inter-cell interference. The scheduler in the serving cell has no knowledge about the interference to neighboring cells due to the scheduling decisions taken. For example, the load in the serving cell may be low and from that perspective, it may be perfectly fine to schedule a high-rate transmission. However, the neighboring cell may not be able to cope with the additional interference caused by the high-rate transmission. Hence, there must be a possibility for the non-serving cell to influence the data rates used. In essence, this can be seen as an 'emergency break' or an 'overload indicator,' commanding non-served UEs to lower their data rate.

Although the name 'relative grant' is used also for the overload indicator, the operation is quite different from the relative grant from the serving cell. First, the overload indicator is a common signal received by all UEs. Since the non-serving cell only is concerned about the total interference level from the neighboring cell, and not which UE that is causing the interference, a common signal is sufficient. Furthermore, as the non-serving cell is not aware of the traffic priorities, etc., of the UEs it is not serving, there would be no use in having dedicated signaling from the non-serving cell.

Second, the overload indicator only takes two, not three, values – 'DTX' and 'down,' where the former does not affect the UE operation. All UEs receiving 'down' from any of the non-serving cells decrease their respective serving grant relative to the previous TTI in the same hybrid ARQ process.

10.2.2.2 *Scheduling information*
For efficient scheduling, the scheduler obviously needs information about the UE situation, both in terms of buffer status and in terms of the available transmission power. Naturally, the more detailed the information is, the better the possibilities for the scheduler to take accurate and efficient decisions. However, at the same time, the amount of information sent in the uplink should be kept low not to consume excessive uplink capacity. These requirements are, to some extent, contradicting and are in Enhanced Uplink addressed by providing two mechanism complementing each other: the out-band 'happy bit' transmitted on the E-DPCCH and in-band scheduling information transmitted on the E-DCH.

Out-band signaling is done through a single bit on the E-DPCCH, the 'happy bit.' Whenever the UE has available power for the E-DCH to transmit at a higher data rate compared to what is allowed by the serving grant, and the number of bits in the buffer would require more than a certain number of TTIs, the UE shall set the bit to 'not happy' to indicate that it would benefit from a higher serving grant. Otherwise, the UE shall declare 'happy.' Note that the happy bit is only transmitted in conjunction with an ongoing data transmission as the E-DPCCH is only transmitted together with the E-DPDCH.

In-band scheduling information provides detailed information about the buffer occupancy, including priority information, and the transmission power available for the E-DCH. The in-band information is transmitted in the same way as user data, either alone or as part of a user data transmission. Consequently, this information benefits from hybrid ARQ with soft combining. As in-band scheduling information is the only mechanism for the unscheduled UE to request resources, the scheduling information can be sent non-scheduled and can therefore be transmitted regardless of the serving grant. Non-scheduled transmissions are not restricted to scheduling information only; the network can configure non-scheduled transmissions also for other data.

10.2.3   E-TFC selection

The E-TFC selection is responsible for selecting the transport format of the E-DCH, thereby determining the data rate to be used for uplink transmission, and to control MAC-e multiplexing. Clearly, the selection needs to take the scheduling decisions taken by the NodeB into account, which is done through the serving grant as previously discussed. MAC-e multiplexing, on the other hand, is handled autonomously by the UE. Hence, while the scheduler handles resource allocation *between* UEs, the E-TFC selection controls resource allocation between flows *within* the UE. The rules for multiplexing of the flows are given by the specification; in principle, high-priority data shall be transmitted before any data of lower priority.

Introduction of the E-DCH needs to take coexistence with DCHs into account. If this is not done, services mapped onto DCHs could be affected. This would be a non-desirable situation as it may require reconfiguration of parameters set for DCH transmission. Therefore, a basic requirement is to serve DCH traffic first and only spend otherwise unused power resources on the E-DCH. Comparisons can be made with HSDPA, where any dedicated channels are served first and the HS-DSCH may use the otherwise unused transmission power. Therefore, TFC selection is performed in two steps. First, the normal DCH TFC selection is performed as in previous releases. The UE then estimates the remaining power and a second TFC selection step is performed where E-DCH can use the remaining power. The overall E-TFC selection procedure is illustrated in Figure 10.9.

Each E-TFC has an associated E-DPDCH-to-DPCCH power offset. Clearly, the higher the data rate, the higher the power offset. When the required transmitter

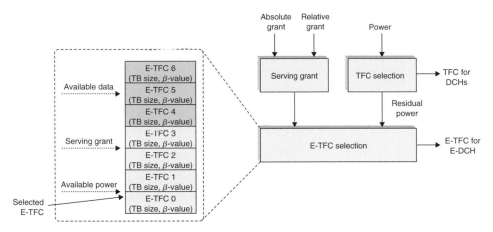

Figure 10.9 *Illustration of the E-TFC selection process.*

power for different E-TFCs has been calculated, the UE can calculate the possible E-TFCs to use from a power perspective. The UE then selects the E-TFC by maximizing the amount of data that can be transmitted given the power constraint and the scheduling grant.

The possible transport block sizes being part of the E-TFCs are predefined in the specifications, similar to HS-DSCH. This reduces the amount of signaling, for example at handover between cells, as there is no need to configure a new set of E-TFCs at each cell change. Generally, conformance tests to ensure the UE obeys the specifications are also simpler the smaller the amount of configurability in the terminal.

To allow for some flexibility in the transport block sizes, there are four tables of E-TFCs specified; for each of the two TTIs specified there is one table optimized for common RLC PDU sizes and one general table with constant maximum relative overhead. Which one of the predefined tables that the UE shall use is determined by the TTI and RRC signaling.

10.2.4   Hybrid ARQ with soft combining

Hybrid ARQ with soft combining for Enhanced Uplink serves a similar purpose as the hybrid ARQ mechanism for HSDPA – to provide robustness against transmission errors. However, hybrid ARQ with soft combining is not only a tool for providing robustness against occasional errors; it can also be used for enhanced capacity as discussed in the introduction. As hybrid ARQ retransmissions are fast, many services allow for a retransmission or two. Combined with incremental redundancy, this forms an implicit rate control mechanism. Thus, hybrid ARQ with soft combining can be used in several (related) ways:

- To provide robustness against variations in the received signal quality.
- To increase the link efficiency by targeting multiple transmission attempts, for example by imposing a maximum number of transmission attempts and operating the outer loop power control on the residual error event after soft combining.

To a large extent, the requirements on hybrid ARQ are similar to HSDPA and, consequently, the hybrid ARQ design for Enhanced Uplink is fairly similar to the design used for HSDPA, although there are some differences as well, mainly originating from the support of soft handover in the uplink.

Similar to HSDPA, Enhanced Uplink hybrid ARQ spans both the MAC layer and the physical layer. The use of multiple parallel stop-and-wait processes for the hybrid ARQ protocol has proven efficient for HSDPA and is used for Enhanced Uplink for the same reasons – fast retransmission and high throughput combined with low overhead of the ACK/NAK signaling. Upon reception of the single transport block transmitted in a certain TTI and intended for a certain hybrid ARQ process, the NodeB attempts to decode the set of bits and the outcome of the decoding attempt, ACK or NAK, is signaled to the UE. To minimize the cost of the ACK/NAK, a single bit is used. Clearly, the UE must know which hybrid ARQ process a received ACK/NAK bit is associated with. Again, this is solved using the same approach as in HSDPA, that is the timing of the ACK/NAK is used to associate the ACK/NAK with a certain hybrid ARQ process. A well-defined time after reception of the uplink transport block on the E-DCH, the NodeB generates an ACK/NAK. Upon reception of a NAK, the UE performs a retransmission and the NodeB performs soft combining using incremental redundancy.

The handling of retransmissions, more specifically when to perform a retransmission, is one of the major differences between the hybrid ARQ operation in the uplink and the downlink (Figure 10.10). For HSDPA, retransmissions are scheduled as any other data and the NodeB is free to schedule the retransmission to the UE at any time instant and using a redundancy version of its choice. It may also address the hybrid ARQ processes in any order, that is it may decide to perform retransmissions for one hybrid ARQ process, but not for another process in the same UE. This type of operation is often referred to as adaptive asynchronous hybrid ARQ. Adaptive since the NodeB may change the transmission format and asynchronous since retransmissions may occur at any time after receiving the ACK/NAK.

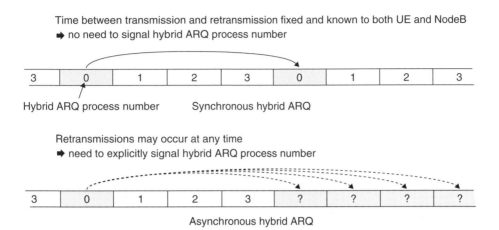

Figure 10.10 *Synchronous vs. asynchronous hybrid ARQ.*

For the uplink, on the other hand, a synchronous, non-adaptive hybrid ARQ operation is used. Hence, thanks to the synchronous operation, retransmissions occur a predefined time after the initial transmission, that is they are not explicitly scheduled. Likewise, the non-adaptive operation implies the transport format and redundancy version to be used for each of the retransmissions is also known from the time of the original transmission. Therefore, neither is there a need to explicitly scheduling the retransmissions nor is there a need for signaling the redundancy version the UE shall use. This is the main benefit of synchronous operation of the hybrid ARQ – minimizing the control signaling overhead. Naturally, the possibility to adapt the transmission format of the retransmissions to any changes in the channel conditions are lost, but as the uplink scheduler in the NodeB has less knowledge of the transmitter status – this information is located in the UE and provided to the NodeB using in-band signaling not available until the hybrid ARQ has successfully decoded the received data – than the downlink scheduler, this loss is by far outweighed by the gain in reduced control signaling overhead.

Apart from the synchronous vs. asynchronous operation of the hybrid ARQ protocol, the other main difference between uplink and downlink hybrid ARQ is the use of soft handover in the former case. In soft handover between different NodeBs, the hybrid ARQ protocol is terminated in multiple nodes, namely all the involved NodeBs. For HSDPA, on the other hand, there is only a single termination point for the hybrid ARQ protocol – the UE. In Enhanced Uplink, the UE therefore needs to receive ACK/NAK from all involved NodeBs. As it, from the UE perspective, is sufficient if at least one of the involved NodeBs receive the transmitted transport block correctly, it considers the data to be successfully delivered to the network if at least one of the NodeBs signals an ACK. This rule is sometimes called 'or-of-the-ACKs.' A retransmission occurs only if all involved NodeBs signal a NAK, indicating that none of them has been able to decode the transmitted data.

As known from the HSDPA description, the use of multiple parallel hybrid ARQ processes cannot itself provide in-sequence delivery and a reordering mechanism is required (Figure 10.11). For HSDPA, reordering is obviously located in the UE. The same aspect with out-of-sequence delivery is valid also for the uplink, which calls for a reordering mechanism also in this case. However, due to the support of soft handover, reordering cannot be located in the NodeB. Data transmitted in one hybrid ARQ process may be successfully decoded in one NodeB, while data transmitted in the next hybrid ARQ process may happen to be correctly decoded in another NodeB. Furthermore, in some situations, several involved NodeBs may succeed in decoding the same transport block. For these reasons, the reordering mechanism needs to have access to the transport blocks delivered from all involved NodeBs and therefore the reordering is located in the RNC. Reordering also removes any duplicates of transport blocks detected in multiple NodeBs.

The presence of soft handover in the uplink has also impacted the design of the control signaling. Similar to HSDPA, there is a need to indicate to the receiving end whether the soft buffer should be cleared, that is the transmission is an initial transmission, or if soft combining with the soft information stored from previous transmissions in this hybrid ARQ process should take place. HSDPA relies on a single-bit new-data indicator, for this purpose. If the NodeB misinterpreted an uplink NAK as an ACK and continues with the next packet, the UE can capture this error event by observing the single-bit 'new-data indicator' which is incremented for each new packet transmission. If the new-data indicator is incremented, the UE will clear the soft buffer, despite its contents were not successfully decoded, and try to decode the new transmission. Although a

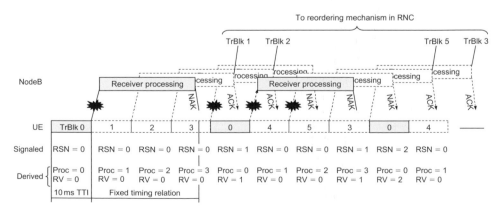

Figure 10.11   *Multiple hybrid ARQ processes for Enhanced Uplink.*

transport block is lost and has to be retransmitted by the RLC protocol, the UE does not attempt to soft combine coded bits originating from different transport blocks and therefore the soft buffer is not corrupted. Only if the uplink NAK *and* the downlink new-data indicator are *both* misinterpreted, which is a rare event, will the soft buffer be corrupted.

For Enhanced Uplink, a single-bit new-data indicator would work in absence of soft handover. Only if the downlink NAK *and* the uplink control signaling both are misinterpreted will the NodeB soft buffer be corrupted. However, in presence of soft handover, this simple method is not sufficient. Instead, a 2-bit *Retransmission Sequence Number* (RSN) is used for Enhanced Uplink. The initial transmission sets RSN to zero and for each subsequent transmission the RSN is incremented by one. Even if the RSN only can take values in the range of 0 to 3, any number of retransmissions is possible; the RSN simply remains at 3 for the third and later retransmissions. Together with the synchronous protocol operation, the NodeB knows when a retransmission is supposed to occur and with what RSN. The simple example in Figure 10.12 illustrates the operation. As the first NodeB acknowledged packet A, the UE continues with packet B, despite that the second NodeB did not correctly decode the packet. At the point of transmission of packet B, the second NodeB expects a retransmission of packet A, but due to the uplink channel conditions at this point in time, the second NodeB does not even detect the presence of a transmission. Again, the first NodeB acknowledge the transmission and the UE continues with packet C. This time, the second

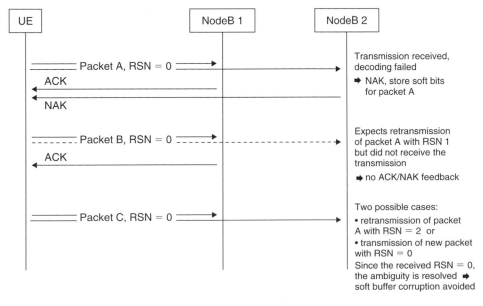

Figure 10.12 *Retransmissions in soft handover.*

NodeB does receive the transmissions and, thanks to the synchronous hybrid ARQ operation, can immediately conclude that it must be a transmission of a new packet. If it were a retransmission of packet A, the RSN would have been equal to two. This example illustrated the improved robustness from a 2-bit RSN together with a synchronous hybrid ARQ operation. A scheme with a single-bit 'new-data indicator,' which can be seen as a 1-bit RSN, would not have been able to handle the fairly common case of a missed transmission in the second NodeB. The new-data indicator would in this case be equal to zero, both in the case of a retransmission of packet A and in the case of an initial transmission of packet C, thereby leading to soft buffer corruption.

Soft combining in the hybrid ARQ mechanism for Enhanced Uplink is based on incremental redundancy. Generation of redundancy versions is done in a similar way as for HSDPA by using different puncturing patterns for the different redundancy versions. The redundancy version is controlled by the RSN according to a rule in the specifications, see further Section 10.3.2.

For Turbo codes, the systematic bits are of higher importance than the parity bits as discussed in Chapter 7. Therefore, the systematic bits should be included in the initial transmission to allow for decodability already after the first transmission attempts. Furthermore, for the best gain with incremental redundancy, the retransmissions should contain additional parity. This leads to a design where the initial transmission is self-decodable and includes all systematic bits as well as some parity bits, while the retransmission mainly contains additional parity bits not previously transmitted.

However, in soft handover, not all involved NodeBs may have received all transmissions. There is a risk that a NodeB did not receive the first transmission with the systematic bits, but only the parity bits in the retransmission. As this would lead to degraded performance, it is preferable if all redundancy versions used when in soft handover are self-decodable and contains the systematic bits. The above-mentioned rule used to map RSN into redundancy versions does this by making all redundancy versions self-decodable for lower data rates, which typically is used in the soft handover region at the cell edge, while using full incremental redundancy for the highest data rates, unlikely to be used in soft handover.

10.2.5   Physical channel allocation

The mapping of the coded E-DCH onto the physical channels is straightforward. As illustrated in Figure 10.13, the E-DCH is mapped to one or several E-DPDCHs, separate from the DPDCH. Depending on the E-TFC selected a different number

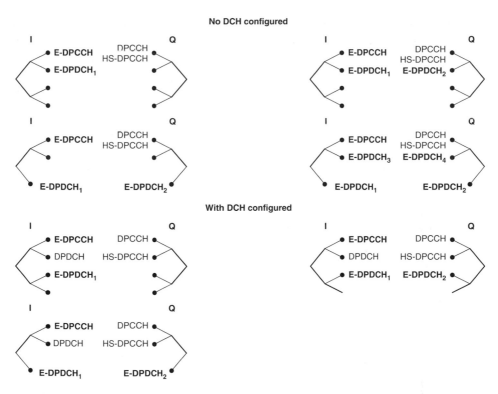

Figure 10.13 *Code allocation in case of simultaneous E-DCH and HS-DSCH operation (note that the code allocation is slightly different when no HS-DPCCH is configured). Channels with SF > 4 are shown on the corresponding SF4 branch for illustrative purposes.*

of E-DPDCHs is used. For the lowest data rates, a single E-DPDCH with a spreading factor inversely proportional to the data rate is sufficient.

To maintain backward compatibility, the mapping of the DPCCH, DPDCH, and HS-DPCCH remains unchanged compared to previous releases.

The order in which the E-DPDCHs are allocated is chosen to minimize the peak-to-average power ratio (PAR) in the UE, and it also depends on whether the HS-DPCCH and the DPDCH are present or not. The higher the PAR, the larger the back-off required in the UE power amplifier, which impact the uplink coverage. Hence, a low PAR is a highly desirable property. PAR is also the reason why SF2 is introduced as it can be shown that two codes of SF2 have a lower PAR than four codes of SF4. For the highest data rates, a mixture of spreading factors, $2 \times SF2 + 2 \times SF4$, is used. The physical channel configurations possible are listed in Table 10.1, and in Figure 10.13 the physical channel allocation with a simultaneous HS-DPCCH is illustrated.

Table 10.1 *Possible physical channel configurations. The E-DPDCH data rates are raw data rate, the maximum E-DCH data rate will be lower due to coding and limitations set by the UE categories.*

#DPCCH	#DPDCH	#HS-DPCCH	#E-DPCCH	#E-DPDCH	Comment
1	1–6	0 or 1	–	–	Rel 5 configurations
1	0 or 1	0 or 1	1	$1 \times SF \geq 4$	0.96 Mbit/s E-DPDCH raw data rate
1	0 or 1	0 or 1	1	$2 \times SF4$	1.92 Mbit/s E-DPDCH raw data rate
1	0 or 1	0 or 1	1	$2 \times SF2$	3.84 Mbit/s E-DPDCH raw data rate
1	0	0 or 1	1	$2 \times SF2 + 2 \times SF4$	5.76 Mbit/s E-DPDCH raw data rate

10.2.6 Power control

The E-DCH power control works in a similar manner as for the DCH and there is no change in the overall power control architecture with the introduction of the E-DCH. A single inner power control loop adjusts the transmission power of the DPCCH. The E-DPDCH transmission power is set by the E-TFC selection relative to the DPCCH power in a similar way as the DPDCH transmission power is set by the TFC selection. The inner loop power control located in the NodeB, bases its decision on the SIR target set by the outer loop power control located in the RNC.

The outer loop in earlier releases is primarily driven by the DCH BLER visible to the RNC. If a DCH is configured, the outer loop, which is an implementation-specific algorithm, may continue to operate on the DCH only. This approach works well as long as there are sufficiently frequent transmissions on the DCH, but the performance is degraded if DCH transmissions are infrequent.

If no DCH is configured, and possibly also if only infrequent transmissions occur at the DCH, information on the E-DCH transmissions need to be taken into account. However, due to the introduction of hybrid ARQ for the E-DCH, the residual E-DCH BLER may not be an adequate input for the outer loop power control. In most cases, the residual E-DCH BLER visible to the RNC is close to zero, which would cause the outer loop to lower the SIR target and potentially

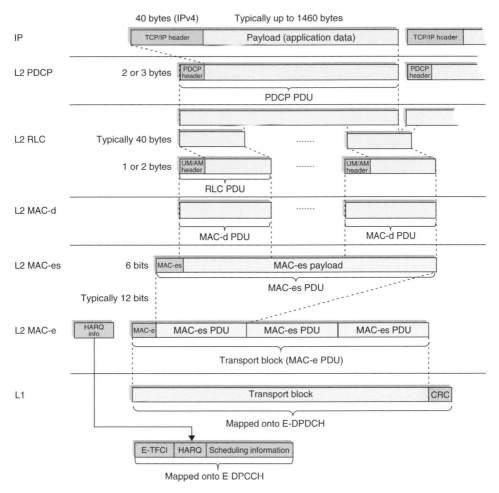

Figure 10.14 *Data flow.*

cause a loss of the uplink DPCCH if the residual E-DCH BLER alone is used as input to the outer loop mechanism. Therefore, to assist the outer loop power control, the number of retransmissions actually used for transmission of a transport block is signaled from the NodeB to the RNC. The RNC can use this information as part of the outer loop to set the SIR target in the inner loop.

10.2.7 Data flow

In Figure 10.14, the data flow from the application through all the protocol layers is illustrated in a similar way as for HSDPA. In this example, an IP service is assumed. The PDCP optionally performs IP header compression. The output from the PDCP is fed to the RLC. After possible concatenation, the RLC SDUs are segmented into smaller blocks of typically 40 bytes and an RLC header is attached. The RLC PDU is passed via the MAC-d, which is transparent for Enhanced Uplink, to the MAC-e. The MAC-e concatenates one or several MAC-d PDUs from one or several MAC-d flows and inserts MAC-es and MAC-e headers to form a transport block, which is forwarded on the E-DCH to the physical layer for further processing and transmission.

10.2.8   Resource control for E-DCH

Similar to HSDPA, the introduction of Enhanced Uplink implies that a part of the radio resource management is handled by the NodeB instead of the RNC. However, the RNC still has the overall responsibility for radio resource management, including admission control and handling of inter-cell interference. Thus, there is a need to monitor and control the resource usage of E-DCH channels to achieve a good balance between E-DCH and non-E-DCH users. This is illustrated in Figure 10.15.

For admission control purposes, the RNC relies on the *Received Total Wideband Power* (RTWP) measurement, which indicates the total uplink resource usage in the cell. Admission control may also exploit the *E-DCH provided bit rate*, which is a NodeB measurement reporting the aggregated served E-DCH bit rate per priority class. Together with the RTWP measurement, it is possible to design an admission algorithm evaluating the E-DCH scheduler headroom for a particular priority class.

To control the load in the cell, the RNC may signal an RTWP target to the NodeB in which case the NodeB should schedule E-DCH transmissions such that the RTWP is within this limit. The RNC may also signal a reference RTWP, which the NodeB may use to improve its estimate of the uplink load in the cell. Note that whether the scheduler uses an absolute measure, such as the RTWP, or

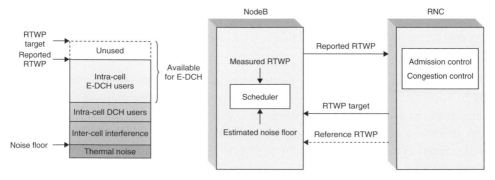

Figure 10.15   *Illustration of the resource sharing between E-DCH and DCH channels.*

a relative measure such as noise rise is not specified. Internally, the NodeB performs any measurements useful to a particular scheduler design.

To provide the RNC with a possibility to control the ratio between inter-cell and intra-cell interference, the RNC may signal a *Target Non-serving E-DCH to Total E-DCH Power Ratio* to the NodeB. The scheduler must obey this limitation when setting the overload indicator and is not allowed to suppress non-serving E-DCH UEs unless the target is exceeded. This prevents a cell to starve users in neighboring cells. If this was not the case, a scheduler could in principle permanently set the overload indicator to 'steal' resources from neighboring cells: a situation which definitely not is desirable.

Finally, the measurement *Transmitted carrier power of all codes not used for HS-PDSCH, HS-SCCH, E-AGCH, E-RGCH, or E-HICH transmission* also includes the E-DCH-related downlink control signaling.

10.2.9 Mobility

Active set management for the E-DCH uses the same mechanisms as for Release 99 DCH, that is the UE measures the signal quality from neighboring cells and informs the RNC. The RNC may then take a decision to update the active set. Note that the E-DCH active set is a subset of the DCH active set. In most cases, the two sets are identical, but in situations where only part of the network support E-DCH, the E-DCH active set may be smaller than the DCH active set as the former only includes cells capable of E-DCH reception.

Changing serving cell is performed in the same way as for HSDPA (see Chapter 9) as the same cell is the serving cell for both E-DCH and HS-DSCH.

10.2.10 UE categories

Similar to HSDPA, the physical layer UE capabilities have been grouped into six categories. Fundamentally, two major physical layer aspects, the number of supported channelization codes and the supported TTI values, are determined by the category number. The E-DCH UE categories are listed in Table 10.2. Support for 10 ms E-DCH TTI is mandatory for all UE categories, while only a subset of the categories support a 2 ms TTI. Furthermore, note that the highest data rate supported with 10 ms TTI is 2 Mbit/s. The reason for this is to limit the amount of buffer memory for soft combining in the NodeB; a larger transport block size translates into a larger soft buffer memory in case of retransmissions. A UE supporting E-DCH must also be able to support HS-DSCH.

Table 10.2   *E-DCH UE categories [99].*

E-DCH category	Max # E-DPDCHs, min SF	Supports 2 ms TTI	Max transport block size	
			10 ms TTI	2 ms TTI
1	1 × SF4	–	7110 (0.7 Mbit/s)	–
2	2 × SF4	Y	14 484 (1.4 Mbit/s)	2798 (1.4 Mbit/s)
3	2 × SF4	–	14 484 (1.4 Mbit/s)	–
4	2 × SF2	Y	20 000 (2 Mbit/s)	5772 (2.9 Mbit/s)
5	2 × SF2	–	20 000 (2 Mbit/s)	–
6	2 × SF4 + 2 × SF2	Y	20 000 (2 Mbit/s)	11 484(5.74 Mbit/s)

10.3   Finer details of Enhanced Uplink

10.3.1   *Scheduling – the small print*

The use of a serving grant as a means to control the E-TFC selection has already been discussed, as has the use of absolute and relative grants to update the serving grant. Absolute grants are transmitted on the shared E-AGCH physical and are used for absolute changes of the serving grant as already stated. In addition to conveying the maximum E-DPDCH-to-DPCCH power ratio, the E-AGCH also contains an activation flag, whose usage will be discussed below. Obviously, the E-AGCH is also carrying the identity of the UE for which the E-AGCH information is intended. However, although the UE receives only *one* E-AGCH, it is assigned *two* identities, one primary and one secondary. The primary identity is UE specific and unique for each UE in the cell, while the secondary identity is a group identity shared by a group of UEs. The reason for having two identities is to allow for scheduling strategies based on both common, or group-wise, scheduling, where multiple terminals are addressed with a single identity and individual per-UE scheduling (Figure 10.16).

Common scheduling means that multiple terminals are assigned the same identity; the secondary identity is common to multiple UEs. A grant sent with the secondary identity is therefore valid for multiple UEs and each of these UEs may transmit up to the limitation set by the grant. Hence, this approach is suitable for scheduling strategies not taking the uplink radio conditions into account, for example CDMA-like strategies where scheduler mainly strives to control the total cell interference level. A low-load condition is one such example. At low cell load, there is no need to optimize for capacity. Optimization can instead focus on minimizing the delays by assigning the same grant level to multiple UEs using the secondary identity. As soon as a UE has data to transmit, the

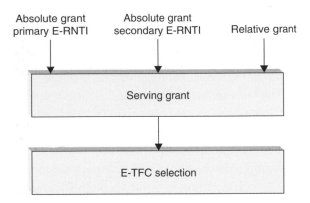

Figure 10.16 *The relation between absolute grant, relative grant and serving grant.*

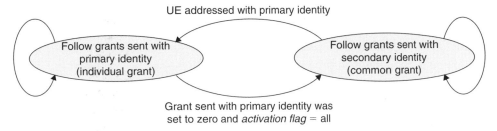

Figure 10.17 *Illustration of UE monitoring of the two identities.*

UE can directly transmit up to the common grant level. There is no need to go through a request phase first, as the UE already has been assigned a non-zero serving grant. Note that multiple UEs may, in this case, transmit simultaneously, which must be taken into account when setting the serving grant level.

Individual per-UE scheduling provides tighter control of the uplink load and is useful to maximize the capacity at high loads. The scheduler determines which user that is allowed to transmit and set the serving grant of the intended user by using the primary identity, unique for a specific UE. In this case, the UEs resource utilization is individually controlled, for example to exploit advantageous uplink channel conditions. The greedy filling strategy discussed earlier is one example of a strategy requiring individual grants.

Which of the two identities, the primary or the secondary, a UE is obeying can be described by a state diagram, illustrated in Figure 10.17. Depending on the state the UE is in, it follows either grants sent with the primary or the secondary identity. Addressing the UE with its unique primary identity causes the UE to stop obeying grants sent using the secondary common identity. There is also a mechanism to force the UE back to follow the secondary, common grant level. The usefulness of this is best illustrated with the example below.

Consider the example in Figure 10.18, illustrating the usage of common and dedicated scheduling. The UEs are all initialized to follow the secondary identity and a suitable common grant level is set using the secondary identity. Any UE that has been assigned a grant level using the secondary identity may transmit using a data rate up to the common grant level; a level that is adjusted as the load in the system varies, for example due to non-scheduled transmissions. As time evolves, UE #1 is in need of a high data rate to upload a huge file. Note that UE #1 may start the upload using the common grant level while waiting for the scheduler to grant a higher level. The scheduler decides to lower the common grant level using the secondary, common identity to reduce the load from other UEs. A large grant is sent to UE #1 using UE #1's primary and unique identity to grant UE #1 a high data rate (or, more accurately, a higher E-DPDCH-to-DPCCH power ratio). This operation also causes UE #1 to enter the 'primary' state in Figure 10.17. At a later point in time, the scheduler decides send a zero grant to UE #1 with activation flag set to *all*, which forces UE #1 back to follow the secondary identity (back to common scheduling).

From this example, it is seen that the two identities each UE is assigned – one primary, UE-specific identity; and one secondary, common identity – facilitates a flexible scheduling strategy.

10.3.1.1 *Relative grants*
Relative grants from the serving cell can take one of the values 'up,' 'down,' and 'no change.' This is used to fine-tune an individual UE's resource utilization as already discussed. To implement the increase (decrease) of the serving grant, the UE maintains a table of possible E-DPDCH-to-DPCCH power ratios as

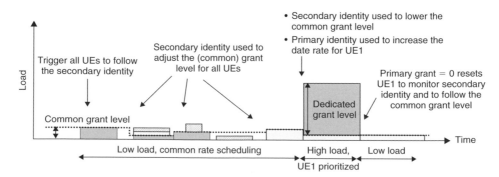

Figure 10.18 *Example of common and dedicated scheduling.*

illustrated in Figure 10.19. The up/down commands corresponds to an increase/ decrease of power ratio in the table by one step compared to the power ratio used in the previous TTI in the same hybrid ARQ process. There is also a possibility to have a larger increase (but not decrease) for small values of the serving grant. This is achieved by (through RRC signaling) configuring two thresholds in the E-DPDCH-to-DPCCH power ratio table, below which the UE may increase the serving grant by three and two steps, respectively, instead of only a single step. The use of the table and the two thresholds allow the network to increase the serving grant efficiently without extensive repetition of relative grants for small data rates (small serving grants) and at the same time avoiding large changes in the power offset for large serving grants.

The 'overload indicator' (relative grant from non-serving cells) is used to control the inter-cell interference (in contrast to the grants from the serving cell which provide the possibility to control the intra-cell interference). As previously described, the overload indicator can take two values: 'down' or 'DTX,' where the latter does not affect the UE operation. If the UE receives 'down' from any of the non-serving cells, the serving grant is decreased relative to the previous TTI in the same hybrid ARQ process.

'Ping-pong effects' describe a situation where the serving grant level in a UE starts to oscillate. One example when this could happen is if a non-serving cell

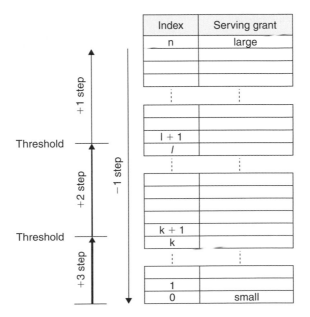

Figure 10.19   *Grant table.*

requests the UEs to lower their transmission power (and hence data rate) due to a too high interference level in the non-served cell. When the UE has reacted to this, the serving cell will experience a lower interference level and may decide to increase the grant level in the UE. The UE utilizes this increased grant level to transmit at a higher power, which again triggers the overload indicator from the non-serving cell and the process repeats.

To avoid 'ping-pong effects,' the UE ignores any 'up' commands from the serving cell for one hybrid ARQ roundtrip time after receiving an 'overload indicator.' During this time, the UE shall not allow the serving grant to increase beyond the limit resulting from the 'overload indicator.' This avoids situations where the non-serving cell reduces the data rate to avoid an overload situation in the non-serving cell, followed by the serving cell increasing the data rate to utilize the interference headroom suddenly available, thus causing the overload situation to reappear. The serving cell may also want to be careful with immediately increasing the serving grant to is previous level as the exact serving grant in the UE is not known to the serving cell in this case (although it may partly derive it by observing the happy bit). Furthermore, to reduce the impact from erroneous relative grants, the UE shall ignore relative grants from the serving cell during one hybrid ARQ roundtrip time after having received an absolute grant with the primary identity.

10.3.1.2   Per-process scheduling

Individual hybrid ARQ processes can be (de)activated, implying that not all processes in a UE are available for data transmission as illustrated in Figure 10.20. Process (de)activation is only possible for 2 ms TTI, for 10 ms TTI all processes are permanently enabled. The reason for process deactivation is mainly to be able to limit the data rate in case of 2 ms E-DCH TTI (with 320-bit RLC PDUs and 2 ms TTI, the minimum data rate is 160 kbit/s unless certain processes are disabled), but it can also be used to enable TDMA-like operation in case of all UEs uplink transmissions being synchronized. Activation of individual processes can either be done through RRC signaling or by using the activation flag being part of

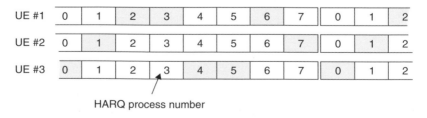

Figure 10.20   *Example of activation of individual hybrid ARQ processes.*

the absolute grant. The activation flag indicates whether only the current process is activated or whether all processes not disabled by RRC signaling are activated.

Non-scheduled transmission can be restricted to certain hybrid ARQ processes. The decision is taken by the serving NodeB and informed to the other NodeBs in the active set through the RNC. Normally, the scheduler needs to operate with a certain margin to be able to handle any non-scheduled transmissions that may occur and restricting non-scheduled transmissions to certain processes can therefore allow the scheduler to operate with smaller margins in the remaining processes.

10.3.1.3 Scheduling requests

The scheduler needs information about the UE status and, as already discussed, two mechanisms are defined in Enhanced Uplink to provide this information: the in-band *scheduling information* and the out-band *happy bit*.

In-band scheduling information can be transmitted either alone or in conjunction with uplink data transmission. From a baseband perspective, scheduling information is no different from uplink user data. Hence, the same baseband processing and hybrid ARQ operation is used.

In case of a standalone scheduling information, the E-DPDCH-to-DPCCH power offset to be used is configured by RRC signaling. To ensure that the scheduling information reaches the scheduler, the transmission is repeated until an ACK is received from the *serving* cell (or the maximum number of transmission attempts is reached). This is different from a normal data transmission, where an ACK from *any* cell in the active set is sufficient.

In case of simultaneous data transmission, the scheduling information is transmitted using the same hybrid ARQ profile as the highest-priority MAC-d flow in the transmission (see Section 10.3.1.5 for a discussion on hybrid ARQ profiles). In this case, periodic triggering will be relied upon for reliability. Scheduling information can be transmitted using any hybrid ARQ process, including processes deactivated for data transmission. This is useful to minimize the delay in the scheduling mechanism.

The in-band scheduling information consists of 18 bits, containing information about:

- Identity of the highest-priority logical channel with data awaiting transmission, 4 bits.

- Buffer occupancy, 5 bits indicating the total number of bytes in the buffer and 4 bits indicating the fraction of the buffer occupied with data from the highest-priority logical channel.
- Available transmission power relative to DPCCH, 5 bits.

Since the scheduling information contains information about both the total number of bits and the number of bits in the highest-priority buffer, the scheduler can ensure that UEs with high-priority data is served before UEs with low-priority data, a useful feature at high loads. The network can configure for which flows scheduling information should be transmitted.

Several rules when to transmit scheduling information are defined. These are:

- If padding allows transmission of scheduling information. Clearly, it makes sense to fill up the transport block with useful scheduling information rather than dummy bits.
- If the serving grant is zero or all hybrid ARQ processes are deactivated and data enters the UE buffer. Obviously, if data enters the UE but the UE has no valid grant for data transmission, a grant should be requested.
- If the UE does have a grant, but incoming data has higher priority than the data currently in the buffer. The presence of higher-priority data should be conveyed to the NodeB as it may affect its decision to scheduler the UE in question.
- Periodically as configured by RRC signaling (although scheduling information is not transmitted if the UE buffer size equals zero).
- At cell change to provide the new cell with information about the UE status.

10.3.1.4   NodeB hardware resource handling in soft handover

From a NodeB internal hardware allocation point of view, there is a significant difference between the serving cell and the non-serving cells in soft handover: the serving cell has information about the scheduling grant sent to the UE and, therefore, knowledge about the maximum amount of hardware resources needed for processing transmissions from this particular UE, information that is missing in the non-serving cells. Internal resource management in the non-serving cells therefore requires some attention when designing the scheduler. One possibility is to allocate sufficient resources for the highest possible data rate the UE is capable of. Obviously, this does not imply any restrictions to the data rates the serving cell may schedule, but may, depending on the implementation, come at a cost of less efficient usage of processing resources in the non-serving NodeBs. To reduce this cost, the highest data rates the scheduler is allowed to assign could be limited by the scheduler design. Alternatively, the non-serving NodeB may under-allocate processing resources, knowing that it may not be able to decode the first few TTIs of a UE transmission. Once the UE starts to transmit at a high data rate, the non-serving NodeB can reallocate resources to this UE, assuming that it will continue to transmit for some time. Non-serving cells may also try to listen to the scheduling requests from the UE to the serving cell to get some information about the amount of resources the UE may need.

10.3.1.5 Quality-of-service support

The scheduler operates per UE, that is it determines which UE that is allowed to transmit and at what maximum resource consumption. However, each UE may have several different flows of different priority. For example, VoIP and RRC signaling typically have a higher priority than a background file upload. Since the scheduler operates per UE, the control of different flows within a UE is *not* directly controlled by the scheduler. In principle, this could be possible, but it would increase the amount of downlink control signaling. For Enhanced Uplink, an E-TFC-based mechanism for quality-of-service support has been selected. Hence, as described earlier, the scheduler handles resource allocation *between UEs,* while the E-TFC selection controls resource allocation between flows *within the UE.*

The basis for QoS support is so-called hybrid ARQ profiles, one per MAC-d flow in the UE. A hybrid ARQ profile consists of a power offset attribute and a maximum number of transmissions allowed for a MAC-d flow.

The power offset value is used to determine the hybrid ARQ operating point, which is directly related to the number of retransmissions. In many cases, several retransmissions may fit within the allowed delay budget. Exploiting multiple transmission attempts together with soft combining is useful to reduce the cost of transmitting at a certain data rate as discussed in conjunction with hybrid ARQ.

However, for certain high-priority MAC-d flows, the delays associated with multiple hybrid ARQ retransmissions may not be acceptable. This could, for example, be the case for RRC signaling such as handover messages for mobility. Therefore, for these flows, it is desirable to increase the E-DPDCH transmission power, thereby increasing the probability for the data to be correctly received at the first transmission attempt. This is achieved by configuring a higher power offset for hybrid ARQ profiles associated with high-priority flows. Of course, the transmission must be within the limits set by the serving grant. Therefore, the payload is smaller when transmitting high-priority data with a larger power offset than when transmitting low-priority data with a smaller power offset.

The power needed for the transmission of an E-DCH transport block is calculated including the power offset obtained from the hybrid ARQ profile for flow

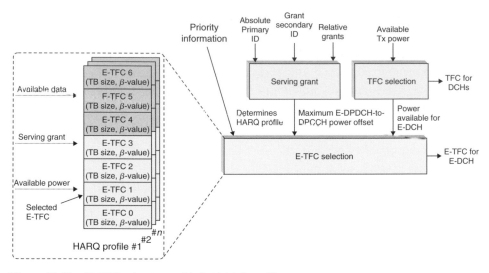

Figure 10.21   *E-TFC selection and hybrid ARQ profiles.*

to be transmitted. The required transmit power for each possible transport block size can then be calculated by adding (in dB) the E-DPDCH-to-DPCCH power offset given by the transport block size and the power offset associated with the hybrid ARQ profile. The UE then selects the largest possible payload, taking these power offsets into account, which can be transmitted within the power available for the E-DCH (Figure 10.21).

Absolute priorities for logical channels are used, that is the UE maximizes the data rate for high-priority data and only transmits data from a low priority in a TTI if all data with higher priority has been transmitted. This ensures that any high-priority data in the UE is served before any low-priority data.

If data from more than one MAC-d flow is included in a TTI, the power offset associated with the MAC-d flow with the logical channel with the highest priority shall be used in the calculation. Therefore, if multiple MAC-d flows are multiplexed within a given transport block, the low-priority flows will get a 'free ride' in this TTI when multiplexed with high-priority data.

There are two ways of supporting guaranteed bit rate services: scheduled and non-scheduled transmissions. With scheduled transmission, the NodeB schedules the UE sufficiently frequent and with sufficiently high bit rate to support the guaranteed bit rate. With non-scheduled transmission, a flow using non-scheduled transmission is defined by the RNC and configured in the UE through RRC signaling. The UE can transmit data belonging to such a flow without first receiving any scheduling grant. An advantage with the scheduled approach is that the network has more control of the interference situation and the power required for downlink ACK/NAK signaling and may, for example, allocate a high bit rate during a fraction of the time while still maintaining the guaranteed bit rate in average. Non-scheduled transmissions, on the other hand, are clearly needed at least for transmitting the scheduling information in case the UE does not have a valid scheduling grant.

10.3.2 Further details on hybrid ARQ operation

Hybrid ARQ for Enhanced Uplink serves a similar purpose as for the HSDPA – to provide robustness against occasional transmissions errors. It can also, as already discussed, be used to increase the link efficiency by targeting multiple hybrid ARQ retransmission attempts.

The hybrid ARQ for the E-DCH operates on a single transport block, that is whenever the E-DCH CRC indicates an error, the MAC-e in the NodeB can request a retransmission representing the same information as the original transport block. Note that there is a single transport block per TTI. Thus it is not possible to mix initial transmission and retransmissions within the same TTI.

Incremental redundancy is used as the basic soft combining mechanism, that is retransmissions typically consists of a different set of coded bits than the initial transmission. Note that, per definition, the set of information bits must be identical between the initial transmission and the retransmissions. For Enhanced Uplink, this implies that the E-DCH transport format, which is defined by the transport block size and includes the number of physical channels and their spreading factors, remains unchanged between transmission and retransmission. Thus, the number of channel bits is identical for the initial transmission and the retransmissions. However, the rate matching pattern will change in order to implement incremental redundancy. The transmission power may also be different for different transmission attempts, for example, due to DCH activity.

The physical layer processing supporting the hybrid ARQ operation is similar to the one used for HS-DSCH, although only a single rate matching stage is used. The reason for two-stage rate matching for HS-DSCH was to handle memory limitations in the UE, but for the E-DCH, any NodeB memory limitations can be handled by proper network configuration. For example, the network could restrict the number of E-TFCs in the UE such that the UE cannot transmit more bits than the NodeB can buffer.

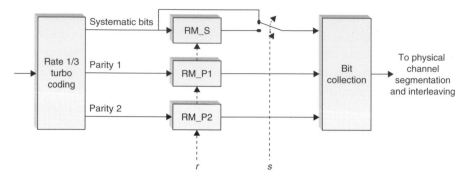

Figure 10.22   *E-DCH rate matching and the r and s parameters. The bit collection procedure is identical to the QPSK bit collection for HS-DSCH.*

The purpose of the E-DCH rate matching, illustrated in Figure 10.22, is twofold:

- To match the number of coded bits to the number of physical channel bits on the E-DPDCH available for the selected E-DCH transport format.
- To generate different sets of coded bits for incremental redundancy as controlled by the two parameters r and s as described below.

The number of physical channel bits depends on the spreading factor and the number of E-DPDCHs allocated for a particular E-DCH transport format. In other words, part of the E-TFC selection is to determine the number of E-DPDCHs and their respective spreading factors. From a performance perspective, coding is always better than spreading and, preferably, the number of channelization codes should be as high as possible and their spreading factor as small as possible. This would avoid puncturing and result in full utilization of the rate 1/3 mother Turbo code. At the same time, there is no point in using a lower spreading factor than necessary to reach rate 1/3 as this only would lead to excessive repetition in the rate matching block. Furthermore, from an implementation perspective, the number of E-DPDCHs should be kept as low as possible to minimize the processing cost in the NodeB receiver as each E-DPDCH requires one set of de-spreaders.

To fulfill these, partially contradicting requirements, the concept of *Puncturing Limit* (PL) is used to control the maximum amount of puncturing the UE is allowed to perform. The UE will select an as small number of channelization codes and as high spreading factor as possible without exceeding the puncturing limits, that is not puncture more than a fraction of $(1 - \text{PL})$ of the coded bits. This is illustrated in Figure 10.23, where it is seen that puncturing is allowed up to a limit before additional E-DPDCHs are used. Two puncturing limits, PL_{max} and $\text{PL}_{\text{non-max}}$, are defined. The limit PL_{max} is determined by the UE category and is used if the number of E-DPDCHs and their spreading factor is equal to the UE capability and the UE therefore cannot increase the number of E-DPDCHs.

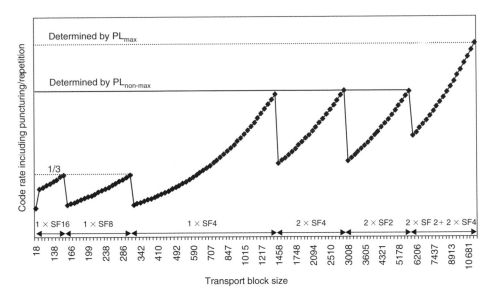

Figure 10.23 *Amount of puncturing as a function of the transport block size.*

Otherwise, PL$_{non-max}$, which is signaled to the UE at the setup of the connection, is used. The use of two different puncturing limits, instead of a single one as for the DCH, allows for a higher maximum data rate as more puncturing can be applied for the highest data rates. Typically, additional E-DPDCHs are used when the code rate is larger than approximately 0.5. For the highest data rates, on the other hand, a significantly larger amount of puncturing is necessary as it is not possible to further increase the number of codes.

The puncturing (or repetition) is controlled by the two parameters r and s in the same way as for the second HS-DSCH rate matching stage (Figure 10.22). If $s = 1$, systematic bits are prioritized and an equal amount of puncturing is applied to the two streams of parity bits, while if $s = 0$, puncturing is primarily applied to the systematic bits. The puncturing pattern is controlled by the parameter r. For the initial transmission attempt r is set to zero and is increased for the retransmissions. Thus, by varying r, multiple, possibly partially overlapping, sets of coded bits representing the same set of information bits can be generated. Note that a change in r also affects the puncturing pattern, even if r is unchanged, as different amounts of systematic and parity bits will be punctured for the two possible values of s.

Equal repetition for all three streams is applied if the number of available channel bits is larger than the number of bits from the Turbo coder, otherwise puncturing is applied. Unlike the DCH, but in line with the HS-DSCH, the E-DCH rate matching may puncture the systematic bits as well and not only the parity bits. This is used for incremental redundancy, where some retransmissions contain mainly parity bits.

The values of s and r are determined from the Redundancy Version (RV), which in turn is linked to the Retransmission Sequence Number (RSN). The RSN is set to zero for the initial transmission and incremented by one for each retransmission as described earlier.

Compared to the HS-DSCH, one major difference is the support for soft handover on the E-DCH. As briefly mentioned above, not all involved cells may receive all transmission attempts in soft handover. Self-decodable transmissions, $s = 1$, are therefore beneficial in these situations as the systematic bits are more important than the parity bits for successful decoding. If full incremental redundancy is used in soft handover, there is a possibility that the first transmission attempt, containing the systematic bits ($s = 1$), is not reliably received in a cell, while the second transmission attempt, containing mostly parity bits ($s = 0$), is received. This could result in degraded performance. However, the data rates in soft handover are typically somewhat lower (the code rate is lower) as the UE in most cases are far from the base station when entering soft handover. Therefore, the redundancy versions are defined such that all transmissions are self-decodable ($s = 1$) for transport formats where the initial code rate is less than 0.5, while the remaining transport formats include retransmissions that are not self-decodable. Thus, thanks to this design, self-decodability 'comes for free' when in soft handover. The design is also well matched to the fact that incremental redundancy (i.e., $s = 0$ for some of the retransmissions) provides most of the gain when the initial code rate is high.

The mapping from RSN via RV to the r and s parameters, illustrated in Figure 10.24, is mandated in the specification and is not configurable with the exception that higher layer signaling can be used to mandate the UE to always use RV = 0,

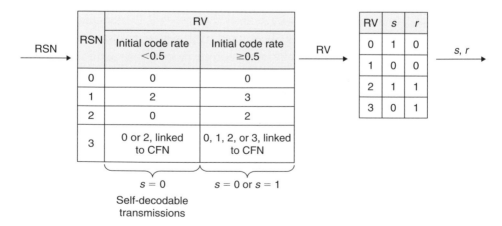

Figure 10.24   *Mapping from RSN via RV to s and r.*

regardless of the RSN. This implies that the retransmission consists of exactly the same coded bits as the initial transmission (Chase combining), and can be used if the memory capabilities of the NodeB are limited. Note that, for RSN = 3, the RV is linked to the (sub)frame number. The reason is to allow for variations in the puncturing pattern even for situations when more than three retransmissions are used.

10.3.2.1 Protocol operation

The hybrid ARQ protocol uses multiple stop-and-wait hybrid ARQ processes similar to HS-DSCH. The motivation is to allow continuous transmission, which cannot be achieved with a single stop-and-wait scheme, while at the same time having some of the simplicity of a stop-and-wait protocol.

As already touched upon several times, the support for soft handover is one of the major differences between the uplink and the downlink. This has also impacted the number of hybrid ARQ processes. For HSDPA, this number is configurable to allow for different NodeB implementations. Although the same approach could be taken for Enhanced Uplink, soft handover between two NodeBs from different vendors would be complicated. In soft handover, all involved NodeBs need to use the same number of hybrid ARQ processes, which partially removes the flexibility of having a configurable number as all NodeBs must support at least one common configuration. To simplify the overall structure, a fixed number of hybrid ARQ processes to be used by all NodeBs are defined. The number of processes is strongly related to the timing of the ACK/NAK transmission in the downlink (see the discussion on control signaling timing for details). For the two TTIs of 10 and 2 ms, the number of processes, N_{HARQ}, is 4 and 8, respectively. This results in a total hybrid ARQ roundtrip time of 4.10 = 40 and 8.2 = 16 ms, respectively.

The use of a synchronous hybrid ARQ protocol is a distinguishing feature compared to HSDPA. In a synchronous scheme, the hybrid ARQ process number is derived from the (sub)frame number and is not explicitly signaled. This implies that the transmissions in a given hybrid ARQ process can only be made once every N_{HARQ} TTI. This also implies that a retransmission (when necessary) always occur N_{HARQ} TTIs after the previous (re)transmission. Note that this does not affect the delay until a first transmission can be made since a data transmission can be started in any available process. Once the transmission of data in a process has started, retransmissions will be made until either an ACK is received or the maximum number of retransmissions has been reached (the maximum number of retransmissions is configurable by the RNC via RRC signaling). The retransmissions are done without the need for scheduling grants; only the initial transmission needs to be scheduled. As the scheduler in the NodeB is aware of whether a retransmission is expected or not, the interference from the (non-scheduled) retransmissions can be taken into account when forming the scheduling decision for other users.

10.3.2.2 In-sequence delivery

Similar to the case for HS-DSCH, the multiple hybrid ARQ processes of E-DCH cannot, in themselves, ensure in-sequence delivery, as there is no interaction between the processes. Also, in soft handover situations, data is received independently in several NodeBs and can therefore be received in the RNC in a different order than transmitted. In addition, differences in Iub/Iur transport delay can cause out-of sequence delivery to RLC. Hence, in-sequence delivery must be implemented on top of the MAC-e entity and a reordering entity in the RNC has been defined for this purpose in a separate MAC entity, the MAC-es. In E-DCH, the reordering is always performed per logical channel such that all data for a logical channel is delivered in-sequence to the corresponding RLC entity. This can be compared to HS-DSCH where the reordering is performed in configurable reordering queues.

The actual mechanism to perform reordering in the RNC is implementation specific and not standardized, but typically similar principles as specified for the HS-DSCH are used. Therefore, each MAC-es PDU transmitted from the UE includes a *Transmission Sequence Number* (TSN), which is incremented for each transmission on a logical channel. By ordering the MAC-es PDUs based on TSN, in-sequence delivery to the RLC entities is possible.

To illustrate the reordering mechanism consider the situation shown in Figure 10.25. The MAC-es PDUs 0, 2, 3, and 5 have been received in the RNC while MAC-es PDUs 1 and 4 have not yet been received. The RNC can in this situation not know why PDUs 1 and 4 are missing and needs to store PDUs 2, 3, and 5 in the reordering buffer. As soon as PDU 1 arrives, PDU 1, 2, and 3 can be delivered to RLC.

The reordering mechanism also needs to handle the situation where PDUs are permanently lost due to, for example, loss over Iub, errors in the hybrid ARQ

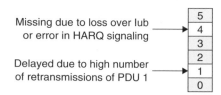

Figure 10.25   *Reordering mechanism.*

signaling, or in case the maximum number of retransmissions was reached without successful decoding. In those situations a stall avoidance mechanism is needed, that is a mechanism to prevent that the reordering scheme waits for PDUs that never will arrive. Otherwise, PDU 5 in Figure 10.25 would never be forwarded to RLC.

Stall avoidance can be achieved with a timer similar to what is specified for the UE in HS-DSCH. The stall avoidance timer delivers packets to the RLC entity if a PDU has been missing for a certain time. If the stall avoidance mechanism delivers PDUs to the RLC entity too early, it may result in unnecessary RLC retransmissions when the PDU is only delayed, for example, due to too many hybrid ARQ retransmissions. If, on the other hand, the PDUs are kept too long in the reordering buffer, it will also degrade the performance since the delay will increase.

To improve the stall avoidance mechanism, the NodeB signals the time (frame and subframe number) when each PDU was correctly decoded to the RNC, as well as how many retransmissions were needed before the PDU was successfully received. The RNC can use this information to optimize the reordering functionality. Consider the example in Figure 10.25. If PDU 5 in the example above needed 4 retransmissions and the maximum number of retransmission attempts configured equals 5, the RNC knows that if PDU 4 has not arrived within one hybrid ARQ roundtrip time (plus some margin to consider variations in Iub delay) after PDU 5, it is permanently lost. In this case, the RNC only have to wait one roundtrip time before delivering PDU 5 to RLC.

10.3.2.3 MAC-e and MAC-es headers

To support reordering and de-multiplexing of the PDUs from different MAC-d flows, the appropriate information is signaled in-band in the form of MAC-es and MAC-e headers. The structures of the MAC-e/es headers are illustrated in Figure 10.26.

Several MAC-d PDUs of the same size and from the same logical channel are concatenated. The *Data Description Indicator* (DDI) provides information about the logical channel from which the MAC-d PDUs belong, as well as the size of the MAC-d PDUs. The number of MAC-d PDUs is indicated by N. The Transmission Sequence Number (TSN), used to support reordering as described in the previous section, is also attached to the set of MAC-d PDUs.

The MAC-e header consists of a number of *DDI* and N pairs. A mapping is provided by RRC from the *DDI* field to a MAC-d PDU size, logical channel ID and MAC-d flow ID. The logical channel also uniquely identifies the reordering queue since reordering in E-DCH is performed per logical channel.

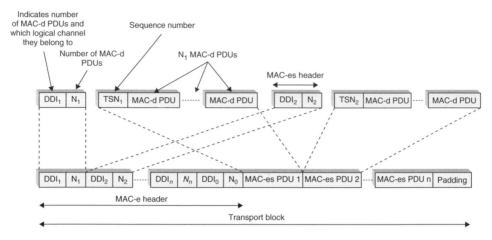

Figure 10.26   *Structure and format of the MAC-e/es PDU.*

The sequence of *DDI* and *N* fields is completed with a predefined value of *DDI* to indicate the end of the MAC-e header. After the MAC-e header follows a number of MAC-es PDUs, where the number of MAC-es PDUs is the same as the number of *DDI* and *N* pairs in the MAC-e header (not counting the predefined DDI value indicating the end of the MAC-e header). After the last MAC-es PDU, the MAC-e PDU may contain padding to fit the current transport block size.

When appropriate, the MAC-e header also includes 18 bits of scheduling information using a special DDI value.

10.3.3   Control signaling

To support E-DCH transmissions in the uplink, three downlink channels carrying out-band control signaling are defined:

- The E-HICH is a dedicated physical channel transmitted from each cell in the active set and used to carry the hybrid ARQ acknowledgments.
- The E-AGCH is a shared physical channel transmitted from the serving cell only and used to carry the absolute grants.
- The E-RGCH carries relative grants. From the serving cell, the E-RGCH is a dedicated physical channel, carrying the relative grants. From non-serving cells, the E-RGCH is a common physical channel, carrying the overload indicator.

Thus, a single UE will receive multiple downlink physical control channels. From the serving cell, the UE receives the E-HICH, E-AGCH, and E-RGCH. From each of the non-serving cells, the UE receives the E-HICH and the E-RGCH.

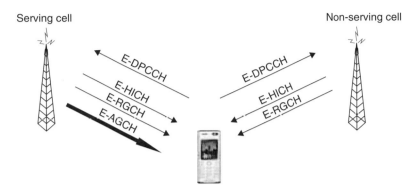

Figure 10.27 *E-DCH-related out-band control signaling.*

Out-band uplink control signaling is also required to indicate the E-TFC the UE selected, the RSN, and the happy bit. This information is carried on the uplink E-DPCCH.

In addition to the E-DCH-related out-band control signaling, downlink control signaling for transmission of power control bits is required. This is no different from WCDMA in general and is carried on the (F-)DPCH. Similarly, the DPCCH is present in the uplink to provide a phase reference for coherent demodulation as well. The overall E-DCH-related out-band control signaling is illustrated in Figure 10.27.

10.3.3.1 E-HICH

The E-HICH is a downlink dedicated physical channel, carrying the binary hybrid ARQ acknowledgments to inform the UE about the outcome of the E-DCH detection at the NodeB. The NodeB transmits either ACK or NAK, depending on whether the decoding of the corresponding E-DCH transport block was successful or a retransmission is requested. To not unnecessarily waste downlink transmission power, nothing is transmitted on the E-HICH if the NodeB did not detect a transmission attempt, that is no energy was detected on the E-DPCCH or the E-DPDCH.

Despite the fact that the ACK/NAK is a single bit of information, the ACK/NAK is transmitted with a duration of 2 or 8 ms, depending on the TTI configured.[3] This ensures that a sufficient amount of energy can be obtained to satisfy the relatively stringent error requirements of the ACK/NAK signaling, without requiring a too high peak power for the E-HICH.

---

[3] The reason for 8 ms and not 10 ms is to provide some additional processing time in the NodeB. See the timing discussion for further details.

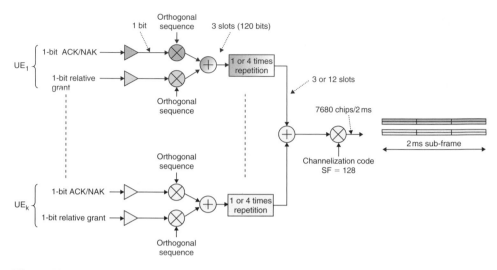

Figure 10.28 *E-HICH and E-RGCH structures (from the serving cell).*

To save channelization codes in the downlink, multiple ACK/NAKs are transmitted on a single channelization code of spreading factor 128. Each user is assigned an orthogonal signature sequence to generate a signal spanning 2 or 8 ms as illustrated in Figure 10.28. The single-bit ACK/NAK is multiplied with a signature sequence of length 40 bits,[4] which equals one slot of bits at the specified spreading factor of 128. The same procedure is used for 3 or 12 slots, depending on the E-DCH TTI, to obtain the desired signaling interval of 2 or 8 ms. This allows multiple UEs to share a single channelization code and significantly reduces amount of channelization codes that needs to be assigned for E-HICH.

As the mutual correlation between different signature sequences varies with the sequence index, signature sequence hopping is used to average out these differences. With hopping, the signature sequence of a certain UE changes from slot to slot using a hopping pattern[5] as illustrated in Figure 10.29.

Both the E-HICH and the E-RGCH use the same structure and to simplify the UE implementation, the E-RGCH and E-HICH for a certain UE shall be allocated the same channelization code and scrambling code. Thus, with length 40 signature sequences, 20 users, each with 1 E-RGCH and 1 E-HICH, can share a single channelization code. Note that the power for different users' E-HICH and E-RGCH can be set differently, despite the fact that they share the same code.

---

[4] In essence, this is identical to defining a spreading factor of $40 \cdot 128$ is $= 5120$.

[5] The use of hopping could also be expressed as a corresponding three-slot-long signature sequence.

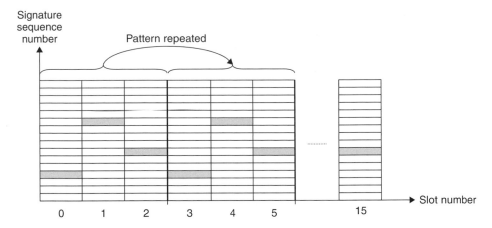

Figure 10.29 *Illustration of signature sequence hopping.*

When a single NodeB is handling multiple cells and a UE is connected to several of those cells, that is, the UE is in softer handover between these cells, it is reasonable to assume that the NodeB will transmit the same ACK/NAK information to the UE in all these cells. Hence, the UE shall perform soft combining of the E-HICH in this case and the received signal on each of the E-HICH-es being received from the same NodeB shall be coherently added prior to decoding. This is the same approach as used for combining of power control bits already from the first release of WCDMA.

The modulation scheme used for the E-HICH is different for the serving and the non-serving cells. In the serving radio link set, BPSK is used, while for non-serving radio link sets, On–Off Keying (OOK) is used such that NAK is mapped to DTX (no energy transmitted). The reason for having different mappings is to minimize downlink power consumption. Generally, BPSK is preferable if ACK is transmitted for most of the transmissions, while the average power consumption is lower for OOK when NAK is transmitted more than 75% of the time as no energy is transmitted for the NAK. When the UE is not in soft handover, there is only the serving cell in the active set and this cell will detect the presence of an uplink transmission most of the time. Thus, BPSK is preferred for the serving cells. In soft handover, on the other hand, at most one cell is typically able to decode the transmission, implying that most of the cells will transmit a NAK, making OOK attractive. However, note that the NodeB will only transmit an ACK or a NAK in case it detected the presence of an uplink transmission attempts. If not even the presence of a data transmission is detected in the NodeB, nothing will be transmitted as described above. Hence, the E-HICH receiver in the UE must be able to handle the DTX case as well, although from a protocol point of view only the values ACK and NAK exist.

10.3.3.2   E-AGCH

The E-AGCH is a shared channel, carrying absolute scheduling grants consisting of:

- The maximum E-DPDCH/DPCCH power ratio the UE is allowed to use for the E-DCH (5 bits).
- An activation flag (1 bit), used for (de)activating individual hybrid ARQ processes.
- An identity that identifies the UE (or group of UEs) for which the E-AGCH information is intended (16 bits). The identity is not explicitly transmitted but implicitly included in the CRC calculation.

Rate 1/3 convolutional coding is used for the E-AGCH and the coded bits are rate matched to 60 bits, corresponding to 2 ms duration at the E-AGCH spreading factor of 256 (Figure 10.30). In case of a 10 ms E-DCH TTI, the 2 ms structure is repeated 5 times. Note that a single channelization code can handle a cell with both TTIs and therefore it is not necessary to reserve two channelization codes in a cell with mixed TTIs. UEs with 2 ms TTI will attempt to decode each subframe of a 10 ms E-AGCH without finding its identity. Similarly, a 10 ms UE will combine five subframes before decoding and the CRC check will fail unless the grant was 10 ms long. For group-wise scheduling, it is unlikely that both 2 and 10 ms UEs will be given the same grant (although the above behavior might be exploited) and the absolute grants for these two groups of UEs can be sent separated in time on the same channelization code.

Each E-DCH-enabled UE receives one E-AGCH (although there may be one or several E-AGCH configured in a cell) from the serving cell. Although the UE is required to monitor the E-AGCH for valid information every TTI, a typical scheduling algorithm may only address a particular UE using the E-AGCH occasionally. The UE can discover whether the information is valid or not by checking the ID-specific CRC.

10.3.3.3   E-RGCH

Relative grants are transmitted on the E-RGCH and the transmission structure used for the E-RGCH is identical to that of the E-HICH. The UE is expected to receive one relative grant per TTI from each of the cells in its active set. Thus, relative grants can be transmitted from both the serving and the non-serving cells.

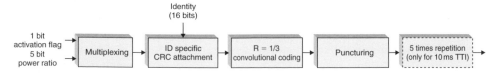

Figure 10.30   *E-AGCH coding structure.*

From the serving cell, the E-RGCH is a dedicated physical channel and the signaled value can be one of $+1$, DTX, and -1, corresponding to UP, HOLD, and DOWN, respectively. Similar to the E-HICH, the duration of the E-RGCH equals 2 or 8 ms, depending on the E-DCH TTI configured.

From the non-serving cells, the E-RGCH is a common physical channel, in essence a common 'overload indicator' used to limit the amount of inter-cell interference. The value on the E-RGCH from the non-serving cells (overload indicator) can only take the values DTX and -1, corresponding to 'no overload' and DOWN, respectively. E-RGCH from the non-serving cells span 10 ms, regardless of the E-DCH TTI configured. Note that Figure 10.28 is representative for the serving cell as each UE is assigned a separate relative grant (from the non-serving cell the E-RGCH is common to multiple UEs).

10.3.3.4 Timing

The timing structure for the E-DCH downlink control channels (E-AGCH, E-RGCH, E-HICH) is designed to fulfill a number of requirements. Additional timing bases in the UE are not desirable from a complexity perspective, and hence the timing relation should either be based on the common pilot or the downlink DPCH as the timing of those channels anyway needs to be handled by the UE.

Common channels, the E-RGCH from the non-serving cell and the E-AGCH, are monitored by multiple UEs and must have a common timing. Therefore, the timing relation of these channels is defined as an offset relative to the common pilot. The duration of the E-AGCH is equal to the E-DCH TTI for which the UE is configured. For the E-RGCH from the non-serving cell, the duration is always 10 ms, regardless of the TTI. This simplifies mixing UEs with different TTIs in a single cell while providing sufficiently rapid inter-cell interference control.

Dedicated channels, the E-RGCH from the serving cell and the E-HICH, are unique for each UE. To maintain a similar processing delay in the UE and NodeB, regardless of the UE timing offset to the common pilot, their timing is defined relative to the downlink DPCH.

The structure of the E-HICH, where multiple E-HICHs share a common channelization code, has influenced the design of the timing relations. To preserve orthogonality between users sharing the same channelization code, the (sub)frame structure of the E-HICHs must be aligned. Therefore, the E-HICH timing is derived from the downlink DPCH timing, adjusted to the closest 2 ms subframe not violating the smallest UE processing requirement.

The number of hybrid ARQ processes directly affects the delay budget in the UE and NodeB. The smaller the number of hybrid ARQ processes, the better from

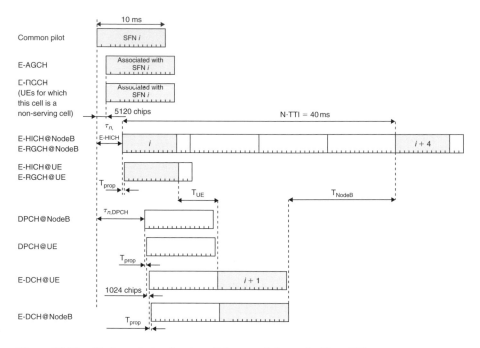

Figure 10.31   *Timing relation for downlink control channels, 10 ms TTI.*

a roundtrip time perspective but also the tighter the implementation requirements. The number of hybrid ARQ processes for E-DCH is fixed to four processes in case of a 10 ms TTI and ten processes in case of a 2 ms TTI. The total
delay budget is split between the UE and the NodeB as given by the expressions
relating the downlink DPCH timing to the corresponding E-HICH subframe.
To allow for 2 ms extra NodeB processing delays, without tightening the UE
requirements, the E-HICH duration is 8 ms, rather than 10 ms in case of a 10 ms
E-DCH TTI. Note that the acceptable UE and NodeB processing delays vary in
a 2 ms interval depending on the downlink DPCH timing configuration. For the
UE, this effect is hard to exploit as it has no control over the network configuration and the UE design therefore must account for the worst case. The NodeB,
on the other hand, may, at least in principle, exploit this fact if the network is
configured to obtain the maximum NodeB processing time.

For simplicity, the timing of the E-RGCH from the serving cell is identical to
that of the E-HICH. This also matches the interpretation of the relative grant
in the UE as it is specified relative to the previous TTI in the same hybrid ARQ
process, that is the same relation that is valid for the ACK/NAK.

The downlink timing relations are illustrated in Figure 10.31 for 10 ms E-DCH
TTI and in Figure 10.32 for 2 ms TTI. An overview of the approximate processing delays in the UE and NodeB can be found in Table 10.3.

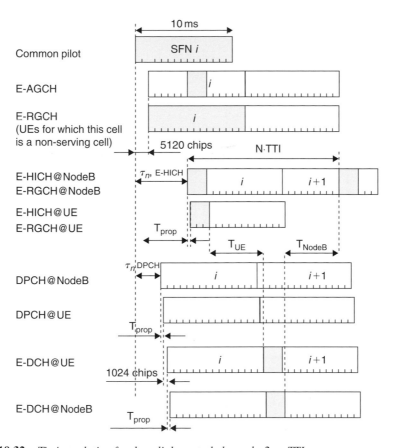

Figure 10.32 *Timing relation for downlink control channels, 2 ms TTI.*

Table 10.3 *Minimum UE and NodeB processing time. Note that the propagation delay has to be included in the NodeB timing budget.*

	10 ms E-DCH TTI	2 ms E-DCH TTI
Number of hybrid ARQ processes	4	8
Minimum UE processing time	5.56 ms	3.56 ms
Minimum NodeB processing time	14.1 ms	6.1 ms

10.3.3.5   Uplink control signaling: E-DPCCH

The uplink E-DCH-related out-band control signaling, transmitted on the E-DPCCH physical channel, consist of:

- 2-bit RSN,
- 7-bit E-TFCI,
- 1-bit rate request ('happy bit').

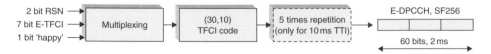

Figure 10.33   *E-DPCCH coding.*

The E-DPCCH is transmitted in parallel to the uplink DPCCH on a separate channelization code with spreading factor 256. In this way, backward compatibility is ensured in the sense that the uplink DPCCH has retained exactly the same structure as in earlier WCDMA releases. An additional benefit of transmitting the DPCCH and the E-DPCCH in parallel, instead of time multiplexed on the same channelization code, is that it allows for independent power-level setting for the two channels. This is useful as the NodeB performance may differ between implementations.

The complete set of 10 E-DPCCH information bits are encoded into 30 bits using a second-order Reed-Müller code (the same block code as used for coding of control information on the DPCCH). The 30 bits are transmitted over three E-DPCCH slots for the case of 2 ms E-DCH TTI (Figure 10.33). In case of 10 ms E-DCH TTI, the 2 ms structure is repeated 5 times. The E-DPCCH timing is aligned with the DPCCH (and consequently the DPDCH and the E-DPDCH).

To minimize the interference generated in the cell, the E-DPCCH is only transmitted when the E-DPDCH is transmitted. Consequently, the NodeB has to detect whether the E-DPCCH is present or not in a certain subframe (DTX detection) and, if present, decode the E-DPCCH information. Several algorithms are possible for DTX detection, for example, comparing the E-DPCCH energy against a threshold depending on the noise variance.

11

MBMS: Multimedia Broadcast Multicast Services

In the past, cellular systems have mostly focused on transmission of data intended for a single user and not on broadcast services. Broadcast networks, exemplified by the radio and TV broadcasting networks, have on the other hand focused on covering very large areas and have offered no or limited possibilities for transmission of data intended for a single user. *Multimedia Broadcast and Multicast Services*, (MBMS), introduced for WCDMA in Release 6, supports multicast/broadcast services in a cellular system, thereby combining multicast and unicast transmissions within a single network.

With MBMS, the same content is transmitted to multiple users located in a specific area, the *MBMS service area*, in a unidirectional fashion. The MBMS service area typically covers multiple cells, although it can be made as small as a single cell.

Broadcast and multicast describe different, although closely related scenarios:

- In *broadcast,* a point-to-multipoint radio resource is set up in each cell being part of the MBMS broadcast area and all users subscribing to the broadcast service simultaneously receive the same transmitted signal. No tracking of users' movement in the radio access network is performed and users can receive the content without notifying the network. Mobile TV is an example of a service that could be provided through MBMS broadcast.
- In *multicast,* users request to join a multicast group prior to receiving any data. The user movements are tracked and the radio resources are configured to match the number of users in the cell. Each cell in the MBMS multicast area may be configured for point-to-point or point-to-multipoint transmission. In sparsely populated cells with only one or a few users subscribing to the MBMS service, point-to-point transmission may be appropriate, while in cells with a larger number of users, point-to-multipoint transmission is better

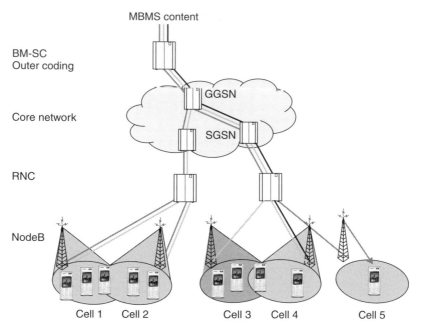

Figure 11.1 *Example of MBMS services. Different services are provided in different areas using broadcast in cells 1–4. In cell 5, unicast is used as there is only single user subscribing to the MBMS service.*

suited. Multicast therefore allows the network to optimize the transmission type in each cell.

To a large extent, MBMS affects mainly the nodes above the radio-access network. A new node, the *Broadcast Multicast Service Center* (BM-SC), illustrated in Figure 11.1, is introduced. The BM-SC is responsible for authorization and authentication of content provider, charging, and the overall configuration of the data flow through the core network. It is also responsible for application-level coding as discussed below.

As the focus of this book is on the radio-access network, the procedures for MBMS will only be briefly described. In Figure 11.2, typical phases during an MBMS session are illustrated. First, the service is announced. In case of broadcast, there are no further actions required by the user; the user simply 'tunes' to the channel of interest. In case of multicast, a request to join the session has to be sent to become member of the corresponding MBMS service group and, as such, receive the data. Before the MBMS transmission can start, the BM-SC sends a session-start request to the core network, which allocates the necessary internal resources and request the appropriate radio resources from the radio-access network. All terminals of the corresponding MBMS service group are

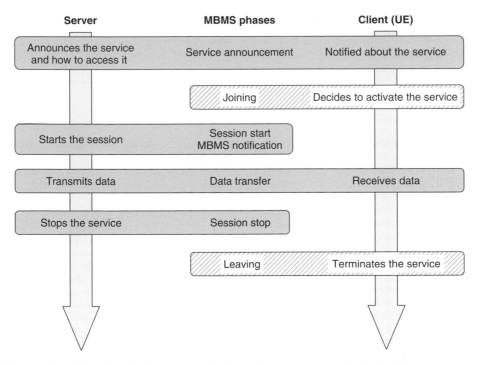

Figure 11.2 *Example of typical phases during an MBMS session. The dashed phases are only used in case of multicast and not for broadcast.*

also notified that content delivery from the service will start. Data will then be transmitted from the content server to the end users. When the data transmission stops, the server will send a session-stop notification. Also, users who want to leave an MBMS multicast service can request to be removed from the MBMS service group.

One of the main benefits brought by MBMS is the resource savings in the network as a single stream of data may serve multiple users. This is seen in Figure 11.1, where three different services are offered in different areas. From the BM-SC, data streams are fed to each of the NodeBs involved in providing the MBMS services. As seen in the figure, the data stream intended for multiple users is not split until necessary. For example, there is only a single stream of data sent to all the users in cell 3. This is in contrast to previous releases of UTRAN, where one stream per user has to be configured throughout both the core network and the radio access network.

In the following, the principles behind MBMS in the radio access network and their introduction into WCDMA will be discussed. The focus is on point-to-multipoint transmission as this requires some new features in the radio interface.

Point-to-point transmission uses either dedicated channels or HS-DSCH and are, from a radio-interface perspective, not different from any other transmission.

A description of MBMS from a specification perspective is found in [102] and the references therein.

11.1 Overview

As discussed above, one of the main benefits with MBMS is resource savings in the network as multiple users can share a single stream of data. This is valid also from a radio-interface perspective, where a single transmitted signal may serve multiple users. Obviously, point-to-multipoint transmission puts very different requirements on the radio interface than point-to-point unicast. User-specific adaptation of the radio parameters, such as channel-dependent scheduling or rate control, cannot be used as the signal is intended for multiple users. The transmission parameters such as power must be set taking the worst case user into account as this determines the coverage for the service. Frequent feedback from the users, for example, in the form of CQI reports or hybrid ARQ status reports, would also consume a large amount of the uplink capacity in cells where a large number of users simultaneously receive the same content. Imagine, for example, a sports arena with thousands of spectators watching their home team playing, all of them simultaneously wanting to receive results from games in other locations whose outcome might affect their home team. Clearly, user-specific feedback would consume a considerable amount of capacity in this case.

From the above discussion, it is clear that MBMS services are power limited and maximizing the diversity without relying on feedback from the users is of key importance. The two main techniques for providing the diversity for MBMS services are

- Macro-diversity by combining of transmissions from multiple cells.
- Time-diversity against fast fading through a long 80 ms TTI and application-level coding.

Fortunately, MBMS services are not delay sensitive and the use of a long TTI is not a problem from the end-user perspective. Additional means for providing diversity can also be applied in the network, for example open-loop transmit diversity. Receive diversity in the terminal also improves the performance, but as the 3GPP UE requirements for Release 6 are set assuming single-antenna UEs, it is hard to exploit this type of diversity in the planning of MBMS coverage. Also, note that application-level coding provides additional benefits, not directly related to diversity, as discussed below.

11.1.1 Macro-diversity

Combining transmissions of the same content from multiple cells (macro-diversity) provides a significant diversity gain [80], in the order of 4–6 dB reduction in transmission power compared to single-cell reception, as illustrated in Figure 11.3. Two combining strategies are supported for MBMS, *soft combining* and *selection combining*, the principles of both illustrated in Figure 11.4.

Soft combining, as the term indicates, combines the soft bits received from the different radio links prior to (Turbo) decoding. In principle, the UE descrambles

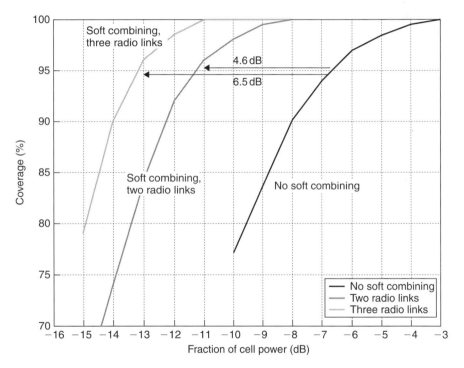

Figure 11.3 *The gain with soft combining and multi-cell reception in terms of coverage vs. power for 64 kbit/s MBMS service (vehicular A, 3 km/h, 80 ms TTI, single receive antenna, no transmit diversity, 1% BLER).*

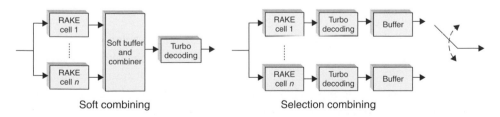

Figure 11.4 *Illustration of the principles for (a) soft combining and (b) selection combining.*

and RAKE combines the transmission from each cell individually, followed by soft combining of the different radio links. Note that, in contrast to unicast, this *macro-diversity* gain comes 'for free' in the sense that the signal in the neighboring cell is anyway present. Therefore, as discussed in Chapter 4, it is better to exploit this signal rather than treat it as interference. However, as WCDMA uses cell-specific scrambling of all data transmissions, the soft combining needs to be performed by the appropriate UE processing. This processing is also responsible for suppressing the interference caused by (non-MBMS) transmission activity in the neighboring cells. To perform soft combining, the physical channels to be combined should be identical. For MBMS, this implies the same physical channel content and structure should be used on the radio links that are soft combined.

Selection combining, on the other hand, decodes the signal received from each cell individually and for each TTI selects one (if any) of the correctly decoded data blocks for further processing by higher layers. From a performance perspective, soft combining is preferable as it provides not only diversity gains, but also a power gain as the received power from multiple cells is exploited. Relative to selection combining, the gain is in the order of 2–3 dB [80].

The reason for supporting two different combining strategies is to handle different levels of asynchronism in the network. For soft combining, the soft bits from each radio link have to be buffered until the whole TTI is received from all involved radio links and the soft combining can start, while for selection combining, each radio link is decoded separately and it is sufficient to buffer the decoded information bits from each link. Hence, for a large degree of asynchronism, selection combining requires less buffering in the UE at the cost of an increase in Turbo decoding processing and loss of performance. The UE is informed about the level of synchronism and can, based upon this information and its internal implementation, decide to use any combination scheme as long as it fulfills the minimum performance requirements mandated by the specifications. With similar buffering requirements as for a 3.6 Mbit/s HSDPA terminal, which is the basis for the definition of the UE MBMS requirements, soft combining is possible provided the transmissions from the different cells are synchronized within approximately 80 ms, which is likely to be realistic in most situations.

As mentioned above, the UE capabilities are set assuming similar buffering requirements as for a 3.6 Mbit/s HSDPA terminal. This result in certain limitations in the number of radio links a terminal is required to be able to soft combine for different TTI values and different data rates. This is illustrated in Table 11.1, from which it is also seen that all MBMS-capable UEs can support data rates up to 256 kbit/s. It is worth noting that there is a single MBMS UE capability – either the UE supports MBMS or not. As network planning has to be done

Table 11.1 *Requirements on UE processing for MBMS reception [99].*

Data rate (on MTCH)	Soft combining		Selection combining	
	Maximum number of RLs	TTI	Maximum number of RLs	TTI
256 kbit/s	3	40	2	40
	≤2	80	1	80
128 kbit/s	≤3	80	3	40
			2	80
			1	80
≤64 kbit/s	≤3	80	≤3	80

assuming a certain set of UE capabilities in terms of soft combining, etc., exceeding these capabilities cannot be exploited by the operator. The end user may of course benefit from a more advanced terminal, for example through the possibility for receiving multiple services simultaneously.

11.1.2 Application-level coding

Many end-user applications require very low error probabilities, in the order of 10^{-6}. Providing these low error probabilities on the transport channel level can power-wise be quite costly. In point-to-point communications, some form of (hybrid) ARQ mechanism is therefore used to retransmit erroneous packets. HSDPA, for example, uses both a hybrid-ARQ mechanism (see Chapter 9) and RLC retransmissions. In addition, the TCP protocol itself also performs retransmissions to provide virtually error-free packet delivery. However, as previously discussed, broadcast typically cannot rely on feedback, and, consequently, alternative strategies need to be used. For MBMS, application-level forward error-correcting coding is used to address this problem. The application-level coding resides in the BM-SC and is thus not part of the radio-access network, but is nevertheless highly relevant for a discussion of the radio-access-network design for support of MBMS. With application-level coding, the system can operate at a transport-channel block-error rate in the order of 1–10% instead of fractions of a percent, which significantly lowers transmit power requirement. As the application-level coding resides in the BM-SC, it is also effective against occasional packet losses in the transport network, for example due to temporary overload conditions.

Systematic Raptor codes [63] have been selected for the application-level coding in MBMS [105], operating on packets of constant size (48–512 bytes). Raptor codes belongs to a class of *Fountain codes*, and as many encoding packets as needed can be generated on-the-fly from the source data. For the decoder to be

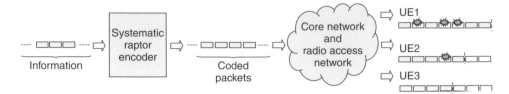

Figure 11.5 *Illustration of application-level coding. Depending on their different ratio conditions, the number of coded packets required for the UEs to be able to reconstruct the original information differs.*

able to reconstruct the information, it only needs to receive sufficiently many coded packets. It does not matter which coded packets it received, in what order they are received, or if certain packets were lost (Figure 11.5).

In addition to provide additional protection against packet losses and to reduce the required transmission power, the use of application-level coding also simplifies the procedures for UE measurements. For HSDPA, the scheduler can avoid scheduling data to a given UE in certain time intervals. This allows the UE to use the receiver for measurement purposes, for example to tune to a different frequency and possible also to a different radio access technology. In a broadcast setting, scheduling measurement gaps is cumbersome as different UEs may have different requirements on the frequency and length of the measurement gaps. Furthermore, the UEs need to be informed when the measurement gaps occur. Hence, a different strategy for measurements is adopted in MBMS. The UE measurements are done autonomously, which could imply that a UE sometimes miss (part of) a coded transport block on the physical channel. In some situations, the inner Turbo code is still able to decode the transport channel data, but if this is not the case, the outer application-level code will ensure that no information is lost.

11.2   Details of MBMS

One requirement in the design of MBMS was to reuse existing channels to the extent possible. Therefore, the FACH transport channel and the S-CCPCH physical channel are reused without any changes. To carry the relevant MBMS data and signaling, three new logical channels are added to Release 6:

- *MBMS Traffic Channel* (MTCH), carrying application data.
- *MBMS Control Channel* (MCCH), carrying control signaling.
- *MBMS Scheduling Channel* (MSCH), carrying scheduling information to support discontinuous reception in the UE.

All these channels use FACH as the transport channel type and the S-CCPCH as the physical channel type. In addition to the three new logical channels, one new physical channel is introduced to support MBMS – MBMS Indicator Channel (MICH), used to notify the UE about an upcoming change in MCCH contents.

11.2.1 MTCH

The MTCH is the logical channel used to carry the application data in case of point-to-multipoint transmission (for point-to-point transmission, DTCH, mapped to DCH or HS-DSCH, is used). One MTCH is configured for each MBMS service and each MTCH is mapped to one FACH transport channel. The S-CCPCH is the physical channel used to carry one (or several) FACH transport channels.

The RLC for MTCH is configured to use unacknowledged mode as no RLC status reports can be used in point-to-multipoint transmissions. To support selective combining (discussed in Section 11.1.1), the RLC has been enhanced with support for in-sequence delivery using the RLC PDU sequence numbers and the same type of mechanism as employed in MAC-hs (see Chapter 9). This enables the UE to do reordering up to a depth set by the RLC PDU sequence number space in case of selection combining.

In Figure 11.6, an example of the flow of application data through RLC, MAC, and physical layer is illustrated. The leftmost part of the figure illustrates the case of point-to-point transmission, while the middle and rightmost parts illustrates the case of point-to-multipoint transmission using the MTCH. In the middle part, one RLC entity is used with multiple MAC entities. This illustrates a typical situation where selection combining is used, where multiple cells are loosely time aligned and the same data may be transmitted several TTIs apart in the different cells. Finally, the rightmost part of the figure illustrates a typical case where soft combining can be used. A single RLC and MAC entity is used for transmission in multiple cells. To allow for soft combining, transmissions from the different cells need to be aligned within 80.67 ms (assuming 80 ms TTI).

11.2.2 MCCH and MICH

The MCCH is a logical channel type used to convey control signaling necessary for MTCH reception. One MCCH is used in each MBMS-capable cell and it can carry control information for multiple MTCHs. The MCCH is mapped to FACH (note, a different FACH than used for MTCH), which in turn is transmitted on an S-CCPCH physical channel. The same S-CCPCH as for the MTCH may be used, but if soft combining is allowed for MTCH, different S-CCPCHs for MTCH and MCCH should be used. The reason for using separate S-CCPCHs in this case is that no selection or soft combining is used for the MCCH, and the UE receives the

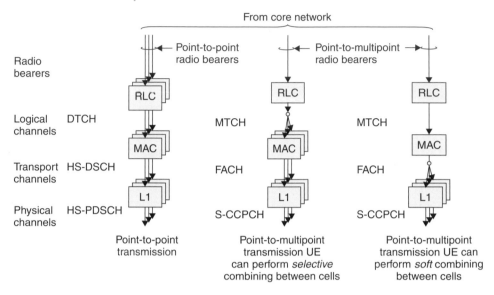

Figure 11.6 *Illustration of data flow through RLC, MAC, and L1 in the network side for different transmission scenarios.*

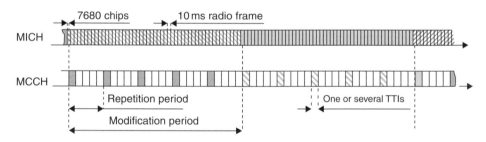

Figure 11.7 *MCCH transmission schedule. Different shades indicate (potentially) different MCCH content, e.g. different combinations of services.*

MCCH from a single cell only. The RLC is operated in unacknowledged mode for MCCH. Where to find the MCCH is announced on the BCCH (the BCCH is the logical channel used to broadcast system configuration information).

Transmission on the MCCH follows a fixed schedule as illustrated in Figure 11.7. The MCCH information is transmitted using a variable number of consecutive TTIs. In each *modification period*, the critical information remains unchanged[1] and is periodically transmitted based on a *repetition period*. This is useful to support mobility between cells; a UE entering a new cell or a UE which missed the first transmission does not have to wait until the start of a new modification period to receive the MCCH information.

---

[1]The MBMS access information may change during a modification period, while the other MCCH information is considered as critical and only may change at the start of a modification period.

The MCCH information includes information about the services offered in the modification period and how the MTCHs in the cell are multiplexed. It also contains information about the MTCH configuration in the neighboring cells to support soft or selective combining of multiple transmissions. Finally, it may also contain information to control the feedback from the UEs in case counting is used.

Counting is a mechanism where UEs connect to the network to indicate whether they are interested in a particular service or not and is useful to determine the best transmission mechanism for a given service. For example, if only a small number of users in a cell are interested in a particular service, point-to-point transmission may be preferable over point-to-multipoint transmission. To avoid the system being heavily loaded in the uplink as a result of counting responses, only a fraction of the UEs transmit the counting information to the network. The MCCH counting information controls the probability with which a UE connects to the network to transmit counting information. Counting can thus provide the operator with valuable feedback on where and when a particular service is popular, a benefit typically not available in traditional broadcast networks.

To reduce UE power consumption and avoid having the UE constantly receiving the MCCH, a new physical channel, the MICH (MBMS Indicator Channel), is introduced to support MBMS. Its purpose is to inform UEs about upcoming changes in the critical MCCH information and the structure is identical to the paging indicator channel. In each 10 ms radio frame, 18, 36, 72, or 144 MBMS indicators can be transmitted, where an indicator is a single bit, transmitted using on–off keying and related to a specific group of services.

By exploiting the presence of the MICH, UEs can sleep and briefly wake up at predefined time intervals to check whether an MBMS indicator is transmitted. If the UE detects an MBMS indicator for a service of interest, it reads the MCCH to find the relevant control information, for example when the service will be transmitted on the MTCH. If no relevant MBMS indicator is detected, the UE may sleep until the next MICH occasion.

11.2.3 MSCH

The purpose of the MSCH is to enable UEs to perform discontinuous reception of the MTCH. Its content informs the UE in which TTIs a specific service will be transmitted. One MSCH is transmitted in each S-CCPCH carrying the MTCH and the MSCH content is relevant for a certain service and a certain S-CCPCH.

12

HSPA Evolution

Although HSPA as defined in Release 6 is a significant enhancement to the packet-data functionality in WCDMA, the performance is further enhanced in Release 7 and beyond. Herein, this is referred to as 'HSPA Evolution' and consists of both the introduction of new major features, such as MIMO, and many smaller improvements to existing structures which, when taken together as a package, represents a major increase in performance and capabilities. In the following sections, some of these enhancements are discussed.

12.1 MIMO

MIMO is one of the major new features in Release 7, introduced to increase the peak data rates through multi-stream transmission. Strictly speaking, MIMO, Multiple Input Multiple Output, in its general interpretation denotes the use of multiple antennas at both transmitter and receiver. This can be used to obtain a diversity gain and thereby increase the carrier-to-interference ratio at the receiver. However, the term is commonly used to denote the transmission of multiple layers or multiple streams as a mean to increase the data rate possible in a given channel. Hence, MIMO, or spatial multiplexing, should mainly be seen as a tool to improve the end-user throughput by acting as a 'data-rate booster.' Naturally, an improved end-user throughput will to some extent also result in an increased system throughput.

As discussed in Chapter 6, MIMO schemes are designed to exploit certain properties in the radio propagation environment to attain high data rates by transmitting multiple data streams in parallel. However, to achieve these high data rates, a correspondingly high carrier-to-interference ratio at the receiver is required. Spatial multiplexing therefore is mainly applicable in smaller cells or close to the base station, where high carrier-to-interference ratios are common. In situations where a sufficient high carrier-to-interference ratios cannot be achieved, the multiple receive antennas, which a MIMO-capable UE is equipped with, can be used for receive diversity also for a single-stream transmission. Hence, a MIMO-capable UE will offer higher cell-edge data rates also in large cells, compared to a corresponding single-antenna UE.

The scheme used for HSDPA-MIMO is sometimes referred to as *dual-stream transmit adaptive arrays* (D-TxAA) [16], which is a multi-codeword scheme with rank adaptation and pre-coding. The scheme can be seen as a generalization of closed-loop mode-1 transmit diversity [97], present already in the first WCDMA release.

Transmission of up to two streams is supported by HSDPA-MIMO. Each stream is subject to the same physical-layer processing in terms of coding, spreading and modulation as the corresponding single-layer HSDPA case. After coding, spreading, and modulation, linear pre-coding is used before the result is mapped to the two transmit antennas. As discussed in Chapter 6, there are several reasons for this pre-coding. Even if only a single stream is transmitted, it can be beneficial to exploit both transmit antennas by using transmit diversity. Therefore, the pre-coding in the single-stream case is similar to closed-loop transmit diversity mode-1 (the difference is mainly in the details for signaling and the update rate as will be discussed below). In essence, this can be seen as a simple form of beam-forming. Furthermore, the pre-coding attempts to pre-distort the signal such that the two streams are (close to) orthogonal at the receiver. This reduces the interference between the two streams and lessens the burden on the receiver processing.

Mainly the physical-layer processing is affected by the introduction of MIMO; the impact to the protocol layer is small and MIMO is to a large extent only visible as a higher data rate.

12.1.1 HSDPA-MIMO data transmission

To support dual-stream transmission, the HS-DSCH is modified to support up to two transport blocks per TTI. Each transport block represents one stream. A CRC is attached to each of the transport blocks and each transport block is individually coded. This is illustrated in Figure 12.1 (compare with Chapter 9 for the non-MIMO case). Since two transport blocks are used in case of multi-stream transmission, HSDPA-MIMO is a multi-codeword scheme (see Chapter 6 for a discussion on single vs multi-codeword schemes) and allows for a successive-interference-cancellation receiver in the UE.

The physical-layer processing for each stream is identical to the single-stream case, up to and including spreading [95, 96]. To avoid wasting channelization-code

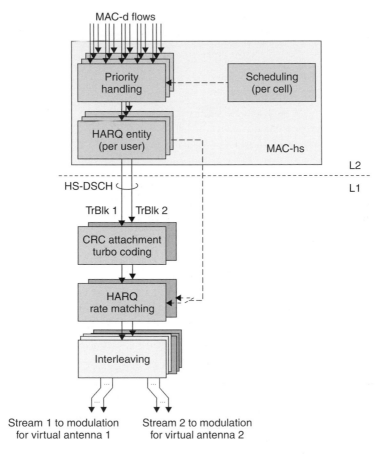

Figure 12.1 *HS-DSCH processing in case of MIMO transmission.*

resources, the same set of channelization codes should be used for the two streams. At the receiver, the two streams are separated by the appropriate receiver processing, for example by interference cancellation (see Chapter 6).

The signal after individual spreading of a stream can be seen as the signal on a virtual antenna. Before each of the virtual-antenna signals are fed to the physical antennas,[1] linear pre-coding is used as illustrated in Figure 12.2. For each of the streams, the pre-coder is simply a pair of weights. Stream i is multiplied with the complex weight w_{ij} before being fed to physical antenna j.

Using pre-coding is beneficial for several reasons, especially in the case of single-stream transmission. In this case, pre-coding provides both *diversity gain*

---

[1] The term 'physical antenna' here denotes the antenna as seen by the specifications, that is each antenna from the specification perspective has a separate common pilot. The actual antenna arrangement on the base station may take any form, although the only antennas visible from the UE are those with a unique common pilot.

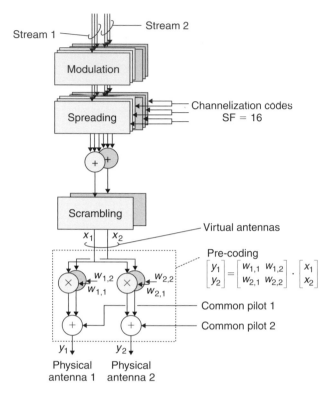

Figure 12.2 *Modulation, spreading, scrambling and pre-coding for two dual-stream MIMO.*

and *array gain* as both transmit antennas are used and the weights are selected such that the signals from the two antennas add coherently at the receiver. This results in a higher received carrier-to-interference ratio than in absence of pre-coding, thus increasing the coverage for a certain data rate. Furthermore, if a separate power amplifier is used for each physical antenna, pre-coding ensures both of them are used also in case of single-stream transmissions, thereby increasing the total transmit power. The pre-coding weights in case of single-stream transmission are chose to be identical to those used for Release 99 closed-loop transmit diversity,

$$w_{1,1} = 1/\sqrt{2},$$
$$w_{2,1} \in \left\{ \frac{(1+j)}{2}, \frac{(1-j)}{2}, \frac{(-1+j)}{2}, \frac{(-1-j)}{2} \right\}. \qquad (4.1)$$

In case of dual-stream transmission, pre-coding can be used to aid the receiver in separating the two streams. If the weights for the second stream are chosen as the (orthogonal) eigenvectors of the covariance matrix at the receiver, the two streams will not interfere with each other at the transmitter. Hence, once the weights for the first stream are selected, the weights $w_{1,2}$, $w_{2,2}$ used for the second stream are given by the requirement to make the columns of the pre-coding matrix orthogonal.

$$W = \begin{bmatrix} w_{1,1} & w_{1,2} \\ w_{2,1} & w_{2,2} \end{bmatrix} \qquad (4.2)$$

As there are four different values possible for $w_{2,1}$, it follows that there are four different values of pre-coding matrix W possible. The pre-coding matrix is constant over one subframe. The setting of the weights are up to the NodeB implementation but is typically based on *Pre-coding Control Indication* (PCI) feedback from the UE.

To demodulate the transmitted data, the UE requires estimates of the channels between each of the base station virtual antennas and each of UEs physical antennas. Hence, a total of four channels need to be estimated. One possibility would be to transmit a common pilot signal on each of the virtual antennas. However, this would not be backwards compatible as UEs assume the primary common pilot to be transmitted from the primary antenna. Demodulation of non-MIMO channels, for example the control channels necessary for the operation of the system, would not be possible either as these channels are not transmitted using any pre-coding. Therefore, the common pilot channels are transmitted in the same way as for transmit diversity. On each physical antenna, a common pilot channel is transmitted. Either a primary common pilot channel is configured on each antenna, using the same channelization and scrambling code on all antennas, or a primary common pilot is configured on one antenna and a secondary common pilot on the other antenna. In case a primary common pilot is configured on both antennas, different mutually orthogonal pilot patterns are used on the different common pilots (all zeros for the first physical antenna as in the single-antenna case, and a sequence of zero and ones for the second physical antenna). Both these schemes enable the UE to estimate the channel from each of the physical transmit antennas to each of its receive antennas. Given knowledge of the pre-coding matrix the NodeB used, the UE can form an estimate of the effective channel from each of the *virtual* antennas to each of the physical receiver antennas as $\hat{H}W$, where

$$\hat{H} = \begin{bmatrix} \hat{h}_{1,1} & \hat{h}_{1,2} \\ \hat{h}_{2,1} & \hat{h}_{2,2} \end{bmatrix} \qquad (4.3)$$

and $\hat{h}_{i,j}$ is the estimate of the channel between the physical antenna i at the base station and the physical antenna j at the UE. For this reason, the pre-coding matrix is signaled to the UE on the HS-SCCH. Explicit signaling of the pre-coding matrix greatly simplifies the UE implementation compared to estimation of the antenna weights which is the case for closed-loop transmit diversity in Release 99 [97].

12.1.2   Rate control for HSDPA-MIMO

Rate control for each stream is similar to the single-stream case. However, the rate-control mechanism also needs to determine the number of streams to transmit and the pre-coding matrix to use. Hence, for each TTI, the number of streams to transmit, the transport-block sizes for each of the streams, the number of channelization codes, the modulation scheme, and the pre-coding matrix is determined by the rate-control mechanism. This information is provided to the UE on the HS-SCCH, similar to the non-MIMO case. As the scheduler controls the size of the two transport blocks in case of multi-stream transmission, the data rate of the two streams can be individually controlled.

Multi-stream transmission is beneficial only at relatively high carrier-to-interference ratios and will consequently only be used for the highest data rates. For lower data rates, single-stream transmission should be used. In this case, the two physical antennas are used for diversity transmission and only one of the virtual antennas is carrying user data.

Similar to Release 6, the rate-control mechanism typically relies on UE feedback of the instantaneous channel quality. In case of dual-stream transmission, information about the supported data rate on each of the streams is required. However, as the NodeB scheduler is free to select the number of streams transmitted to a UE, the supported data rate in case of single-stream transmission is also of interest. The CQI reports are therefore extended to cover both these cases, as well as indicating the preferred pre-coding matrix.

12.1.3   Hybrid-ARQ with soft combining for HSDPA-MIMO

For each stream, the physical-layer hybrid-ARQ processing and the use of multiple hybrid-ARQ processes are identical to the single-stream case. However, as the multiple streams are transmitted over different antennas, one stream may be correctly received while another stream require retransmission of the payload.[2] Therefore, one hybrid-ARQ acknowledgement per stream is sent from the UE to the NodeB.

---

[2] Typically, if the first stream is erroneously received, decoding of the second stream is likely to fail if successive interference cancellation is used.

12.1.4   Control signaling for HSDPA-MIMO

To support MIMO, the out-band control signaling is modified accordingly. No changes to the in-band control signaling in the form of the MAC-hs header are required as reordering and priority queue selection is not affected by the introduction of MIMO. However, to efficiently support the higher data rates provided by MIMO, the MAC, and RLC layers are updated with flexible segmentation as discussed in Section 12.5.

The downlink out-band control signaling is carried on the HS-SCCH. In case MIMO support is enabled in the UE, an alternative format of the HS-SCCH (type 3) is used to accommodate the additional information required (Figure 12.3). The division of the HS-SCCH into two parts (see Chapter 9) is maintained. Part one is extended to include information about the number of streams transmitted to the UE, one or two, and their respective modulation scheme as well as which of the four pre-coding matrices the NodeB used for the transmission. For the second part of the HS-SCCH, the format used depends on whether one or two streams were scheduled to the UE. In the latter case, additional bits are transmitted on part two to convey the hybrid-ARQ and transport-block size information also for the second stream. Despite the slight increase of the number of bits on the HS-SCCH in case MIMO support is enabled in the UE, the HS-SCCH spreading factor is kept at 128. The additional bits are fitted onto the physical channel by adjusting the rate matching of the two parts appropriately. Furthermore, the new-data indicator, the redundancy version, and the constellation-rearrangement fields are combined in to a joint field, eliminating some less-frequently used combinations from previous releases of HSPA. Finally, the restriction that the same HS-SCCH must be used between subframes in case of continuous transmission to a UE (see Chapter 9) has been removed for all Release 8 terminals to provide additional flexibility to the scheduler.

Uplink out-band control signaling consists of hybrid-ARQ acknowledgements, PCI, and CQI, and is transmitted on the HS-DPCCH. In case of single-stream

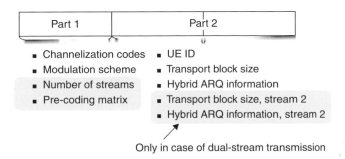

Figure 12.3   *HS-SCCH information in case of MIMO support. The gray shaded information is added compared to Release 5.*

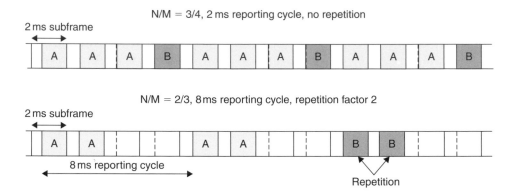

Figure 12.4   *Example of type A and type B PCI/CQI reporting for a UE configured for MIMO reception.*

transmission, only a single hybrid-ARQ acknowledgement bit is transmitted and the format is identical to Release 5. In case of dual-stream transmission, the two hybrid-ARQ acknowledgement bits are jointly coded to 10 bits and transmitted in one slot on the HS-DPCCH.

The Pre-coding Control Information (PCI), consists of 2 bits, indicating which of the four pre-coding matrices that best matches the channel conditions at the UE.

The Channel Quality Indicator (CQI), indicates the data rate the UE recommends in case transmission is done using the recommended PCI. Both single-stream and dual-stream CQI reports are useful as the scheduler may decide to transmit a single stream, even if the channel conditions permit two streams, in some situations, for example if the amount of data to transmit is small. In this case, a single-stream CQI report is required, while in the case of dual-stream transmission obviously a CQI report per stream is useful. As it is not straightforward to deduce the single-stream quality from a dual-stream report, two types of CQI reports are defined:

1. *Type A reports*, containing the PCI and the recommended number of streams, one or two, as well as the CQI for each of these streams.
2. *Type B reports*, containing the PCI and the CQI in case of single-stream transmission.

For the type B reports, the same 5-bit CQI report as in earlier releases is used, while for type A the CQI consists of 8 bits. In both cases, the PCI and CQI reports are concatenated and coded into 20 bits using a block code.

Similar to the non-MIMO case, the PCI/CQI reports are transmitted in two slots on the HS-DPCCH. To allow for a flexible adaptation to different propagation environments, the first N out of M PCI/CQI reports are type A reports and the remaining *M–N* reports are type B reports. The ratio *N/M* is configured by RRC signaling. Two examples are shown in Figure 12.4.

12.1.5 UE capabilities

To allow for a wide range of different UE implementations, MIMO support is not mandatory for all UEs. Furthermore, as multi-stream transmission mainly is a tool for increasing the supported peak data rates, MIMO is mainly relevant for the high-end UE categories. Therefore, UEs supporting MIMO are always capable of receiving 15 channelization codes [99]. Furthermore, some capabilities may differ depending on whether the terminal is configured in MIMO mode or not; as an example some categories of MIMO-capable terminals may support 64QAM only when operating in non-MIMO mode.

12.2 Higher-order modulation

The introduction of MIMO, discussed in the previous section, allows for a substantial increase in peak data rate. This increase is achieved by exploiting certain propagation conditions in the channel through multi-stream transmission. However, in some situations the UE can experience a high carrier-to-interference ratio at the same time as the channel does not support multi-stream transmission, for example in case of line-of-sight propagation. As discussed in Chapter 3, higher-order modulation is useful in such cases. Higher-order modulation is also useful as it allows for high data rates in cases when the UE or NodeB are not equipped with multiple antennas. Therefore, Release 7 increases the highest modulation order to 64QAM in the downlink and 16QAM in the uplink [96] and Release 8 takes this one step further by providing simultaneous support for 64QAM and MIMO in the downlink. The peak rates with higher order modulation and MIMO are listed in Table 12.1.

Supporting 64QAM in the downlink and 16QAM in the uplink is based on the same principles as already specified for Release 6. However, to fulfill the transmitter RF requirements, a somewhat larger power amplifier back-off may be required for these modulation schemes.

Table 12.1 *Peak rates in downlink and uplink with higher order modulation and MIMO.*

Downlink peak rate, Mbit/s				Uplink peak rate, Mbit/s	
Non-MIMO		MIMO			
16QAM	64QAM	16QAM	64QAM	BPSK/QPSK	16QAM
14	21	28	42	5.7	11

12.3 Continuous packet connectivity

Packet-data traffic is often highly bursty with occasional periods of transmission activity. Clearly, from a user performance perspective, it is advantageous to keep the HS-DSCH and E-DCH configured to rapidly be able to transmit any user data. At the same time, maintaining the connection in uplink and downlink comes at a cost. From a network perspective, there is a cost in uplink interference from the DPCCH transmission even in absence of data transmission. From a UE perspective, power consumption is the main concern; even when no data is received the UE needs to transmit the DPCCH and monitor the HS-SCCH.

To reduce UE power consumption, WCDMA, like most other cellular systems, has several states: URA_PCH, CELL_PCH, CELL_FACH, and CELL_DCH. The different states are illustrated in Figure 12.5.

The lowest power consumption is achieved when the UE is in one of the two paging states specified for WCDMA, namely CELL_PCH and URA_PCH. In these states, the UE sleeps and only occasionally wakes up to check for paging messages. The paging mechanism is mainly intended for longer periods of inactivity. For exchange of data, the UE need to be moved to the CELL_FACH or CELL_DCH states.

In CELL_FACH, the UE can transmit small amounts of data as part of the random-access procedure. The UE also monitors common downlink channels for small amounts of user data and RRC signaling from the network.

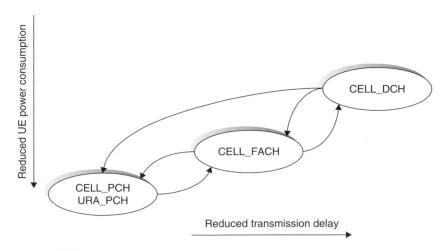

Figure 12.5 *WCDMA state model.*

The high-transmission-activity state is known as CELL_DCH. In this state, the UE can use HS-DSCH and E-DCH for exchanging data with the network as described in Chapters 9 and 10, respectively. This state allows for rapid transmission of large amounts of user data, but also has the highest UE power consumption.

RRC signaling is used to move the UE between the different states. Thus, as discussed above, from a delay perspective it is preferable to keep the UE in CELL_DCH, while from interference and power-consumption perspective, one of the paging states is preferred.

To improve the packet-data support in HSPA, a set of features known as *Continuous Packet Connectivity* (CPC), is introduced in Release 7. CPC consists of three building blocks:

1. *Discontinuous transmission* (DTX), to reduce the uplink interference and thereby increase the uplink capacity, as well as to save battery power.
2. *Discontinuous reception* (DRX), to allow the UE to periodically switch off the receiver circuitry and save battery power.
3. *HS-SCCH-less operation* to reduce the control signaling overhead for small amounts of data, as will be the case for services such as VoIP.

The intention with these features is to provide an 'always-on' experience for the end user by keeping the UE in CELL_DCH for a longer time and avoiding frequent state changes to the low-activity states, as well as improving the capacity for services, such as VoIP. Since they mainly relate to packet-data support, they are only supported in combination with HSPA; thus if a DCH is configured, the CPC features cannot be used. In the following, the three building blocks and the interaction between them is described.

12.3.1 DTX–reducing uplink overhead

The shared resource in the uplink is, as discussed in Chapter 10, the interference headroom in the cell. During the periods when no data transmission is ongoing in the uplink, the interference generated by a UE is due to the uplink DPCCH, which is continuously transmitted as long as the E-DCH is configured. Any reduction in unnecessary DPCCH activity would therefore directly reduce the uplink interference, thereby lowering the cost in terms of system capacity of keeping the UE connected. Clearly, from an interference reduction perspective, the best approach would be to completely switch off the DPCCH when no data transmission is taking place. However, this would have a serious impact on the possibility to maintain uplink synchronization, as well as negatively impact the power control operation. Therefore, occasional slots of DPCCH activity, even if there is no data to transmit, are beneficial to maintain uplink synchronization and to maintain a reasonably accurate power control. This is the basic idea behind Uplink Discontinuous Transmission (uplink DTX). Obviously, the burstier the data traffic, the larger the benefits with discontinuous transmission.

Basically, if there is no E-DCH transmission in the uplink, the UE automatically stops continuous DPCCH transmission and regularly transmits a DPCCH burst according to a UE DTX cycle. The UE DTX cycle, configured in the UE and the NodeB by the RNC, defines when to transmit the DPCCH even if there is no E-DCH activity. This is illustrated in Figure 12.6. The length of the DPCCH burst can be configured. Note that the DPCCH is transmitted whenever there is activity on the uplink E-DPDCH, regardless of the UE DTX cycle. There is also a possibility to set UE-specific offsets to spread the DPCCH transmission occasions from different UEs in time.

To adapt the UE DTX cycle to the traffic properties, two different cycles are defined, UE DTX cycle 1 and UE DTX cycle 2, where the latter is an integer multiple of the former. After a certain configurable period of inactivity on the E-DCH, the UE switches from UE DTX cycle 1 to UE DTX cycle 2, which has less frequent DPCCH transmission instants.

Discontinuous reception in the NodeB is possible thanks to the use of uplink DTX and can be useful to save processing resources in the NodeB as it does not have to continuously process the received signal from all users. To enable this possibility, the network can configure the UE to allow E-DCH transmissions to start only in certain (sub)frames. A certain time after the last E-DCH transmission, the restriction takes effect and the UE can only transmit in the uplink according to the MAC DTX Cycle.

During slots where the DPCCH is not transmitted, the NodeB cannot estimate the uplink signal-to-interference ratio for power-control purposes and there is no

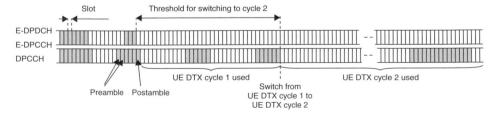

Figure 12.6   *Example of uplink DTX.*

reason for transmitting a power control bit in the downlink. Consequently, the UE shall not receive any power control commands on the F-DPCH in downlink slots corresponding to inactive uplink DPCCH slots. For improved channel-estimation performance and more accurate power control, preambles and postambles are used. For UE DTX cycle 1, the UE starts DPCCH transmission two slots prior to the start of E-DPDCH, as well as ends the DPCCH transmission one slot after the E-DPDCH transmission. This can be seen in Figure 12.6. For UE DTX cycle 2, the preamble can be extended to 15 slots. The preamble and postamble is used also for the DPCCH bursts due to data transmission as well as any HS-DPCCH transmission activity as discussed below.

Until now, the discussion has concerned user-data transmission on the E-DCH and not the control signaling on the HS-DPCCH, which also represents a certain overhead. With CPC enabled, the hybrid-ARQ operation remains unchanged and the UE transmits a hybrid-ARQ acknowledgment after each HS-DSCH reception, regardless of the UE DTX cycle. Clearly, this is sensible as hybrid-ARQ acknowledgement signaling is important for the HS-DSCH performance. It also does not conflict with the possibilities for NodeB discontinuous reception as the NodeB knows when to expect any acknowledgments.

For the CQI reports, the transmission of those reports depends on whether there has been a recent HS-DSCH transmission or not. If any HS-DSCH transmission has been directed to the UE within at most *CQI DTX Timer* subframes, where *CQI DTX Timer* is configured via RRC signaling, the CQI reports are transmitted according to the configured CQI feedback cycle in the same way as in Release 5 and 6. However, if there has not been any recent HS-DSCH transmission, CQI reports are only transmitted if they coincide with the DPCCH bursts. Expressed differently, the uplink DTX pattern overrides the CQI reporting pattern in this case (Figure 12.7).

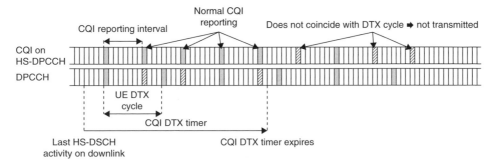

Figure 12.7 *CQI reporting in combination with uplink DTX.*

12.3.2 DRX–reducing UE power consumption

In 'normal' HSDPA operation, the UE is required to monitor up to four HS-SCCHs in each subframe. Although this allows for full scheduling flexibility, it also requires the UE to continuously have its receiver circuitry switched on, leading to a non-negligible power consumption. Therefore, to reduce the power consumption, CPC introduces the possibility for Downlink Discontinuous Reception (downlink DRX). With discontinuous reception, which always is used in combination with discontinuous transmission, the network can restrict in which subframes the UE shall monitor the downlink HS-SCCH, E-AGCH, and E-RGCH by configuring a UE DRX cycle to be used after a certain period of HS-DSCH inactivity. Note that, in this case, the UE can only be scheduled in a subset of all the subframes, which limits the scheduling flexibility somewhat, but for many services such as VoIP with regular packet arrival approximately once per 20 ms, this is not a major problem.

The E-HICH is not subject to DRX as this obviously would not make sense. Hence, whenever the UE has transmitted data in the uplink, it shall monitor the E-HICH in the corresponding downlink subframe to receive the acknowledgment (or negative acknowledgment).

For proper power-control operation, the UE needs to receive the power control bits on the F-DPCH in all downlink slots corresponding to uplink slots where the UE does transmit. This holds, regardless of any UE DRX cycle in the downlink. Therefore, to fully benefit from downlink DRX operation, the network should use uplink DTX in combination with downlink DRX and configure the UE DTX and UE DRX cycles to match each other. An example of simultaneous use of DTX and DRX is shown in Figure 12.8.

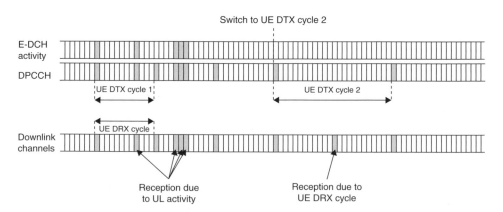

Figure 12.8 *Example of simultaneous use of uplink DTX and downlink DRX.*

12.3.3 HS-SCCH-less operation: downlink overhead reduction

In the downlink, each user represents a certain overhead for the network in terms of code usage and transmission power. The fractional DPCH, F-DPCH, introduced already in Release 6 addresses this issue by significantly reducing the channelization code space overhead. Another source of overhead is the HS-SCCH, used for downlink scheduling. In case of medium-to-large payloads on the HS-DSCH, the HS-SCCH overhead is small relative to the payload; however, for services such as VoIP with frequent transmissions of small payloads, the overhead compared to the actual payload may not be insignificant. Therefore, to address this issue and increase the capacity for VoIP, the possibility for *HS-SCCH-less operation* is introduced in Release 7. The basic idea with HS-SCCH-less operation is to perform HS-DSCH transmissions *without* any accompanying HS-SCCH. As the UE in this case is not informed about the transmission format, it has to revert to blind decoding of the transport format used on the HS-DSCH.

When HS-SCCH-less operation is enabled, the network configures a set of predefined formats that can be used on the HS-DSCH. To limit the complexity of the blind detections in the UE, the number of formats is limited to four and all formats are limited to QPSK and at most two channelization codes. This is well matched to the small transport-block sizes, in the order of a few hundred bits, for which HS-SCCH-less operation is intended. Furthermore, the UE knows which channelization code(s) that may be used for HS-SCCH-less transmission.

In each subframe where the UE has not received any HS-SCCH control signaling, the UE tries to decode the signal received according to each of the preconfigured formats. If decoding of one of the formats is successful, the UE transmits a positive hybrid-ARQ acknowledgement on the HS-DPCCH and delivers the transport block to higher layers. If the decoding was not successful, the UE stores the received soft bits in a soft buffer for potential later retransmissions. Note that no explicit NAK is transmitted in this case. Clearly, this would not be possible as the UE does not know whether the unsuccessful decoding was the result of the UE being addressed, but the transmission received in error, or the UE not being addressed at all. In 'normal' operation, these two cases can be differentiated as there is an HS-SCCH transmission detected in the former but not in the latter case, but in HS-SCCH-less operation this is obviously not possible.

Normally, the HS-SCCH carries the identity of the UE being scheduled. However, in case of HS-SCCH-less operation, this is obviously not possible and the identity of the scheduled UE must be conveyed elsewhere. This is solved by masking the 24-bit CRC on the HS-DSCH with the UE ID using the same general procedure as for the HS-SCCH. Since the UE knows its identity, it can take this into account when checking the CRC and will thus discard transmissions intended for other UEs.

It is possible to mix HS-SCCH-less operation with 'normal' transmissions. If the UE receives the HS-SCCH in a subframe for an initial transmission, it obeys the HS-SCCH and does not try to perform blind decoding. Only if no HS-SCCH directed to this UE is detected will the UE attempt to blindly decode the data. For backward compatibility reasons the same procedure as in previous releases is used for CRC attachment; only for HS-SCCH-less operation is the HS-DSCH CRC masked with the UE ID.

Unlike the initial transmissions discussed so far, hybrid-ARQ retransmissions are accompanied with an HS-SCCH. The HS-SCCH is transmitted using the same structure as for normal HS-DSCH transmissions; however, the bits are reinterpreted to provide the UE with:

- an indication that this is a retransmission of a previous HS-SCCH-less transmission;
- whether it is the first or second retransmission;
- the channelization code set and transport-block size;
- a pointer to the previous transmission attempt the retransmission should be soft combined with.

The reason for this information is to guide the UE in how to perform soft combining; if this information would not have been provided to the UE, the UE would have been forced to blindly try different soft combining strategies and take a hit in complexity. Furthermore, to reduce complexity, at most two retransmissions are supported and the redundancy version to use for each of them is preconfigured.

To be able to perform soft combining, the UE needs to store the soft bits from the previous attempts. With a maximum of three transmissions, one initial and two retransmissions, a total of 13 subframes of soft buffering memory is required. Keeping the amount of soft buffering to a reasonable size is one of the reasons for limiting the number of retransmission attempts to a maximum of two and limiting the payload sizes for HS-SCCH-less operation.

HS-SCCH-less operation in combination with retransmissions is illustrated in Figure 12.9.

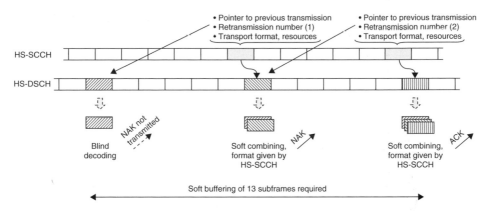

Figure 12.9 *Example of retransmissions with HS-SCCH-less operation.*

12.3.4 Control signaling

Higher-layer signaling is the primary way of setting up and controlling the CPC features. UE DTX and UE DRX cycles are configured and activated by RRC signaling. However, they are not activated immediately after call setup, but only after a configurable time (known as the *Enabling Delay*) to allow synchronization and power control loops to stabilize. HS-SCCH-less operation, on the other hand, can be activated immediately at call setup.

In addition to RRC signaling, there is also a possibility for the serving NodeB to switch on or off uplink DTX and downlink DRX by using reserved HS-SCCH bit patterns, not used for normal scheduling operation. Although this mechanism is typically not used, it provides the scheduler with the possibility of overriding the DTX/DRX operation for additional flexibility. If a UE receives a DTX/DRX activation or deactivation order on HS-SCCH, it responds by sending an acknowledgment on the HS-DPCCH. HS-SCCH activation orders can also be used to activate or deactivate HS-SCCH-less operation.

12.4 Enhanced CELL_FACH operation

The purpose of continuous packet connectivity is, as discussed in the previous sections, to provide an 'always-on' user experience by keeping the UE in the active state (known as CELL_DCH in WCDMA) while still providing mechanisms for reduced power consumption. However, eventually the UE will be switched to CELL_FACH if there has been no transmission activity for a certain period of time. Once the UE is in to CELL_FACH, signaling on the *Forward Access Channel* (FACH), a low-rate common downlink transport channel, is required to move the UE to CELL_DCH prior to any data exchange on HS-DSCH and E-DCH can take place. The physical resources to which the

FACH is mapped is semi-statically configured by the RNC and, to maximize the resources available for HS-DSCH and other downlink channels, the amount of resources (and thus the FACH data rate) is typically kept small, in the order of a few tens of kbit/s.

To reduce the latency associated with state changes, Release 7 improves the performance by allowing HS-DSCH to be used also in the CELL_FACH state. This is often referred to as *Enhanced CELL_FACH operation*. Using the HS-DSCH also in CELL_FACH allows for a significant reduction in the delays associated with switching to CELL_DCH state. Instead of using a low-rate FACH, the signaling from the network to the UE can be carried on the high-rate HS-DSCH. This can result in a significant reduction in call-setup delay and a corresponding improvement in the user perception.

In enhanced CELL_FACH operation, the UE monitors the HS-SCCH for scheduling information using the same principles as described in Chapter 9. However, one major difference compared to the HS-DSCH procedures described in Chapter 9 is that no dedicated uplink is present in the CELL_FACH state. Consequently, no CQI reports are available for rate adaptation and channel-dependent scheduling, nor is it possible to transmit any hybrid-ARQ feedback. Therefore, rate adaptation and channel-dependent scheduling has to be based on long-term measurements, transmitted as part of the random-access procedure used to initiate the state change. To account for the lack of hybrid-ARQ feedback, the network can blindly retransmit the downlink data a preconfigured number of times to ensure reliable reception at the UE.

If the same MAC header format as described in Section 12.5 below is used, it is even possible to start transmitting user data to the mobile terminal while carrying out the switch from CELL_FACH to CELL_DCH. This result in a significant improvement in the user perception compared to the approach used prior to Release 7, where data transmission is suspended during the state change.

Furthermore, HS-DSCH reception is also supported in the paging states. This allows for rapid switching also from the paging states and is similar to the approach taken by LTE for paging as described in Chapter 17.

In Release 8, the CELL_FACH enhancements are taken one step further by activating E-DCH in the uplink in CELL_FACH to reduce the delay before the E-DCH can be used. In the conventional WCDMA/HSPA random-access procedure, the power of a short random-access preamble is successively increased until it is detected by the NodeB as indicated by downlink signaling on the *Acquisition Indicator Channel* (AICH). If the UE receives a positive acknowledgement on the AICH, it proceeds by transmitting the actual message on the uplink *Random-Access Channel* (RACH), while a negative acknowledgement implies that the NodeB is not ready to receive the random-access message and requests the UE to make another random access attempt at a later time. As the RACH is intended for small amounts of data, it has a limited data rate and can only support transmission of a single transport block. Thus, terminals with a larger amount of uplink data either need to rely on multiple random-access attempts to transmit the data or have to switch from CELL_FACH to CELL_DCH, which introduces delays.

One way of reducing these delays is to transmit data on the E-DCH already in CELL_FACH. In Release 8, this is solved by pre-configuring default parameters for the E-DCH-related channels. Upon reception of an ACK on the AICH, the UE transmits the message not on RACH but on E-DCH with its significantly higher data rate. There is also the possibility to configure additional sets of E-DCH parameters. In case the NodeB decides not to use the default set, it can use an *extended AICH* (E-AICH) to signal an offset indicating which set to use. If an NAK is signaled on the AICH and the E-AICH is configured, the UE uses the offset provided on the E-AICH.

Contention on the E-DCH in case multiple terminals tried to perform random-access at the same time is resolved by means of UE identities in the E-DCH transmission. Using the enhanced MAC header formats described in Section 12.5, it is, similarly to the downlink, possible to start data transmission already in CELL_FACH. Data transmissions can continue without interruption, even during the state switch from CELL_FACH to CELL_DCH.

12.5 Layer 2 protocol enhancements

To fully benefit from the high data rates supported by HS-DSCH, especially in combination with 64QAM and MIMO, Release 7 introduces enhancements to the RLC and MAC-hs protocols in additions to the physical-layer enhancements. In releases prior to Release 7, the RLC PDU size is semi-statically configured. This is appropriate for the low-to-medium data rates, but for the high data rates targeted by HSPA Evolution, the RLC PDU size, the RLC roundtrip time, and the RLC window size may limit the peak data rates and cause the RLC protocol to stall [55]. One possibility to avoid this is to increase the RLC PDU size, but for Release 7 a somewhat more advanced solution has been adopted, *flexible RLC*. The flexible RLC is based on ideas such as those in [18].

Segmentation of RLC PDUs into smaller MAC PDUs, matched to the instantaneous radio conditions, is introduced. This allows the RLC size to be sufficiently large to keep the overhead from RLC headers small while at the same time keeping the padding overhead modest. It would appear natural that RLC directly creates RLC PDUs with a size adapted to the radio conditions. This is also the approach taken by the RLC in LTE as described in Chapter 15 where the RLC and the scheduler are located in the same node. For HSPA, the situation is different. Since the RLC and the scheduler in this case are located in the RNC and NodeB, respectively, and the instantaneous radio conditions are not known to the RNC, this is not possible for HSPA. However, segmenting the RLC PDUs into smaller MAC PDUs in the NodeB, where the size depends on the instantaneous radio conditions, is a good approximation to fully adaptive RLC PDU sizes.

Furthermore, RLC SDUs are segmented if the SDU size exceeds a certain limit. This increases the RLC retransmission efficiency in case the MAC hybrid-ARQ mechanism fails, triggering an RLC retransmission.

The restriction of not allowing multiplexing data from different radio bearers into the same transport block is also removed in Release 7. This increases the resource efficiency for mixed-service scenarios.

12.6 Advanced receivers

There are many ways to enhance performance in terms of, for example, data throughput and coverage without modifications to the specifications. Many of these enhancements are based on more advanced receiver algorithms and are thus implemented in software in the baseband processing. Other enhancements require more 'hardware' in terms of antennas and RF components, for example, receiver antenna diversity and beam-forming techniques. Advanced receivers are possible for both base stations and mobile devices (UEs).

For the single receiver, the enhancement is manifested by a decrease in the signal-to-noise ratio (E_b/N_0) required for a specific quality of service. The improved receiver performance enables improved quality of service in terms of, for example, end-user data rates. If a large number of the user devices have receiver enhancements, it will lead to improved system performance in terms of, for example, system wide data throughput.

The standards developed in 3GPP do in principle not specify the receiver structure to be used. The specifications define performance requirements for demodulation of the different physical channels. What type of receiver implementation that is used to meet those requirements is not specified, there is full freedom for a UE vendor to use any implementation, as long as the 3GPP requirements are met.

It is for this reason not possible to mandate use of certain receivers through the 3GPP specifications, if the freedom of implementation is to be kept. Most performance requirements are however developed with a baseline receiver in mind. The performance of the baseline receiver is simulated and an agreed 'implementation margin' is added to the results to model (additional) receiver imperfections not included in the simulations. Once the agreed performance limit is entered into the specification, it is to be fulfilled regardless of what receiver has been implemented.

12.6.1 Advanced UE receivers specified in 3GPP

The typical receiver for CDMA is the so-called RAKE receiver [50]. It assumes that noise is uncorrelated between the so-called RAKE taps that independently demodulate propagation components received with different delays.

As described above, the advanced receivers are not specified and mandated as such in the 3GPP specifications. Instead there are multiple 'types' of requirements defined, each based on a different baseline receiver. The UE vendor declares which type of requirements that the UE conforms to. There are four types of enhanced receiver-performance requirements defined in 3GPP specifications, see [92]; see also Table 12.2. Each type of requirement below is optional:

1. *Type 1*: Performance requirements which are based on UEs utilizing receiver diversity.
2. *Type 2*: Performance requirements which are based on UEs utilizing a *Linear Minimum Mean Square Error* (LMMSE) chip-level equalizer receiver structure.
3. *Type 3*: Performance requirements which are based on UEs utilizing both receiver diversity and a chip-level equalizer structure.
4. *Type 3i*: Performance requirements which are based on UEs utilizing both receiver diversity and an interference-aware chip-level equalizer structure.

12.6.2 Receiver diversity (type 1)

Receiver diversity usually means two antennas at the UE, but as explained above the exact implementation is not in any way mandated. A second antenna can be integral or external to the device. Placement of two antennas on or inside a UE can often not be made in an equal fashion, leading to gain imbalance between the antennas paths. Even with such an imbalance, the gain with antenna diversity can still be substantial, especially in an interference limited scenario [54].

Table 12.2   *Advanced receiver requirements in the 3GPP UE performance specification [92].*

3GPP requirements in TS 25.101 (basis for requirements)	Dedicated channels	HSDPA	Enhanced uplink	MBMS
Type 1 (Rx diversity)	DPCH	HS-DSCH	E-RGCH	S-CCPCH
	F-DPCH	HS-SCCH	E-AGCH E-HICH	MICH
Type 2 (chip-level equalizer)	–	HS-DSCH	–	–
Type 3 (chip-level equalizer and Rx diversity)	–	HS-DSCH	–	–
Type 3i (interference-aware chip-level equalizer and Rx diversity)	–	HS-DSCH	–	–

There are 3GPP requirements for type 1 receivers defined for HSDPA [92], specifically for demodulation of HS-DSCH and HS-SCCH, both for QPSK and 16QAM modulation. The reason that type 1 requirements were first developed for HSDPA is that the gain is directly visible for HSDPA as an efficient means to reach higher end-user data rates, through use of more codes and higher-order modulation. There is however also gain for other services such as the ones based on the Release 99 dedicated channels. Additional requirements for type 1 receivers are also developed for DCH, MBMS, and Enhanced Uplink (E-RGCH, E-AGCH, and E-HICH).

12.6.3   Chip-level equalizers and similar receivers (type 2)

Another way to improve downlink throughput for HSDPA in general and for higher-order modulation specifically is to introduce more advanced receivers in the UE, as described in Chapter 5. In highly dispersive radio environments, the main factor limiting performance is self-interference from multipath propagation, which limits the obtainable carrier-to-interference ratio and reduces the number of occasions for which 16QAM or 64QAM can be used. In time-dispersive scenarios, performance (capacity) losses can be partially compensated for by using advanced receivers that suppress self-interference [112]. The use of rate adaptation provides terminal manufacturers an incentive to implement more advanced receivers since those receivers will result in higher end-user data rates than standard receivers.

Multiple strategies for interference-suppressing receivers are possible [29, 113]. The 3GPP type 2 requirements are based on a reference receiver architecture being an LMMSE chip-level equalizer with a 1/2 chip tap spacing and a length of 20 chips (40 taps) [61]. In a UE implementation, any receiver structure can however be used, for example the G-RAKE receiver [29]. There are 3GPP requirements for type 2 receivers defined for HS-DSCH [92].

12.6.4 Combination with antenna diversity (type 3)

Combining an advanced receiver such as G-RAKE with receive antenna diversity gives possibilities for further performance gains as shown in Figure 12.10. While antenna diversity gives performance gain over the whole cell for all geometries, the G-RAKE receiver results in additional performance gains when the own-cell interference dominates, that is for higher geometries closer to the base station. With G-RAKE and antenna diversity combined, there is a substantial performance gain for all geometries over the whole cell.

The rationale for introducing combined advanced receivers and antenna diversity in the 3GPP standards has been the need for high performance of high-end UEs supporting ten or more HS-PDSCH channelization codes that need good receiver performance to efficiently utilize the supported high data rates. It is

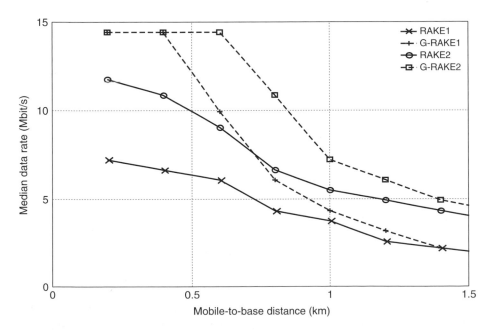

Figure 12.10 *Median HSDPA data rate in a mildly dispersive propagation channel for UEs with 15 channelization codes (from [112]). Numbers are with default RAKE receiver (RAKE1), with antenna diversity (RAKE2 = Type 1), G-RAKE (G-RAKE1 = Type 2), and combined G-RAKE and antenna diversity (G-RAKE2 = Type 3).*

also driven by the ongoing development of high-end UE platforms that support more and more advanced features, implying that a UE with antenna diversity will often also have advanced receivers and will therefore be a type 1 + 2 UE.

Such requirements have also been introduced in 3GPP as type 3 requirements. The reference receiver for type 3 is a Linear Minimum Mean Square Error (LMMSE) chip-level equalizer with two-branch diversity, but any implementation and receiver structure can be used. There are 3GPP requirements for type 3 receivers in [92] defined for HS-DSCH.

12.6.5 Combination with antenna diversity and interference cancellation (type 3i)

As was demonstrated in Figure 12.10 for the downlink, antenna diversity provides gain in both high and low geometry cases, while advanced receivers give performance gains mainly in high geometry scenarios when the intra-cell interference dominates. It is identified in 3GPP standardization that further improvements in performance are possible for lower geometry cases, where other-cell interference dominates. Such improvements can be based on cancellation or rejection of other-cell interference.

Interference cancellation techniques have been studied for a long time [113], where one example is the joint detection scheme developed for UTRA TDD receivers. Another example where substantial gain was shown is the demonstration of parallel interference cancellation in a live WCDMA system, indicating up to 40% multi-cell uplink capacity gain [5]. Implementing interference cancellation for the downlink is however different since the algorithms need to be implemented in each UE that is to benefit from the cancellation. The complexity of the algorithm will then be of higher importance than for an uplink scheme implemented in a base station.

To achieve higher suppression of other-cell interfering signals in a UE, methods such as projection or even explicit demodulation of another base-station signal can be used [113]. The gains of projection-based interference methods have to be weighed against the loss of power and orthogonality that can result from the projection. Explicit demodulation is very complex and may not be suitable for UEs. Both methods rely on a single interfering base station being the dominant interferer, which is the case in approximately 30% of the cell in a multi-cell layout.

With two receive antennas in a UE there are more possibilities to cancel interference as discussed in Chapter 6. The antenna weights can be selected to strongly mitigate a single interfering other-cell base-station signal. This is also called spatio-temporal equalization.

For a UE, both single-branch and dual-branch advanced receivers (types 2 and 3) already cancel some of the own-cell interference by accounting for the correlation of the interference in the receiver process. By also accounting for the correlation of other-cell interference, receivers with a G-RAKE or an LMMSE can achieve some suppression of other-cell interference [136]. The degree to which cancellation is possible will depend on whether the other-cell interference is dominated by a few strong BS signals, since this increases the correlation.

Such requirements with two antenna branches are introduced in 3GPP for HS-DSCH as type 3i requirements. The reference receiver for type 3i is an *interference-aware* Linear Minimum Mean Square Error (LMMSE) chip-level equalizer with two-branch diversity, but any implementation and receiver structure can be used. The requirements in [92] are based on an interference scenario where 70% of the power comes from two dominating base stations' signals and the rest is modeled as AWGN. The chosen power ratios are derived from a system analysis for a lower geometry scenario [136].

12.7   MBSFN operation

The basic principle behind MBMS, described in Chapter 11, is to transmit the same signal from multiple base stations. However, as WCDMA/HSPA uses cell-specific scrambling, the combining needs to be carried out *by processing in the terminal* as described in Chapter 11. To improve the performance of MBMS, true single-frequency operation, also referred to as *Multicast-Broadcast Single Frequency Network* (MBSFN) is advantageous. In MBSFN, the cells are time-synchronized and identical copies of the signal are transmitted from all the cells. This was described already in Chapter 4 and in the framework of WCDMA/HSPA MBMS, it implies the use of the *same* scrambling code in all the cells involved in MBSFN operation. Combined with advanced interference-suppressing receivers in the terminal, very high signal-to-noise ratios can be obtained. To exploit this, Release 7 adds support for 16QAM on FACH and time-multiplexed pilots.

12.8   Conclusion

In the previous chapters the evolution of WCDMA has been discussed. HSPA, consisting of improved packet-data support in the downlink and uplink, has been thoroughly described. By adapting the transmission parameters to rapid variations in the radio-channel quality, as well as traffic variations, a significant performance gain in terms of higher peak data rates, reduced latencies, and improved system capacity can be achieved. Technologies such as channel-dependent scheduling, rate adaptation, and hybrid ARQ with soft combining are used to rapidly adjust the transmission parameters.

The broadcast performance of WCDMA has also been considerably improved through the introduction of MBMS and its enhancements. Transmissions from multiple cells are combined to obtain diversity, which is of key importance for efficient broadcast performance. Furthermore, the use of application-level coding provides additional diversity as the terminal is able to reconstruct the source data even if some packets are missing.

HSPA is also subject to a continuous evolution and in this chapter, some of these steps have been described. Through the use of MIMO, multiple antennas at both the NodeB and the UE can be use to further increase the peak data rates. The use of Continuous Packet Connectivity provides an 'always-on' experience for the end user.

No doubt will this evolution of HSPA continue beyond the steps described in this chapter. For example, HSPA can be extended to bandwidths beyond 5 MHz by aggregating multiple 5 MHz HSPA carriers. This would allow for even higher data rates with a corresponding improvement in end-user experience. Improvements to reduce the interruption time when moving between cells is another possibility for enhancements. However, although the enhancements described in this chapter aim at improving the packet-data experience in a cellular network, it is important to stress that HSPA, MBMS, MIMO, etc. are part of an *evolution* of WCDMA. These enhancements all build upon the basic WCDMA structure defined by Release 99. Terminals using a later release of the specifications are able to coex-ist with terminals from previous releases on the same carrier. Naturally, this sets some constraints on what is possible to introduce, but also offers the significant benefit for an operator to gradually improve the network capacity. In the next chapters, the long-term evolution will be discussed, which puts less emphasis on the backwards compatibility in order to push the performance even further.

Part IV
LTE and SAE

13

LTE and SAE: Introduction and design targets

In Chapters 8–12, HSPA and the evolution thereof were described. As explained in these chapters, HSPA is an evolution of WCDMA, building upon the basic WCDMA structure and with a strong requirement on backwards compatibility to leverage on already deployed networks. In parallel to evolving HSPA, 3GPP has specified a new radio access technology, known as Long-Term Evolution (LTE). As described in Chapter 2, LTE targets more complex spectrum situations and has fewer restrictions on backwards compatibility. Thus, 3G evolution consists of two parallel tracks, both having their respective merits, for the radio access evolution. Figure 13.1 illustrates the relation between HSPA and LTE.

To support the new packet-data capabilities provided by the LTE radio interfaces, an evolved core network has been developed. The work on specifying the core network is commonly known as System Architecture Evolution (SAE).

Part IV of this book describes LTE and SAE, based on the specification work in 3GPP. The drivers behind LTE and SAE were explained in Chapter 2. Prior to starting the work on LTE and SAE, 3GPP agreed on a set of requirements, or design targets, to be taken as a basis for the development of LTE and SAE. Naturally, many of the SAE and LTE requirements overlap in terms of scope,

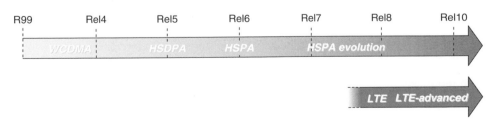

Figure 13.1 *LTE and HSPA Evolution.*

many being the same. These requirements are outlined in the remaining part of this chapter to form an understanding of the background upon which LTE has been developed.

The following chapter (Chapter 14) provides an introductory technical overview. This chapter describes the most important technologies used by LTE to support the requirements, including transmission schemes, scheduling, multi-antenna support, and spectrum flexibility. The generic technologies for the schemes were described in Part II while Part IV reveals their specific application to LTE. The chapter can either be read on its own to get a high-level overview of LTE, or as an introduction to the following chapters.

Chapter 15 describes the LTE protocol structure, including RLC, MAC, and the physical layer, explaining the logical and physical channels, and the related data flow. The LTE physical layer is described in detail in Chapters 16 and 17, including details on processing and control signaling for the OFDM downlink and the Single-Carrier FDMA uplink. Chapter 18 gives the details of the access procedures, including cell search, random access, and paging. Chapter 19 discusses the LTE transmission procedures. Finally, Chapter 20 provides details on the flexible bandwidth in LTE.

The system architecture is the topic for Chapter 20. In this chapter, details are provided of the HSPA system architecture and the SAE, including the different nodes and interfaces of the radio access network and the core network, their individual functionality, and the functional split between them.

13.1   LTE design targets

As discussed in Chapter 2, the initial 3GPP activity on 3G evolution was setting the objectives, requirements, and targets for LTE. These targets/requirements are documented in 3GPP TR 25.913 [86]. It should be noted that the capabilities, system performance, and other aspects as outlined below and in TR25.913 are the *targets* set out in the initial phase of the LTE standards development. The final capabilities and performance reached are different and do in many cases exceed the targets set at the beginning of the LTE development. More details on the performance of the final LTE standard can be found in Chapter 23.

The requirements for LTE were divided into seven different areas:

- capabilities,
- system performance,

- deployment-related aspects,
- architecture and migration,
- radio resource management,
- complexity, and
- general aspects.

Below, each of these groups is discussed.

13.1.1 Capabilities

The targets for downlink and uplink peak data-rate requirements are 100 Mbit/s and 50 Mbit/s, respectively, when operating in 20 MHz spectrum allocation. For narrower spectrum allocations, the peak data rates are scaled accordingly. Thus, the requirements can be expressed as 5 bit/s/Hz for the downlink and 2.5 bit/s/Hz for the uplink. As will be discussed below, LTE supports both FDD and TDD operation. Obviously, for the case of TDD, uplink and downlink transmission cannot, by definition, occur simultaneously. Thus the peak data rate requirement cannot be met simultaneously. For FDD, on the other hand, the LTE specifications should allow for simultaneous reception and transmission at the peak data rates specified above.

The latency requirements are split into control-plane requirements and user-plane requirements. The control-plane latency requirements address the delay for transiting from different non-active terminal states to an active state where the mobile terminal can send and/or receive data. There are two measures: one measure is expressed as the transition time from a camped state such as the Release 6 idle[1] mode state, where the requirement is 100 ms; The other measure is expressed as the transition time from a dormant state such as Release 6 Cell_ PCH[2] state where the requirement is 50 ms. For both these requirements, any sleep mode delay and non-RAN signaling are excluded.

The user-plane latency requirement is expressed as the time it takes to transmit a small IP packet from the terminal to the RAN edge node or vice versa measured on the IP layer. The one-way transmission time should not exceed 5 ms in an unloaded network, that is, no other terminals are present in the cell.

---

[1] Release 6 idle mode is a state where the terminal is unknown to the radio access network, that is, the radio access network does not have any context of the terminal and the terminal does not have any radio resources assigned. The terminal may be in sleep mode, that is, only listening to the network at specific time intervals.

[2] Release 6 Cell_ PCH state is a state where the terminal is known to the radio access network. Furthermore, the radio access network knows in which cell the terminal is in, but the terminal does not have any radio resources assigned. The terminal may be in sleep mode.

As a side requirement to the control-plane latency requirement, LTE should support at least 200 mobile terminals in the active state when operating in 5 MHz. In wider allocations than 5 MHz, at least 400 terminals should be supported. The number of inactive terminals in a cell is not explicitly stated, but should be significantly higher.

13.1.2 System performance

The LTE system performance design targets address user throughput, spectrum efficiency, mobility, coverage, and further enhanced MBMS.

In general, the LTE performance requirements in [86] are expressed relative to a baseline system using Release 6 HSPA as described in Part III in this book. For the base station, one transmit and two receive antennas are assumed, while the terminal has a maximum of one transmit and two receive antennas. However, it is important to point out that the more advanced features described in Chapter 12 as part of the evolution of HSPA are not included in the baseline reference. Hence, albeit the terminal in the baseline system is assumed to have two receive antennas, a simple RAKE receiver is assumed (referred to as Type 1 in Chapter 12). Similarly, spatial multiplexing is not assumed in the baseline system.

The LTE user throughput requirement is specified at two points: at the average and at the fifth percentile of the user distribution (where 95 percent of the users have better performance). A spectrum efficiency target has also been specified, where in this context, spectrum efficiency is defined as the system throughput per cell in bit/s/ MHz /cell. These design targets are summarized in Table 13.1.

The mobility requirements focus on the mobile terminals speed. Maximal performance is targeted at low terminal speeds, 0–15 km/h, whereas a slight degradation is allowed for higher speeds. For speeds up to 120 km/h, LTE should provide high performance and for speeds above 120 km/h, the system should

Table 13.1 *LTE user throughput and spectrum efficiency requirements.*

Performance measure	Downlink target relative to baseline	Uplink target relative to baseline
Average user throughput (per MHz)	3× − 4×	2× − 3×
Cell-edge user throughput (per MHz, 5th percentile)	2× − 3×	2× − 3×
Spectrum efficiency (bit/s/Hz/cell)	3× − 4×	2× − 3×

be able to maintain the connection across the cellular network. The maximum speed to manage in an LTE system is set to 350 km/h (or even up to 500 km/h depending on frequency band). Special emphasis is put on the voice service that LTE needs to provide with equal quality as supported by WCDMA/HSPA.

The coverage requirements focus on the cell range (radius), that is the maximum distance from the cell site to a mobile terminal in a cell. The requirement for non-interference-limited scenarios is to meet the user throughput, the spectrum efficiency, and the mobility requirements for cells with up to 5 km cell range. For cells with up to 30 km cell range, a slight degradation of the user throughput is tolerated and a more significant degradation of the spectrum efficiency is acceptable relative to the requirements. However, the mobility requirements should be met. Cell ranges up to 100 km should not be precluded by the specifications, but no performance requirements are stated in this case.

The further enhanced MBMS requirements address both broadcast mode and unicast mode. In general, LTE should provide MBMS services better than what is possible with Release 6. The requirement for the broadcast case is a spectral efficiency of 1 bit/s/Hz, corresponding to around 16 mobile-TV channels using in the order of 300 kbit/s each in a 5 MHz spectrum allocation. Furthermore, it should be possible to provide the MBMS service as the only service on a carrier, as well as mixed with other, non-MBMS services. Naturally, simultaneously voice calls and MBMS services should be possible with the LTE specifications.

13.1.3 Deployment-related aspects

The deployment-related requirements include deployment scenarios, spectrum flexibility, spectrum deployment, and coexistence and interworking with other 3GPP radio access technologies such as GSM and WCDMA /HSPA.

The requirement on the deployment scenario includes both the case when the LTE system is deployed as a stand-alone system and the case when it is deployed together with WCDMA/HSPA and/or GSM. Thus, this requirement is not in practice limiting the design criteria. The requirements on the spectrum flexibility and deployment are outlined in more detail in the section 13.1.3.1.

The coexistence and interworking with other 3GPP systems and their respective requirements set the requirement on mobility between LTE and GSM, and between LTE and WCDMA/HSPA for mobile terminals supporting those technologies. Table 13.2 lists the interruption requirements, that is, longest acceptable interruption in the radio link when moving between the different radio-access

Table 13.2   *Interruption time requirements, LTE – GSM and LTE – WCDMA.*

	Non-real-time (ms)	Real-time (ms)
LTE to WCDMA	500	300
LTE to GSM	500	300

technologies, for both real-time and non-real-time services. It is worth noting that these requirements are very loose for the handover interruption time and significantly better values are expected in real deployments.

The coexistence and interworking requirement also address the switching of multicast traffic from being provided in a broadcast manner in LTE to being provided in unicast manner in either GSM or WCDMA, albeit no numbers are given.

13.1.3.1   *Spectrum flexibility and deployment*
The basis for the requirements on spectrum flexibility is the requirement for LTE to be deployed in existing IMT-2000 frequency bands, which implies coexistence with the systems that are already deployed in those bands, including WCDMA/HSPA and GSM. A related part of the LTE requirements in terms of spectrum flexibility is the possibility to deploy LTE -based radio access in both paired and unpaired spectrum allocations, that is LTE should support both Frequency Division Duplex (FDD), and Time Division Duplex (TDD).

The duplex scheme or duplex arrangement is a property of a radio access technology. However, a given spectrum allocation is typically also associated with a specific duplex arrangement. FDD systems are deployed in paired spectrum allocations, having one frequency range intended for downlink transmission and another for uplink transmission. TDD systems are deployed in unpaired spectrum allocations. LTE should be able to operate in unpaired as well as paired spectrum. It should also be possible to deploy LTE in different frequency bands. The supported frequency bands should be specified based on 'release independence,' which means that the first release of LTE does not have to support all bands from the start.

An example is the IMT-2000 spectrum at 2 GHz, that is, the IMT-2000 'core band.' As shown in Figure 13.2, it consists of the paired frequency bands 1920–1980 MHz and 2110–2170 MHZ intended for FDD-based radio access, and the two frequency bands 1910–1920 MHz and 2010–2025 MHz intended for TDD-based radio access. Note that through local and regional regulation the use of the IMT-2000 spectrum may be different than shown here.

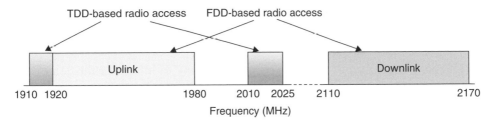

Figure 13.2 *The original IMT -2000 'core band' spectrum allocations at 2 GHz.*

A very likely scenario in many frequency bands is that an operator wants to migrate to LTE in a spectrum allocation which is almost fully deployed with a 2G or 3G technology, such as GSM or CDMA2000. Such a migration can usually take place only in steps, in order to guarantee a maintained good service for existing users. This requirement for gradual deployment in allocations that can initially be quite small puts a requirement on spectrum flexibility for LTE in terms of a scalable transmission bandwidth.

Furthermore, [86] also addresses coexistence and co-siting with GSM and WCDMA on adjacent frequencies, as well as coexistence between operators on adjacent frequencies and networks in different countries using overlapping spectrum. There is also a requirement that no other system should be required in order for a terminal to access LTE, that is, LTE is supposed to have all the necessary control signaling required for enabling access.

13.1.4 Architecture and migration

A few guiding principles for the LTE RAN architecture design as stated by 3GPP are listed in [86]:

- A single LTE RAN architecture should be agreed.
- The LTE RAN architecture should be packet based, although real-time and conversational class traffic should be supported.
- The LTE RAN architecture should minimize the presence of 'single points of failure' without additional cost for backhaul.
- The LTE RAN architecture should simplify and minimize the introduced number of interfaces.
- Radio Network Layer (RNL) and Transport Network Layer (TNL) interaction should not be precluded if in the interest of improved system performance.
- The LTE RAN architecture should support an end-to-end QoS. The TNL should provide the appropriate QoS requested by the RNL.

- QoS mechanism(s) should take into account the various types of traffic that exists to provide efficient bandwidth utilization: *Control-Plane* traffic, *User-Plane* traffic, *O&M* traffic, etc.
- The LTE RAN should be designed in such a way to minimize the delay variation (jitter) for traffic needing low jitter, for example, TCP/IP.

13.1.5   Radio resource management

The radio resource management requirements are divided into enhanced support for end-to-end QoS, efficient support for transmission of higher layers, and support of load sharing and policy management across different radio access technologies.

The enhanced support for end-to-end QoS requires an 'improved matching of service, application and protocol requirements (including higher layer signaling) to RAN resources and radio characteristics.'

The efficient support for transmission of higher layers requires that the LTE RAN should 'provide mechanisms to support efficient transmission and operation of higher layer protocols over the radio interface, such as IP header compression.'

The support of load sharing and policy management across different radio access technologies requires consideration of reselection mechanisms to direct mobile terminals toward appropriate radio access technologies in all types of states as well as that support for end-to-end QoS during handover between radio access technologies.

13.1.6   Complexity

The LTE complexity requirements address the complexity of the overall system as well as the complexity of the mobile terminal. Essentially, these requirements imply that the number of options should be minimized with no redundant mandatory features. This also leads to a minimized number of necessary test cases.

13.1.7   General aspects

The section covering general requirements on LTE address the cost-and service-related aspects. Obviously, it is desirable to minimize the cost while maintaining the desired performance for all envisioned services. Specific to the cost, the backhaul and operation and maintenance is addressed. Thus not only the radio interface, but also the transport to the base-station sites and the management system should be addressed by LTE. A strong requirement on multi-vendor interfaces also falls into this category of requirements. Furthermore, low complexity and low power consuming mobile terminals are required.

13.2 SAE design targets

The SAE objectives were outlined in the study item description of SAE, and some very high-level targets are set in [88] produced by TSG SA WG 1. The SAE targets are divided into several areas:

- high-level user and operational aspects,
- basic capabilities,
- multi-access and seamless mobility,
- man–machine interface aspects,
- performance requirements for the evolved 3GPP system,
- security and privacy, and
- charging aspects.

Although the SAE requirements are many and split into the subgroups above, the SAE requirements are mainly non-radio access related. Thus, this section tries to summarize the most important SAE requirements that have an impact on either the radio access network or the SAE architecture.

The SAE system should be able to operate with more than the LTE radio access network and there should be mobility functions allowing a mobile terminal to move between the different radio-access systems. In fact, the requirements do not limit the mobility between radio access networks, but opens up for mobility to fixed-access network. The access networks need not to be developed by 3GPP, other non-3GPP access networks should also be considered.

As always in 3GPP, roaming is a very strong requirement for SAE, including inbound and outbound roaming to other SAE networks and legacy networks. Furthermore, interworking with legacy packet-switched and circuit-switched services is a requirement. However, it is not required to support the circuit switched services from the circuit-switched domain of the legacy networks.

The SAE requirements also list performance as an essential requirement but do not go into the same level of details as the LTE requirements. Different traffic scenarios and usage are envisioned, for example user to user and user to group communication. Furthermore, resource efficiency is required, especially radio resource efficiency (cf. spectrum efficiency requirement for LTE). The SAE resource efficiency requirement is not as elaborated as the LTE requirement. Thus it is the LTE requirement that is the design requirement.

Of course, the SAE requirements address the service aspects and require that the traditional services such as voice, video, messaging, and data file exchange should be supported, and in addition multicast and broadcast services. In fact, with the requirement to support IPv4 and IPv6 connectivity, including mobility between access networks supporting different IP versions as well as communication between terminals using different versions, any service based on IP will be supported, albeit perhaps not with optimized quality of service.

The quality of service requirement of SAE is well elaborated upon in [88]. The SAE system should for example, provide no perceptible deterioration of audio quality of a voice call during and following handover between dissimilar circuit-switched and packet-switched access networks. Furthermore, the SAE should ensure that there is no loss of data as a result of a handover between dissimilar fixed and mobile access systems. A particular important requirement for the SAE QoS concept is that the SAE QoS concept should be backwards compatible with the pre- SAE QoS concepts of 3GPP. This is to ensure smooth mobility between different 3GPP accesses (LTE, WCDMA/HSPA and GSM).

The SAE system should provide advanced security mechanisms that are equivalent to or better than 3GPP security for WCDMA/HSPA and GSM. This means that protection against threats and attacks including those present on the Internet should be part of SAE. Furthermore, the SAE system should provide information authenticity between the mobile terminal and the network, but at the same time enable lawful interception of the traffic.

The SAE system has strong requirements on user privacy. Several levels of user privacy should be provided, for example communication confidentiality, location privacy, and identity protection. Thus, SAE -based systems will hide the identity of the users from unauthorized third parties, protect the content, origin and destination of a particular communication from unauthorized parties, and protect the location of the user from unauthorized parties. Authorized parties are normally government agencies, but the user may give certain parties the right to know about the location of the mobile terminal. One example hereof is fleet management for truck dispatchers.

Several charging models, including calling party pays, flat rate, and charging based on QoS is required to be supported in SAE. Charging aspects are sometimes visible in the radio access networks, especially those charging models that are based on delivered QoS or delivered data volumes. However, most charging schemes are only looking at information available in the core network.

14

LTE radio access: An overview

In the previous chapter, the targets of LTE were discussed. In this chapter, an overview of some of the most important technology components and features of LTE will be provided. Chapters 15–20 will then provide a more detailed description of the LTE radio access.

As already mentioned, in parallel to the development of LTE, there has also been an evolution of the overall 3GPP architecture, a work known as the *System Architecture Evolution* (SAE). A description of SAE and the guiding principles behind the SAE design is found in Chapter 21.

14.1 LTE transmission schemes: Downlink OFDM and uplink DFTS-OFDM/SC-FDMA

The LTE downlink transmission scheme is based on OFDM. As discussed in Chapter 4, OFDM is an attractive downlink transmission scheme for several reasons. Due to the relatively long OFDM symbol time in combination with a cyclic prefix, OFDM provides a high degree of robustness against channel frequency selectivity. Although signal corruption due to a frequency-selective channel can, in principle, be handled by equalization at the receiver side, the complexity of such equalization starts to become unattractively high for implementation in a mobile terminal at bandwidths exceeding 5 MHz. Therefore, OFDM with its inherent robustness to channel frequency selectivity is attractive for the downlink when extending the bandwidth beyond 5 MHz.

Additional benefits with OFDM include:

- OFDM provides access to the frequency domain, thereby enabling an additional degree of freedom to the channel-dependent scheduler compared to HSPA for which only time-domain scheduling is possible.

- Flexible transmission bandwidth to support operation in spectrum allocations of different size is straightforward with OFDM, at least from a baseband
  perspective, by varying the number of OFDM subcarriers used for transmission. Note, however, that support a flexible transmission bandwidth also
  requires flexible RF filtering, etc., for which the exact transmission scheme
  is irrelevant. Nevertheless, maintaining the same baseband-processing structure, regardless of the bandwidth, eases terminal development, design, and
  implementation.
- Broadcast/multicast transmission, where the same information is transmitted
  from multiple base stations, is straightforward with OFDM as described
  already in Chapter 4.

For the LTE uplink, single-carrier transmission based on DFT-spread OFDM
(DFTS-OFDM), as described in Chapter 5, is used. The use of single-carrier
modulation in the uplink is motivated by the lower peak-to-average ratio of the
transmitted signal compared to multi-carrier transmission such as OFDM. The
smaller the peak-to-average ratio of the transmitted signal, the higher the average transmission power can be for a given power amplifier. Single-carrier transmission therefore allows for more efficient usage of the power amplifier, which
translates into an increased coverage and reduced terminal power consumption.
At the same time, the equalization required to handle corruption of the single-
carrier signal due to frequency-selective fading is less of an issue in the uplink
due to the feasibility of more powerful signal processing resources at the base-
station side, compared to the mobile terminal.

In contrast to the non-orthogonal WCDMA/HSPA uplink, which is also based
on single-carrier transmission, the LTE uplink is based on orthogonal separation of uplink transmissions in the time and/or frequency domain[1]. Orthogonal
separation is in many cases beneficial as it avoids intra-cell interference.
However, as discussed in Chapter 5, allocating a very large instantaneous bandwidth resource for transmission from a single terminal is not an efficient strategy
in situations where the data rate is mainly limited by the available terminal
transmit power rather than bandwidth. In such situations, a terminal is instead
allocated only a part of the total available bandwidth and other terminals can
transmit in parallel on the remaining part of the spectrum. Thus, as the LTE
uplink contains a frequency-domain multiple-access component, the LTE uplink
transmission scheme is sometimes also referred to as Single-Carrier FDMA
(SC-FDMA).

---

[1] In principle, orthogonal user separation can be achieved in the time domain only by assigning the entire
uplink transmission bandwidth to one user at a time (this is possible already with enhanced uplink as described
in Chapter 11).

14.2 Channel-dependent scheduling and rate adaptation

At the core of the LTE transmission scheme is the use of *shared-channel transmission*, with the overall time-frequency resource dynamically shared between users. This is similar to the approach taken in HSDPA, although the realization of the shared resource differs between the two – time and frequency in case of LTE vs. time and channelization codes in case of HSDPA. The use of shared-channel transmission is well matched to the rapidly varying resource requirements posed by packet data and also enables several of the other key technologies used by LTE.

The *scheduler* controls, for each time instant, to which users the shared resources should be assigned. The scheduler also determines the data rate to be used for each link, that is *rate adaptation* can be seen as a part of the scheduler. The scheduler is thus a key element and to a large extent determines the overall downlink system performance, especially in a highly loaded network. Both downlink and uplink transmissions are subject to tight scheduling. From Chapter 7 it is well known that a substantial gain in system capacity can be achieved if the channel conditions are taken into account in the scheduling decision, so-called *channel-dependent scheduling*. This is exploited already in HSPA, where the downlink scheduler transmits to a user when its channel conditions are advantageous to maximize the data rate, and is, to some extent, also possible for the Enhanced Uplink. However, LTE has, in addition to the time domain, also access to the frequency domain, due to the use of OFDM in the downlink and DFTS-OFDM in the uplink. Therefore, the scheduler can, for each frequency region, select the user with the best channel conditions. In other words, scheduling in LTE can take channel variations into account not only in the time domain, as HSPA, but also in the frequency domain. This is illustrated in Figure 14.1.

The possibility for channel-dependent scheduling in the frequency domain is particularly useful at low terminal speeds, in other words when the channel is varying slowly in time. As discussed in Chapter 7, channel-dependent scheduling relies on channel-quality variations between users to obtain a gain in system capacity. For delay-sensitive services, a time-domain only scheduler may be forced to schedule a particular user, despite the channel quality not being at its peak. In such situations, exploiting channel-quality variations also in the frequency domain will help improving the overall performance of the system. For LTE, scheduling decisions can be taken as often as once every 1 ms and the granularity in the frequency domain is 180 kHz. This allows for relatively fast channel variations to be tracked and utilized by the scheduler.

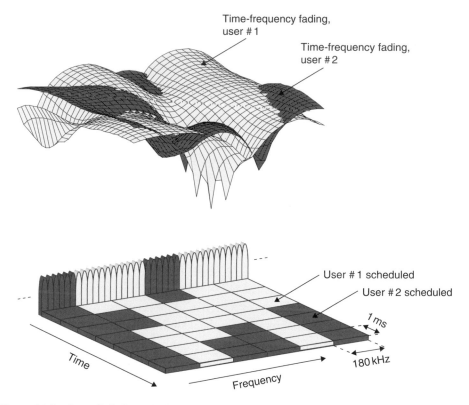

Figure 14.1 *Downlink channel-dependent scheduling in time and frequency domains.*

14.2.1 Downlink scheduling

To support downlink scheduling, a terminal may provide the network with *channel-status* reports indicating the instantaneous downlink channel quality in both the time and frequency domain. The channel status can, for example, be obtained by measuring on a reference signal transmitted on the downlink and used also for demodulation purposes. Based on the channel-status report, the downlink scheduler can assign resources for downlink transmission to different mobile terminals, taking the channel quality into account in the scheduling decision. In principle, a scheduled terminal can be assigned an arbitrary combination of 180 kHz wide resource blocks in each 1 ms scheduling interval.

14.2.2 Uplink scheduling

The LTE uplink is based on orthogonal separation of different uplink transmissions and it is the task of the uplink scheduler to assign resources in both time and frequency domain (combined TDMA/FDMA) to different mobile terminals. Scheduling decisions, taken once per 1 ms, control what set of mobile terminals

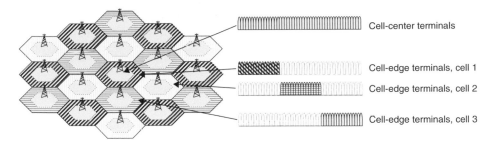

Figure 14.2 *Example of inter-cell interference coordination.*

are allowed to transmit within a cell during a given time interval and, for each terminal, on what frequency resources the transmission is to take place and what uplink data rate (transport format) to use.

Channel conditions can also be taken into account in the uplink scheduling process, similar to the downlink scheduling. However, as will be discussed in more detail in the following chapters, obtaining information about the uplink channel conditions is a non-trivial task. Therefore, different means to obtain uplink diversity are important as a complement in situations where uplink channel-dependent scheduling is not suitable.

14.2.3 Inter-cell interference coordination

LTE provides orthogonality between users within a cell in both uplink and downlink, that is at least in principle there is no interference between transmissions within one cell (no intra-cell interference). Hence, LTE performance in terms of spectrum efficiency and available data rates is, relatively speaking, more limited by interference from other cells (inter-cell interference) compared to WCDMA/HSPA. Means to reduce or control the inter-cell interference can therefore, potentially, provide substantial benefits to LTE performance, especially in terms of the service (data rates, etc.) that can be provided to users at the cell edge.

Inter-cell interference coordination is a scheduling strategy in which the cell-edge data rates are increased by taking inter-cell interference into account. Basically, inter-cell interference coordination implies certain (frequency-domain) restrictions to the uplink and downlink schedulers in order to control the inter-cell interference. By restricting the transmission power of parts of the spectrum in one cell, the interference seen in the neighboring cells in this part of the spectrum will be reduced. This part of the spectrum can then be used to provide higher data rates for users in the neighboring cell. In essence, the frequency reuse factor is different in different parts of the cell (Figure 14.2).

Note that inter-cell interference coordination is mainly a scheduling strategy, taking the situation in neighboring cells into account. Thus, inter-cell interference coordination is to a large extent an implementation issue and hardly visible in the specifications. This also implies that interference coordination can be applied to only a selected set of cells, depending on the requirements set by a particular deployment. To aid the implementation of various intercell-interference coordination schemes, LTE supports exchange of interference indicators between base stations.

14.3   Hybrid ARQ with soft combining

Fast hybrid ARQ with soft combining is used in LTE for very similar reasons as in HSPA, namely to allow the terminal to rapidly request retransmissions of erroneously received transport blocks and to provide a tool for implicit rate adaptation. The underlying protocol is also similar to the one used for HSPA– multiple parallel stop-and-wait hybrid ARQ processes. Retransmissions can be rapidly requested after each packet transmission, thereby minimizing the impact on end-user performance from erroneously received packets. Incremental redundancy is used as the soft combining strategy and the receiver buffers the soft bits to be able to do soft combining between transmission attempts.

14.4   Multiple antenna support

LTE already from the beginning supports multiple antennas at both the base station and the terminal as an integral part of the specifications. In many respects, the use of multiple antennas is the key technology to reach the aggressive LTE performance targets. As discussed in Chapter 6, multiple antennas can be used in different ways for different purposes:

- Multiple receive antennas can be used for receive diversity. For uplink transmissions, this has been used in many cellular systems for several years. However, as dual receive antennas is the baseline for all LTE terminals, the downlink performance is also improved. The simplest way of using multiple receive antennas is classical receive diversity to collect additional energy and suppress fading, but additional gains can be achieved in interference-limited scenarios if the antennas also are used not only to provide diversity, but also to suppress interference as discussed in Chapter 6.
- Multiple transmit antennas at the base station can be used for transmit diversity and different types of beam-forming. The main goal of beam-forming is to improve the received SNR and/or SIR and, eventually, improve system capacity and coverage.

- *Spatial multiplexing*, sometimes referred to as MIMO, using multiple antennas at both the transmitter and receiver is supported by LTE. Spatial multiplexing results in an increased data rate, channel conditions permitting, in bandwidth-limited scenarios by creating several parallel 'channels' as described in Chapter 6.

In general, the different multi-antenna techniques are beneficial in different scenarios. As an example, at relatively low SNR and SIR, such as at high load or at the cell edge, spatial multiplexing provides relatively limited benefits. Instead, in such scenarios multiple antennas at the transmitter side should be used to raise the SNR/SIR by means of beam-forming. On the other hand, in scenarios where there already is a relatively high SNR and SIR, for example in small cells, raising the signal quality further provides relatively minor gains as the achievable data rates are then mainly bandwidth limited rather than SIR/SNR limited. In such scenarios, spatial multiplexing should be used instead to fully exploit the good channel conditions. The multi-antenna scheme used is under control of the base station, which therefore can select a suitable scheme for each transmission.

14.5 Multicast and broadcast support

Multi-cell broadcast implies transmission of the same information from multiple cells as described in Chapter 4. By exploiting this at the terminal, effectively using signal power from multiple cell sites at the detection, a substantial improvement in coverage (or higher broadcast data rates) can be achieved. This is already exploited in WCDMA where, in case of multi-cell broadcast/multicast, a mobile terminal may receive signals from multiple cells and actively *soft combine* these within the receiver as described in Chapter 11.

LTE takes this one step further to provide highly efficient multi-cell broadcast. By transmitting not only identical signals from multiple cell sites (with identical coding and modulation), but also synchronize the transmission timing between the cells, the signal at the mobile terminal will appear exactly as a signal transmitted from a single cell site and subject to multi-path propagation. Due to the OFDM robustness to multi-path propagation, such multi-cell transmission, also referred to as *Multicast–Broadcast Single-Frequency Network* (MBSFN) transmission, will then not only improve the received signal strength, but also eliminate the inter-cell interference as described in Chapter 4. Thus, with OFDM, multi-cell broadcast/multicast throughput may eventually be limited by noise only and can then, in case of small cells, reach extremely high values.

It should be noted that the use of MBSFN transmission for multi-cell broadcast/multicast assumes the use of tight synchronization and time alignment of the signals transmitted from different cell sites.

14.6   Spectrum flexibility

As discussed in Chapter 13, a high degree of spectrum flexibility is one of the main characteristics of the LTE radio access. The aim of this spectrum flexibility is to allow for the deployment of the LTE radio access in diverse spectrum with different characteristics, including different duplex arrangements, different frequency-bands-of-operation, and different sizes of the available spectrum. Chapter 20 outlines further details of how the spectrum flexibility is achieved in LTE.

14.6.1   Flexibility in duplex arrangement

One important part of the LTE requirements in terms of spectrum flexibility is the possibility to deploy LTE-based radio access in both paired *and* unpaired spectrum. Therefore, LTE supports both frequency- and time-division-based duplex arrangements. *Frequency Division Duplex* (FDD) as illustrated to the left in Figure 14.3 implies that downlink and uplink transmission take place in different, sufficiently separated, frequency bands. *Time Division Duplex* (TDD), as illustrated to the right in Figure 14.3, implies that downlink and uplink transmission take place in different, non-overlapping time slots. Thus, TDD can operate in unpaired spectrum, whereas FDD requires paired spectrum. The required flexibility and resulting requirements to support LTE operation in different paired and unpaired frequency arrangements are further discussed in Chapter 20.

Support for both paired and unpaired spectrum is part of the 3GPP specifications already from Release 99 through the use of FDD-based WCDMA/HSPA radio access as described in Part III in paired allocations and TDD-based TD-CDMA/TD-SCDMA radio access (see Chapter 24) in unpaired allocations. However, this is achieved by means of, at least in the details, relatively different radio-access

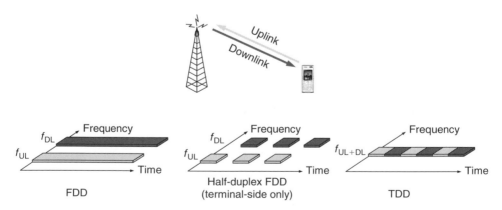

Figure 14.3   *Frequency- and time-division duplex.*

technologies and, as a consequence, terminals capable of both FDD and TDD operations are fairly uncommon. LTE, on the other hand, supports both FDD and TDD *within a single radio-access technology*, leading to a minimum of deviation between FDD and TDD for LTE-based radio access. As a consequence of this, the overview of the LTE radio access provided in the following chapters is, to a large extent, valid for both FDD and TDD. In case of differences between FDD and TDD, these differences will be explicitly indicated.

LTE also supports *half-duplex* FDD at the terminal (illustrated in the middle of Figure 14.3). In half-duplex FDD, transmission and reception *at a specific terminal* are separated in both frequency and time. The base station still uses full duplex as it simultaneously may schedule *different* terminals in uplink and downlink; this is similar to, for example, GSM operation. The main benefit with half-duplex FDD is the reduced terminal complexity as no duplex filter is needed in the terminal, which is especially beneficial in case of multi-band terminals which otherwise would need multiple sets of duplex filters.

14.6.2 Flexibility in frequency-band-of-operation

LTE is envisioned to be deployed on a per-need basis when and where spectrum can be made available, either by the assignment of new spectrum for mobile communication, such as the 2.6 and 3.5 GHz band, or by the migration to LTE of spectrum currently used for other mobile-communication technologies, such as GSM or cdma2000 systems, or even non-mobile radio technologies such as in current broadcast spectrum. As a consequence, it is required that the LTE radio access should be able to operate in a wide range of frequency bands, from as low as 450 MHz band up to, at least, 3.5 GHz.

14.6.3 Bandwidth flexibility

Related to the possibility to deploy the LTE radio access in different frequency bands is the possibility of being able to operate LTE with different transmission bandwidths on both downlink and uplink. The main reason for this is that the amount of spectrum being available for LTE may vary significantly between different frequency bands and also depending on the exact situation of the operator. Furthermore, the possibility to operate in different spectrum allocations gives the possibility for gradual migration of spectrum from other radio access technologies to LTE.

LTE supports operation in a wide range of spectrum allocations, achieved by a flexible transmission bandwidth being part of the LTE specifications. To efficiently support very high data rates when spectrum is available, a wide transmission bandwidth is necessary as discussed in Chapter 3. However, a sufficiently large amount of spectrum may not always be available, either due to the band-of-operation or due to a gradual migration from another radio-access technology, in which case LTE can be operated with a more narrow transmission bandwidth. Obviously, in such cases, the maximum achievable data rates will be reduced correspondingly.

The LTE physical-layer specifications [106–109] are bandwidth-agnostic and do not make any particular assumption on the supported transmission bandwidths beyond a minimum value. As will be seen in the following, the basic radio-access specification including the physical-layer and protocol specifications, allows for any transmission bandwidth ranging from roughly 1 MHz up to around 20 MHz. At the same time, at an initially stage, radio-frequency requirements are only specified for a limited subset of transmission bandwidth, corresponding to what is predicted to be relevant spectrum-allocation sizes and relevant migration scenarios. Thus, in practice LTE radio access supports a limited set of transmission bandwidths, but additional transmission bandwidths can easily be supported by updating only the RF specifications.

15

LTE radio interface architecture

Similar to WCDMA/HSPA, as well as to most other modern communication systems, the processing specified for LTE is structured into different protocol layers. Although several of these layers are similar to those used for WCDMA/HSPA, there are some differences, for example due to the differences in the overall architecture between WCDMA/HSPA and LTE. This chapter contains an overview of these protocol layers and their interaction. A detailed description of the LTE architecture is found in Chapter 21, where the location of the different protocol entities in the different network nodes is discussed. For the discussion in this chapter, it suffices to note that the LTE radio-access architecture consists of a single node – the eNodeB.[1] The eNodeB communicates with one or several mobile terminals, also known as UEs.

A general overview of the LTE protocol architecture for the downlink is illustrated in Figure 15.1. As this will become clear in the subsequent discussion, not all the entities illustrated in Figure 15.1 are applicable in all situations. For example, neither MAC scheduling nor hybrid ARQ with soft combining is used for broadcast of system information. Furthermore, the LTE protocol structure related to uplink transmissions is similar to the downlink structure in Figure 15.1, although there are some differences with respect to, for example, transport-format selection and multi-antenna transmission.

Data to be transmitted in the downlink enters the processing chain in the form of IP packets on one of the *SAE bearers*. Prior to transmission over the radio interface,

---

[1] The term eNodeB is introduced in LTE to indicate the additional functionality placed in the eNodeB compared to the functionality in the NodeB in WCDMA/HSPA.

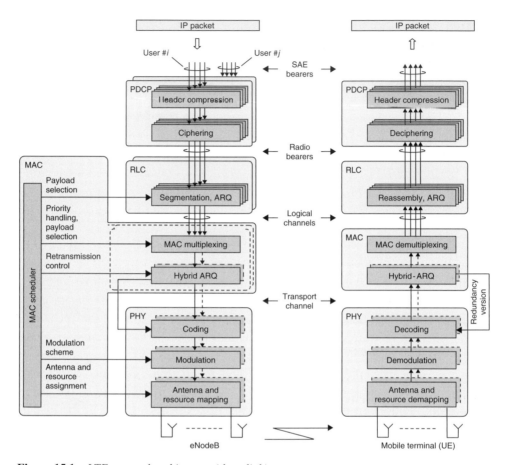

Figure 15.1   *LTE protocol architecture (downlink).*

incoming IP packets are passed through multiple protocol entities, summarized below and described in more detail in the following sections:

- *Packet Data Convergence Protocol* (PDCP) performs IP header compression to reduce the number of bits to transmit over the radio interface. The header-compression mechanism is based on Robust Header Compression (ROHC) [64], a standardized header-compression algorithm also used in WCDMA as well as several other mobile-communication standards. PDCP is also responsible for ciphering and integrity protection of the transmitted data. At the receiver side, the PDCP protocol performs the corresponding deciphering and decompression operations. There is one PDCP entity per SAE bearer configured for a mobile terminal.
- *Radio Link Control* (RLC) is responsible for segmentation/concatenation, retransmission handling, and in-sequence delivery to higher layers. Unlike WCDMA, the RLC protocol is located in the eNodeB since there is only a single type of node in the LTE radio-access-network architecture. The RLC

offers services to the PDCP in the form of *radio bearers*. There is one RLC entity per radio bearer configured for a terminal.

- *Medium Access Control* (MAC) handles hybrid-ARQ retransmissions and uplink and downlink scheduling. The scheduling functionality is located in the eNodeB, which has one MAC entity per cell, for both uplink and downlink. The hybrid-ARQ protocol part is present in both the transmitting and receiving end of the MAC protocol. The MAC offers services to the RLC in the form of *logical channels*.
- *Physical Layer* (PHY) handles coding/decoding, modulation/demodulation, multi-antenna mapping, and other typical physical layer functions. The physical layer offers services to the MAC layer in the form of *transport channels*.

The remaining of the chapter contains an overview of the RLC, MAC, and physical layers. A more detailed description of the LTE physical layer is given in Chapters 16 (downlink) and 17 (uplink), followed by an overview of LTE access procedures in Chapter 18 and transmission procedures in Chapter 19.

15.1 Radio link control

Similar to WCDMA/HSPA, LTE RLC is responsible for segmentation/concatenation of (header-compressed) IP packets, also known as RLC SDUs, from the PDCP into suitably sized RLC PDUs.[2] It also handles retransmission of erroneously received PDUs, as well as duplicate removal of received PDUs. Finally, the RLC ensures in-sequence delivery of SDUs to upper layers. Depending on the type of service, the RLC can be configured in different modes to perform some or all of these functions.

Segmentation and concatenation, one of the main RLC functions, is illustrated in Figure 15.2. Depending on the scheduler decision, a certain amount of data is selected for transmission from the RLC SDU buffer and the SDUs are segmented/concatenated to create the RLC PDU. Thus, for LTE the RLC PDU size varies *dynamically*, whereas WCDMA/HSPA prior to Release 7 uses a semi-static PDU size.[3] For high data rates, a large PDU size results in a smaller relative overhead, while for low data rates, a small PDU size is required as the payload would otherwise be too large, leading to extensive padding. Hence, as the LTE data rates may range from a few kbit/s up to 300 Mbit/s, dynamic PDU sizes are motivated for

---

[2] In general, the data entity from/to a higher protocol layer is known as a Service Data Unit (SDU) and the corresponding entity to/from a lower protocol layer entity is denoted Protocol Data Unit (PDU).

[3] The possibility to segment RLC PDUs is introduced in WCDMA/HSPA Release 7 as described in Chapter 12, providing similar benefits as a dynamic PDU size.

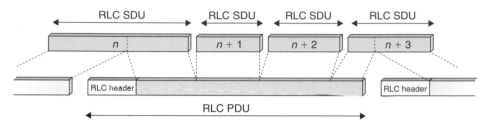

Figure 15.2 *RLC segmentation and concatenation.*

LTE. Since the RLC, scheduler, and rate adaptation mechanisms are all located in the eNodeB, dynamic PDU sizes are easily supported for LTE. In each RLC PDU, a header is included, containing, among other things, a sequence number used for in-sequence delivery and by the retransmission mechanism.

The RLC retransmission mechanism is also responsible for providing error-free delivery of data to higher layers. To accomplish this, a retransmission protocol operates between the RLC entities in the receiver and transmitter. By monitoring the sequence numbers of the incoming PDUs, the receiving RLC can identify missing PDUs. Status reports are then fed back to the transmitting RLC entity, requesting retransmission of missing PDUs. Based on the received status report, the RLC entity at the transmitter can take the appropriate action and retransmit the missing PDUs if needed.

Although the RLC is capable of handling transmission errors due to noise, unpredictable channel variations, etc., error-free delivery is in most cases handled by the MAC-based hybrid-ARQ protocol. The use of a retransmission mechanism in the RLC may therefore seem superfluous at first. However, as will be discussed in Section 15.2.3, this is not the case and the use of both RLC- and MAC-based retransmission mechanisms is in fact well motivated by the differences in the feedback signaling.

The details of RLC are further described in Chapter 19.

15.2 Medium access control

The MAC layer handles logical-channel multiplexing, hybrid-ARQ retransmissions, and uplink and downlink scheduling. In contrast to HSPA, which uses uplink macro-diversity and therefore defines both serving and non-serving cells (see Chapter 10), LTE only defines a serving cell as there is no uplink macro-diversity. The serving cell is the cell the mobile terminal is connected to and which is responsible for scheduling decisions and hybrid-ARQ operation for the mobile terminal.

15.2.1 Logical channels and transport channels

The MAC offers services to the RLC in the form of *logical channels*. A logical channel is defined by the *type* of information it carries and is generally classified as a *control channel*, used for transmission of control and configuration information necessary for operating an LTE system, or as a *traffic channel*, used for the user data. The set of logical-channel types specified for LTE includes:

- *Broadcast Control Channel* (BCCH), used for transmission of *system information* from the network to all mobile terminals in a cell. Prior to accessing the system, a mobile terminal needs to acquire the system information to find out how the system is configured and, in general, how to behave properly within a cell.
- *Paging Control Channel* (PCCH), used for paging of mobile terminals whose location on cell level is not known to the network. The paging message therefore needs to be transmitted in multiple cells.
- *Common Control Channel* (CCCH), used for transmission of control information in conjunction with random access.
- *Dedicated Control Channel* (DCCH), used for transmission of control information to/from a mobile terminal. This channel is used for individual configuration of mobile terminals such as different handover messages.
- *Multicast Control Channel* (MCCH), used for transmission of control information required for reception of the MTCH, see below.
- *Dedicated Traffic Channel* (DTCH), used for transmission of user data to/from a mobile terminal. This is the logical channel type used for transmission of all uplink and non-MBSFN downlink user data.
- *Multicast Traffic Channel* (MTCH), used for downlink transmission of MBMS services.

A similar logical-channel structure is used for WCDMA/HSPA. However, compared to WCDMA/HSPA, the LTE logical-channel structure is somewhat simplified, with a reduced number of logical-channel types.

From the physical layer, the MAC layer uses services in the form of *Transport Channels*. A transport channel is defined by *how* and *with what characteristics* the information is transmitted over the radio interface. Following the notation from HSPA, which has been inherited for LTE, data on a transport channel is organized into *transport blocks*. In each *Transmission Time Interval* (TTI), at most one transport block of a certain size is transmitted over the radio interface to/from a mobile terminal in absence of spatial multiplexing. In case of spatial multiplexing ('MIMO'), there can be up to two transport blocks per TTI.

Associated with each transport block is a *Transport Format* (TF), specifying *how* the transport block is to be transmitted over the radio interface. The transport format includes information about the transport-block size, the modulation scheme, and the antenna mapping. Together with the resource assignment, the resulting code rate can then be derived from the transport format. By varying the transport format, the MAC layer can thus realize different data rates. Rate control is therefore also known as *transport-format selection*.

The following transport channels are defined for LTE:

- *Broadcast Channel* (BCH) has a fixed transport format, provided by the specifications. It is used for transmission of parts of the BCCH system information, more specifically the so-called *Master Information Block* (MIB), as described in Chapter 18.
- *Paging Channel* (PCH) is used for transmission of paging information from the PCCH logical channel. The PCH supports *discontinuous reception* (DRX) to allow the mobile terminal to save battery power by waking up to receive the PCH only at predefined time instants. The LTE paging mechanism is described in Chapter 18.
- *Downlink Shared Channel* (DL-SCH) is the main transport channel used for transmission of downlink data in LTE. It supports key LTE features such as dynamic rate adaptation and channel-dependent scheduling in the time and frequency domains, hybrid ARQ with soft combining, and spatial multiplexing. It also supports DRX to reduce mobile-terminal power consumption while still providing an always-on experience. The DL-SCH is also used for transmission of the parts of the BCCH system information not mapped to the BCH and for single-cell MBMS services.
- *Multicast Channel* (MCH) is used to support MBMS. It is characterized by a semi-static transport format and semi-static scheduling. In case of multi-cell transmission using MBSFN, the scheduling and transport format configuration is coordinated among the cells involved in the MBSFN transmission.
- *Uplink Shared Channel* (UL-SCH) is the uplink counterpart to the DL-SCH, that is the uplink transport channel used for transmission of uplink data.

In addition, the *Random Access Channel* (RACH) is also defined as a transport channel although it does not carry transport blocks.

Part of the MAC functionality is multiplexing of different logical channels and mapping of the logical channels to the appropriate transport channels. The

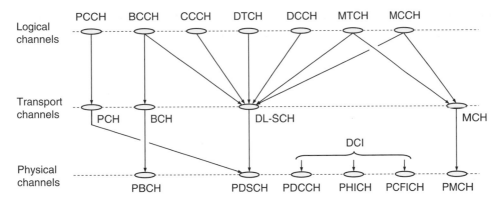

Figure 15.3 *Downlink channel mapping.*

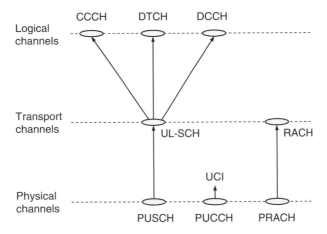

Figure 15.4 *Uplink channel mapping.*

supported mappings between logical channels and transport channels are given in Figure 15.3 for the downlink and Figure 15.4 for the uplink. The figures clearly indicate how DL-SCH and UL-SCH are the main downlink and uplink transport channels, respectively. In the figures, the corresponding physical channels, described further below, are also included and the mapping of transport channels to physical channels is illustrated.

15.2.2 Scheduling

One of the basic principles of the LTE radio access is shared-channel transmission, that is time–frequency resources are dynamically shared between users. The *scheduler* is part of the MAC layer and controls the assignment of uplink and downlink resources. The basic operation of the scheduler is so-called *dynamic* scheduling, where the eNodeB in each 1 ms TTI makes a scheduling decision and sends scheduling information to the selected set of terminals. However, there is also a possibility for semi-persistent scheduling to reduce the control-signaling overhead.

Uplink and downlink scheduling are separated in LTE and uplink and downlink scheduling decisions can be taken independently of each other (within the limits set by the uplink/downlink split in case of half-duplex FDD operation). The terminal follows scheduling commands from a single cell only, the serving cell. This is in contrast to HSPA Enhanced Uplink, where the terminal may follow scheduling information also from non-serving cells in order to control the inter-cell interference. For LTE, inter-cell coordination between different eNodeBs relies on inter-eNodeB signaling over the X2 interface.

The downlink scheduler is responsible for dynamically controlling the terminal(s) to transmit to and, for each of these terminals, the set of resource blocks upon which the terminal's DL-SCH should be transmitted. Transport-format selection (selection of transport-block size, modulation scheme, and antenna mapping) and logical-channel multiplexing for downlink transmissions are controlled by the eNodeB as illustrated in Figure 15.5a. The basic time–frequency unit in the scheduler is a so-called *resource block.* Resource blocks are described in more detail in Chapter 16 in conjunction with the mapping of data to physical resources, but in principle a resource block is a unit spanning 180 kHz in the frequency domain. In each 1 ms scheduling interval, the scheduler assigns resource blocks for DL-SCH

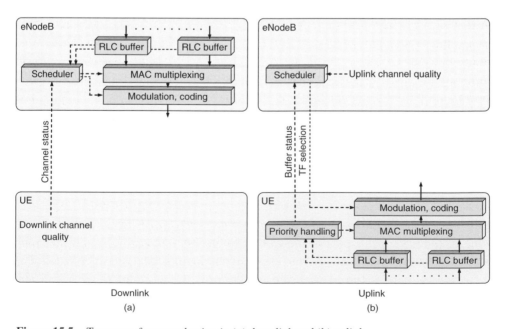

Figure 15.5   *Transport-format selection in (a) downlink and (b) uplink.*

transmission to a terminal, an assignment used by the physical-layer processing as described in Chapter 16. As a consequence of the scheduler controlling the data rate, the RLC segmentation and MAC multiplexing will also be affected by the scheduling decision. The outputs from the downlink scheduler can be seen in Figure 15.1.

The uplink scheduler serves a similar purpose, namely to dynamically control which mobile terminals are to transmit on their UL-SCH and on which uplink time/frequency resources. Despite the fact that the eNodeB scheduler determines the transport format for the mobile terminal, it is important to point out that the uplink scheduling decision is taken *per mobile terminal* and not per radio bearer. Thus, although the eNodeB scheduler controls the payload of a scheduled mobile terminal, the terminal is still responsible for selecting *from which radio bearer(s)* the data is taken. Thus, the mobile terminal autonomously handles logical-channel multiplexing. This is illustrated in Figure 15.5b, where the eNodeB scheduler controls the transport format and the mobile terminal controls the logical-channel multiplexing. The radio-bearer multiplexing in the mobile terminal is done according to rules, the parameters of which can be configured by RRC signaling from the eNodeB. Each radio bearer is assigned a priority and a prioritized data rate. The mobile terminal performs the radio-bearer multiplexing such that the radio bearers are served in priority order up to their prioritized data rate. Remaining resources, if any, after fulfilling the prioritized data rate are given to the radio bearers in priority order.

Although the scheduling strategy is implementation specific and not specified by 3GPP, the overall goal of most schedulers is to take advantage of the channel variations between mobile terminals and preferably schedule transmissions to a mobile terminal on resources with advantageous channel conditions. In this respect, operation of the LTE scheduler is in principle similar to the scheduler in HSPA. However, LTE can exploit channel variations in both frequency *and* time domains, while HSPA can only exploit time-domain variations. This was mentioned already in Chapter 14 and was illustrated in Figure 14.1. For the larger bandwidths supported by LTE, where a significant amount of frequency-selective fading often will be experienced, the possibility for the scheduler to exploit also frequency-domain channel variations becomes increasingly important compared to exploiting time-domain variations only. Especially at low speeds, where the variations in the time domain are relatively slow compared to the delay requirements set by many services, the possibility to exploit also frequency-domain variations is beneficial.

Channel-dependent scheduling is typically used for the downlink. To support this, the mobile terminal transmits *channel-status reports* reflecting the instantaneous channel quality in the time and frequency domains, in addition to information necessary to determine the appropriate antenna processing in case of spatial multiplexing. In principle, channel-dependent scheduling can be used also for the uplink. Channel-quality estimates are in this case based on a *sounding reference signal* transmitted from each mobile terminal for which the eNodeB wants to estimate the uplink channel quality. Such a sounding reference signal is supported by LTE and further described in Chapter 17, but comes at a cost in terms of overhead. Therefore, means to provide uplink diversity as an alternative to uplink channel-dependent scheduling are also supported within LTE. To aid the uplink scheduler in its decisions, the mobile terminal can transmit buffer-status information to the eNodeB using a MAC message. Obviously, this information can only be transmitted if the mobile terminal has been given a valid scheduling grant. For situations when this is not the case, an indicator that the mobile terminal needs uplink resources is provided as part of the uplink L1/L2 control-signaling structure, see further Chapter 17.

Interference coordination, which tries to control the inter-cell interference on a slow basis as mentioned in Chapter 14, is also part of the scheduler. As the scheduling strategy is not mandated by the specifications, the interference-coordination scheme (if used) is vendor specific and may range from simple higher-order reuse deployments to more advanced schemes. The mechanisms used to support inter-cell interference coordination are discussed in Chapter 19.

15.2.3   Hybrid ARQ with soft combining

Hybrid ARQ with soft combining serves a similar purpose for LTE as for HSPA – to provide robustness against transmission errors. As hybrid-ARQ retransmissions are fast, many services allow for one or multiple retransmissions, thereby forming an implicit (closed loop) rate-control mechanism. In the same way as for HSPA, the hybrid-ARQ protocol is part of the MAC layer, while the actual soft combining is handled by the physical layer.[4]

Obviously, hybrid ARQ is not applicable for all types of traffic. For example, broadcast transmissions, where the same information is intended for multiple users, typically do not rely on hybrid ARQ. Hence, hybrid ARQ is only supported for the DL-SCH and the UL-SCH.

The LTE hybrid-ARQ protocol is similar to the corresponding protocol used for HSPA, that is multiple parallel stop-and-wait processes are used. Upon reception

---

[4] The soft combining is done before or as part of the channel decoding which is clearly a physical-layer functionality.

of a transport block, the receiver makes an attempt to decode the transport block and informs the transmitter about the outcome of the decoding operation through a single acknowledgement bit indicating whether the decoding was successful or if a retransmission of the transport block is required. Further details on transmission of hybrid-ARQ acknowledgements are found in Chapters 16, 17, and 19. Clearly, the receiver must know to which hybrid-ARQ process a received acknowledgement is associated. This is solved using the same approach as in HSPA, namely to use the timing of the acknowledgement for association with a certain hybrid-ARQ process. Note that, in case of TDD operation, the time relation between the reception of data in a certain hybrid-ARQ process and the transmission of the acknowledgement is also affected by the uplink/downlink allocation.

The use of multiple parallel hybrid-ARQ processes, illustrated in Figure 15.6, for each user can result in data being delivered from the hybrid-ARQ mechanism out-of-sequence. For example, transport block 5 in the figure was successfully decoded before transport block 1, which required retransmissions. In-sequence delivery of data is therefore ensured by the RLC layer. In contrast, HSPA, which is an add-on to WCDMA, handles reordering in the MAC layer as the RLC was kept unchanged for compatibility reasons as discussed in Chapter 9. For LTE, on the other hand, the protocol layers are all designed jointly, implying fewer restrictions in the design.

Similarly to HSPA, an asynchronous protocol is the basis for downlink hybrid-ARQ operation. Hence, downlink retransmissions may occur at any time after the initial transmission and an explicit hybrid-ARQ process number is used to indicate which process is being addressed. In an asynchronous hybrid-ARQ protocol, the retransmissions are in principle scheduled similarly to the initial transmissions. Uplink retransmissions, on the other hand, are based on a synchronous protocol and the retransmission occurs at a predefined time after the initial transmission and the process number can be implicitly derived. In a synchronous protocol the time instant for the retransmissions is fixed once the initial transmission has been scheduled, which must be accounted for in the scheduling operation. However, note that the scheduler knows from the hybrid-ARQ entity in the eNodeB whether a mobile terminal will do a retransmission or not.

The hybrid-ARQ mechanism will rapidly correct transmission errors due to noise or unpredictable channel variations. As discussed above, the RLC is also capable of requesting retransmissions, which at first sight may seem unnecessary. However, the reason for having two retransmission mechanisms on top of each other can be seen in the feedback signaling – hybrid ARQ provides fast retransmissions but due to errors in the feedback the residual error rate is typically too

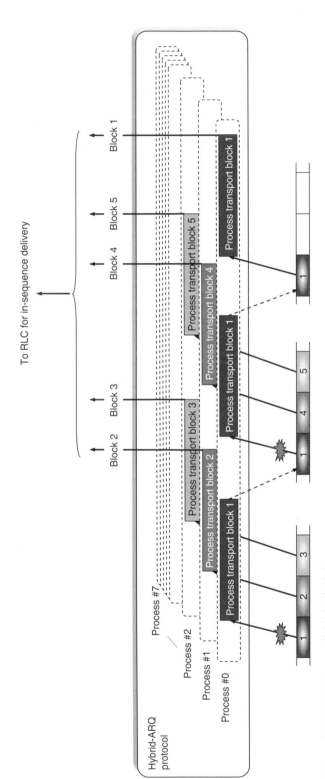

Figure 15.6 *Multiple parallel hybrid-ARQ processes.*

high for good TCP performance, while RLC ensures (almost) error-free data delivery but slower retransmissions than the hybrid-ARQ protocol. Hence, the combination of hybrid ARQ and RLC attains a good combination of small round-trip time and reliable data delivery where the two components complement each other. Furthermore, as the RLC and hybrid ARQ are located in the same node, tight interaction between the two is possible as discussed in Chapter 19.

15.3 Physical layer

The physical layer is responsible for coding, physical-layer hybrid-ARQ processing, modulation, multi-antenna processing, and mapping of the signal to the appropriate physical time–frequency resources. It also handles mapping of transport channels to physical channels as shown in Figure 15.4. A simplified overview of the processing for the DL-SCH is given in Figure 15.7.

As mentioned already in the introduction, the physical layer offers services to the MAC layer in the form of transport channels. In the downlink, the DL-SCH is the main channel for data transmission and is assumed in the description below but the processing for PCH and MCH is similar. In each TTI, there is at most one (two in case of spatial multiplexing) transport blocks.

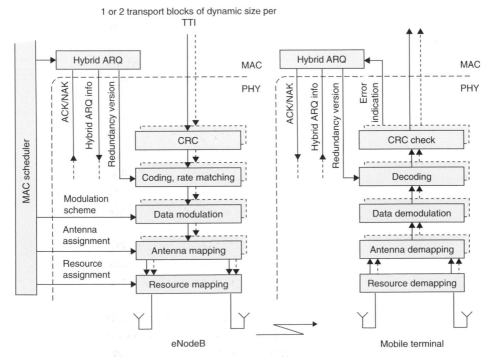

Figure 15.7 *Simplified physical-layer processing for DL-SCH.*

To the transport block(s) to transmit on the DL-SCH, a CRC, used for error detection in the receiver, is attached, followed by Turbo coding for error correction. In case of spatial multiplexing, the processing is duplicated for each of the transport blocks. Rate matching is used not only to match the number of coded bits to the amount of resources allocated for the DL-SCH transmission, but also to generate the different redundancy versions as controlled by the hybrid-ARQ protocol.

After rate matching, the coded bits are modulated using QPSK, 16QAM, or 64QAM, followed by antenna mapping. The antenna mapping can be configured to provide different multi-antenna transmission schemes including transmit diversity, beam-forming, and spatial multiplexing. Finally, the output of the antenna processing is mapped to the physical resources used for the DL-SCH. The resources, as well as the transport-block size and the modulation scheme, are under control of the scheduler.

The physical-layer processing for the UL-SCH follows closely the processing for the DL-SCH. However, note that the MAC scheduler in the eNodeB is responsible for selecting the mobile-terminal transport format and resources to be used for uplink transmission as described in Section 15.2.2. Furthermore, the uplink does not support spatial multiplexing and consequently there is no antenna mapping in the uplink.

A *physical channel* corresponds to the set of time–frequency resources used for transmission of a particular transport channel and each transport channel is mapped to a corresponding physical channel as shown in Figures 15.3 and 15.4. In addition to the physical channels with a corresponding transport channel, there are also physical channels without a corresponding transport channel. These channels, known as L1/L2 control channels, are used for downlink control information (DCI), providing the terminal with the necessary information for proper reception and decoding of the downlink data transmission, and uplink control information (UCI) used for providing the scheduler and the hybrid-ARQ protocol with information about the situation in the terminal.

The physical-channel types defined in LTE include the following:

- *Physical Downlink Shared Channel* (PDSCH) is the main physical channel used for unicast transmission, but also for transmission of paging information.
- *Physical Broadcast Channel* (PBCH) caries part of the system information, required by the terminal in order to access the network.
- *Physical Multicast Channel* (PMCH) is used for MBSFN operation.

- *Physical Downlink Control Channel* (PDCCH) is used for downlink control information, mainly scheduling decisions, required for reception of PDSCH and for scheduling grants enabling transmission on the PUSCH.
- *Physical Hybrid-ARQ Indicator Channel* (PHICH) carries the hybrid-ARQ acknowledgement to indicate to the terminal whether a transport block should be retransmitted or not.
- *Physical Control Format Indicator Channel* (PCFICH) is a channel providing the terminals with information necessary to decode the set of PDCCHs. There is only one PCFICH in each cell.
- *Physical Uplink Shared Channel* (PUSCH) is the uplink counterpart to the PDSCH. There is at most one PUSCH per terminal.
- *Physical Uplink Control Channel* (PUCCH) is used by the terminal to send hybrid-ARQ acknowledgements, indicating to the eNodeB whether the downlink transport block(s) was successfully received or not, to send channel-status reports aiding downlink channel-dependent scheduling, and for requesting resources to transmit uplink data upon. There is at most one PUCCH per terminal.
- *Physical Random Access Channel* (PRACH) is used for random access as described in Chapter 18.

The mapping between transport channels and physical channels was illustrated in Figures 15.3 and 15.4. Note that some of the physical channels, more specifically the channels used for downlink control information (PCFICH, PDCCH, PHICH) and uplink control information (PUCCH), do not have a corresponding transport channel.

The remaining downlink transport channels are based on the same general physical-layer processing as the DL-SCH, although with some restrictions in the set of features used. For the broadcast of system information on the BCH, a mobile terminal must be able to receive this information channel as one of the first steps prior to accessing the system. Consequently, the transmission format must be known to the terminals a priori and there is no dynamic control of any of the transmission parameters from the MAC layer in this case.

For transmission of paging messages on the PCH, dynamic adaptation of the transmission parameters can to some extent be used. In general, the processing in this case is similar to the generic DL-SCH processing. The MAC can control modulation, the amount of resources, and the antenna mapping. However, as an uplink has not yet been established when a mobile terminal is paged, hybrid ARQ cannot be used as there is no possibility for the mobile terminal to transmit a hybrid-ARQ acknowledgement.

The MCH is used for MBMS transmissions, typically with single-frequency network operation as described in Chapter 4 by transmitting from multiple cells on the same resources with the same format at the same time. Hence, the scheduling of MCH transmissions must be coordinated between the involved cells and dynamic selection of transmission parameters by the MAC is not possible.

15.4 Terminal states

In LTE, a mobile terminal can be in two different states as illustrated in Figure 15.8. RRC_CONNECTED is the state used when the mobile terminal is active. In this state, the mobile terminal is connected to a specific cell within the network. One or several IP addresses have been assigned to the mobile terminal, as well as an identity of the terminal, the *Cell Radio-Network Temporary Identifier* (C-RNTI), used for signaling purposes between the mobile terminal and the network. Although expressed differently in the specifications, RRC_CONNECTED can be said to have two substates, IN_SYNC and OUT_OF_SYNC, depending on whether the uplink is synchronized to the network or not. Since LTE uses an orthogonal FDMA/TDMA-based uplink, it is necessary to synchronize the uplink transmission from different mobile terminals such that they arrive at the receiver at (approximately) the same time. The procedure for obtaining and maintaining uplink synchronization is described in Chapter 19 but in short the receiver measures the arrival time of the transmissions from each actively transmitting mobile terminal and sends timing-correction commands in the downlink. As long as the uplink is synchronized, uplink transmission of user data and L1/L2 control signaling is possible. In case no uplink transmission has taken place within a given time window, timing alignment is obviously not possible and the uplink is declared to be non-synchronized. In this case, the mobile terminal needs to perform a random-access procedure to restore uplink synchronization.

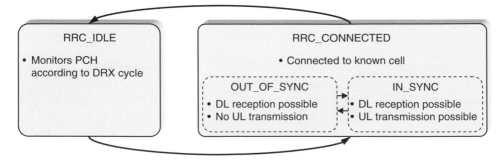

Figure 15.8 *LTE states.*

RRC_IDLE is a low activity state in which the mobile terminal sleeps most of the time in order to reduce battery consumption. Uplink synchronization is not maintained and hence the only uplink transmission activity that may take place is random access to move to RRC_CONNECTED. In the downlink, the mobile terminal can periodically wake up in order to be paged for incoming calls as described in Chapter 18. The mobile terminal keeps its IP address(es) and other internal information in order to rapidly move to RRC_CONNECTED when necessary.

15.5 Data flow

To summarize the flow of downlink data through all the protocol layers, an example illustration for a case with three IP packets, two on one radio bearer and one on another radio bearer, is given in Figure 15.9. The data flow in case of uplink transmission is similar. The PDCP performs (optional) IP header compression, followed by ciphering. A PDCP header is added, carrying information required for deciphering in the mobile terminal. The output from the PDCP is fed to the RLC.

The RLC protocol performs concatenation and/or segmentation of the PDCP SDUs and adds an RLC header. The header is used for in-sequence delivery (per logical channel) in the mobile terminal and for identification of RLC PDUs in case of retransmissions. The RLC PDUs are forwarded to the MAC layer, which takes a number of RLC PDUs, assembles those into a MAC SDU, and attaches the MAC header to form a transport block. The transport-block size depends on the instantaneous data rate selected by the link adaptation mechanism. Thus, the link adaptation affects both the MAC and RLC processing. Finally, the physical layer attaches a CRC to the transport block for error-detection purposes, performs coding and modulation, and transmits the resulting signal over the air.

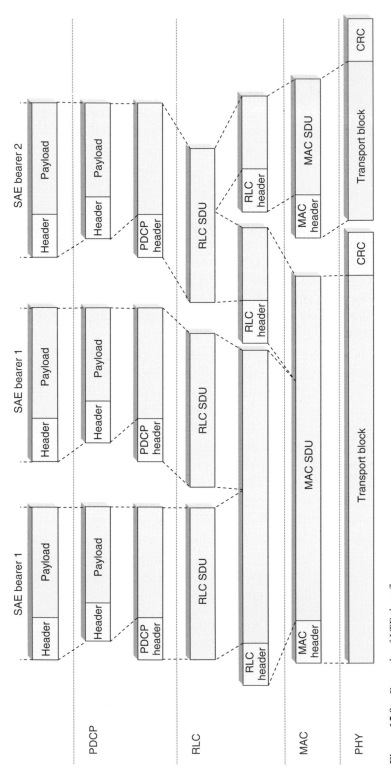

Figure 15.9 *Example of LTE data flow.*

16

Downlink transmission scheme

In Chapter 15, the LTE radio-interface architecture was discussed with an overview of the functions and characteristics of the different protocol layers. This chapter will provide a more detailed overview of the LTE physical-layer transmission scheme, more specifically, the *downlink* transmission scheme. Chapter 17 will provide a corresponding overview of the LTE *uplink* transmission scheme. The following two chapters will then go further into the details of some important LTE access and transmission procedures.

However, before going into the details of the LTE transmission scheme, a brief overview of the overall LTE time-domain structure and the duplex alternatives for LTE will be provided.

16.1   Overall time-domain structure and duplex alternatives

Figure 16.1 illustrates the high-level time-domain structure for LTE transmission with each (*radio*) *frame* of length 10 ms consisting of ten equally sized *sub-frames* of length 1 ms. On an even higher level, each frame is identified by its *System Frame Number* (SFN). The SFN is used to control different transmission cycles that may have a period longer than one frame, such as paging sleep-mode cycles and periods for channel-status reporting.

To provide consistent and exact timing definitions, different time intervals within the LTE radio-access specification are defined as multiples of a basic time unit $T_s = 1/30\,720\,000$.[1] The time intervals outlined in Figure 16.1 can thus also be expressed as $T_{\text{frame}} = 307\,200 \cdot T_s$ and $T_{\text{subframe}} = 30\,720 \cdot T_s$.

LTE can operate in both FDD and TDD as illustrated in Figure 16.2. Although the time-domain structure is, in most respects, the same for FDD and TDD there

---

[1] The reason for this exact value of T_s will be clarified in the next section.

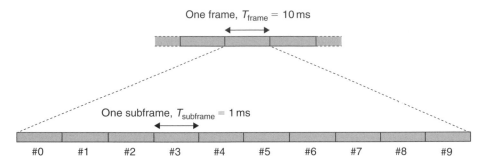

Figure 16.1   *LTE high-level time-domain structure.*

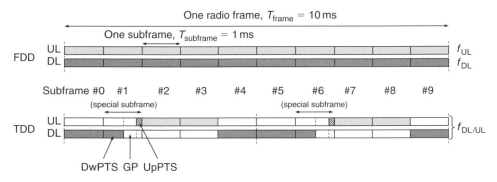

Figure 16.2   *Uplink/downlink time/frequency structure in case of FDD and TDD.*

are some differences between the two duplex modes, most notably the presence of a *special subframe* in case of TDD. The special subframe is used to provide the necessary guard time for downlink-to-uplink switching as discussed below.

In case of FDD operation (upper part of Figure 16.2), there are two carrier frequencies, one for uplink transmission (f_{UL}) and one for downlink transmission (f_{DL}). During each frame, there are thus ten uplink subframes and ten downlink subframes and uplink and downlink transmission can occur simultaneously within a cell.

Even if uplink and downlink transmission can occur simultaneously within a cell in case of FDD operation, a terminal may be capable of *full-duplex* operation or only *half-duplex* operation depending on whether or not it is capable of simultaneous transmission/reception. In case of full-duplex capability, transmission and reception may occur simultaneously at a terminal, whereas a terminal only capable of half-duplex operation cannot transmit and receive simultaneously. As mentioned in Chapter 14, supporting only half-duplex operation allows for simplified terminal implementation due to relaxed duplex-filter requirements. This is especially the case for certain frequency bands with narrow duplex gap. At the same time, half-duplex operation has an impact on the data rates that can be provided to/from a single mobile terminal. It should be noted that full/half-duplex capability may be frequency-band dependent such that a terminal may support only half-duplex operation in certain frequency bands while being capable of full-duplex operation in the remaining supported bands.

In case of TDD operation (lower part of Figure 16.2), there is only a single carrier frequency and uplink and downlink transmissions are always separated in time also on a cell basis. As the same carrier frequency is used for uplink and downlink transmission, both the base station and the mobile terminals need to switch from transmission to reception and vice versa. An essential aspect of any TDD system is to provide the possibility for a sufficiently large *guard time* where neither downlink nor uplink transmissions occur. This is required to avoid interference between uplink and downlink transmissions as elaborated upon in Chapter 19. For LTE, this guard time is provided by special subframes (subframe 1 and, in some cases, subframe 6), which are split into three parts: a downlink part (DwPTS), a guard period (GP), and an uplink part (UpPTS). The remaining subframes are either allocated to uplink or downlink transmission.

LTE TDD allows for different asymmetries in terms of the amount of resources allocated for uplink and downlink transmission, respectively, by means of seven supported downlink/uplink configurations as shown in Figure 16.3. As seen in the figure, subframes 0 and 5 are always allocated for downlink transmission while subframe 2 is always allocated for uplink transmissions. To avoid severe interference between downlink and uplink transmissions between the cells, neighbor cells should have the same downlink/uplink configuration. Therefore, the downlink/uplink asymmetry cannot vary dynamically on, for example, a frame-by-frame basis. However, it can be changed on a slower basis to, for example, match different traffic characteristics such as differences and variations in the downlink/uplink traffic asymmetry. Furthermore, for reasons that will be further elaborated on in Chapter 19, the duration of the DwPTS and UpPTS fields in the special subframe can also be configured. Information about the downlink/uplink configuration, as well as the duration of the fields of the special subframe, is provided to the terminals as part of the cell system information.

16.2 The downlink physical resource

As already mentioned in the overview of the LTE radio access provided in Chapter 14, LTE downlink transmission is based on OFDM. The basic LTE downlink physical resource can thus be seen as a time-frequency resource grid

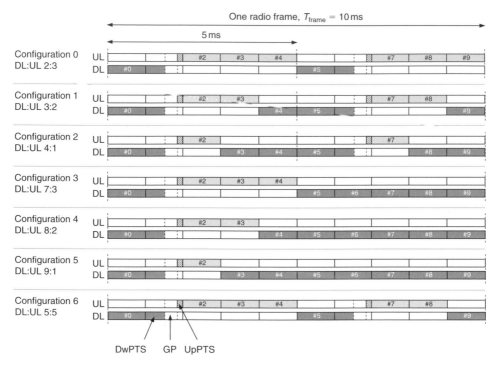

Figure 16.3 *Different downlink/uplink configurations in case of TDD.*

(Figure 16.4), where each *resource element* corresponds to one OFDM subcarrier during one OFDM symbol interval.[2]

For LTE, the OFDM subcarrier spacing has been chosen to $\Delta f = 15\,\text{kHz}$. Assuming an FFT-based transmitter/receiver implementation, this corresponds to a sampling rate $f_s = 15\,000 \cdot N_{\text{FFT}}$, where N_{FFT} is the FFT size. The basic time unit T_s defined in the previous section can thus be seen as the sampling time of an FFT-based transmitter/receiver implementation with an FFT size equal to 2048. It is important to understand though that the time unit T_s is introduced in the LTE radio-access specifications purely as a tool to define different time intervals and does not impose any specific transmitter and/or receiver implementation constraints (e.g. a certain sampling rate). In practice, an FFT-based transmitter/receiver implementation with an FFT size equal to 2048 and a corresponding sampling rate of 30.72 MHz is suitable for the wider LTE transmission bandwidths, such as bandwidths in the order of 15 MHz and above. However, for smaller transmission bandwidths, a smaller FFT size and a correspondingly lower sampling rate can very well be used. As an example, for transmission bandwidths in the order of 5 MHz, an FFT size equal to 512 and a corresponding sampling rate of 7.68 MHz may be sufficient.

---

[2] In case of multi-antenna transmission, there will be one resource grid per antenna.

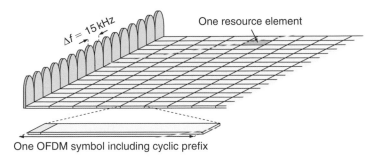

Figure 16.4   *The LTE downlink physical resource.*

Assuming a power-of-two FFT size and a subcarrier spacing of 15 kHz, the sampling rate $\Delta f \cdot N_{FFT}$ will be a multiple or submultiple of the WCDMA/HSPA chip rate (3.84 Mcps). This relation can be utilized when implementing multi-mode terminals supporting both WCDMA/HSPA and LTE.

In addition to the 15 kHz subcarrier spacing, a *reduced subcarrier spacing* $\Delta f_{low} = 7.5$ kHz with twice as long OFDM symbol time is also defined for LTE. The reduced subcarrier spacing specifically targets MBSFN-based multicast/broadcast transmissions (see Section 16.7 for more discussions on MBSFN transmission in LTE). The remaining discussions within this and the following chapters will assume the 15 kHz subcarrier spacing unless explicitly stated otherwise.

As illustrated in Figure 16.5, in the frequency domain the downlink subcarriers are grouped into *resource blocks*, where each resource block consists of 12 consecutive subcarriers.[3] In addition, there is an unused *DC-subcarrier* in the center of the downlink band. The reason why the DC-subcarrier is not used for downlink transmission is that it may be subject to un-proportionally high interference, for example, due to local-oscillator leakage.

The LTE physical-layer specification allows for a downlink carrier to consist of any number of resource blocks, ranging from a minimum of 6 resource blocks up to a maximum of 110 resource blocks. This corresponds to an overall downlink transmission bandwidth ranging from roughly 1 MHz up to in the order of 20 MHz with very fine granularity and thus allows for a very high degree of LTE bandwidth flexibility, at least from a physical-layer-specification point-of-view. However, as mentioned in Chapter 14, LTE radio-frequency requirements are, at least initially, only specified for a limited set of transmission bandwidths, corresponding to a limited set of possible values for the number of resource blocks within a carrier.

Figure 16.6 outlines the more detailed time-domain structure for LTE downlink transmission. Each 1 ms subframe consists of two equally sized *slots* of length

---

[3] The resource blocks are actually two-dimensional (time-frequency) units with a size of 12 subcarriers in the frequency domain and one 0.5 ms slot (seven/six OFDM symbols) in the time domain (see Figure 16.7).

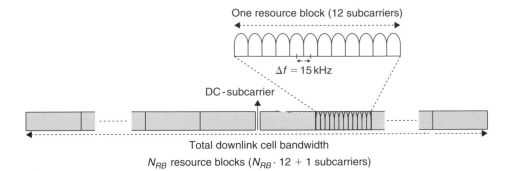

Figure 16.5 *Frequency-domain structure for LTE downlink.*

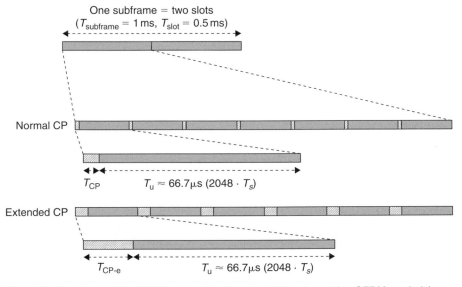

$T_{CP} : 160 \cdot T_s \approx 5.1\,\mu s$ (first OFDM symbol), $144 \cdot T_s \approx 4.7\,\mu s$ (remaining OFDM symbols)
$T_{CP\text{-}e} : 512 \cdot T_s \approx 16.7\,\mu s$

Figure 16.6 *Detailed time-domain structure for LTE downlink transmission.*

$T_{slot} = 0.5\,ms$ ($15\,360 \cdot T_s$). Each slot then consists of a number of OFDM symbols including cyclic prefix.[4]

According to Chapter 4, a subcarrier spacing of 15 kHz corresponds to a useful symbol time of approximately 66.7 μs. The overall OFDM symbol time is then the sum of the useful symbol time and the cyclic-prefix length. As illustrated in Figure 16.6, LTE defines two cyclic-prefix lengths, the normal cyclic prefix and an *extended*

---

[4] Figure 16.6 is valid for the 'normal' downlink subframes. As described in Section 16.1, the special subframe is not divided into two slots but rather into three fields. However, the number of (DFTS-)OFDM symbols in the special subframe is the same as for the normal subframes illustrated in Figure 16.6.

cyclic prefix, corresponding to seven and six OFDM symbols per slot, respectively. The exact cyclic-prefix lengths, expressed in the basic time unit T_s, are given in Figure 16.6. It can be noted that, in case of the normal cyclic prefix, the cyclic-prefix length for the first OFDM symbol of a slot is somewhat larger, compared to the remaining OFDM symbols. The reason for this is simply to fill the entire 0.5 ms slot as the number of basic time units T_s per slot (15 360) is not dividable by seven.

The reasons for defining two cyclic-prefix lengths for LTE are twofold:

- A longer cyclic prefix, although less efficient from a cyclic-prefix-overhead point-of-view, may be beneficial in specific environments with very extensive delay spread, for example in very large cells. It is important to have in mind, though, that a longer cyclic prefix is not necessarily beneficial in case of large cells, even if the delay spread is very extensive in such cases. If, in large cells, link performance is limited by noise rather than by signal corruption due to residual time dispersion not covered by the cyclic prefix, the additional robustness to radio-channel time dispersion, due to the use of a longer cyclic prefix, may not justify the corresponding loss in terms of reduced received signal energy.
- As already discussed in Chapter 4, in case of MBSFN-based multicast/broadcast transmission, the cyclic prefix should not only cover the main part of the actual channel time dispersion but also the timing difference between the transmissions received from the cells involved in the MBSFN transmission. In case of MBSFN operation, the extended cyclic prefix is therefore often needed.

Thus, the main use of the extended cyclic prefix can be expected to be MBSFN-based transmission. It should be noted that different cyclic-prefix lengths may be used for different subframes within a frame. As an example, as will be further discussed in Section 16.7, MBSFN-based multicast/broadcast transmission is typically confined to certain subframes in which case the use of the extended cyclic prefix, with its associated additional cyclic-prefix overhead, may only be applied to these subframes.[5]

Taking into account also the downlink time-domain structure, the *resource blocks* mentioned above consist of 12 subcarriers during a 0.5-ms slot, as illustrated in Figure 16.7. Each resource block thus consists of 84 resource elements in case of normal cyclic prefix and 72 resource elements in case of extended cyclic prefix.

Although resource blocks are defined over one slot, the basic time-domain unit for dynamic scheduling in LTE is one subframe, consisting of two consecutive slots. The reason to define the resource blocks over one slot is that *distributed downlink transmission* as described in Section 16.5.8, as well as *uplink frequency hopping* (see Section 17.5), is defined on a slot basis. The minimum scheduling

---

[5]The extended cyclic prefix is then actually applied only to the so-called *MBSFN part* of the MBSFN subframes, see Section 16.7.

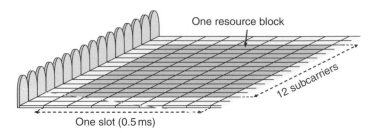

Figure 16.7   *Downlink resource block assuming normal cyclic prefix (i.e. seven OFDM symbols per slot). With extended cyclic prefix there are six OFDM symbols per slot.*

unit consisting of two resource blocks within one subframe (one resource block per slot) is sometimes referred to as a *resource-block pair*.

16.3   Downlink reference signals

To carry out coherent demodulation of different downlink physical channels, a mobile terminal needs estimates of the downlink channel. More specifically, in case of OFDM transmission, the terminal needs an estimate of the complex channel of each subcarrier.

As described in Chapter 4, one way to enable channel estimation in case of OFDM transmission is to insert known *reference symbols* into the OFDM time–frequency grid. In LTE, these reference symbols are jointly referred to as *downlink reference signals*. More specifically, three types of reference signals are defined for the LTE downlink:

- *Cell-specific downlink reference signals* are transmitted in every downlink subframe, and span the entire downlink cell bandwidth. The cell-specific reference signals can be used for channel estimation for coherent demodulation of any downlink transmission except when so-called *non-codebook-based beam-forming* (see Section 16.6.3) is used.
- A *UE-specific reference signal* is specifically intended for channel estimation for coherent demodulation of downlink-shared-channel (DL-SCH) transmissions for which non-codebook-based beam-forming *has* been applied. The term *UE-specific* relates to the fact that each such reference signal is typically intended to be used for channel estimation by one specific terminal. The UE-specific reference signals are then only transmitted within the resource blocks assigned for DL-SCH transmission to that specific terminal.
- *MBSFN reference signals* are used for channel estimation for coherent demodulation of signals being transmitted by means of MBSFN. MBSFN transmission, including more details on MBSFN reference signals, will be further discussed in Section 16.7.

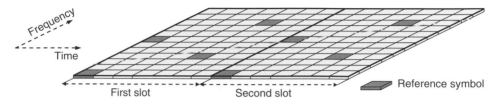

Figure 16.8 *Structure of cell-specific reference signal within a pair of resource blocks (normal cyclic prefix).*

16.3.1 Cell-specific downlink reference signals

The basic structure for cell-specific downlink reference signals is illustrated in Figure 16.8. A cell-specific reference signal consists of known *reference symbols* inserted within the first and third last[6] OFDM symbol of each slot and with a frequency-domain spacing of six subcarriers. Furthermore, there is a frequency-domain staggering of three subcarriers for the reference symbols within the third last OFDM symbol. Within each resource block, consisting of 12 subcarriers during one slot, there are thus four reference symbols. As will be discussed further in Section 16.7, in so-called *MBSFN subframes*, only the reference symbols in the first OFDM symbol of the subframe are actually transmitted.

To estimate the channel over the entire time–frequency grid as well as reducing the noise in the channel estimates, the mobile terminal should carry out interpolation/ averaging over multiple reference symbols. Thus, when estimating the channel for a certain resource block, the mobile terminal may not only use the reference symbols within that resource block but also, in the frequency domain, neighbor resource blocks, as well as reference symbols of previously received slots/subframes. However, the extent to which the mobile terminal can average over multiple resource blocks in the frequency and/or time domain depends on the channel characteristics. In case of high channel frequency selectivity, the possibility for averaging in the frequency domain is limited. Similarly, the possibility for time-domain averaging, that is the possibility to use reference symbols in previously received slots/subframes, is limited in case of fast channel variations, for example, due to high mobile-terminal velocity. It should also be noted that, in case of TDD, the possibility for time averaging may be further limited, as previous subframes may not even be assigned for downlink transmission.

In general, the complex values of the reference symbols will vary between different reference-symbol positions and also between different cells. Thus, the reference signal of a cell can be seen as a *cell-specific two-dimensional* sequence. The period of this sequence equals one frame. Furthermore, regardless of the cell bandwidth, the reference-signal sequence is defined assuming the maximum

---

[6]This corresponds to the fifth and fourth OFDM symbols of the slot in case of normal and extended cyclic prefix, respectively.

possible LTE cell bandwidth corresponding to 110 resource blocks in the frequency domain. In case of a more narrow cell bandwidth, only the reference symbols within the actual bandwidth are then transmitted. Thus, the reference symbols in the center part of the band will always be the same, regardless of the actual cell bandwidth. This allows for the terminal to estimate the channel corresponding to the center part of the band, where, for example, the basic system information of the cell is transmitted using the BCH transport channel, without knowing the actual cell bandwidth. Information about the downlink cell bandwidth is then provided on the BCH.

There are 504 different reference-signal sequences defined for LTE, where each sequence corresponds to one out of 504 different *physical-layer cell identities*. As will be described in more detail in Chapter 18, during the so-called *cell-search* procedure, the terminal detects the physical-layer identity of the cell, as well as the cell frame timing. Thus, from the cell-search procedure, the terminal knows both what reference-signal sequence is used within the cell as well as the start of the reference-signal sequence.

The set of reference-symbol positions outlined in Figure 16.8 is only one out of six possible *frequency shifts* of the reference symbols (Figure 16.9). What shift to use in a cell depends on the physical-layer cell identity such that each shift corresponds to 84 different cell identities, that is the six shifts jointly cover all 504 cell identities. Thus, by properly selecting the physical-layer cell identities to be assigned to different cells, different reference-signal frequency shifts can be used in neighbor cells. This can be beneficial, for example, if the reference symbols are transmitted with higher energy compared to other resource elements, also referred to as *reference-signal (power) boosting*. If reference signals of neighbor cells were transmitted using the same time/frequency resource, the boosted reference symbols of one cell would be interfered by boosted reference symbols of neighbor cells,[7] implying no gain in the reference-signal SIR. However, if different frequency shifts are used for the reference-signal transmissions of neighbor cells, the reference symbols of one cell will be interfered by non-reference symbols of neighbor cells, implying an improved reference-signal SIR in case of reference-signal boosting.

In case of downlink multi-antenna transmission the mobile terminal should be able to estimate the downlink channel corresponding to each transmit antenna. To enable this, there is one downlink reference signal transmitted from each antenna or, more exactly, from each *antenna port*. The LTE radio-access specification talks about *antenna ports* rather than *antennas* to emphasize that what is referred to does not necessarily correspond to a single physical antenna. Actually, an antenna

---

[7]This assumes that the cells are time synchronized.

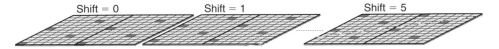

Figure 16.9 *Different reference-signal frequency shifts.*

port is *defined* by the presence of an *antenna-port-specific reference signal*. Thus, if identical reference signals are transmitted from several physical antennas, these antennas cannot be distinguished from each other from a mobile-terminal point-of-view and the antennas can be jointly seen as a single antenna port.

Figure 16.10 illustrates the reference-signal structure for each antenna port in case of multiple antenna ports within a cell:

- In case of two antenna ports (Figure 16.10a), the reference symbols of the second antenna port are frequency multiplexed with the reference symbols of the first antenna port, with a frequency-domain offset of three subcarriers.
- In case of four antenna ports (Figure 16.10b), the reference symbols for the third and fourth antenna ports are frequency multiplexed within the *second* OFDM symbol of each slot. Note that the reference symbols for antenna port three and four are only transmitted within one OFDM symbol of each slot.

It can also be noted that in a resource element carrying a reference symbol for a certain antenna port, nothing is being transmitted on the other antenna ports. Thus, the reference symbols of a certain antenna port are not interfered by transmissions from other antenna ports within the cell. Multi-antenna transmission schemes, such as spatial multiplexing, to a large extent rely on good channel estimates to suppress interference between the transmissions from the different antenna ports. However, in the channel estimation itself there is then obviously no such suppression and avoiding interference to the reference signals of one antenna port from the other antennas ports is therefore important in order to allow for good channel estimation.

Clearly, in case of four antenna ports the time-domain reference-symbol density of the third and fourth antenna ports is reduced compared to the first and second antenna ports. This is done in order to limit the reference-signal overhead in case of four antenna ports. At the same time, this has a negative impact on the possibility to track very fast channel variations. However, this can be justified on the basis of an expectation that, for example, four-antenna spatial multiplexing will mainly be applied to scenarios with low mobility.

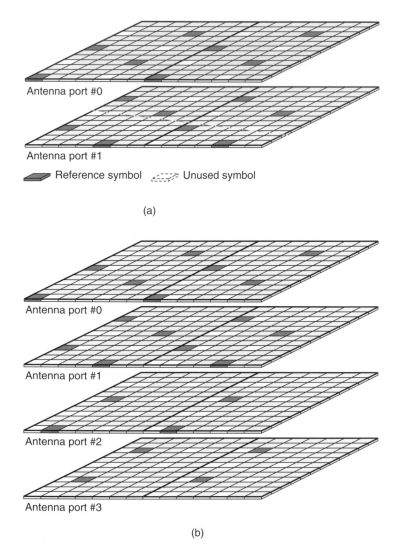

Figure 16.10 *Cell-specific reference signals in case of multi-antenna transmission: (a) two antenna ports and (b) four antenna ports.*

16.3.2   UE-specific reference signals

The downlink reference signals discussed above are referred to as *cell-specific reference signals* indicating that:

- Different reference signals (different reference-signal sequences and possibly also different reference-signal frequency shifts) are used in neighbor cells.
- The reference signals are common to all terminals in a cell in the sense that any terminal can use the reference signals to estimate the channel for the different antenna ports and then use these estimates for the coherent demodulation of

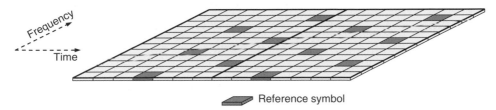

Reference symbol

Figure 16.11 *Structure of UE-specific reference signal within a pair of resource blocks (normal cyclic prefix).*

any physical channel being transmitted using these antenna ports. This includes transmission using the multi-antenna transmission schemes explicitly specified for LTE, that is, transmit diversity and spatial multiplexing (see Section 16.6). It also includes beam-forming as a special case of spatial multiplexing.

However, as will be described in Section 16.6, LTE also allows for more general beam-forming. In order to allow for channel estimation also for such transmissions, additional reference signals, beam-formed in the same way as the data transmission, are needed. As such a reference signal can only be used by the specific terminal to which the beam-formed transmission is intended, it is referred to as a *UE-specific reference signal*. Transmissions for which channel estimation is assumed to be based on a UE-specific reference signal is also referred to as transmission using *antenna port 5*.[8]

A UE-specific reference signal is only transmitted within the resource blocks that are used for the corresponding beam-formed transmission. Figure 16.11 illustrates more exactly how reference symbols of a UE-specific reference signal are inserted into a pair of consecutive resource blocks corresponding to one sub-frame. It should be noted that the UE-specific reference symbols are not inserted into OFDM symbols in which the cell-specific reference signals are transmitted (compare Figure 16.8). Thus the UE-specific reference symbols never collide with the cell-specific reference signals. Also, the UE-specific reference symbols are only transmitted within the data part of the assigned resource blocks. As L1/L2 control signaling is never transmitted using antenna port 5, no UE-specific reference symbols need to be transmitted within the control part. This also implies that only the terminal for which beam-formed downlink-shared-channel data is being transmitted within a resource block needs to know that UE-specific reference symbols are inserted in that resource block.

---

[8] Antenna port 0–3 corresponds to the (up to) four cell-specific reference signals. Transmission using antenna port 4 corresponds to MBSFN transmission, see Section 16.7.

The terminal is informed by the network if downlink-shared-channel transmission is done using the cell-specific antenna ports (antenna ports 0–3) or using antenna port 5. In the latter case, the terminal knows that some of the resource elements within the assigned resource blocks are used for UE-specific reference symbols. The terminal also knows that it should use these reference symbols for channel estimation for coherent demodulation of the beam-formed data transmission, rather than the cell-specific reference signals.

16.4 Downlink L1/L2 control signaling

To support the transmission of downlink and uplink transport channels, there is a need for certain *associated downlink control signaling*. This control signaling is often referred to as the *downlink L1/L2 control signaling*, indicating that the corresponding information partly originates from the physical layer (Layer 1) and partly from Layer 2 MAC. The downlink L1/L2 control signaling consists of downlink scheduling assignments including information required for the terminal to be able to properly receive, demodulate, and decode the DL-SCH,[9] uplink scheduling grants informing the terminal about the resources and transport format to use for uplink (UL-SCH) transmission, and hybrid-ARQ acknowledgements in response to UL-SCH transmissions. In addition, the downlink control signaling can also be used for the transmission of power-control commands for power control of uplink physical channels.

As illustrated in Figure 16.12, the downlink L1/L2 control signaling is transmitted within the first part of each subframe. Thus, each subframe can be said to be divided into a *control region* followed by a *data region*, where the control region corresponds to the part of the subframe in which the L1/L2 control signaling is transmitted. To simplify the overall design, the control region always occupies an integer number of OFDM symbols, more specifically one, two, or three OFDM symbols.[10]

The size of the control region, that is, the number of OFDM symbols that the control region spans, can be dynamically varied on a per-subframe basis. Thus, the amount of radio resources used for control signaling can be dynamically adjusted to match the instantaneous traffic situation. In case of a small number of users being scheduled in a subframe, the required amount of control signaling

---

[9]L1/L2 control signaling is also needed for the reception, demodulation, and decoding of the PCH transport channel.

[10]For narrow cell bandwidths, the control region consists of two, three, or four OFDM symbols to allow for a sufficient amount of control signaling.

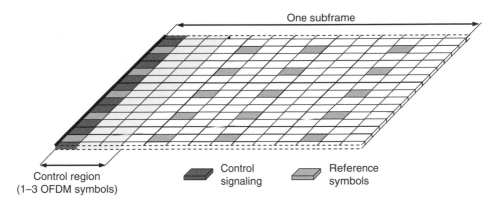

Figure 16.12 *LTE time/frequency grid illustrating the split of the subframe into (variable-sized) control and data regions.*

is small and a larger part of the subframe can be used for data transmission (larger data region).

The reason for transmitting the control signaling at the beginning of the subframe is to allow for terminals to decode downlink scheduling assignments as early as possible. Processing of the data region, that is, demodulation and decoding of the DL-SCH transmission, can then begin before the end of the subframe. This reduces the delay in the DL-SCH decoding and thus the overall downlink transmission delay. Furthermore, by transmitting the L1/L2 control channel at the beginning of the subframe, that is, by allowing for early decoding of the L1/L2 control information, mobile terminals that are not scheduled in the subframe may power down the receiver circuitry for a part of the subframe, allowing for a reduced terminal power consumption.

The downlink L1/L2 control signaling corresponds to three different physical-channel types:

- The *Physical Control Format Indicator Channel* (PCFICH), informing the terminal about the size of the control region (one, two, or three OFDM symbols). There is one and only one PCFICH in each cell.
- The *Physical Downlink Control Channel* (PDCCH), used to signal downlink scheduling assignments and uplink scheduling grants. Each PDCCH carries signaling for a single terminal (or a group of terminals). There are typically multiple PDCCHs in each cell.
- The *Physical Hybrid-ARQ Indicator Channel* (PHICH), used to signal hybrid-ARQ acknowledgements in response to uplink UL-SCH transmissions. There are multiple PHICHs in each cell.

The maximum size of the control region is normally three OFDM symbols (four in case of narrow cell bandwidths) as mentioned above. However, there are a few exceptions to this rule. When operating in TDD mode, the control region in subframes 1 and 6 is restricted to at most two OFDM symbols since, in case of TDD, the primary synchronization signal (see Chapter 18) occupies the third OFDM symbol in those subframes. Similarly, for MBSFN subframes (see Section 16.7), the control region is restricted to a maximum of two OFDM symbols.

16.4.1 Physical Control Format Indicator Channel

The PCFICH indicates the size of the control region in terms of number of OFDM symbols, that is, indirectly where in the subframe the data region starts. Correct decoding of the PCFICH information is thus essential. If the PCFICH is incorrectly decoded, the terminal will neither know where to find the control channels nor where the data region starts for the corresponding subframe.

The PCFICH consists of two bits of information, corresponding to a control-region size of one, two, or three OFDM symbols,[11] which is coded into a 32-bit codeword. The coded bits are scrambled with a cell- and subframe-specific scrambling code to randomize inter-cell interference, QPSK-modulated, and mapped to 16 resource elements. As the size of the control region is unknown until the PCFICH is decoded, the PCFICH is always mapped to the first OFDM symbol of each subframe.

The mapping of the PCFICH to resource elements in the first OFDM symbol in the subframe is done in groups of four resource elements with the four groups being well separated in frequency to obtain good diversity. Furthermore, to avoid collisions between PCFICH transmissions in neighbor cells, the location of the four groups in the frequency domain depends on the physical-layer cell identity.

The transmission power of the PCFICH is under control of the eNodeB. If necessary for coverage in a certain cell, the power of the PCFICH can be set higher than for other channels by 'borrowing' power from, for example, simultaneously transmitted PDCCHs. Obviously, increasing the power of the PCFICH to improve the performance in an interference-limited system depends on the neighboring cells not increasing their transmit power on the interfering resource elements. Otherwise, the interference would increase as much as the signal power, implying no gain in received SIR. However, as the PCFICH-to-resource-element mapping depends on the cell identity, the probability of (partial)

---

[11] The fourth combination is reserved for future use.

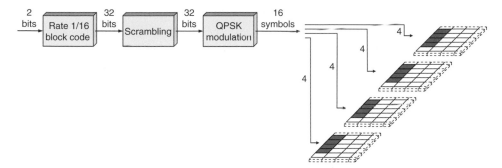

Figure 16.13 *Overview of the PCFICH processing.*

collisions with PCFICH in neighboring cells in synchronized networks is reduced, thereby improving the performance of PCFICH power boosting as a tool to control the error rate.

The overall PCFICH processing is illustrated in Figure 16.13.

To describe the mapping of the PCFICH, and L1/L2 control signaling in general, to resource elements, some terminology is required. As mentioned above, the mapping is specified in terms of groups of four resource elements, so-called *resource-element groups*. To each resource-element group, a *symbol quadruplet* consisting of four (QPSK) symbols is mapped. The main motivation behind this, instead of simply mapping the symbols one-by-one, is the support of transmit diversity. As will be further discussed in Section 16.6, LTE supports several downlink multi-antenna transmission schemes for up to four different antenna ports. Among these transmission schemes, only transmit diversity is applicable to the physical channels used for L1/L2 control signaling. Transmit diversity for four antenna ports is specified in terms of groups of four symbols (resource elements), and, consequently, also the L1/L2 control-channel processing is defined in terms of symbol quadruplets. The transmit diversity scheme applied to the L1/L2 control signaling is identical to the scheme applied to the BCH transport channel.

The definition of the resource-element groups assumes that reference symbols corresponding to two antenna ports are present in the first OFDM symbol, regardless of the actual number of antenna ports configured in the cell. This simplifies the definition and reduces the number of different structures to handle. Thus, as illustrated in Figure 16.14, in the first OFDM symbol there are two resource-element groups per resource block as every third resource element is reserved for reference signals (or non-used resource elements corresponding to reference symbols on the other antenna port). As also illustrated in

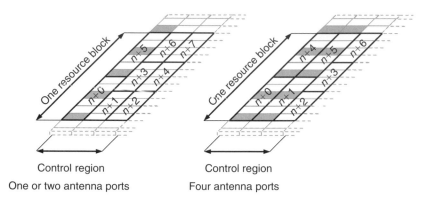

Control region   Control region

One or two antenna ports   Four antenna ports

Figure 16.14 *Numbering of resource-element groups in the control region (assuming a size of three OFDM symbols).*

Figure 16.14, in the second OFDM symbol (if part of the control region) there are two or three resource-element groups depending on the number of antenna ports configured. Finally, in the third OFDM symbol (if part of the control region) there are always three resource-element groups per resource block. Figure 16.14 also illustrates how resource-element groups are numbered in a time-first manner within the size of the control region.

Returning to the PCFICH, four resource-element groups are used for the transmission of the 16 QPSK symbols. To obtain good frequency diversity the resource-element groups should be well spread in frequency and cover the full downlink cell bandwidth. Therefore, the four resource-element groups are separated by one quarter of the downlink cell bandwidth in the frequency domain with the starting position given by physical-layer cell identity. This is illustrated in Figure 16.15, where the PCFICH mapping to the first OFDM symbol in a subframe are shown for three different physical-layer cell identities in case of a downlink cell bandwidth of eight resource blocks. As seen in the figure, the PCFICH mapping depends on the physical-layer cell identity to reduce the risk of inter-cell PCFICH collisions. The cell-specific shifts of the reference symbols, described in Section 16.3.1, are also seen in the figure.

16.4.2 *Physical Hybrid-ARQ Indicator Channel*

The PHICH is used for transmission of hybrid-ARQ acknowledgements in response to UL-SCH transmission.

For proper operation of the hybrid-ARQ protocol as discussed in Chapter 19, the error rate of the PHICH should be sufficiently low. The operating point of the

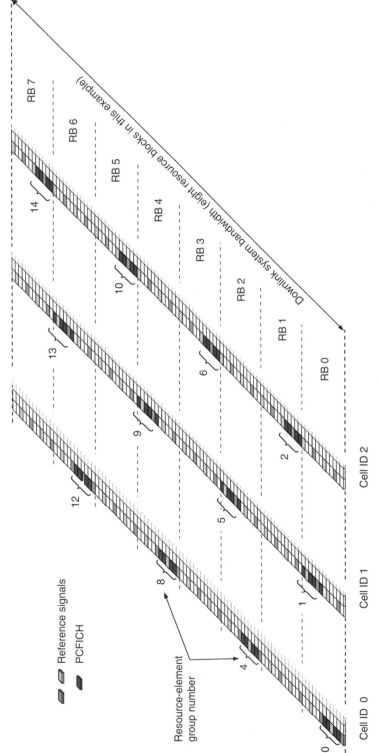

Figure 16.15 *Example of PCFICH mapping in the first OFDM symbol for three different physical-layer cell identities.*

PHICH is not specified but is up to the network operator to decide on, but typically ACK-to-NAK and NAK-to-ACK error rates in the order of 10^{-2} and 10^{-3} to 10^{-4}, respectively, are targeted. The reason for the asymmetric error rates is that an NAK-to-ACK error would imply a loss of a transport block at the MAC level, a loss that has to be recovered by RLC retransmissions with the associated delays, while an ACK-to-NAK error rate only implies an unnecessary retransmission of an already correctly decoded transport block. To meet these error-rate targets without excessive power, it is beneficial to control the PHICH transmission power as a function of the radio-channel quality of the terminal to which the PHICH is directed. This has influenced the design of the PHICH structure.

In principle, a PHICH could be mapped to a set of resource elements exclusively used by this PHICH. However, taking the dynamic PHICH power setting into account, this could result in significant variations in transmission power between resource elements, which can be challenging from an RF implementation perspective. Therefore, it is preferable to spread each PHICH on multiple resource elements to reduce the power differences while at the same time providing the energy necessary for accurate reception. To fulfill this, a structure where several PHICHs are code multiplexed onto a set of resource elements is used for LTE. The hybrid-ARQ acknowledgement (one single bit of information) is repeated three times, followed by BPSK modulation on either the I or the Q branch and spreading with a length-four orthogonal sequence. A set of PHICHs transmitted on the same set of resource-elements is denoted a PHICH group, where a PHICH group consists of eight PHICHs in case of normal cyclic prefix. An individual PHICH can thus be uniquely represented by the number of the PHICH group, the number of the orthogonal sequence within the group, and information whether the I or the Q branch is used.[12]

For extended cyclic prefix, which is typically used in time-dispersive environments, the radio channel may not be flat over the frequency spanned by a length-four sequence. A non-flat channel would negatively impact the orthogonality between the sequences. Hence, for extended cyclic prefix, orthogonal sequences of length two are used for spreading implying only four PHICHs per PHICH group. However, the general structure remains the same as for the normal cyclic prefix.

After forming the composite signal representing the PHICHs in a group, the signal is scrambled and the twelve scrambled symbols are mapped to three resource-element groups. Similarly to the other L1/L2 control channels, the mapping is described using resource-element groups to be compatible with the transmit diversity schemes defined for LTE.

---

[12] The specifications lump the sequence number and the IQ mapping together into a single index number.

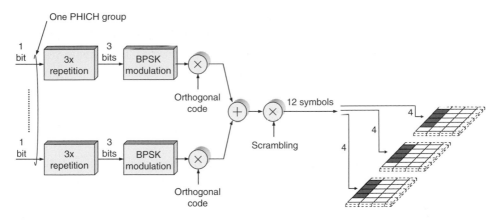

Figure 16.16 *PHICH structure.*

The overall PHICH processing is illustrated in Figure 16.16.

The requirements on the mapping of PHICH groups to resource elements is similar to those for the PCFICH, namely to obtain good frequency diversity and to avoid collisions between neighboring cells in synchronized networks. Hence, each PHICH group is mapped to three resource-element groups, separated by approximately one-third of the downlink cell bandwidth. In the first OFDM symbol in the control region, resources are first allocated to the PCFICH, the PHICH is mapped to resource elements not used by the PCFICH, and finally, the PDCCH is mapped to the remaining resource elements.

Typically, the PHICH is transmitted in the first OFDM symbol only which allows the terminal to attempt to decode the PHICH even if it failed decoding of the PCFICH. This is advantageous as the error requirements on the PHICH typically are stricter than for PCFICH. However, in some propagation environments, having a PHICH span of a single OFDM symbol would unnecessarily restrict the coverage. To alleviate this, it is possible to *semi-statically* configure a PHICH duration of three OFDM symbols. In this case, the control region is three OFDM symbols long in all subframes to fulfill the general principle of separating the control region from data in the time domain only. Obviously, the value transmitted on the PCFICH will be fixed (and can be ignored) in this case.

In order to minimize the overhead and not introduce any additional signaling in the uplink grants, the PHICH which the terminal will expect the hybrid-ARQ acknowledgement upon, that is the PHICH group number, the orthogonal sequence, and the I/Q branch, is given by the number of the first resource block upon which the current uplink PUSCH transmission occurred. This principle is also compatible with semi-persistently scheduled transmission as well as retransmissions. In addition, the resources used for a particular PHICH further depend on the reference-signal phase rotation signaled as part of the uplink grant. In this way, multiple terminals scheduled simultaneously on the same set of resource blocks in the uplink, sometimes referred to as multi-user MIMO, will use different PHICH resources as their reference signals are assigned different phase rotations through the corresponding field in the uplink grant.

The time-domain association between an uplink subframe with data and the corresponding downlink PHICH subframe is given by a fixed rule as described in Chapter 19. For FDD, an uplink transport block received in subframe n should be acknowledged on the PHICH in subframe $n + 4$. For TDD a similar rule is used, although the uplink–downlink configuration is taken into account. Furthermore, as there is not necessarily a one-to-one relation between uplink and downlink subframes for TDD, multiple uplink subframes may need to be acknowledged in one downlink subframe. This is the case for TDD configuration zero (see Section 16.1 for an overview of the downlink–uplink asymmetries), and a terminal consequently needs to monitor multiple PHICHs. Consequently, the amount of resources for PHICH may differ between subframes in case of TDD.

The PHICH configuration is part of the system information transmitted on the PBCH; one bit indicates whether the duration is one or three OFDM symbols and two bits indicate the amount of resources in the control region reserved for PHICHs, expressed as a fraction of the downlink cell bandwidth in terms of resource blocks. Having the possibility to configure the amount of PHICH resources is useful as the PHICH capacity depends on, for example, whether the network uses multi-user MIMO or not. The PHICH configuration must reside on the PBCH, as it needs to be known in order to properly process the PDCCHs for reception of the part of the system information on the DL-SCH. For TDD, the PHICH information provided on the PBCH is not sufficient for the terminal to know the exact set of resources used by PHICH, as there is a dependency on the uplink/downlink allocation as well. In order to receive the system information on the DL-SCH, which contains the uplink/downlink allocation, the terminal has to blindly process the PDCCHs under different PHICH configuration hypotheses.

16.4.3 Physical Downlink Control Channel

The PDCCH is used to carry DCI such as scheduling decisions and power-control commands. More specifically, the DCI includes:

- Downlink scheduling assignments, including PDSCH resource indication, transport format, hybrid-ARQ information, and control information related to spatial multiplexing (if applicable). A downlink scheduling assignment also includes a command for power control of the PUCCH uplink physical channel.

- Uplink scheduling grants, including PUSCH resource indication, transport format, and hybrid-ARQ-related information. An uplink scheduling grant also includes a command for power control of the PUSCH uplink physical channel.
- Power-control commands for a set of terminals as a complement to the commands included in the scheduling assignments/grants.

The different types of control information above typically correspond to different DCI message sizes. For example, supporting spatial multiplexing with non-contiguous allocation of resource blocks in the frequency domain requires a larger scheduling message in comparison with an uplink grant allowing for frequency-contiguous allocations only. The DCI is therefore categorized into different *DCI formats*, where a format corresponds to a certain message size and usage. The DCI formats are summarized in Table 16.1 together with their respective relative message sizes. The actual message size depends on the cell bandwidth as, for larger bandwidths, a larger number of bits are required to indicate the resource-block allocation. This will be discussed further below; at this stage it suffices to note that some DCI formats have the same message size.

One PDCCH carries one message with one of the DCI formats above. As multiple terminals can be scheduled simultaneously, on both downlink and uplink, there must be a possibility to transmit multiple scheduling messages within each subframe. Each scheduling message is transmitted on a separate PDCCH, and consequently there are typically multiple simultaneous PDCCH transmissions

Table 16.1 *DCI formats.*

	Usage		
Relative DCI size	Uplink grant	Downlink assignment	Power control
Small	–	1C Small contiguous allocations	–
	0	1A Contiguous allocations only	3, 3A
…	–	1B Contiguous allocations with spatial multiplexing	–
	–	1 Flexible allocations, no spatial multiplexing	–
Large	–	2 Flexible allocations, full spatial multiplexing	–

within each cell. Furthermore, to support different radio-channel conditions, link adaptation, where the code rate of the PDCCH is selected to match the radio-channel conditions, can be used.

16.4.4 Downlink scheduling assignment

Downlink scheduling assignments are valid for the same subframe in which they are transmitted. The scheduling assignments use one of DCI formats 1C, 1A, 1B, 1, and 2, ordered in increasing order of the message size. The reason for supporting multiple formats with different message sizes for the same purpose is to allow for a trade-off in control-signaling overhead and scheduling flexibility. Parts of the contents are the same for the different DCI formats as seen in Figure 16.17, but obviously there are also differences due to the different capabilities.

DCI format 1C is used for various special purposes such as random-access response, paging, and transmission of system information. Common for these applications is simultaneous reception of a relatively small amount of information by *multiple* users. Hence, DCI format 1C supports QPSK only, has no support for hybrid-ARQ retransmissions, and does not support closed-loop spatial multiplexing. Consequently, the message size for DCI format 1C is small, which is beneficial for coverage and efficiency of the type of system messages for which it is intended. Furthermore, as only a small number of resource blocks can be indicated, the size of the corresponding indication field in DCI format 1C is independent of the cell bandwidth.

DCI format 1A, also known as the 'compact' downlink assignment, supports allocation of frequency-contiguous resource blocks only. Contiguous allocations reduce the payload size and are also used in the uplink grants in DCI format 0 as a consequence of the single-carrier properties.

Codebook-based beam-forming described in Section 16.6.2, with a low control-signaling overhead, is supported by DCI format 1B. The content is similar to DCI format 1A with the addition of bits for signaling of the precoding matrix. As codebook-based beam-forming can be used to improve the data rates for cell-edge terminals, it is important to keep the related DCI message size small not to unnecessarily limit the coverage.

Unlike DCI format 1A, DCI format 1 supports non-contiguous allocations of resource blocks. Thus, it not only offers increased flexibility in the resource

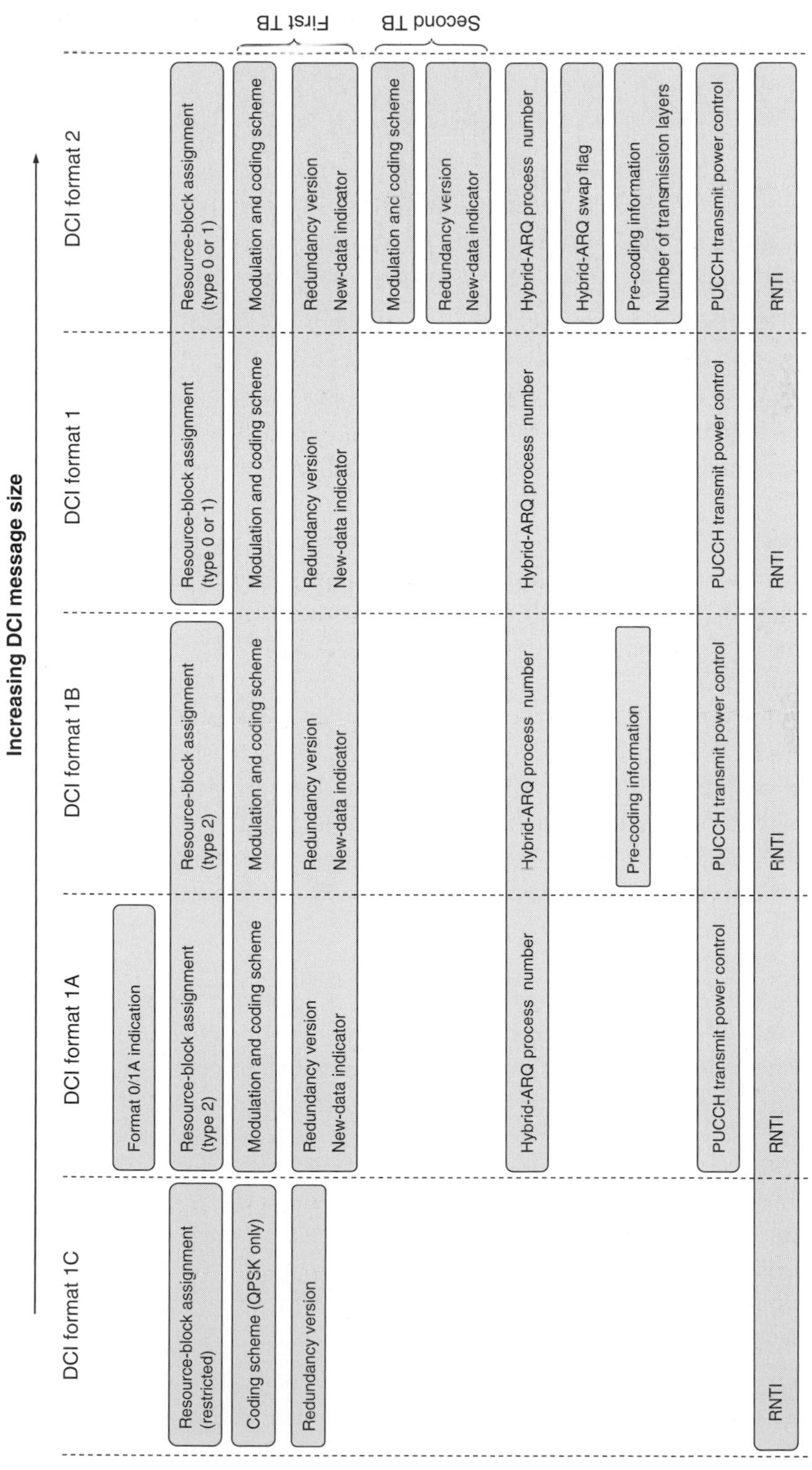

Figure 16.17 *Overview of DCI formats for downlink scheduling (FDD).*

allocation but also implies a larger control-signaling overhead. Apart from supporting non-contiguous allocations, most of the information fields in DCI format 1 are the same as for DCI format 0.

DCI format 2, which is the largest of the DCI formats for downlink scheduling assignments, can be seen as an extension for DCI format 1 supporting spatial multiplexing. Thus, fields for indicating the number of transmission layers are included and some of the fields in DCI format 1 have been duplicated to handle the two transport blocks transmitted in parallel in case of spatial multiplexing.

The information fields in the different DCI formats are, as already mentioned, common among several of the formats. The contents of the different DCI formats are summarized in Figure 16.17 and explained in more detail below:

- DCI format 0/1A indication [1 bit], used to differentiate between DCI formats 1A and 0 as the two formats have the same message size. This field is present in DCI formats 0 and 1A only. DCI formats 3 and 3A, which have the same size, are separated from DCI formats 0 and 1A through the use of a different RNTI.
- Distributed transmission flag [1 bit], determines whether downlink distributed transmission as described in Section 16.5.8 is used or not. This field is present only in DCI format 1A as distributed transmission only applies to transmissions scheduled with this DCI format.
- Resource-block allocation. This field indicates the resource blocks upon which the terminal should receive the PDSCH and the size of this fields depends on the DCI format:
  – DCI format 1C supports a small frequency-contiguous allocation only. Furthermore, the allocation is limited to a relatively small number of resource blocks, thereby keeping the control-signaling overhead small.
  – DCI formats 1A and 1B support a frequency-contiguous allocation of resource blocks only, that is the resource allocation type 2 as described below is used. The size of this field depends on the downlink cell bandwidth.
  – DCI formats 1 and 2 support non-contiguous allocations of resource blocks using resource allocation type 0 or 1 as described below. The size of this field depends on the downlink cell bandwidth.
- For the first (or only) transport block:
  – Modulation and coding scheme [5 bit], used to provide the terminal with information about the modulation scheme, the code rate, and the transport-block size as described further below. DCI format 1C has a restricted size of this field as only QPSK is supported.

- New-data indicator [1 bit], used to clear the soft buffer for initial transmissions. Not present in DCI format 1C as this format does not support hybrid ARQ.
- Redundancy version [2 bit].
- For the second transport block (present in DCI format 2 only):
  - Modulation and coding scheme.
  - New-data indicator [1 bit].
  - Redundancy version [2 bit].
- Hybrid-ARQ process number [3 bit for FDD, 4 bit for TDD], informing the terminal about the process to use for soft combining. Not present in DCI format 1C.
- Information related to spatial multiplexing (present in DCI format 2 only):
  - Pre-coding information.
  - Number of transmission layers.
  - Hybrid-ARQ swap flag [1 bit], indicating whether the two codewords should be swapped prior to being fed to the hybrid-ARQ processes. Used for averaging the channel quality between the codewords.
- Downlink assignment index [2 bit], informing the terminal about the number of downlink transmissions for which a single hybrid-ARQ acknowledgement shall be generated according to Section 19.1.1.3. Present for TDD only.
- Transmit-power control for PUCCH [2 bit]. Not present in DCI format 1C.
- Identity (RNTI) of the terminal for which the PDSCH transmission is intended [16 bit]. As described in Section 16.4.7, the identity is not explicitly transmitted but implicitly included in the CRC calculation. There are different RNTIs defined depending on the type of transmission (unicast data transmission, paging, power-control commands, etc.).

16.4.4.1 *Signaling of resource-block allocations*

Focusing on the signaling of resource-block allocations, there are three different possibilities, types 0, 1, and 2, as indicated in Figure 16.17. Resource-block allocation types 0 and 1 both support non-contiguous allocations of resource block in the frequency domain, whereas type 2 supports contiguous allocations only. A natural question is why multiple ways of signaling the resource-block allocations are supported; a question where the answer lies in the number of bits required for the signaling. The most flexible way of indicating the resource blocks, which the terminal is supposed to receive the downlink transmission upon, is to include a bitmap with the size equal to the number of resource blocks in the cell bandwidth. This would allow for an arbitrary combination of resource blocks to be scheduled for transmission to the terminal but would, unfortunately, also result in a very large bitmap for the larger cell bandwidths. For example, in case of a downlink cell bandwidth corresponding to 100 resource blocks, the downlink PDCCH would require 100 bits for the bitmap alone to which the other pieces of information need to be added. Not only would this result in a large control-signaling overhead, but it could also result in downlink coverage problems as more than 100 bits in one OFDM symbol corresponds to a data rate exceeding 1.4 Mbit/s. Consequently, there is a need for a resource allocation scheme requiring a smaller number of bits while keeping sufficient allocation flexibility.

In resource allocation type 0, the size of the bitmap has been reduced by pointing not to individual resource blocks in the frequency domain but to groups of contiguous resource blocks as shown at the top of Figure 16.18. The size of such a group is determined by the downlink cell bandwidth; for the smallest bandwidths there is only a single resource block in a set implying that an arbitrary set of resource blocks can be scheduled, whereas for the largest cell bandwidths groups of four resource blocks are used (in the example in Figure 16.18, the cell bandwidth is 25 resource blocks, implying a group size of two resource blocks). Thus, the bitmap for the system with a downlink cell bandwidth of 100 resource blocks is reduced from 100 to 25 bits. A drawback is that the scheduling granularity is reduced; single resource blocks cannot be scheduled for the largest cell bandwidths using allocation type 0.

However, also in large cell bandwidths a frequency resolution of a single resource block is sometimes useful, for example, to support small payloads. Resource allocation type 1 address this by dividing the total number of resource blocks in the frequency domain into dispersed subsets as shown in the middle of Figure 16.18. The number of subsets is given from the cell bandwidth with the number of subsets in type 1 being equal to the group size in type 0. Thus, in Figure 16.18, there are two subsets, whereas for a cell bandwidth of 100 resource blocks there would have been four different subsets. Within a subset, a bitmap indicates the resource blocks in the frequency domain upon which the downlink transmission occurs.

To inform the terminal whether resource allocation type 0 or 1 is used, the resource allocation field includes a flag for this purpose, denoted 'type' to the left in Figure 16.18. For type 0, the only additional information is the bitmap discussed above. For type 1, on the other hand, in addition to the bitmap itself, information about the subset for which the bitmap relates is also required. As one of the requirements on the design of resource allocation type 1 was to maintain the same number of bits in the allocation as for type 0 without adding unnecessary overhead, the bitmap in resource allocation type 1 is smaller than in type 0 to allow for the signaling of the subset number. However, a

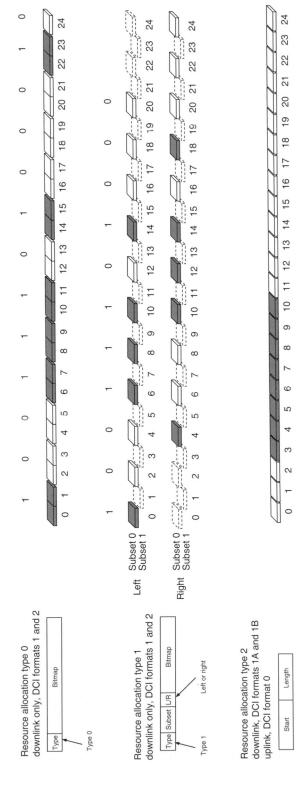

Figure 16.18 *Illustration of resource-block allocation types (cell bandwidth corresponding to 25 resource blocks used in this example).*

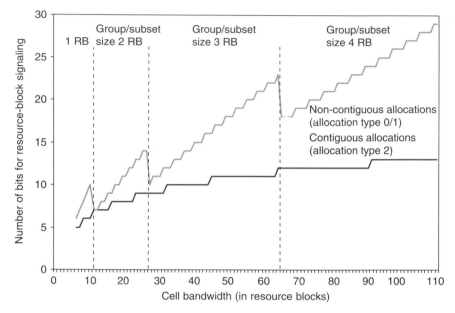

Figure 16.19   *Number of bits used for resource allocation signaling for allocation types 0/1 and 2.*

consequence of a smaller bitmap is that not all resource blocks in the subset can be addressed simultaneously. To be able to address all resources with the bitmap, there is a flag indicating whether the bitmap relates to the 'left' or 'right' part of the resource blocks as depicted in the middle part of Figure 16.18.

Unlike the other two types of resource-block allocation signaling, type 2 does not rely on a bitmap. Instead, it encodes the resource allocation as a start position and length of the resource-block allocation. Thus, it does not support arbitrary allocations of resource blocks but only frequency-contiguous allocations, thereby reducing the number of bits required for signaling the resource-block allocation. The number of bits required for resource-signaling type 2 compared to type 0 or 1 is shown in Figure 16.19 and, as shown, the difference is fairly large for the larger cell bandwidths.

16.4.4.2   *Signaling of transport-block sizes*

Proper reception of a downlink transmission requires, in addition to the set of resource blocks, knowledge about the modulation scheme and the transport-block size; information (indirectly) provided by a five-bit field in the different DCI formats. Out of the 32 combinations, 29 are used to signal the modulation-and-coding scheme whereas three are reserved, the purpose of which is described later. Together, the modulation-and-coding scheme and the number of resource blocks assigned provide the transport-block size on the DL-SCH. Thus, the possible transport-block sizes can be described as a table with 29 rows and 110 columns, one column for each number of resource blocks possible to transmit upon (assuming the maximum downlink cell bandwidth of 110 resource blocks).

Each modulation-and-coding scheme represents a particular combination of modulation scheme and channel-coding rate, or, equivalently, a certain spectral efficiency measured in the number of information bits per modulation symbol. Although the 29-by-110 table of transport-block sizes in principle could be filled directly from the modulation-and-coding scheme and the number of resource blocks, this would result in arbitrary transport-block sizes which is not desirable. First, as all the protocol layers are byte aligned, the resulting transport-block sizes should be an integer number of bytes. Secondly, common payloads, for example, RRC signaling messages and VoIP, should be possible to transmit without padding. Aligning with the QPP interleaver sizes is also beneficial, as this would avoid the use of filler bits (see Section 16.5.3). Finally, the same transport-block size should ideally appear for several different resource-block allocations as this allows the number of resource blocks to be changed between retransmission attempts providing increased scheduling flexibility. Therefore, a 'mother table' of transport-block sizes is first defined. Each entry in the 29-by-110 table is picked from the mother table such that the resulting spectral efficiency is as close as possible to the spectral efficiency of the signaled modulation-and-coding scheme. The mother table spans the full range of transport-block sizes possible with an approximately constant worst-case padding, which is similar to HSPA.

From a simplicity perspective, it is desirable if the transport-block sizes do not vary with the configuration of the system. The set of transport-block sizes is therefore independent of the actual number of antenna ports and the size of the control region. The design of the table assumes a control region of three OFDM symbols and two antenna ports, the 'reference configuration.' If the actual configuration is different, the resulting code rate for the DL-SCH will be slightly different as a result from the rate-matching procedure. However, the difference is small and of no practical concern. Also, if the actual size of the control region is smaller than the three-symbol assumption in the reference configuration, the spectral efficiencies will be somewhat smaller than the range indicated by the modulation-and-coding scheme signaled as part of the DCI. Thus, information about the modulation scheme used is obtained directly from the modulation-and-coding scheme whereas the exact code rate and rate matching is obtained from the implicitly signaled transport-block size together with the number of resource elements used for DL-SCH transmission.

For bandwidths smaller than the maximum of 110 resource blocks, a subset of the table is used. More specifically, in case of a cell bandwidth of N resource blocks, the first *N* columns of the table are used. Also, in case of a single transport block being mapped to two transmission layers in case of spatial multiplexing, the transport-block size is in principle twice the size indicated by the table.

The 29 combinations of modulation-and-coding schemes each represent a reference spectral efficiency in the approximate range of 0.2–5.6 bits/s/symbol.[13] There is some overlap in the combinations in the sense that some of the 29 combinations represent the same spectral efficiency. The reason is that the best combination for realizing a specific spectral efficiency depends on the channel properties; sometimes higher-order modulation with a low code rate is preferable over lower-order modulation with a higher code rate, sometimes the situation is opposite. With the overlap, the eNodeB can select the best combination, given the propagation scenario. As a consequence of the overlap, two of the rows in the 29-by-110 table are duplicates, that is result in the same spectral efficiency but with different modulation schemes, and there are only 27 unique rows of transport-block sizes.

Returning to the three reserved combinations in the modulation-and-coding field mentioned at the beginning, those entries can be used for retransmissions only. In case of a retransmission, the transport-block size is per definition unchanged and fundamentally there is no need to signal this piece of information. Instead, the three reserved values represent the modulation scheme, QPSK, 16QAM or 64QAM, which allows the scheduler to use an (almost) arbitrary combination of resource blocks for the retransmission. Obviously, using any of the three reserved combinations assume that the terminal properly received the control signaling for the initial transmission; if this is not the case, the retransmission should explicitly indicate the transport-block size.

The derivation of the transport-block size from the modulation-and-coding scheme and the number of scheduled resource blocks is illustrated in Figure 16.20.

16.4.5   Uplink scheduling grants

Uplink scheduling grants use DCI format 0. Unlike the downlink, neither noncontiguous resource-block allocations nor spatial multiplexing from a single terminal is supported for the LTE uplink. Hence a single DCI format, format 0, is sufficient for the uplink grants.

---

[13] In order to not limit the possibility to fully exploit the possibilities offered by the physical layer, a few entries in the 29-by-110 table are slightly modified, resulting in a spectral efficiency of more than 5.6 bits/s/Hz.

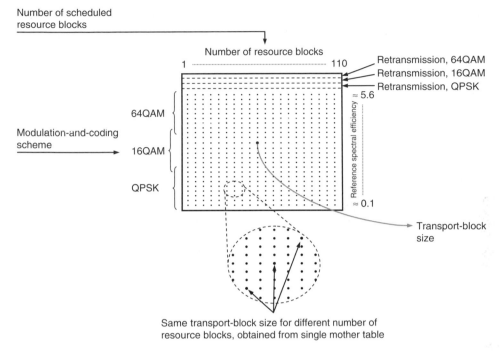

Figure 16.20 *Computing the transport-block size.*

DCI format 0 has the same size of the control-signaling message as the 'compact' downlink assignment[14] (DCI format 1A). A flag in the message is used to inform the terminal whether the message is an uplink scheduling grant (DCI format 0) or a downlink scheduling assignment (DCI format 1A). The information fields of DCI format 1A includes:

- DCI format 0/1A indication [1 bit], used to differentiate between DCI formats 1A and 0 as the two formats have the same message size.
- Hopping flag [1 bit], indicating whether or not uplink frequency hopping as described in Section 17.5 is to be applied to the uplink PUSCH transmission.
- Resource-block allocation. This field indicates the resource blocks upon which the terminal should transmit the PUSCH. Since the single-carrier properties of the uplink implies that resource blocks must be contiguous in frequency, resource allocation type 2, described in Section 16.4.4.1, is used for the uplink scheduling grants.
- Modulation-and-coding scheme including redundancy version [5 bit], used to provide the terminal with information about the modulation scheme, the

---

[14]The smaller of DCI formats 0 and 1A is padded to ensure the same payload size. This is mainly relevant for the case of different uplink and downlink cell bandwidths; in case of identical uplink and downlink bandwidths in a cell there is a single bit of padding in format 0.

code rate, and the transport-block size. The signaling of the transport-block size use the same transport-block table as for the downlink, that is the modulation-and-coding scheme together with the number of scheduled resource blocks provides the transport-block size. However, as the support of 64QAM in the uplink is not mandatory for all terminals (and eNodeBs), those terminals use 16QAM even if 64QAM is indicated in the modulation-and-coding field. The use of the three reserved combinations is slightly different than for the downlink; the three reserved values are used for implicit signaling of the redundancy version as described below.

- New-data indicator [1 bit], used to clear the soft buffer for initial transmissions.
- Phase rotation of the uplink demodulation reference signal [3 bit]. Multiple terminals can be scheduled to transmit upon the same set of resource blocks, but if the uplink demodulation reference signals are distinguishable through different phase rotations as described in Chapter 17, the eNodeB can estimate the uplink channel response from each terminal and suppress the inter-terminal interference by the appropriate processing. This is sometimes referred to as multi-user MIMO.
- Channel-status request flag [1 bit]. The network can explicitly request an aperiodic channel-status report, discussed in Chapter 19, by setting this bit in the uplink grant.
- Uplink index [2 bit]. This field is present only when operating in TDD and is used to signal for which uplink subframe the grant is valid as described below.
- Transmit-power control for PUSCH [2 bit].
- Identity (RNTI) of the terminal for which the PDSCH transmission is intended [16 bit]. As described in Section 16.4.7, the identity is not explicitly transmitted but implicitly included in the CRC calculation.

The resource allocation field indicates the set of resource blocks to use for uplink transmission on PUSCH. As single-carrier transmission is used for the uplink, resource blocks must be contiguous in frequency. Hence, resource-block allocation type 2, described in the previous section, is used. The exact set of resource blocks to use in the two slots of a subframe is controlled by the hopping flag as described in Section 17.5.

There is no explicit signaling of the redundancy version in the uplinks scheduling grants. This is motivated by the use of a synchronous hybrid-ARQ protocol in the uplink; retransmissions are normally triggered by a negative acknowledgement on the PHICH and not explicitly scheduled as for downlink data transmissions. However, as described in Chapter 19, there is a possibility to explicitly schedule retransmissions. This is useful in a situation where the network explicitly will move the retransmission in the frequency domain by using the PDCCH instead of the PHICH. Three values of the modulation-and-coding field are reserved to mean redundancy versions one, two, and three. If one of those values are signaled, the terminal should assume that the same modulation and coding as the original transmission is used. The remaining entries are used to signal the modulation-and-coding scheme to use and also imply that redundancy version zero should be used. The difference in usage of the reserved values compared to the downlink scheduling assignments implies that the modulation scheme, unlike the downlink case, cannot change between uplink (re)transmission attempts.

The time between reception of an uplink scheduling grant on a PDCCH and the corresponding transmission on the UL-SCH is fixed. For FDD, the time relation is the same as for PHICH, that is, an uplink grant received in downlink subframe n applies to uplink subframe $n + 4$. The reason for having the same timing relation as for PHICH is the possibility to override the acknowledgement on PHICH with a scheduling grant on PDCCH as described in Chapter 19. For TDD, such a time relation is not possible as subframe $n + 4$ may not be an uplink subframe. Therefore, for some downlink–uplink configurations, the delay between the reception of an uplink scheduling grant and the actual transmission differs between subframes, depending on the subframe in which the uplink scheduling grant was received. Furthermore, in one of the downlink–uplink asymmetries for TDD, configuration 0, there are more uplink subframes than downlink subframes, three-versus-two. Therefore, there is a need to be able to schedule multiple uplink subframes in one downlink subframe; if this was not possible, not all uplink subframes would be possible to schedule. Consequently, a two-bit uplink index field is part of the uplink scheduling grant. The index field specifies which uplink subframe a grant received in a downlink subframe applies to. For example, as illustrated in Figure 16.21, an uplink scheduling grant received in downlink subframe 0 applies to one or both of the uplink subframes 4 and 7, depending on the contents of the index field.

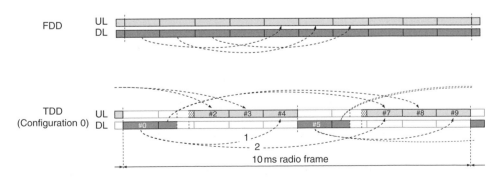

Figure 16.21 *Timing relation for uplink grants in FDD and TDD configuration 0.*

16.4.6 Power-control commands

As a complement to the power-control commands sent as part of the downlink scheduling assignments and the uplink scheduling grants, there is a possibility to transmit a power-control command using DCI formats 3 (single-bit command per terminal) or 3A (two-bit command per terminal). The power-control message is directed to a group of terminals using an RNTI specific for that group. Each terminal can be allocated two power-control RNTIs, one for PUCCH power control and the other for PUSCH power control. Although the power-control RNTIs are common to a group of terminals, each terminal is informed through RRC signaling which bit(s) in the DCI message it should follow.

16.4.7 PDCCH processing

Having discussed the contents of the L1/L2 control signaling in terms of the different DCI formats, the transmission of the DCI message on a PDCCH can be described. The processing of downlink control signaling is illustrated in Figure 16.22. A CRC is attached to each DCI message payload. Similarly to HSPA, the identity of the terminal (or terminals) addressed, that is, the RNTI, is included in the CRC calculation but not explicitly transmitted. Depending on the purpose of the DCI message (unicast data transmission, power-control command, random-access response, etc.), different RNTIs are used; for normal unicast data transmission, the terminal-specific C-RNTI is used.

Upon reception of DCI, the terminal will check the CRC using its set of assigned RNTIs. If the CRC checks, the message is declared to be correctly received and intended for the terminal. Thus, the identity of the terminal that is supposed to receive the DCI message is implicitly encoded in the CRC and not explicitly transmitted. This reduces the amount of bits necessary to transmit on the PDCCH as, from a terminal point-of-view, there is no difference between a corrupt message whose CRC will not check and a message intended for another terminal.

After CRC attachment, the bits are coded with a rate-1/3 tail-biting convolutional code, and the rate is matched to fit the amount of resources used for PDCCH transmission. Tail-biting convolutional coding is similar to conventional convolutional coding with the exception that no tail bits are used. Instead, the convolutional encoder is initialized with the last bits of the message prior to the encoding process. Thus, the starting and ending states in the trellis in a MLSE (Viterbi) decoder are identical.

To allow for simple yet efficient processing of the control channels in the terminal, the mapping of PDCCHs to resource elements are subject to a certain

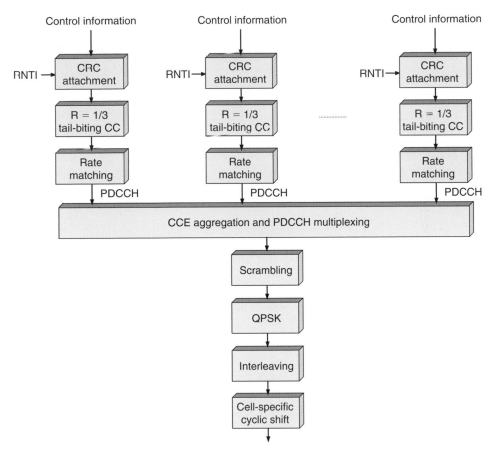

Figure 16.22 *Processing of L1/L2 control signaling.*

structure. This structure is based on so-called *Control-Channel Elements* (CCEs), which in essence is a convenient name for a set of 36 useful resource elements (9 resource-element groups as defined in Section 16.4.1). The number of CCEs, one, two, four, or eight, required for a certain PDCCH depends on the payload size of the control information (DCI payload) and the channel-coding rate. This is used to realize link adaptation for the PDCCH; if the channel conditions for the terminal to which the PDCCH is intended are disadvantageous a larger number of CCEs need to be used compared to the case of advantageous channel conditions.

The number of CCEs available for PDCCHs depends on the size of the control region, the cell bandwidth, the number of downlink antenna ports, and the amount of resources occupied by PHICH. The size of the control region can vary dynamically from subframe to subframe as indicated by the PCFICH whereas the other quantities are semi-statically configured. The CCEs available for PDCCH transmission can be numbered from zero and upward as illustrated in Figure 16.23. A specific PDCCH can thus be identified by the numbers of the corresponding CCEs in the control region.

As the number of CCEs for each of the PDCCIIs may vary and is not signaled, the terminal has to blindly determine the number of CCEs used for the PDCCH it is addressed upon. To reduce the complexity of this process somewhat, certain restrictions on the aggregation of CCEs have been specified. For example, an aggregation of eight CCEs can only start on CCE numbers evenly divisible by eight as illustrated in Figure 16.23. The same principle is applied to the other aggregation levels. Furthermore, some combinations of DCI formats and CCE aggregations that result in excessively high channel-coding rates are removed.

The sequence of CCEs should match the amount of resources available for PDCCH transmission in a given subframe, that is the number of CCEs varies according to the value transmitted on the PCFICH. This implies that, in many cases, not all the PDCCHs that are possible to transmit in the control region are used. Nevertheless, unused PDCCHs are part of the interleaving and mapping process in the same way as any other PDCCH. At the terminal, the CRC will not check for those 'dummy' PDCCHs. Preferably, the transmission power is set to zero for those unused PDCCHs; the power can be used by other control channels.

After the PDCCHs to be transmitted in a given subframe have been allocated to the desired CCEs as described above, the sequence of bits corresponding to all the CCEs to be transmitted in the subframe, including the unused CCEs, is scrambled by a cell- and subframe-specific scrambling sequence to randomize inter-cell interference, followed by QPSK modulation and mapping to resource elements.

The mapping of the modulated composite control information is, for the same reason as for the other control channels, described in terms of symbol quadruplets being mapped to resource-element groups. Thus, the first step of the mapping stage is to group the QPSK symbols into symbol quadruplets, each consisting of four consecutive QPSK symbols. In principle, the sequence of quadruplets could be mapped directly to the resource elements in sequential order. However, this would not exploit all the frequency diversity available in the channel and diversity is important for good performance. Furthermore, if the same CCE-to-resource-element mapping is used in all neighboring cells, a given PDCCH will persistently collide with one and the same PDCCH in the neighboring cells

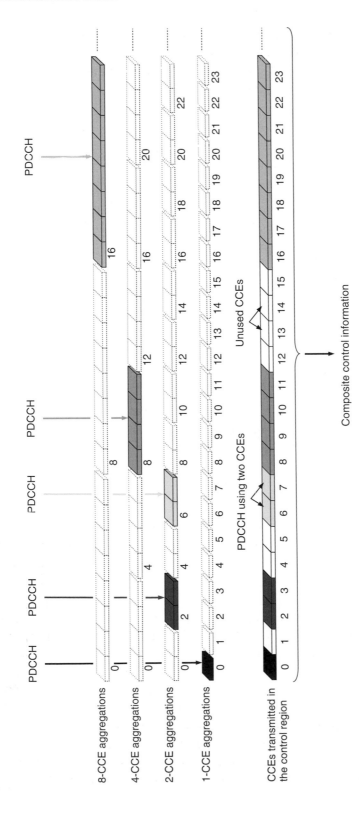

Figure 16.23 *CCE aggregation and PDCCH multiplexing.*

assuming a fixed PDCCH format and inter-cell synchronization. In practice, the number of CCEs per PDCCH varies in the cell as a function of the scheduling decisions, which gives some randomization to the interference, but further randomization is desirable to obtain a robust control channel design. Therefore, the sequence of quadruplets are first interleaved using a block interleaver to allow exploitation of the frequency diversity, followed by a cell-specific cyclic shift to randomize the interference between neighboring cells. The output from the cell-specific shift is mapped to resource-element groups in a time-first manner as illustrated in Figure 16.14, skipping resource-element groups used for PCFICH and PHICH. Time-first mapping preserves the interleaving properties; with frequency-first over multiple OFDM symbols, resource-element groups that are spread far apart after the interleaving process may end up close in frequency although on different OFDM symbols.

The interleaving operation described above, in addition to enabling exploitation of the frequency diversity and randomizing the inter-cell interference, also serves the purpose of ensuring that each CCE spans virtually all the OFDM symbols in the control region. This is beneficial for coverage as it allows flexible power balancing between the PDCCHs to ensure good performance for each of the terminals addressed. In principle, the energy available in the OFDM symbols in the control region can be balanced arbitrarily between the PDCCHs. The alternative of restricting each PDCCH to a single OFDM symbol would imply that power cannot be shared between PDCCHs in different OFDM symbols.

Similarly to the PCFICH, the transmission power of each PDCCH is under the control of the eNodeB. Power adjustments can therefore be used as a complementary link adaptation mechanism in addition to adjusting the code rate. Relying on power adjustments alone might seem a tempting solution and, although possible in principle, it can result in relatively large power differences between PDCCHs. This may have implications on the RF implementation and may violate the out-of-band emission masks specified. Hence, to keep the power differences between the PDCCHs reasonable, link adaptation through adjusting the channel code rate, or equivalently, the number of CCEs aggregated for a PDCCH, is necessary. The two mechanisms for link adaptation, power adjustments and different code rates, complement each other.

To summarize and to illustrate the mapping of PDCCHs to resource elements in the control region, consider the example shown in Figure 16.24. In this example, the size of the control region in the subframe considered equals three OFDM symbols. Two downlink antenna ports are configured (but as explained above, the mapping would be identical in case of a single antenna port). One PHICH

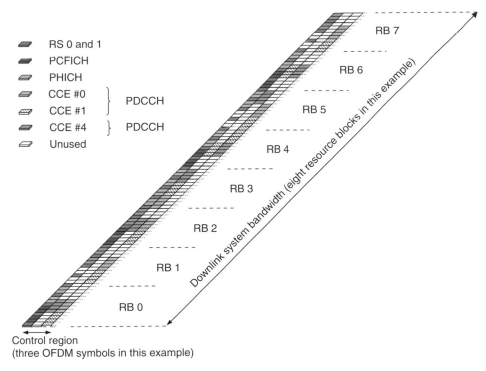

Figure 16.24 *Example of mapping of PCFICH, PHICH, and PDCCH.*

group is configured and three resource-element groups are therefore used by the PHICHs. The cell identity is assumed to be identical to zero in this case.

The mapping can then be understood as follows: first, the PCFICH is mapped to four resource-element groups, followed by allocating the resource-element groups required for the PHICH. The resource-element groups left after the PCFICH and PHICH are used for the different PDCCHs in the system. In this particular example, one PDCCH is using CCE numbers 0 and 1, while another PDCCH is using CCE number 4. Consequently, there is a relatively large amount of unused resource-element groups in this example; either they can be used for additional PDCCHs or the power otherwise used for the unused CCEs could be allocated to the PDCCHs in use.

16.4.8   Blind decoding of PDCCHs

As described above, each PDCCH supports multiple formats and the format used is a priori unknown to the terminal. Therefore, the terminal needs to blindly detect the format of the PDCCHs. The CCE structure described in the previous section helps in reducing the number of blind decoding attempts, but is not sufficient. Hence, it is required to have mechanisms to limit the CCE aggregations that the terminal is supposed to monitor. Clearly, from a scheduling point-of-view, restrictions in the allowed CCE aggregations are undesirable as they may influence the scheduling flexibility and requires additional processing at the transmitter side. At the same time, requiring the terminal to monitor all possible CCE aggregations, also for the larger cell bandwidths, is not attractive from a terminal-complexity point-of-view. To impose as few restrictions as possible on the scheduler while at the same time limit the maximum number of blind decoding attempts in the terminal to 44 per subframe, LTE defines so-called *search spaces* which describes the set of CCEs the terminal is supposed to monitor.

A search space is a set of candidate control channels formed by CCEs on a given aggregation level, which the terminal is supposed to attempt to decode. As there are multiple aggregation levels, corresponding to one, two, four, and eight CCEs, a terminal has multiple search spaces. In each subframe, the terminals will attempt to decode all the PDCCHs that can be formed from the CCEs in each of its search spaces. If the CRC checks, the content of the control channel is declared as valid for this terminal and the terminal processes the information (scheduling assignment, scheduling grants, etc.). Clearly, the network can only address a terminal if the control information is transmitted on a PDCCH formed by the CCEs in one of the terminal's search spaces. For example, terminal A in Figure 16.25 cannot be addressed on a PDCCH starting on CCE number 20 whereas terminal B can. Furthermore, if terminal B is using CCEs 16–23 and another terminal, terminal C, is scheduled using CCEs 8–15, terminal A cannot be addressed on aggregation level four as all CCEs in its level-four search space are blocked by the use for the other terminals. From this it can be intuitively understood that for efficient utilization of the CCEs in the system, the search spaces should differ between terminals. Each terminal in the system therefore has a *terminal-specific* search space at each aggregation level.

The DCI formats to decode in the terminal-specific search space depend on the transmission mode but at least DCI format 1A is always decoded. Transmission modes are described in Section 16.6 and, in principle, corresponds to different multi-antenna configurations. As an example, there is no need to attempt to decode DCI format 2 when the terminal has not been configured for spatial multiplexing.

As the terminal-specific search space is typically smaller than the number of PDCCHs the network could transmit at the corresponding aggregation level,

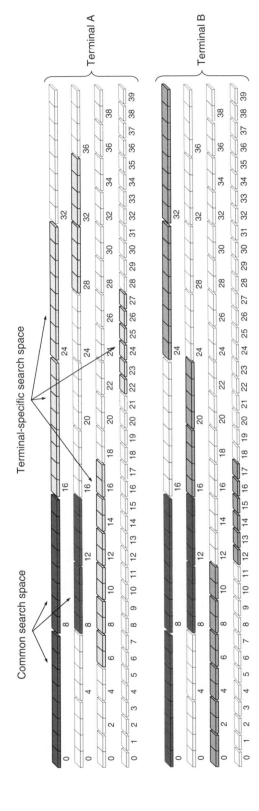

Figure 16.25 *Principal illustration of search spaces in two terminals.*

there must be a mechanism determining the set of CCEs in the terminal-specific search space for each aggregation level. One possibility would be to let the network configure the terminal-specific search space in each terminal. However, this would require explicit signaling to each of the terminals and possibly reconfiguration at handover. Instead, the terminal-specific search spaces in LTE are defined without explicit signaling through a function of the terminal identity and implicitly the subframe number. Dependence on the subframe number results in the terminal-specific search spaces being time varying, which helps resolving blocking between terminals. If a given terminal cannot be scheduled in a subframe as all the CCEs that the terminal is monitoring have already been used for scheduling other terminals in the same subframe, the time-varying definition of the terminal-specific search spaces is likely to resolve the blocking in the next subframe.

In several situations, there is a need to address a group of, or all, terminals in the system. One example hereof is dynamic scheduling of system information, another is transmission of paging messages, both described in Chapter 18. Transmission of explicit power-control commands to a group of terminals is a third example. To allow all terminals to be addressed at the same time, LTE has defined *common search spaces* in addition to the terminal-specific search spaces. A common search space is, as the name implies, common, and all terminals in the cell monitor the CCEs in the common search spaces for control information. Although the motivation for the common search space is primarily transmission of various system messages, it can be used to schedule individual terminals as well. Thus, it can be used to resolve situations where scheduling of one terminal is blocked due to lack of available resources in the terminal-specific search space. Unlike unicast transmissions, where the transmission parameters of the control signaling can be tuned to match the channel conditions of a specific terminal, system messages typically need to reach the cell border. Furthermore, the data rate of the associated DL-SCH transmission is typically modest. Consequently, the common search spaces are only defined for aggregation levels of four and eight CCE and only for the smallest DCI formats, 0/1A/3/3A and 1C. There is no support for DCI formats with spatial multiplexing in the common search space. This helps reducing the number of blind decoding attempts in the terminal used for monitoring the common search space.

Figure 16.25 illustrates the terminal-specific and common search spaces in two terminals in a certain subframe. The terminal-specific search spaces are different in the two terminals and will, as described above, vary from subframe to subframe. Furthermore, the terminal-specific search spaces partially overlap between the two terminals in this subframe (CCEs 24–31 on aggregation level 8) but, as the terminal-specific search space varies between subframes, the overlap in the next subframe is most likely different.

16.5 Downlink transport-channel processing

As discussed in Chapter 15, the LTE physical layer interfaces to higher layers, more specifically to the MAC layer, by means of *Transport Channels*. LTE has inherited the basic principle of WCDMA/HSPA that data are delivered to the physical layer in the form of *Transport Blocks* of a certain size. In terms of the more detailed transport-block structure, LTE has adopted a similar approach as was adopted for HSPA:

- In case of single-antenna transmission there is a single transport block of dynamic size for each TTI.
- In case of multi-antenna transmission, there can be up to two transport blocks of dynamic size for each TTI, where each transport block corresponds to one *codeword* in case of downlink spatial multiplexing. This implies that, although LTE supports downlink spatial multiplexing with up to four layers, the number of codewords and thus also the number of transport blocks is still limited to two. More details on LTE downlink spatial multiplexing are provided in Section 16.6.

With this transport-block structure in mind, the LTE downlink transport-channel processing, more specifically the processing for DL-SCH transmission, can be outlined according to Figure 16.26 with two, mainly separated, processing chains, each corresponding to the processing of a single transport block. The second processing chain, corresponding to the second transport block, is thus only present in the case of downlink spatial multiplexing. In this case, the two transport blocks of, in general, different sizes are combined as part of the *antenna mapping* in the lower part of Figure 16.26.

This processing chain is also to a large extent applicable to the PCH and MCH transport channels, although with certain restrictions. In contrast, as will be clear from Chapter 18, the processing for the BCH transport channel differs from that of DL-SCH in several respects.

16.5.1 CRC insertion per transport block

In the first step of the transport-channel processing, a 24-bit CRC is calculated for and appended to each transport block. The CRC allows for receiver-side detection of errors in the decoded transport block. The corresponding error indication is then, for example, used by the downlink hybrid-ARQ protocol as a trigger for requesting retransmissions.

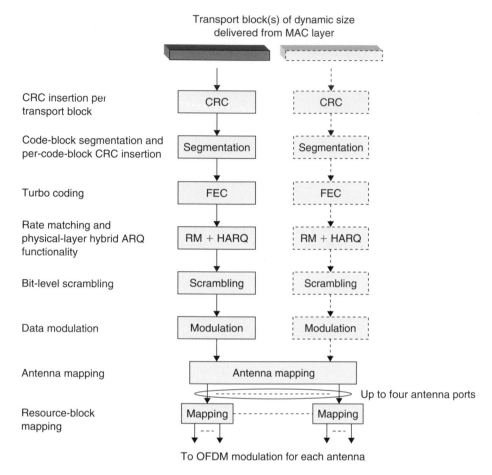

Figure 16.26   *LTE downlink transport-channel processing. Dashed parts are only present in case of spatial multiplexing, that is when two transport blocks are transmitted in parallel within a TTI.*

16.5.2   *Code-block segmentation and per-code-block CRC insertion*

The LTE Turbo-coder internal interleaver is only defined for a limited number of code-block sizes with a maximum block size of 6144 bits. In case the transport block, including the transport-block CRC, exceeds this maximum code-block size, *code-block segmentation* as illustrated in Figure 16.27 is applied before Turbo coding. Code-block segmentation implies that the transport block is segmented into smaller *code blocks* that match the set of code-block sizes defined for the Turbo coder.

In order to ensure that the size of each code block is matched to the set of available code-block sizes, *filler bits* may have to be inserted at the head of the first

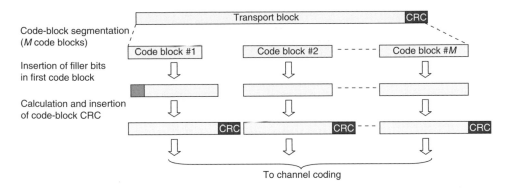

Figure 16.27 *Code-block segmentation and per-code-block CRC insertion.*

code block. Note that filler bits may be needed also if there is no actual code-block segmentation, that is if the transport-block size does not exceed the maximum code-block size.[15]

As can be seen in Figure 16.27 code-block segmentation also implies that an additional (24 bits) CRC is calculated for and appended to each code block.[16] Having a CRC per code block allows for early detection of correctly decoded code blocks and corresponding early termination of the iterative decoding of that code block. This can be used to reduce the terminal processing effort and power consumption. It should be noted that, in case of no code-block segmentation, that is in case of a single code block, no additional code-block CRC is applied.

One could argue that, in case of code-block segmentation, the transport-block CRC is redundant as the set of code-block CRCs should indirectly provide information about the correctness of the complete transport block. However, code-block segmentation is only applied to large transport blocks, in which case the extra overhead of the additional, and partly redundant, transport-block CRC is insignificant. The transport-block CRC also adds additional error-detection capabilities and thus reduces the risk for undetected errors in the decoded transport block.

16.5.3 Turbo coding

The first releases of the WCDMA radio-access specifications (before HSPA) allowed for both convolutional coding and Turbo coding to be applied to transport channels. For HSPA, channel coding was simplified in the sense that only Turbo

---

[15] The supported transport-block sizes of LTE have been chosen so that, in the majority of cases, no filler bits are needed.

[16] The code-block CRC is based on a different CRC polynomial compared to the transport-block CRC.

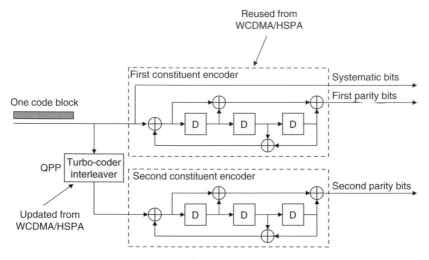

Figure 16.28   *LTE Turbo encoder.*

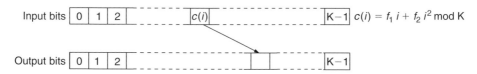

Figure 16.29   *Principles of QPP-based interleaving.*

coding can be applied to the HSPA-related transport channels (HS-DSCH for the downlink and E-DCH for the uplink). The same is true for the LTE downlink shared channel, that is, only Turbo coding can be applied in case of DL-SCH transmission. This also holds for the PCH and MCH transport channels.

The overall structure of the LTE Turbo encoding is illustrated in Figure 16.28. The Turbo encoding reuses the two WCDMA/HSPA rate-1/2, eight-state constituent encoders, implying an overall code rate of 1/3. However, the WCDMA/HSPA Turbo encoder internal interleaver has, for LTE, been replaced by QPP-based[17] interleaving [72]. As illustrated in Figure 16.29, the QPP interleaver provides a mapping from the input (non-interleaved) bits to the output (interleaved) bits according to the function:

$$c(i) = f_1 \cdot i + f_2 \cdot i^2 \bmod K$$

where i is the index of the bit at the output of the interleaver, $c(i)$ is the index of the same bit at the input of the interleaver, and K is the code-block/interleaver

---

[17] QPP = Quadrature Permutation Polynomial.

size. The values of the parameters f_1 and f_2 depend on the code-block size K. The LTE specification lists all supported code-block sizes, ranging from a minimum of 40 bits to a maximum of 6144 bits, together with the associated values for the parameters f_1 and f_2.

In contrast to the WCDMA/HSPA turbo-code interleaver, a QPP-based interleaver is maximum *contention free* [73], implying that the decoding can be straightforwardly parallelized without the risk for contention when the different parallel processes are accessing the interleaver memory. For the very high data rates to be supported by LTE, the improved possibilities for parallel processing offered by the QPP-based interleaving will substantially simplify the Turbo-encoder/decoder implementation.

16.5.4 *Rate-matching and physical-layer hybrid-ARQ functionality*

The task of the rate-matching and physical-layer hybrid-ARQ functionality is to extract, from the blocks of code bits delivered by the channel encoder, the exact set of bits to be transmitted within a given TTI.

As illustrated in Figure 16.30, the outputs of the Turbo encoder (systematic bits, first parity bits, and second parity bits) are first separately interleaved. The interleaved bits are then inserted into what can be described as a circular buffer with the systematic bits inserted first, followed by alternating insertion of the first and second parity bits.

The bit selection then extracts consecutive bits from the circular buffer to the extent that fits into the assigned resource. The set of bits to extract depends on

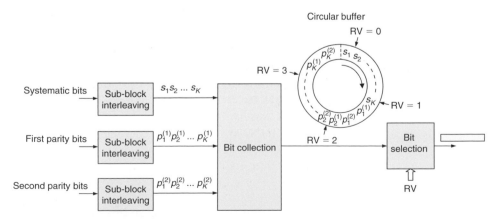

Figure 16.30 *Rate-matching and hybrid-ARQ functionality.*

the *redundancy version* corresponding to different starting points for the extraction of coded bits from the circular buffer. As can be seen, there are four different alternatives for the redundancy version.

16.5.5   Bit-level scrambling

LTE downlink scrambling implies that the block of code bits delivered by the hybrid-ARQ functionality is multiplied (*exclusive-or* operation) by a bit-level *scrambling sequence*. In general, scrambling of the coded data helps to ensure that the receiver-side decoding can fully utilize the processing gain provided by the channel code. Without downlink scrambling, the channel decoder at the mobile terminal could, at least in principle, be equally matched to an interfering signal as to the target signal, thus not being able to properly suppress the interference. By applying different scrambling sequences for neighbor cells, the interfering signal(s) after descrambling are randomized, ensuring full utilization of the processing gain provided by the channel code.

In contrast to WCDMA/HSPA, where downlink scrambling is applied to the complex-valued chips after spreading (*chip-level scrambling*), LTE applies downlink scrambling to the coded bits of each transport channel (*bit-level scrambling*). Chip-level scrambling is necessary for WCDMA/HSPA to ensure that the processing gain provided by the spreading can be efficiently utilized. On the other hand, scrambling of code bits rather than complex-valued modulation symbols implies somewhat lower implementation complexity, with no negative impact on performance in case of LTE.

In LTE, downlink scrambling is applied to all transport channels. As described in Section 16.4, scrambling is also applied to the downlink *L1/L2 control signaling*. For all downlink transport channels except the MCH, as well as for the L1/L2 control signaling, the scrambling sequences should be different for neighbor cells (*cell-specific scrambling*) to ensure interference randomization between the cells. This is achieved by having the scrambling depend on the physical-layer cell identity. In contrast, in case of MBSFN-based transmission using the MCH transport channel, the same scrambling should be applied to all cells taking part in the MBSFN transmission (*cell-common scrambling*). This is achieved by having the scrambling depend on the so-called *MBSFN area* identity (see Section 16.7).

16.5.6   Data modulation

The downlink data modulation transforms the block of scrambled bits to a corresponding block of complex modulation symbols. The set of modulation schemes supported for the LTE downlink includes QPSK, 16QAM, and 64QAM, corresponding to two, four, and six bits per modulation symbol, respectively. All these modulation schemes are applicable to the DL-SCH, PCH, and MCH transport channels. As will be described in Chapter 18, only QPSK modulation can be applied to the BCH transport channel.

16.5.7 Antenna mapping

The *Antenna Mapping* jointly processes the modulation symbols corresponding to, in the general case, two transport blocks, and maps the result to the different antenna ports. As can be seen from Figure 16.26, LTE supports up to four transmit antenna ports. The antenna mapping can be configured in different ways to provide different multi-antenna schemes including *transmit diversity* and *spatial multiplexing.* More details about the antenna mapping and, in general, about LTE downlink multi-antenna transmission, are provided in Section 16.6.

16.5.8 Resource-block mapping

The resource-block mapping maps the symbols to be transmitted on each antenna port to the resource elements of the set of resource blocks assigned by the MAC scheduler for transmission of the transport block(s) to the terminal. Each resource block consists of 84 resource elements (12 subcarriers during 7 OFDM symbols)[18] as described in Section 16.2. However, some of the resource elements within a resource block will not be available as they are occupied by

- Downlink reference symbols including non-used resource elements corresponding to reference symbols of other antenna ports as discussed in Section 16.3.
- Downlink L1/L2 control signaling (one, two, or three OFDM symbols at the head of each subframe) as discussed in Section 16.4.

Furthermore, as will be described in Chapter 18, within some resource blocks, additional resource elements are used for the transmission of the BCH transport channels and the so-called *synchronization signals.* The physical resource, that is, the set of resource elements to which the DL-SCH is mapped, is in the LTE specifications referred to as the *Physical Downlink Shared Channel* (PDSCH).

As discussed already in Chapter 14, when deciding what set of resource blocks to use for transmission to a specific terminal, the network may take the downlink

---

[18] $12 \times 6 = 72$ resource elements in case of extended cyclic prefix.

channel conditions in both the time and frequency domain into account. Such time/frequency-domain channel-dependent scheduling, taking channel variations, for example, due to frequency-selective fading into account, may significantly improve system performance in terms of achievable data rates and overall cell throughput.

However, in some cases downlink channel-dependent scheduling is not suitable to use or even practically possible:

- For low-rate services such as voice, the signaling associated with channel-dependent scheduling may lead to extensive relative overhead.
- At high mobility (high terminal speed), it may be difficult or even practically impossible to track the instantaneous channel conditions to the accuracy required for channel-dependent scheduling to be efficient.

In such situations, an alternative means to handle radio-channel frequency selectivity is to achieve frequency diversity by distributing a downlink transmission in the frequency domain.

One way to distribute a downlink transmission in the frequency domain, and thereby achieve frequency diversity, is to assign multiple non-frequency-contiguous resource blocks for the transmission to a terminal. As described in Section 16.4, LTE allows for such *distributed resource-block allocation* by means of resource allocation types 0 and 1. However, although being sufficient in many cases, distributed resource-block allocation by means of resource allocation types 0 and 1 has certain drawbacks:

- For both types of resource allocations, the minimum size of the allocated resource can be as large as four resource-block pairs. Thus resource allocation types 0 and 1 may not be suitable when resource allocations of limited size are needed, especially in case of large cell bandwidths.[19]
- In general, resource allocation types 0 and 1 are associated with a relatively large PDCCH payload, once again especially in case of large cell bandwidths.

In contrast, resource allocation type 2 always allows for the allocation of a single resource-block pair and is also associated with a relatively small PDCCH payload size. However, resource allocation type 2 only allows for the allocation of resource blocks that are contiguous in the frequency domain. In addition, regardless of the type of resource allocation, frequency diversity by means of distributed resource-block allocation will only be achieved in case of resource allocations larger than one resource-block pair.

---

[19] In case of narrow cell bandwidths, the minimum resource allocation is one resource-block pair for types 0 and 1 resource allocations.

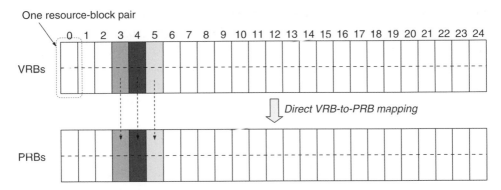

Figure 16.31 *VRB-to-PRB mapping in case of localized VRBs. Figure assumes a cell bandwidth corresponding to 25 resource blocks.*

In order to provide the possibility for distributed resource-block allocation in case of resource allocation type 2, as well as to allow for distributing the transmission of a single resource-block pair in the frequency domain, the notion of a *Virtual Resource Block* (VRB) has been introduced for LTE.

What is being provided in the resource allocation is the resource allocation in terms of VRB pairs. The key to distributed transmission then lies in the mapping from VRB pairs to *Physical Resource Block* (PRB) pairs, that is, to the actual physical resource used for transmission.

The LTE specification defines two types of VRBs: *localized* VRBs and *distributed* VRBs.

In case of localized VRBs, there is a direct mapping from VRB pairs to PRB pairs as illustrated in Figure 16.31.

However, in case of distributed VRBs, the mapping from VRB pairs to PRB pairs is more elaborate in the sense that

- consecutive VRBs are not mapped to PRBs that are consecutive in the frequency domain,
- even a single VRB pair is distributed in the frequency domain.

The basic principle of distributed transmission is outlined in Figure 16.32 and consists of two steps:

- A mapping from VRB pairs to PRB pairs such that consecutive VRB pairs are not mapped to frequency-consecutive PRB pairs (first step of Figure 16.32).

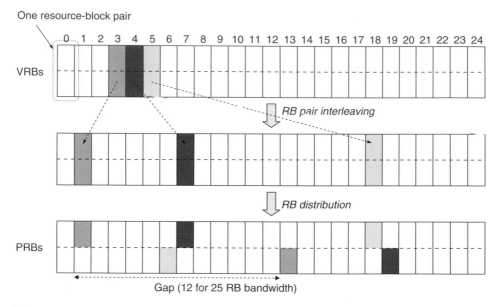

Figure 16.32 *VRB-to-PRB mapping in case of distributed VRBs. Figure assumes a cell bandwidth corresponding to 25 resource blocks.*

This provides frequency diversity between consecutive VRB pairs. The spreading in the frequency domain is done by means of a block-based 'interleaver' operating on resource-block pairs.

• A split of each resource-block pair such that the two resource blocks of the resource-block pair are transmitted with a certain frequency gap in between (second step of Figure 16.32). This provides frequency diversity also for a single VRB pair. This step can be seen as the introduction of frequency hopping on a slot basis.

Whether the VRBs are localized, and thus mapped according to Figure 16.31, or distributed (mapped according to Figure 16.32) is indicated on the associated PDCCH in case of type 2 resource allocation. Thus it is possible to dynamically switch between distributed and localized transmission and also mix distributed and localized transmission for different terminals within the same subframe.

The exact size of the frequency gap in Figure 16.32 depends on the overall downlink cell bandwidth according to Table 16.2. These gaps have been chosen based on two criteria:

1. The gap should be in the order of half the downlink cell bandwidth in order to provide good frequency diversity also in the case of a single VRB pair.

Table 16.2 *Gap size for different cell bandwidths (number of resource blocks).*

Bandwidth	6	7–8	9–10	11	12–19	20–26	27–44	45–63	64–79	80–110
P	1	1	1	2	2	2	3	3	4	4
Gap size	3	4	5	4	8	12	18	27	32	48

Table 16.3 *Second gap size for different cell bandwidth (only applicable to bandwidths ≥50 RBs).*

Bandwidth	50–63	64–110
Gap size	9	16

2. The gap should be a multiple of P^2, where P is the size of a *resource-block group* as defined in Section 16.4.4. The reason for this constraint is to ensure a smooth coexistence in the same subframe between distributed transmission as described above and transmissions based on downlink allocation types 0 and 1.

Due to the constraint that the gap size should be a multiple of P^2, the gap size will in most cases deviate from exactly half the cell bandwidth. In these cases, not all resource blocks within the cell bandwidth can be used for distributed transmission. As an example, for a cell bandwidth corresponding to 25 resource blocks (the example in Figure 16.32) and a corresponding gap size equal to 12 according to Table 16.2, resource-block pair 24 cannot be used for distributed transmission. As another example, for a cell bandwidth corresponding to 50 resource blocks (gap size equal to 27 according to Table 16.2) only 46 resource blocks would be available for distributed transmission.

In addition to the gap size outlined in Table 16.2, for wider cell bandwidths (50 RBs and beyond), there is a possibility to use a second, smaller frequency gap of a size in the order of one-fourth of the cell bandwidth (see Table 16.3). The use of the smaller gap allows for restricting the distributed transmission to only a part of the overall cell bandwidth. Selection between the larger gap according to Table 16.2 and the smaller gap according to Table 16.3 is indicated by an additional bit in the resource allocation on PDCCH.

16.6 Multi-antenna transmission

In this section, a more detailed overview of LTE downlink multi-antenna transmission will be provided. As outlined in the previous section (Figure 16.26), the

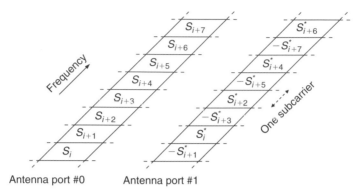

Figure 16.33 *Two-antenna-port transmit diversity – SFBC.*

antenna mapping operates on the block of modulation symbols (one block per each transport block) also referred to as a *codeword*.

LTE supports the following multi-antenna transmission schemes or *transmission modes*, in addition to single-antenna transmission:

- Transmit diversity
- Closed-loop spatial multiplexing including codebook-based beam-forming
- Open-loop spatial multiplexing.

All these multi-antenna transmission schemes rely on the cell-specific reference signals described in Section 16.3.1 for channel estimation. In addition, LTE also supports more general beam-forming, not based on the cell-specific reference signals but on the transmission of a UE-specific reference signal as described in Section 16.3.2.

16.6.1 Transmit diversity

In case of two antenna ports, LTE transmit diversity is based on *Space Frequency Block Coding* (SFBC) as described in Chapter 6. As can be seen from Figure 16.33, SFBC implies that consecutive modulation symbols S_i and S_{i+1} are mapped directly on adjacent subcarriers on the first antenna port. On the second antenna port, the swapped and transformed symbols $-S_{i+1}^*$ and S_i^* are transmitted on the corresponding subcarriers.

In case of four antenna ports, LTE transmit diversity is based on a combination of SFBC and *Frequency Shift Transmit Diversity* (FSTD). As can be seen in

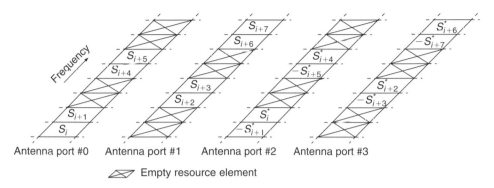

Antenna port #0   Antenna port #1   Antenna port #2   Antenna port #3

◿◺ Empty resource element

Figure 16.34 *Four-antenna-port transmit diversity – combined SFBC/FSTD.*

Figure 16.34, combined SFBD/FSTD implies that pairs of modulation symbols are transmitted by means of SFBC with transmission alternating between pairs of antenna ports (antenna ports 0 and 2 and antenna ports 1 and 3, respectively). For the subcarriers where transmission is on one pair of antenna ports, there is no transmission on the other pair of antenna ports. Thus combined SFBC/FSTD in some sense operates on groups of four symbols as well as groups of four resource elements on each antenna port. As mentioned in Section 16.4.1, this is the reason for the use of *resource-element groups*, each consisting of four frequency-consecutive resource elements, when defining the mapping of the L1/L2 control signaling to the physical resource.

Transmit diversity is the only multi-antenna transmission scheme that can be applied to the BCH and PCH transport channels. It is also the only multi-antenna transmission scheme that can be applied to the downlink L1/L2 control channels PCFICH, PHICH, and PDSCH. Actually, if two antenna ports are available within a cell, SFBC *must* be used for BCH as well as for the downlink control channels. Similarly, if four antenna ports are available within the cell, combined SFBC/FSTD *must* be used for these channels.

16.6.2 *Spatial multiplexing*

As described in Chapter 6, spatial multiplexing implies that multiple streams or 'layers' are transmitted in parallel, thereby allowing for higher data rates within a given bandwidth. LTE spatial multiplexing allows for the transmission of a variable number of layers, up to a maximum of N_A layers, where N_A is the number of antenna ports.

The LTE spatial multiplexing may operate in two different modes: *closed-loop spatial multiplexing* and *open-loop spatial multiplexing* where closed-loop spatial multiplexing relies on more extensive feedback from the mobile terminal.

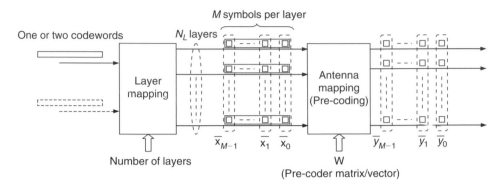

Figure 16.35   *The basic structure of LTE closed-loop spatial multiplexing.*

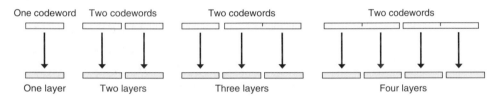

Figure 16.36   *Codeword-to-layer mapping for spatial multiplexing.*

The basic structure of LTE closed-loop spatial multiplexing is illustrated in Figure 16.35. One or two codewords, corresponding to one or two transport blocks, are mapped to the N_L layers. The number of N_L layers may range from a minimum of one layer up to a maximum number of layers equal to the number of antenna ports. As described in Section 16.5, there is one transport block, and thus also one codeword, in case of a single layer ($N_L = 1$) and two transport blocks (two codewords) in case of two or more layers ($N_L \geq 2$). The mapping to layers is such that the number of modulation symbols on each layer is the same. Thus, in case of three layers, the second codeword, which is mapped to the second and third layer, is twice as large as the first codeword (mapped to the first layer). This is guaranteed by the set of supported transport-block sizes in combination with the structure of the code-block-segmentation and rate-matching functionality. In case of four layers, the first codeword is mapped to the first and second layer while the second codeword is mapped to the third and fourth layer. In this case the two codewords are thus of the same size. The LTE codeword-to-layer mapping in case of spatial multiplexing is illustrated Figure 16.36.

After layer mapping, a set of N_L symbols (one symbol from each layer) is linearly combined and mapped to the N_A antenna ports. This combining/mapping can be described by means of a *pre-coder matrix* **W** of size $N_A \times N_L$. More specifically, the vector $\overline{y}_i$ of size N_A, consisting of one symbol for each antenna port, is given by the vector $\overline{x}_i$ of size N_L, consisting of one symbol from each layer, multiplied by the pre-coder matrix, that is $\overline{y}_i = \mathbf{W}\overline{x}_i$. In case of spatial multiplexing, the number of layers is also often referred to as the *transmission rank*.[20]

In case of a single layer ($N_L = 1$), $\mathbf{W}$ is a vector of size $N_A \times 1$ and the pre-coding provides beam-forming for the single symbol x_i using the N_A antenna ports. Thus, beam-forming can be seen as a special case of closed-loop spatial multiplexing with the number of layers N_L equal to one.

As LTE supports multi-antenna transmission using two or four antenna ports, pre-coding matrices are defined for:

- two antenna ports ($N_A = 2$) and one and two layers, corresponding to pre-coder matrices of size 2×1 and 2×2, respectively;
- four antenna ports ($N_A = 4$) and one, two, three, and four layers, corresponding to pre-coder matrices of size 4×1, 4×2, 4×3, and 4×4, respectively.

As an example, the set of defined pre-coder matrices for the case of two antenna ports is illustrated in Table 16.4. As can be seen, there are four 2×1 pre-coder matrices (actually vectors) for single-layer transmission and two 2×2 pre-coder matrices for two-layer transmission. Similar matrices are defined for the case of four antenna ports (4×1, 4×2, 4×3, and 4×4 matrices for the case of one, two, three, and four layers, respectively).

To assist the network in selecting a suitable pre-coder matrix for transmission, the terminal may report a recommended number of layers (expressed as a *Rank Indication*, RI), as well as a recommended *pre-coder matrix* (*Pre-coder-Matrix Indication*, PMI) corresponding to that number of layers, depending on estimates

Table 16.4   *LTE pre-coder matrices* $\mathbf{W}$ *in case of two antenna ports.*

One layer	$\frac{1}{\sqrt{2}}\begin{bmatrix} +1 \\ +1 \end{bmatrix}$	$\frac{1}{\sqrt{2}}\begin{bmatrix} +1 \\ -1 \end{bmatrix}$	$\frac{1}{\sqrt{2}}\begin{bmatrix} +1 \\ +j \end{bmatrix}$	$\frac{1}{\sqrt{2}}\begin{bmatrix} +1 \\ -j \end{bmatrix}$
Two layers	$\frac{1}{2}\begin{bmatrix} +1 & +1 \\ +1 & -1 \end{bmatrix}$	$\frac{1}{2}\begin{bmatrix} +1 & +1 \\ +j & -j \end{bmatrix}$		

---

[20] In LTE, transmit-diversity schemes are also described as using multiple 'layers.' However, they are then still single-rank transmission schemes.

of the downlink channel conditions. The network may, or may not, follow the terminal recommendation when transmitting to the terminal. When not following the terminal recommendation, the network must explicitly inform the terminal what pre-coder matrix is used for the subsequent downlink transmission. Some more details on these reports are provided in Chapter 19.

In addition to closed-loop spatial multiplexing, LTE also supports *open-loop spatial multiplexing*, also sometimes referred to as *large-delay CDD*.[21] Open-loop spatial multiplexing does not rely on any detailed pre-coder recommendation being fed back from the mobile terminal and does not require any explicit pre-coder information being signaled to the terminal from the network. Open-loop spatial multiplexing is thus suitable for high-mobility scenarios and cases where the additional overhead associated with closed-loop spatial multiplexing are not justifiable.

The structure of large-delay CDD is illustrated in Figure 16.37. As can be seen, the overall pre-coding functionality can in this case be seen as a combination of two pre-coder matrices, a matrix $\mathbf{P}$ of size $N_L \times N_L$ and a matrix $\mathbf{W}$ of size $N_A \times N_L$.

In case of large-delay CDD and two antenna ports, the pre-coder matrix $\mathbf{W}$ is fixed and is given by the first 2×2 matrix in Table 16.4:[22]

$$\mathbf{W} = \frac{1}{2}\begin{bmatrix} +1 & +1 \\ +1 & -1 \end{bmatrix}$$

In case of four antenna ports, the pre-coder matrix is cycling through four of the defined $4 \times N_L$ pre-coder matrixes and is different for consecutive resource elements. It should be noted that, although the pre-coder matrices of Figure 16.37

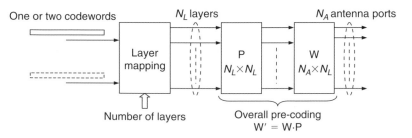

Figure 16.37   *Open-loop spatial multiplexing ('large-delay CDD').*

---

[21] Large-delay CDD is only applicable in case of two or more layers. In case of a single layer, open-loop spatial multiplexing corresponds to transmit diversity, that is SFBC and combined SFBC/FSTD in case of two and four antenna ports, respectively.

[22] Note that large-delay CDD is only defined for the case of two or more layers.

may vary in time and frequency, they vary in a pre-determined way. No information about the pre-coder matrix thus needs to be explicitly provided to the terminal.

The matrix $\mathbf{P}$ in Figure 16.37 can be expressed as a product of two matrices $\mathbf{P} = \mathbf{U} \cdot \mathbf{D}$, where $\mathbf{U}$ is a constant matrix of size $N_L \times N_L$ and $\mathbf{D}(i)$ is matrix of size $N_L \times N_L$ that varies between subcarriers. As an example, the matrices $\mathbf{U}$ and $\mathbf{D}(i)$ for the case of two layers ($N_L = 2$) are given by:

$$\mathbf{U} = \frac{1}{\sqrt{2}}\begin{bmatrix} 1 & 1 \\ 1 & e^{-j2\pi/2} \end{bmatrix} \quad \mathbf{D}(i) = \begin{bmatrix} 1 & 0 \\ 0 & e^{-j2\pi i/2} \end{bmatrix}$$

It can be shown that the second matrix $\mathbf{D}(i)$ is equivalent to CDD in the layer domain, thus the name 'large-delay CDD'. The *'large-delay'* attribute refers to the fact that the cyclic delay is exceptionally large, in the case of two layers equal to half the block length.

The basic idea with the matrix $\mathbf{P}$, that is the 'large-delay CDD' part of the open-loop spatial multiplexing, is to average out any differences in the channel conditions as seen by the different layers.

Spatial multiplexing (closed- and open-loop) is only applicable to the downlink shared channel (DL-SCH).

16.6.3 General beam-forming

As described above, closed-loop spatial multiplexing includes beam-forming as a special case when the number of layers N_L equals one. This kind of beam-forming can be referred to as *codebook-based beam-forming*, indicating that

- the network selects one pre-coding vector (the beam-forming vector) from a set of pre-defined pre-coding vectors (the 'codebook') with the selection, for example, based on the terminal reporting a recommended pre-coding vector;
- if not following the terminal recommendation, the network must explicitly inform the terminal about what pre-coding vector, from the set of predefined vectors, is actually used for transmission to the terminal.

As the terminal knows what pre-coding vector is used for the data transmission, it can use the cell-specific (and not pre-coded/beam-formed) reference signal of each antenna port for channel estimation. Based on these estimated channels for each antenna port, and knowledge about the used pre-coding vector, the terminal can derive an estimate of the channel experienced by the beam-formed data transmission and use this for coherent demodulation of the corresponding data transmission.

In contrast, *non-codebook-based beam-forming* implies that the network can apply an arbitrary beam-forming at the transmitter side and does not need to inform the terminal what beam-forming has been applied. In order to allow for estimation of the channel experienced by the beam-formed transmission, the base station has to transmit a reference signal to which has been applied the same beam-forming as for the data transmission. Clearly this reference signal is UE specific as it is beam-formed with a pre-coder vector selected for transmission to a specific terminal.

As already described in Section 16.3.2, in order to support non-codebook-based pre-coding, LTE allows for the transmission of such UE-specific reference signals. Transmission corresponding to the UE-specific reference signals is referred to as *transmission using antenna port 5*. Transmission using antenna port 5 is seen as an additional transmission mode, in addition to the transmission modes corresponding to antenna ports 0–3 (single-antenna-port transmission, transmit diversity, and closed/open-loop spatial multiplexing).

Transmission using antenna port 5, that is, with non-codebook-based beam-forming and UE-specific reference signals, is only applicable to the downlink shared channel (DL-SCH).

16.7   MBSFN transmission and MCH

As discussed in Chapter 4, OFDM transmission offers some specific benefits in terms of the provisioning of multi-cell multicast/broadcast services. More specifically, OFDM transmission offers the possibility to make synchronous multi-cell multicast/broadcast transmissions appear as a transmission from a single point over a time-dispersive channel. As mentioned in Chapter 14, for LTE this kind of transmission is referred to as *Multicast/Broadcast over Single Frequency Network* (MBSFN).

MBSFN transmission provides several benefits:

*   Increased received signal strength, especially at the border between cells involved in the MBSFN transmission, as the terminal can utilize the signal energy received from multiple cells.
*   Reduced interference level, once again especially at the border between cells involved in the MBSFN transmission, as the signals received from neighbor cells will not appear as interference but as useful signals.

- Additional diversity against fading on the radio channel as the information is received from several, geographically separated locations.

Altogether this allows for significant improvements in the multicast/broadcast reception quality, especially at the border between cells involved in the MBSFN transmission, and, as a consequence, significant improvements in the achievable multicast/broadcast data rates.

LTE supports MBSFN-based multicast/broadcast transmission by means of the MCH transport channel. The MCH is mapped to special subframes, so-called *MBSFN subframes*, to which other downlink transport channels (DL-SCH, PCH, and BCH) cannot be mapped.

Figure 16.38 illustrates the resource-block structure of MBSFN subframes. As can be seen, an MBSFN subframe consists of two parts:

- A unicast part
- An MBSFN part, to which the MCH is mapped.

The unicast part of an MBSFN subframe uses the same cyclic-prefix length as non-MBSFN subframes. In contrast, the MBSFN part of MBSFN subframes *always* uses the extended cyclic prefix. Thus, in case normal cyclic prefix is used for non-MBSFN subframes, and thus also in the unicast part of MBSFN subframes, there will be a small 'hole' between the unicast part and the MBSFN part of MBSFN subframes as illustrated in Figure 16.38. Note that, as the extended cyclic prefix is always used for MBSFN subframes there are only 12 OFDM symbols within the subframe. In case a carrier is used for MBSFN transmission only, also known as *stand-alone MBSFN carrier*, there is no unicast parts as unicast signaling in this case obviously is not needed.

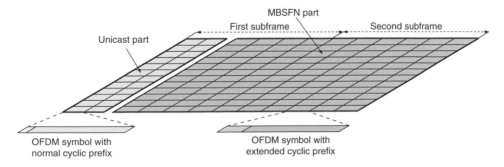

Figure 16.38 *Resource-block structure for MBSFN subframes, assuming normal cyclic prefix for the unicast part.*

The unicast part of MBSFN subframes serves the same purpose as, and is in essence identical to, the control region of non-MBSFN subframes, see Section 16.4, that is, within the unicast part of MBSFN subframes downlink L1/L2 control channels (PDCCH, PCFICH, and PHICH) are transmitted. The reason for transmitting these control channels also in MBSFN subframes, despite the fact that there is no DL-SCH transmission in MBSFN subframes, is that downlink L1/L2 control signaling is still needed to provide scheduling grants, hybrid-ARQ acknowledgements, and power-control commands for uplink transmissions. One difference compared to non-MBSFN subframes is that, in MBSFN subframes, the control-channel part consists of a maximum of two symbols. The reason is that the need for PDCCH transmission capacity is less in MBSFN subframes as there is no transmission of downlink scheduling assignments for DL-SCH transmission in MBSFN subframes. It should be noted that, in practice, the size of the control region for MBSFN subframes cannot be dynamically adjusted as it has to be the same for all cells involved in the MBSFN transmission.[23]

Within the MBSFN part of MBSFN subframes, the MCH is transmitted by means of MBSFN from multiple cells, more specifically from the set of cells that is to provide the specific broadcast service, also referred to as the *MBSFN area* of that service. The radio channel that the MCH has propagated over, as seen from the UE point-of-view, is thus the channels of each cell within the MBSFN area aggregated into a single channel. For channel estimation for coherent demodulation of the MCH, the terminal can thus not rely on the normal cell-specific reference signals transmitted from each cell. Rather, to enable coherent demodulation for MCH, special *MBSFN reference symbols* are inserted within the MBSFN part of the MBSFN subframe as illustrated in Figure 16.39. These reference symbols are transmitted by means of MBSFN over the set of cells that constitute the MBSFN area, that is, they are transmitted at the same time/ frequency position and with the same reference-symbol values from each cell.

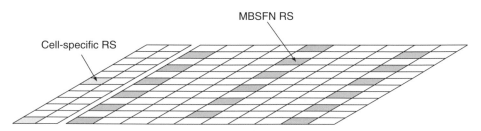

Figure 16.39   *Reference-signal structure for MBSFN subframes.*

---

[23] More specifically, the size of the MBSFN part of the subframe must be the same for all cells involved in the MBSFN transmission. Thus, indirectly, also the size of the unicast part (the control region) must be the same.

Channel estimation using these reference symbols will thus correctly reflect the overall aggregated channel corresponding to the MCH transmissions of all cells being part of the MBSFN area.

MBSFN transmission in combination with specific MBSFN reference signals can be seen as transmission using a specific antenna port. In the LTE specification this is referred to as transmission using *antenna port 4*.

A terminal can assume that all MBSFN transmissions within a given subframe correspond to the same MBSFN area. Thus, a terminal can average over all MBSFN reference symbols within a given MBSFN subframe while estimating the aggregated MBSFN channel. In contrast, MCH transmissions in different subframes may correspond to different MBSFN areas. Thus a terminal cannot necessarily average over multiple subframes when doing channel estimation for MBSFN transmissions.

As can be seen in Figure 16.39, the frequency-domain density of MBSFN reference symbols is higher than the corresponding density of cell-specific reference symbols as, for example, illustrated in Figure 16.8 in Section 16.3.1. This is needed, as the aggregated channel of all cells involved in the MBSFN transmission will be equivalent to a highly time-dispersive or, equivalently, frequency-selective channel. Thus, a higher frequency-domain reference-symbol density is needed.

There is only a single MBSFN reference signal in MBSFN subframes. Thus multi-antenna transmission such as transmit diversity and spatial multiplexing is not supported for MBSFN (MCH) transmission. The main argument for not supporting transmit diversity for MCH transmission is that the high frequency selectivity of the aggregated MBSFN channel in itself provides substantial (frequency) diversity.

Furthermore, as also illustrated in Figure 16.39, within MBSFN subframes cell-specific reference signals are only transmitted within the unicast part of the subframe. Thus, all terminals, including terminals that are not to receive any MCH, need to be aware of what subframes are MBSFN subframes in order to properly carry out channel estimation based on the cell-specific reference signals. Information about what subframes are MBSFN subframes are provided as part of the cell system information.

The transport-channel processing for MCH is, in most respects, the same as that for DL-SCH as illustrated in Figure 16.26, with some exceptions:

- In case of MBSFN transmission, the same data is to be transmitted with the same transport format using the same physical resource from multiple cells typically belonging to different eNodeB. Thus, the MCH transport format and resource allocation cannot be dynamically adjusted by the eNodeB.

- As the MCH transmission is simultaneously targeting multiple mobile terminals, hybrid ARQ is not directly applicable in case of MCH transmission.
- As already mentioned, multi-antenna transmission (transmit diversity and spatial multiplexing) do not apply to MCH transmission.

Furthermore, as also mentioned in Section 16.2.3, the MCH scrambling should be *MBSFN-area specific*, that is identical for all cells involved in the MBSFN transmission.

17

Uplink transmission scheme

In Chapter 16, a detailed overview of the LTE downlink transmission scheme was provided. In this chapter, a similar overview will be provided for the LTE uplink transmission scheme.

17.1   The uplink physical resource

As already mentioned in the overview provided in Chapter 14, LTE uplink transmission is based on so-called DFTS-OFDM transmission. As described in Chapter 5, DFTS-OFDM is a low-PAR 'single-carrier' transmission scheme that allows for flexible bandwidth assignment and orthogonal multiple access not only in the time domain but also in the frequency domain. Thus, the LTE uplink transmission scheme is also sometimes referred to as *Single-Carrier FDMA* (SC-FDMA).

Figure 17.1 recapitulates the basic structure of DFTS-OFDM transmission with a size-*M* DFT being applied to a block of *M* modulation symbols. The output of the DFT is then mapped to selective inputs of an OFDM modulator, typically implemented as an inverse FFT. The DFT size determines the instantaneous bandwidth of the transmitted signal whereas the exact mapping of the DFT output to the input of the OFDM modulator determines the position of the transmitted signal within the overall uplink cell bandwidth. A cyclic prefix is then inserted for each DFT block. As explained in Chapter 5, the use of a cyclic prefix allows for straightforward application of low-complexity frequency-domain equalization at the receiver side. The transmitted signal corresponding to one DFT block, including the cyclic prefix, can be referred to as one *DFTS-OFDM symbol*.

As described in Chapter 5, in the general case both localized and distributed DFTS-OFDM transmissions is possible. However, LTE uplink transmission is limited to localized transmission, that is the output of the DFT is always mapped to consecutive inputs of the OFDM modulator.

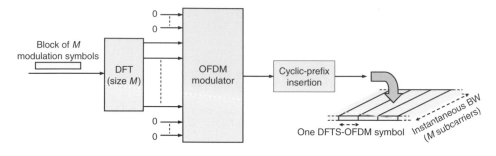

Figure 17.1   *Basic principles of DFTS-OFDM for LTE uplink transmission.*

From a DFT-implementation point-of-view, the DFT size should preferably be constrained to a power of two. However, such a constraint is in direct conflict with a desire to have a high degree of flexibility in terms of the amount of resources (the instantaneous transmission bandwidth) that can be dynamically assigned to a mobile terminal for uplink transmission. From a flexibility point-of-view, all possible DFT sizes should rather be allowed. For LTE, a middle way has been adopted where the DFT size is limited to products of the integers two, three, and five. For example, DFT sizes of 60, 72, and 96[1] are allowed but a DFT size of 84 is not allowed. In this way, the DFT can be implemented as a combination of relatively low-complex radix-2, radix-3, and radix-5 FFT processing.

As DFTS-OFDM can be seen as conventional OFDM transmission combined with DFT-based pre-coding one can speak about a subcarrier spacing also in case of DFTS-OFDM transmission. Furthermore, similar to OFDM, the DFTS-OFDM physical resource can be seen as a time–frequency grid with the additional constraint that the overall time–frequency resource assigned to a mobile terminal for uplink transmission must always consist of a set of *frequency-consecutive* subcarriers.

The basic parameters of the LTE uplink transmission scheme have been chosen to be aligned, as much as possible, with the corresponding parameters of the OFDM-based LTE downlink. Thus, as illustrated in Figure 17.2, the uplink subcarrier spacing equals 15 kHz and resource blocks, consisting of 12 subcarriers, are also defined for the LTE uplink. However, in contrast to the downlink, no unused DC-subcarrier is defined for the LTE uplink. The presence of a DC-carrier in the center of the spectrum would have prevented the assignment of the entire cell bandwidth to a single mobile terminal while still retaining the

---

[1] As uplink resource assignments are always done in terms of resource blocks of size 12 subcarriers, the DFT size is always a multiple of 12.

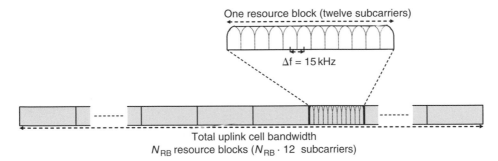

One resource block (twelve subcarriers)

$\Delta f = 15\,\text{kHz}$

Total uplink cell bandwidth
N_{RB} resource blocks ($N_{RB} \cdot 12$ subcarriers)

Figure 17.2 *Frequency-domain structure for LTE uplink.*

assumption of mapping to consecutive inputs of the OFDM modulator, something which is required to keep the low-PAR property of the uplink transmission. Also, due to the DFT-based pre-coding, the impact of any DC interference will be spread over the block of M modulation symbols and will therefore be less harmful compared to normal OFDM transmission.

Similar to the downlink, also for the uplink the LTE physical-layer specification allows for a very high degree of flexibility in terms of the overall cell bandwidth by allowing for, in essence, any number of uplink resource blocks ranging from a minimum of 6 resource blocks up to a maximum of 110 resource blocks. However, also similar to the downlink, there are restrictions in the sense that radio-frequency requirements are, at least initially, only specified for a limited set of uplink cell bandwidths.

Also in terms of the more detailed *time-domain* structure, the LTE uplink is very similar to the downlink, as can be seen from Figure 17.3.[2] Each 1 ms uplink subframe consists of two equally sized slots of length 0.5 ms. Each slot then consists of a number of DFTS-OFDM symbols including cyclic prefix. Also, similar to the downlink, two cyclic-prefix lengths are defined for the uplink; the normal cyclic prefix and an extended cyclic prefix.

17.2 Uplink reference signals

17.2.1 Uplink demodulation reference signals

Similar to the downlink, reference signals for channel estimation are also needed for the LTE uplink to enable coherent demodulation of different uplink physical channels

---

[2] The figure illustrates the structure of 'normal' uplink subframes. As described in Section 16.1, in case of TDD operation there is also a *special subframe* with a special structure including both downlink and uplink transmission as well as a guard period.

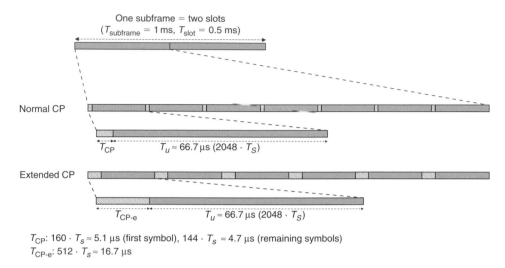

T_{CP}: $160 \cdot T_s \approx 5.1$ µs (first symbol), $144 \cdot T_s \approx 4.7$ µs (remaining symbols)
$T_{CP\text{-}e}$: $512 \cdot T_s \approx 16.7$ µs

Figure 17.3 *Detailed time-domain structure for LTE uplink transmission.*

at the receiver side. These reference signals are more specifically referred to as uplink *demodulation reference signals* (*DRS*) to distinguish them from so-called uplink *sounding reference signals* (SRS) discussed in Section 17.2.2. For simplicity, in this section the term uplink reference signals will refer to the uplink DRS.

Uplink reference signals are needed for coherent demodulation of the *Physical Uplink Shared Channel* (PUSCH) to which the UL-SCH transport channel is mapped. Uplink reference signals are also needed for coherent demodulation of the *Physical Uplink Control Channel* (PUCCH), which carries different types of L1/L2 control signaling and which is described in much more detail in Section 17.3. In many respects the principles for uplink reference-signal transmission are the same for PUSCH and PUCCH, with some specific differences to be highlighted below.

17.2.1.1 Basic principles of uplink DRS transmission

Due to the importance of low power variations for uplink transmissions, the principles for uplink reference-signal transmission are different from those of the downlink. In essence, having reference signals frequency multiplexed with other uplink transmissions from the same mobile terminal is not suitable for the uplink. Instead, certain DFTS-OFDM symbols are exclusively used for reference-signal transmission, that is, the uplink reference signals are *time multiplexed* with other uplink transmissions from the same mobile terminal.

More specifically, in case of PUSCH transmission, a reference signal is transmitted within the fourth symbol of each uplink slot[3] (Figure 17.4). Within each subframe, there are thus two reference-signal transmissions, one in each slot.

---

[3] The third symbol in case of extended cyclic prefix.

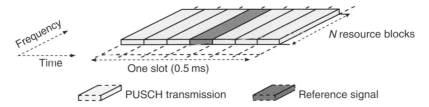

Figure 17.4 *Transmission of uplink reference signals within a slot in case of PUSCH transmission (normal cyclic prefix).*

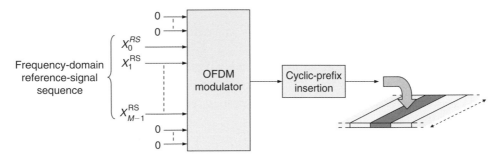

Figure 17.5 *Generation of uplink reference signal from a frequency-domain reference-signal sequence.*

In case of PUCCH transmission, the number of DFTS-OFDM symbols used for reference-signal transmission, as well as the exact position of these symbols within the slot, differs between different PUCCH formats.

Regardless of the kind of uplink transmission (PUSCH or PUCCH), the basic structure of each reference-signal transmission is the same. As illustrated in Figure 17.5, an uplink reference signal is defined as a *frequency-domain reference-signal sequence* applied to consecutive inputs (consecutive subcarriers) of an OFDM modulator. A time-domain cyclic prefix is then inserted, similar to other uplink transmissions. It should be noted though that one could equivalently have defined a time-domain reference-signal sequence, being the inverse DFT of the frequency-domain sequence of Figure 17.5, and applied this sequence to a DFTS-OFDM modulator.

In general, there is no reason to estimate the channel outside the frequency band of the corresponding PUSCH/PUCCH transmission. The bandwidth of the reference signal, corresponding to the length of the reference-signal sequence in Figure 17.5, should thus be equal to the bandwidth of the corresponding PUSCH/PUCCH transmission measured in number of subcarriers. This means that, in case of PUSCH transmission, it should be possible to generate reference-signal sequences of different length, corresponding to the possible bandwidths of a PUSCH transmission. Note, however, that the length of the reference-signal sequence will always be a multiple of 12 as uplink resource allocations for PUSCH transmission are always done in terms of resource blocks of size 12 subcarriers.

In contrast, as will be described in Section 17.3, PUCCH transmission is always carried out over a single resource block in the frequency domain, that is over 12 subcarriers. In case of PUCCH transmission, the length of the reference-signal sequence is thus always equal to 12.

Uplink reference signals should preferably have the following properties:

- Limited power variations in the frequency domain to allow for similar channel-estimation quality for all frequencies.
- Limited power variations in the time domain to allow for high power-amplifier efficiency.

Furthermore, sufficiently many reference-signal sequences of the same length, that is corresponding to the same transmission bandwidth, should be available to avoid an unreasonable planning effort when assigning reference-signal sequences to cells.

So-called *Zadoff–Chu* sequences [123] have the property of constant power in both the frequency and the time domain. The elements of a Zadoff–Chu sequence of length M_{ZC} can be expressed as

$$X_k^{ZC} = e^{-j\pi u\left(k(k+1)/M_{ZC}\right)} \quad 0 \le k < M_{ZC} \tag{17.1}$$

where u is the *index* of the Zadoff–Chu sequence within the set of Zadoff–Chu sequences of length M_{ZC}. From the point-of-view of small power variations in both the frequency and the time domain, Zadoff–Chu sequences would thus be excellent as uplink reference-signal sequences. However, there are two reasons why Zadoff–Chu sequences are not suitable for direct usage as uplink reference-signal sequences in LTE:

- The number of available Zadoff–Chu sequences of a certain length, that is, the number of possible values of the index u in the expression above, equals the number of integers that are relative prime to the sequence length. This implies that to maximize the number of Zadoff–Chu sequences and thus, in the end, to maximize the number of available uplink reference signals, prime-length Zadoff–Chu sequences would be preferred. At the same time, the length of the uplink reference-signal sequences should be a multiple of 12 which is obviously not a prime number.

- For short sequence lengths, corresponding to narrow uplink transmission bandwidths, relatively few reference-signal sequences would be available even if they were based on prime-length Zadoff–Chu sequences.

Instead, for sequence lengths larger than or equal to 36, corresponding to transmission bandwidths larger than or equal to three resource blocks, the reference-signal sequences are defined as *cyclic extensions* of Zadoff–Chu sequences of length M_{ZC}, where M_{ZC} is the largest prime number smaller than or equal to the desired reference-signal sequence length. For example, the largest prime number less than or equal to 36 is 31, implying that reference-signal sequences of length 36 are defined as cyclic extensions of Zadoff–Chu sequences of length 31. The number of available sequences is then equal to 30, that is, one less than the length of the Zadoff–Chu sequence. For larger sequence lengths, more sequences are available. For example, for a reference-signal sequence length equal to 72, there are 70 sequences available.[4]

For sequence lengths equal to 12 and 24, corresponding to transmission bandwidths of one and two resource blocks, respectively, special QPSK-based sequences have instead been found from computer search and are explicitly listed in the LTE specifications. For each of the two sequence lengths, 30 sequences are then available.

17.2.1.2 Phase-rotated reference-signal sequences

In the previous section it was described how different reference-signal sequences (a minimum of 30 sequences for each sequence length) can be derived, primarily by cyclically extending different prime-length Zadoff–Chu sequences. Additional reference-signal sequences can be derived by applying different linear phase rotations to the same basic reference-signal sequence as illustrated in Figure 17.6, where the basic reference-signal sequence is defined according to Section 17.2.1.1 above (cyclically extended Zadoff–Chu sequence for sequence lengths larger than or equal to 36, special sequences for sequence lengths equal to 12 and 24).

Applying a linear phase rotation in the frequency domain is equivalent to applying a cyclic shift in the time domain. Thus, although being *defined* as different *frequency-domain phase rotations*, in the LTE specification this is often referred to as applying different *cyclic shifts*[5] to the same basic reference-signal

---

[4]The largest prime-number smaller than or equal to 72 is 71. The number of sequences is then one less than the length of the Zadoff–Chu sequence (i.e., 70).

[5]Not to be mixed up with the cyclic extension of Zadoff–Chu sequences to generate the basic reference-signal sequences. The cyclic shift discussed here would be a time-domain cyclic shift, equivalent to a frequency-domain phase rotation.

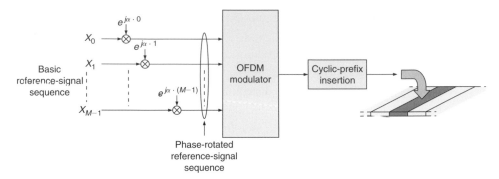

Figure 17.6 *Generation of uplink reference-signal sequence from linear phase rotation of a basic reference-signal sequence.*

sequence. Here the more correct term 'phase rotation' will be used. However, it should then be had in mind that what is here referred to as 'phase rotation' is often, although in some sense somewhat incorrectly, referred to as 'cyclic shift.'

Reference signals defined from different reference-signal sequences typically have relatively low but still non-zero mutual correlation. In contrast, reference signals defined from different phase rotations of the same basic reference-signal sequence can be made completely orthogonal, thus causing no interference to each other, assuming the parameter α in Figure 17.6 takes a value $m\pi/6$ where m ranges from 0 to 11.[6] Up to 12 orthogonal reference signals can thus be defined from each basic reference-signal sequence.

However, to preserve the orthogonality between these reference signals also at the receiver side, the channel should be frequency non-selective over the span of 12 subcarriers (one resource block). If that is not the case, a subset of the available values for α may be used, for example, only the values $\{0, 2\pi/6, 4\pi/6, \ldots, 10\pi/6\}$ or perhaps even less values. Limiting the set of values for α implies orthogonality over a smaller number of subcarriers, that is, less sensitivity to channel frequency selectivity.

Another prerequisite for orthogonality between reference signals defined from different phase rotations of the same basic reference-signal sequence is that the reference signals should be received relatively time aligned. A timing misalignment between the reference signals will, in the frequency domain, appear as a

---

[6]The orthogonality is due to the fact that, for $\alpha = m\pi/6$, there will be an integer number of full-circle rotations over 12 subcarriers, that is, over one resource block.

phase rotation that may counteract the phase rotation applied to separate the reference-signal sequences. The result may be substantial interference between the reference-signal transmissions.

In general for LTE, uplink transmissions from different terminals within the same cell are typically anyway well-time aligned as this is a prerequisite for retaining the orthogonality between different frequency-multiplexed transmissions. Thus, one possible use of reference signals defined from different phase rotations of the same basic reference-signal sequence is for the case when multiple mobile terminals within the same cell simultaneously transmit on the uplink *using the same frequency resource*. As will be seen in Section 17.3, this is, for example, the case for uplink PUCCH transmissions.

Another possible use of reference signals defined from different phase rotations of the same basic reference-signal sequence is for mobile terminals in neighbor cells belonging to the same eNodeB as such cells are, in practice, often tightly synchronized to each other and thus uplink transmissions in these cells are, in practice, relatively well-time aligned. The use of different phase rotations of the same basic reference-signal sequence for such mobile terminals will reduce the interference between the reference-signal transmissions of these terminals. This is then only applicable to PUSCH transmission as, for PUCCH, the different phase rotations are already used within the same cell as mentioned above.

17.2.1.3 Reference-signal assignment to cells

As described above, the design of the LTE reference-signal sequences allows for at least 30 sequences of each sequence length. To assign reference-signal sequences to cells these sequences are first grouped into 30 sequence groups where each group consists of:

- One reference-signal sequence for each sequence length less than or equal to 60, corresponding to a transmission bandwidth of five resource blocks or less.
- Two reference-signal sequences for each sequence length larger than or equal to 72,[7] corresponding to a transmission bandwidth of six resource blocks or more.

The grouping of reference-signal sequences into sequence groups is illustrated in Figure 17.7. It can be noted that the sequence groups do not include any

---

[7]For sequence lengths larger than or equal to 72 there are more than 60 sequences of each length allowing for two sequences per group.

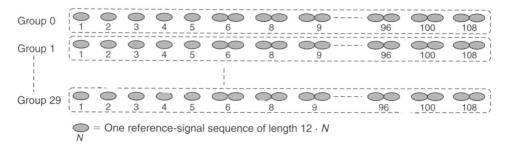

= One reference-signal sequence of length 12 · N

Figure 17.7 *Grouping of reference-signal sequences into sequence groups. The number indicates the corresponding bandwidth in number of resource blocks.*

reference-signal sequences corresponding to seven resource blocks. As described in Section 17.1, the uplink transmission bandwidths, measured in number of resource blocks, should always be a product of the integers two, three, and five.[8] A resource assignment of size seven resource blocks is thus not a valid resource assignment and there is no need to define reference-signal sequences of the corresponding length. For the same reason, the maximum length of the reference-signal sequences corresponds to 108 resource blocks, rather than 110 resource blocks. Similarly, the second largest sequence length corresponds to 100 resource blocks.

In a given time slot, the uplink reference-signal sequences to use within a cell are taken from one specific sequence group. What group to use within a cell may then be the same for all slots (fixed assignment). Alternatively, what group to use within a cell may be varying between slots, also referred to as *group hopping*. If there is a fixed group assignment or group hopping is to be used is indicated by the cell system information.

Regardless of whether a fixed group assignment or group hopping is used, there are certain differences between PUCCH and PUSCH transmission.

In case of a fixed group assignment, the sequence group to use for PUCCH transmission is given by the physical-layer cell identity modulo 30, where the cell identity ranges from 0 to 503. Thus, cell identities 0, 30, 60, …, 480 correspond to sequence group 0, cell identities 1, 31, 61, …, 481 correspond to sequence group 1, and so on.

---

[8] Strictly speaking, the number of subarriers should be a product of the integers two, three, and five. However, as there are $12 = 3 \times 2^2$ subcarriers within one resource block, this is equivalent to a constraint that the number of resource blocks should be a product of the integers two, three, and five.

In contrast, what sequence group to use for PUSCH transmission is explicitly signaled as part of the cell system information. The reason for explicitly indicating what sequence group to use for PUSCH transmission is that it should be possible to use the same sequence group for PUSCH transmission in neighbor cells, despite the fact that such cells typically have different cell identities. In this case, the reference signals for PUSCH transmission would instead be distinguished by different phase rotations as discussed in Section 17.2.1.2.

In case of group hopping, an additional cell-specific group-hopping pattern is added to the sequence group. Thus, in the case of reference signals for PUCCH transmission, the sequence group to use in a given slot is given by the value of the group-hopping pattern for that slot plus the cell identity modulo 30. The group-hopping pattern is also given by the cell identity and identical group-hopping patterns are used for PUSCH and PUCCH within a cell.

Thus, to have the same sequence group for PUSCH in two cells:

- the cell identities of the two cells should be different (to have different sequence groups for PUCCH which should always be the case);
- if group hopping is enabled, the cell identities of the two cells should be such that the same group-hopping pattern is used in the two cells;
- the sequence group for PUSCH (indicated as part of the cell system information) should be the same for the two cells.

17.2.1.4 Sequence hopping

As can be seen from Figure 17.7, for sequence lengths corresponding to six resource blocks and above, there are two reference-signal sequences in each sequence group. What sequence to use may then either be fixed (always the first sequence within the group) or varying between slots, also referred to as *sequence hopping*.[9] Sequence hopping can only be enabled if group hopping is not enabled, that is, if there is a fixed assignment of groups to cells.

17.2.2 Uplink sounding reference signals

The uplink DRS discussed in Section 17.2.1 are used for uplink channel estimation to allow for coherent demodulation of uplink physical channels (PUSCH or PUCCH). They are always transmitted together with and covering the same frequency band as the corresponding physical channel.

---

[9]Not to be mixed with group hopping as described in Section 17.2.1.3.

In contrast, *sounding reference signals* (SRS), are transmitted on the uplink to allow for the network to estimate the uplink *channel quality* at different frequencies. Such quality estimates can then, for example, be used by the network to schedule uplink transmissions on resource blocks of instantaneously good quality. Thus an SRS is not necessarily transmitted together with any physical channel and if transmitted together with, for example, PUSCH, the SRS may cover a different, often larger, frequency span.

A terminal can be configured to transmit SRS at regular intervals ranging from as often as once in every 2 ms (every second subframe) to as infrequently as once in every 160 ms (every 16th frame). When an SRS is to be transmitted within a subframe, it occupies the last symbol of the subframe as illustrated in Figure 17.8.

In the frequency domain, SRS transmissions should cover the frequency band that is of interest for the frequency-domain scheduling. This can be achieved in the following two ways:

- By means of a sufficiently wideband SRS transmission that allows for sounding of the entire frequency band of interest with a single SRS transmission as illustrated in the upper part of Figure 17.9.

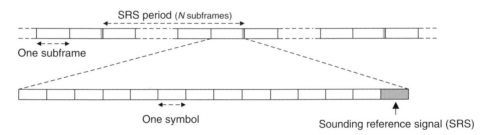

Figure 17.8   *Transmission of SRS.*

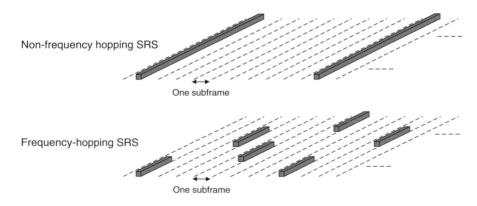

Figure 17.9   *Non-frequency-hopping (wideband) SRS versus frequency-hopping SRS.*

- By means of a more narrowband SRS that is hopping in the frequency domain in such a way that a sequence of SRS transmissions jointly covers the frequency band of interest as illustrated in the lower part of Figure 17.9.

The main benefit of a wideband SRS transmission as illustrated in the upper part of Figure 17.9 is that the entire frequency band of interest can be sounded with a single SRS transmission, that is, within a single DFTS-OFDM symbol. As the DFTS-OFDM symbols in which SRS is transmitted will be unavailable for PUSCH transmission, a wideband SRS transmission is thus more efficient from a resource-utilization point-of-view. However, in case of bad channel conditions, for example, a high uplink path loss, a wideband SRS transmission may lead to a relatively low received power density, which may degrade the channel-quality estimation. In such cases it can instead be preferred to use a more narrowband SRS transmission that hops over the total band to be sounded.

In general, different bandwidths of the SRS transmission can be used within a cell. A narrow SRS bandwidth, corresponding to four resource blocks is always available in all cells, regardless of the uplink cell bandwidth. Up to three additional, more wideband SRS bandwidths may also be defined depending on the cell bandwidth. The SRS bandwidths are, in these cases, always a multiple of four resource blocks. A terminal is then explicitly configured to use one of the SRS bandwidths available in the cell.

If a terminal is transmitting SRS in a certain subframe, then that SRS transmission may very well overlap, in the frequency domain, with PUSCH transmissions from other terminals within the cell. To avoid collision between SRS and PUSCH transmissions from different terminals, all terminals within a cell are aware of in what set of subframes SRS may be transmitted by *any* terminal within the cell. In such subframes, the last DFTS-OFDM symbol of the subframe is then not used for PUSCH transmission by *any* terminal within the cell.

On the more detailed level, the principles for SRS transmission are similar to those of the *demodulation reference signals* (DRS). More specifically, an SRS is also defined as a frequency-domain reference-signal sequence derived as a cyclic extension of prime-length Zadoff–Chu sequences. However, in case of SRS, the reference-signal sequence is mapped to *every second subcarrier* creating a 'comb'-like spectrum as illustrated in Figure 17.10. Taking into account that the bandwidth of the SRS transmission is always a multiple of four resource blocks, the lengths of the reference-signal sequences for SRS are thus always a multiple of 24. The reference-signal sequence to use for SRS transmission within the cell is taken from the same sequence group as the DRS for PUCCH.

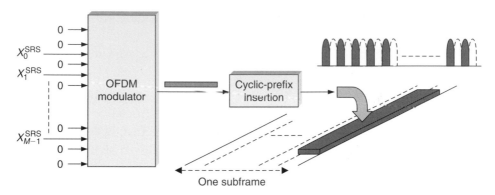

Figure 17.10   *Generation of SRS from a frequency-domain reference-signal sequence.*

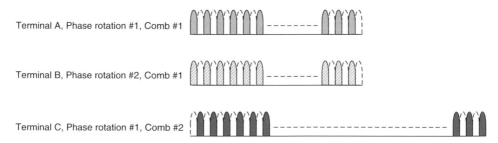

Figure 17.11   *Multiplexing of SRS transmissions from different mobile terminals.*

Similar to DRS, different phase rotations (also for SRS often referred to as 'cyclic shifts') can be used to generate different SRS that are orthogonal to each other. By assigning different phase rotations to different terminals, multiple SRS can thus be transmitted in parallel in the same subframe as illustrated in the upper part of Figure 17.11. However, it is then required that the reference signals span the same frequency band.

Another way to allow for SRS to be simultaneously transmitted from different terminals is to rely on the fact that each SRS only occupies every second subcarrier. Thus, SRS transmissions from two terminals can be *frequency multiplexed* by assigning them to different frequency shifts or 'combs' as illustrated in the lower part of Figure 17.11. In contrast to the multiplexing of SRS transmission by means of different 'cyclic shifts,' frequency multiplexing of SRS transmissions does not require the transmissions to cover identical frequency bands.

17.3   Uplink L1/L2 control signaling

Similar to the LTE downlink, there is also need for the uplink for certain associated control signaling (*'uplink L1/L2 control'*) to support the transmission of downlink and uplink transport channels. The uplink L1/L2 control signaling consists of:

- hybrid-ARQ acknowledgements for received DL-SCH transport blocks;
- terminal reports related to the downlink channel conditions,[10] used as assistance for the downlink scheduling;
- scheduling requests, indicating that a mobile terminal needs uplink resources for UL-SCH transmissions.

In contrast to the downlink, there is no information indicating the UL-SCH transport-format signaled on the uplink. As mentioned in Chapter 15, the eNodeB is in complete control of the uplink UL-SCH transmissions and the mobile terminal always follows the scheduling grants received from the network, including the UL-SCH transport-format specified in those grants. Thus, the network will know the transport-format used for the UL-SCH transmission in advance and there is no need for any explicit transport-format signaling on the uplink. For similar reasons, there is also no explicit uplink signaling of information related to UL-SCH hybrid ARQ.

Uplink L1/L2 control signaling needs to be transmitted on the uplink regardless of whether or not the mobile terminal has any uplink transport-channel (UL-SCH) data to transmit and thus regardless of whether or not the mobile terminal has been assigned any uplink resources for UL-SCH transmission. Hence, two different methods are used for the transmission of the uplink L1/L2 control signaling, depending on whether or not the mobile terminal has been assigned an uplink resource for UL-SCH transmission.

- *No simultaneous transmission of UL-SCH*:
  In case the terminal does not have a valid scheduling grant, that is, no resources have been assigned for UL-SCH in the current subframe, a separate physical channel, the *Physical Uplink Control Channel* (PUCCH), is used for transmission of uplink L1/L2 control signaling.
- *Simultaneous transmission of UL-SCH*:
  In case, the terminal has a valid scheduling grant, that is, resources have been assigned for UL-SCH in the current subframe, the uplink L1/L2 control signaling is time multiplexed with the coded UL-SCH onto the PUSCH prior to DFTS-OFDM modulation. Obviously, as the terminal has been assigned UL-SCH resources, there is no need to support transmission of the scheduling

---

[10]One or several of CQI, PMI, and RI as described in Chapter 19.

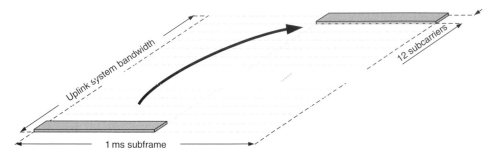

Figure 17.12   *Uplink L1/L2 control signaling transmission on PUCCH.*

request in this case. Instead, scheduling information can be included in the MAC headers as described in Chapter 19.

These two cases are described in more detail in the following.

17.3.1   Uplink L1/L2 control signaling on PUCCH

If the mobile terminal has not been assigned an uplink resource for UL-SCH transmission, the L1/L2 control information (channel-status reports, hybrid-ARQ acknowledgments, and scheduling requests) is transmitted in uplink resources (resource blocks) specifically assigned for uplink L1/L2 control on PUCCH. As illustrated in Figure 17.12, these resources are located at the edges of the total available cell bandwidth. Each such resource consists of 12 'subcarriers' (one resource block) within each of the two slots of an uplink subframe. To provide frequency diversity, these frequency resources are frequency hopping on the slot boundary, that is, one 'frequency resource' consists of 12 subcarriers at the upper part of the spectrum within the first slot of a subframe and an equally sized resource at the lower part of the spectrum during the second slot of the subframe or vice versa. If more resources are needed for the uplink L1/L2 control signaling, for example, in case of very large overall transmission bandwidth supporting a large number of users, then additional resources blocks can be assigned next to the previously assigned resource blocks.

The reasons for locating the PUCCH resources at the edges of the overall available spectrum are twofold:

- Together with the frequency hopping described above, this maximizes the frequency diversity experienced by the control signaling.
- Assigning uplink resources for the PUCCH at other positions within the spectrum, that is, not at the edges, would have fragmented the uplink spectrum,

making it impossible to assign very wide transmission bandwidths to single mobile terminal and still retain the single-carrier property of the uplink transmission.

The bandwidth of one resource block during one subframe is too large for the control signaling needs of a single terminal. Therefore, to efficiently exploit the resources set aside for control signaling, multiple terminals can share the same resource block. This is done by assigning the different terminals to different orthogonal phase rotations of a cell-specific length-12 frequency-domain sequence, where the sequence is identical to a length-12 reference-signal sequence. Furthermore, as described in conjunction with the reference signals in Section 17.2, a linear phase rotation in the frequency domain is equivalent to applying a cyclic shift in the time domain. Thus, although the term 'phase rotation' is used herein, the term cyclic shift is sometimes used with an implicit reference to the time domain.

The resource used by a PUCCH is therefore not only specified in the time–frequency domain by the resource-block pair, but also by the phase rotation applied. Similarly to the case of reference signals, there are up to 12 different phase rotations specified, providing up to 12 different orthogonal sequences from each cell-specific sequence. However, in the case of frequency-selective channels, not all the 12 phase rotations can be used if orthogonality is to be retained. Typically, up to six rotations are considered usable in a cell.

As mentioned earlier, uplink L1/L2 control signaling includes hybrid-ARQ acknowledgements, channel-status reports, and scheduling requests. Different combinations of these types of messages are possible as described later, but to explain the structure for these cases it is beneficial to discuss separate transmission of each of the types first, starting with the hybrid-ARQ acknowledgement and the scheduling request. There are two formats defined for PUCCH, each capable of carrying a different number of bits.

17.3.1.1 PUCCH format 1

Hybrid-ARQ acknowledgements are used to acknowledge the reception of one (or two in case of spatial multiplexing) transport blocks on the DL-SCH. The hybrid-ARQ acknowledgement is only transmitted in case the terminal correctly received control signaling related to DL-SCH transmission intended for this terminal on one of the PDCCHs. If no valid DL-SCH-related control signaling was detected, then nothing is transmitted on the PUCCH (i.e., DTX). Apart from not unnecessarily occupying PUCCH resources that can be used for other purposes, this allows the eNodeB to do three-state detection, ACK, NAK, or DTX, on the

PUCCH. Similarly to HSPA (Chapter 9), this is useful when controlling the redundancy version to use for the retransmission.

Scheduling requests are used to request resources for uplink data transmission. Obviously, a scheduling request should only be transmitted when the terminal is requesting resources, otherwise the terminal should be silent to save battery resources and not create unnecessary interference. Hence, unlike hybrid-ARQ acknowledgements, no explicit information bit is transmitted by the scheduling request; the information is instead conveyed by the presence (or absence) of energy on the corresponding PUCCH. However, the scheduling request, although used for a completely different purpose, shares the same PUCCH format as the hybrid-ARQ acknowledgement. This format is referred to as *PUCCH format 1* in the specifications.[11]

PUCCH format 1 uses the same structure in the two slots of a subframe as illustrated in Figure 17.13. For transmission of a hybrid-ARQ acknowledgement, the single hybrid-ARQ acknowledgement bit is used to generate a BPSK symbol (in case of downlink spatial multiplexing the two acknowledgement bits are used to generate a QPSK symbol). For a scheduling request, in contrast, the BPSK/QPSK symbol is replaced by a constellation point treated as negative acknowledgement at the eNodeB. The modulation symbol is then used to generate the signal to be transmitted in each of the two PUCCH slots.

There are seven OFDM symbols per slot for normal cyclic prefix (six in case of extended cyclic prefix). In each of those seven OFDM symbols, a length-12 sequence, obtained by a phase rotation of the cell-specific sequence as described earlier, is transmitted. Three of the symbols are used as reference signals to enable channel estimation by the eNodeB and the remaining four[12] are modulated by the BPSK/QPSK symbols described earlier. In principle, the BPSK/QPSK modulation symbol could directly modulate the rotated length-12 sequence used to differentiate terminals transmitting on the same time–frequency resource. However, this would result in unnecessarily low capacity on the PUCCH. Therefore, the BPSK/QPSK symbol is multiplied by a length-4 orthogonal cover sequence.[13] Multiple terminals may transmit on the same time–frequency

---

[11] There are actually three variants in the LTE specifications, formats 1, 1a, and 1b, used for transmission of scheduling requests and one or two hybrid-ARQ acknowledgements, respectively. However, for simplicity, they are all referred to as format 1 herein.

[12] The number of symbols used for reference signals and the acknowledgement is a trade-off between channel-estimation accuracy and energy in the information part; three symbols for reference symbols and four symbols for the acknowledgement were found to be a good compromise.

[13] In case of sounding and PUCCH transmissions in the same subframe, a length-3 sequence is used, thereby making one DFTS-OFDM symbol available for the sounding reference signal.

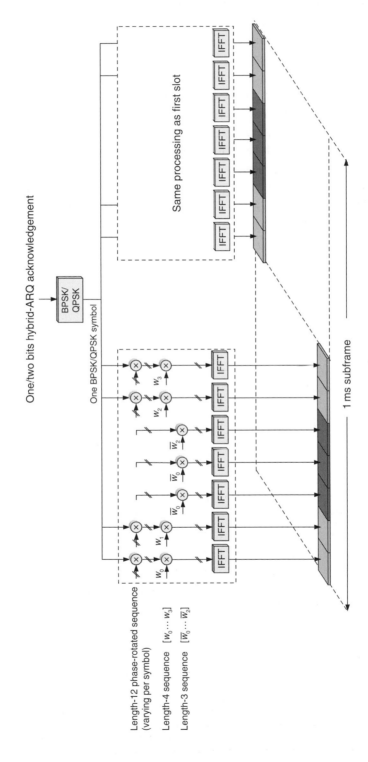

Figure 17.13 *PUCCH format 1 (normal cyclic prefix).*

resource using the same phase-rotated sequence and be separated through different orthogonal covers. To be able to estimate the channels for the respective terminals, the reference signals also employ an orthogonal cover sequence with the only difference being the length of the sequence – three for the case of normal cyclic prefix. Thus, since each cell specific sequence can be used for up to $3 \times 12 = 36$ different terminals (assuming all twelve rotations being available; typically only six of them are used), a threefold improvement in the PUCCH capacity is compared to the case of no cover sequence. The cover sequences are three Walsh sequences of length four for the data part and three DFT sequences of length three for the reference signals (for extended cyclic prefix, the reference-signal cover sequence is of length two).

A PUCCH format 1 resource, used for either a hybrid-ARQ acknowledgement or a scheduling request, is represented by a single scalar resource index. From the index, the phase rotation and the orthogonal cover sequences are derived.

The use of a phase rotation of a cell-specific sequence together with orthogonal sequences as described earlier provides orthogonality between different terminals in the same cell transmitting PUCCH on the same set of resource blocks. Hence, in the ideal case, there will be no intra-cell interference, which helps improving the performance. However, there will typically be inter-cell interference for the PUCCH as the different sequences used in neighboring cells are not orthogonal. To randomize this interference between neighboring cells, the phase rotation of the sequence used in a cell varies on a symbol-by-symbol basis in a slot according to a hopping pattern derived from the physical-layer cell identity.

To further randomize the PUCCH inter-cell interference, slot-level hopping is applied to the orthogonal cover and phase rotation. This is exemplified in Figure 17.14 assuming normal cyclic prefix and six out of twelve rotations used for each cover sequence. To the phase rotation given by the cell-specific hopping a slot-specific offset is added. In cell A, a terminal is transmitting on PUCCH resource number 3, which in this example corresponds to using the (phase rotation, cover sequence) combination (6, 0) in the first slot and (11, 1) in the second slot of this particular subframe. PUCCH resource number 11, used by another terminal in cell A transmitting in the same subframe, corresponds to (9, 1) and (2, 2) in the first and second slot, respectively, of the subframe. In another cell the PUCCH resource numbers are mapped to different sets of (rotation, cover sequence) in the slots. This helps randomizing the inter-cell interference.

Figure 17.13 illustrates the overall structure for the PUCCH format 1 in case of normal cyclic prefix. The same structure is used for the case of extended cyclic

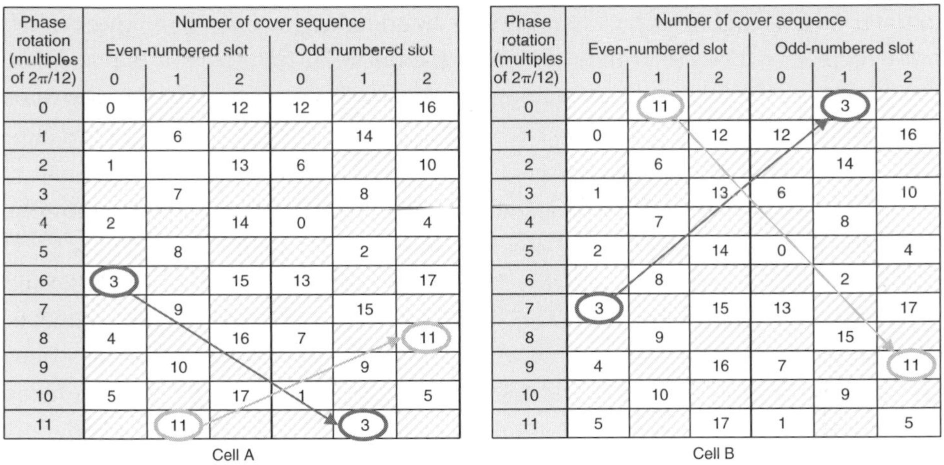

Cell A

Phase rotation (multiples of 2π/12)	Number of cover sequence					
	Even-numbered slot			Odd numbered slot		
	0	1	2	0	1	2
0	0		12	12		16
1		6			14	
2	1		13	6		10
3		7			8	
4	2		14	0		4
5		8			2	
6	(3)		15	13		17
7		9			15	
8	4		16	7		(11)
9		10			9	
10	5		17	1		5
11		(11)			(3)	

Cell B

Phase rotation (multiples of 2π/12)	Number of cover sequence					
	Even-numbered slot			Odd-numbered slot		
	0	1	2	0	1	2
0		(11)			(3)	
1	0		12	12		16
2		6			14	
3	1		13	6		10
4		7			8	
5	2		14	0		4
6		8			2	
7	(3)		15	13		17
8		9			15	
9	4		16	7		(11)
10		10			9	
11	5		17	1		5

Figure 17.14 *Example of phase rotation and cover hopping for two PUCCH resource indices in two different cells.*

prefix with the difference being the number of reference symbols in each slot. In this case, the six OFDM symbols in each slot are divided such that the two middle symbols are used for reference signals and the remaining four symbols used for the information. Thus, the length of the orthogonal sequence used to spread the reference symbols is reduced from three to two and the multiplexing capacity is lower. However, the general principles described above still apply.

As mentioned earlier, a PUCCH resource can be represented by an index. For a hybrid-ARQ acknowledgement, the resource index to use is given as a function of the first CCE in the PDCCH used to schedule the downlink transmission to the terminal. This way, there is no need to explicitly include information about the PUCCH resources in the downlink scheduling assignment, which of course reduces overhead. Furthermore, as described in Chapter 19, hybrid-ARQ acknowledgements are transmitted a fixed time after the reception of a DL-SCH transport block and when to expect a hybrid ARQ on the PUCCH is therefore known to the eNodeB.

In addition to dynamic scheduling by using the PDCCH, there is also, as described in Chapter 19, the possibility to semi-persistently schedule a terminal according to a specific pattern. In this case there is no PDCCH to derive the PUCCH resource index from. Instead, the configuration of the semi-persistent scheduling pattern includes information on the PUCCH index to use for the hybrid-ARQ acknowledgement. In either of these two cases, a terminal is using PUCCH resources only when it has been scheduled in the downlink. Thus, the amount of PUCCH resources required for hybrid-ARQ acknowledgements does not necessarily increase with an increasing number of terminals in the cell, but, for dynamic scheduling, is rather related to the number of CCEs in the downlink control signaling.

Unlike the hybrid-ARQ acknowledgements, whose occurrence is known to the eNodeB from the downlink scheduling decisions, the need for uplink resources for a certain terminal is in principle unpredictable by the eNodeB. One way to handle this would be to have a contention-based mechanism for requesting uplink resources. The random-access mechanism is based on this principle and can, to some extent, be used also for scheduling requests as discussed in Chapter 19. Contention-based mechanisms typically work well for low intensities, but for higher scheduling-request intensities, the collision rate between different terminals simultaneously requesting resources increases. Therefore, LTE provides a contention-free scheduling-request mechanism on the PUCCH, where each terminal in the cell is given reserved resource on which it can transmit a request for uplink resources. The contention-free resource is represented by a PUCCH resource index as described earlier, occurring at every n:th subframe. The more frequent these time instants occur, the lower the scheduling-request delay at the cost of a higher PUCCH resource consumption. As the eNodeB configures all the terminals in the cell, when and on which resources a terminal can request resources is known to the eNodeB.

The discussion earlier has considered transmission of *either* a hybrid-ARQ acknowledgement *or* a scheduling request. However, there are situations when the terminal needs to transmit *both* of them. In such a situation, the hybrid-ARQ acknowledgement is transmitted on the scheduling-request resource, see Figure 17.15. This is possible as the same PUCCH structure is used for both of them and the scheduling request carries no explicit information. By comparing the amount of energy detected on the acknowledgement resource and the scheduling-request resource for a specific terminal, the eNodeB can determine whether the terminal is requesting resources or not. Once the resource used for transmission of the acknowledgement is detected, the hybrid-ARQ acknowledgement can be decoded. Other, more advanced methods jointly decoding hybrid ARQ and scheduling request can also be envisioned.

17.3.1.2   PUCCH format 2
Channel-status reports, the contents of which are discussed in Chapter 19, are used to provide the eNodeB with an estimate of the channel properties at the terminal to aid channel-dependent scheduling. A channel-status report consists of multiple bits per subframe. PUCCH format 1, which is capable of at most two bits of information per subframe, obviously cannot be used for this purpose.

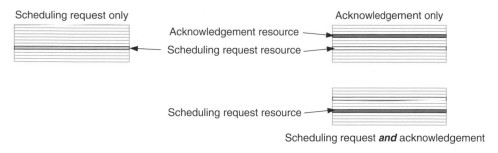

Figure 17.15 *Multiplexing of scheduling request and hybrid-ARQ acknowledgement from a single terminal.*

Transmission of channel-status reports on the PUCCH is instead handled by PUCCH format 2, which is capable of multiple information bits per subframe.[14]

PUCCH format 2, illustrated for normal cyclic prefix in Figure 17.16, is based on a phase rotation of the same cell-specific sequence as format 1. After block coding and QPSK modulation of the channel-status information, there are ten QPSK symbols to transmit in the subframe; the first five symbols are transmitted in the first slot and the remaining five in the last slot.

Assuming normal cyclic prefix, there are seven DFTS-OFDM symbols per slot. Out of the seven DFTS-OFDM symbols in each slot, two[15] are used for reference-signal transmission to allow coherent demodulation at the eNodeB. In the remaining five, the respective QPSK symbol to be transmitted is multiplied by a phase-rotated length-12 cell-specific sequence and the result is transmitted in the corresponding DFTS-OFDM symbol. For extended cyclic prefix, where there are six DFTS-OFDM symbols per slot, the same structure is used but with one reference signal symbols per slot instead of two.

Basing the format 2 structure on phase rotations of the same cell-specific sequence as format 1 is highly beneficial as it allows the two formats to be transmitted in the same resource block. As phase-rotated sequences are orthogonal, one rotated sequence in the cell can be used either for one PUCCH instance using format 2 or three PUCCH instances using format 1. Thus, the 'resource consumption' of one channel-status report is as large as three hybrid-ARQ acknowledgements (assuming normal cyclic prefix). Note that no orthogonal cover sequences are used for format 2.

---

[14] There are actually three variants in the LTE specifications, formats 2, 2a, and 2b, where the last two formats are used for simultaneous transmission of hybrid-ARQ acknowledgements as discussed later in this section. However, for simplicity, they are all referred to as format 2 herein.

[15] Similarly to format 1, the number of symbols used for reference signals and the coded channel-quality information is a trade-off between channel-estimation accuracy and energy in the information part. Two symbols for reference symbols and five symbols for the coded information part in each slot were found to be the best compromise.

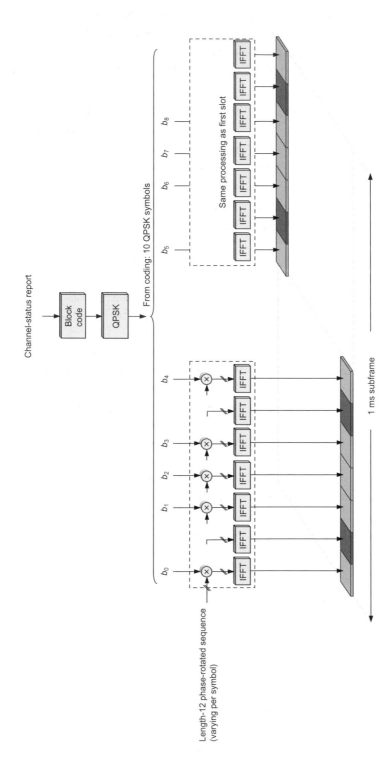

Figure 17.16　*PUCCH format 2 (normal cyclic prefix).*

The rotation angles to use in the different symbols for PUCCH format 2 are hopping in a similar way as for format 1, motivated by interference randomization. Resources for PUCCH format 2 can, similar to format 1, be represented by a scalar index, which can be seen as a 'channel number' and higher-layer signaling is used to configure each terminal with a resource to transmit its channel-status report on, as well as when those reports should be transmitted. Hence, the eNodeB has full knowledge on when and on which resources each of the terminals will transmit channel-status reports on PUCCH. The channel-status reports on PUCCH are also known as *periodic* reports; as will be discussed later there are also *aperiodic* reports but those reports can only be transmitted on PUSCH.

17.3.1.3 Simultaneous transmission of multiple feedback reports

The above description has focused on transmission of a channel-status report alone. However, there is also a need to handle the case when hybrid ARQ or scheduling request needs to be transmitted in the same subframe. Transmitting the two PUCCH formats simultaneously from the same terminal would increase the peak-to-average ratio and violate the 'single-carrier' property. Instead, a single PUCCH structure supporting simultaneous transmission of multiple feedback signals is used.

In principle, the following four situations requiring simultaneous transmission of multiple feedback signals from a single terminal can be envisioned:

1. *Hybrid-ARQ acknowledgement and scheduling request*:
   This is supported by transmitting the hybrid-ARQ acknowledgement on the scheduling-request resource as described earlier.
2. *Scheduling request and channel-status report*:
   The eNodeB is in control of when a terminal may transmit a scheduling request and when the terminal should report the channel status. Hence, this situation can be avoided by proper configuration, but if this is not done the terminal drops the channel-status report and transmits the scheduling request only. Missing a channel-status report is not detrimental and only incurs some degradation in the scheduling and rate-adaptation accuracy whereas the scheduling request is critical for the uplink transmissions.
3. *Scheduling request, hybrid-ARQ acknowledgement, and channel-status report*:
   This is similar to the previous situation; the channel-status report is dropped and multiplexing of hybrid ARQ and scheduling request is handled as described in point 1 above.
4. *Hybrid-ARQ acknowledgement and channel-status report*:
   Simultaneous transmission of acknowledgement and channel-status report from the same terminal is possible as described later. There is also the possibility to configure the terminal to drop the status report and only transmit the acknowledgement.

Transmission of data in the downlink implies transmission of hybrid-ARQ acknowledgements in the uplink. At the same time, since data is transmitted in the downlink, up-to-date channel-status reports are beneficial to optimize the downlink transmissions. Hence, simultaneous transmission of hybrid-ARQ acknowledgements and channel-status reports is supported by LTE. The basis is PUCCH format 2 as described earlier for both normal and extended cyclic prefix, although the detailed solution differs between the two.

For normal cyclic prefix, each slot in PUCCH format 2 has two OFDM symbols used for reference signals. When transmitting a hybrid-ARQ acknowledgement at the same time as a channel-status report, the second reference symbol in each slot is modulated by the acknowledgement as illustrated in Figure 17.17a. Either BPSK or QPSK is used, depending on whether one or two acknowledgement bits are to be fed back. The fact that the acknowledgement is superimposed on the reference signal needs to be accounted for at the eNodeB. One possibility is to decode the acknowledgement bit(s) modulated onto the second reference symbol using the first reference symbol for channel estimation. Once the acknowledgement bit(s) have been decoded, the modulation imposed on the second reference symbol can be removed and channel estimation and decoding of the channel-status report can be handled the same way as in absence of simultaneous hybrid-ARQ acknowledgement. This two-step approach works well for low to medium Doppler frequencies; for higher Doppler frequencies the acknowledgement and channel-status reports are preferably decoded jointly.

For extended cyclic prefix, there is only a single reference symbol per slot. Hence, it is not possible to overlay the hybrid-ARQ acknowledgement on the reference symbol. Instead, the acknowledgement bit(s) are jointly coded with the channel-status report prior to transmission using PUCCH format 2 as illustrated in Figure 17.17b.

The time instances when to expect channel-status reports and hybrid-ARQ acknowledgements are known to the eNodeB which therefore knows whether to expect a hybrid-ARQ acknowledgement along with the channel-status report or not. If the PDCCH assignment is missed by the terminal, then only the channel-status report will be transmitted as the terminal is not aware that it has been scheduled. In absence of a simultaneous channel-status report, the eNodeB can employ DTX detection to discriminate between a missed assignment and a failed decoding of downlink data. However, one consequence of the structures above is that DTX detection is cumbersome if not impossible. This implies that incremental redundancy needs to be operated with some care if the eNodeB has scheduled data such that the acknowledgement occurs at the same time as a channel-status report. As the terminal may have missed the original transmission attempt of downlink, it may be preferable for the eNodeB to select the redundancy version of the retransmission such that systematic bits are also included in the retransmission.

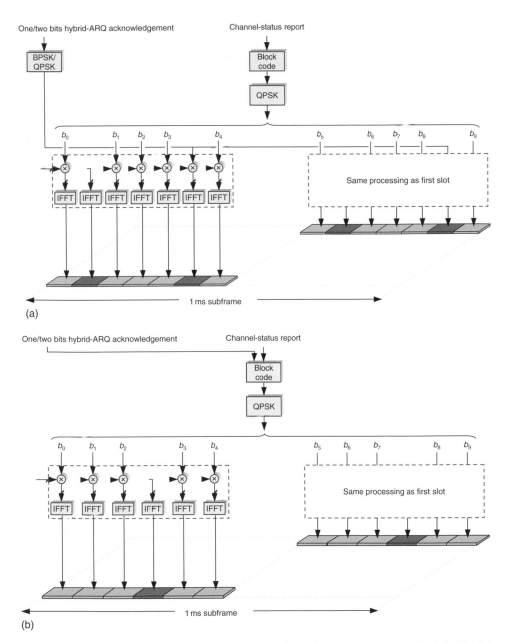

Figure 17.17 *Simultaneous transmission of channel-status reports and hybrid-ARQ acknowledgements: (a) normal cyclic prefix and (b) extended cyclic prefix.*

One possibility to circumvent this is to configure the terminal to drop the channel-status report in case of simultaneous transmission of a hybrid-ARQ acknowledgement. In this case, the eNodeB can detect DTX as the acknowledgement will be transmitted using PUCCH format 1 as described earlier. Obviously, there will be no channel-status report sent in this case, which needs to be taken into account in the scheduling process. Dropping the channel-status report can also be useful for handling coverage-limited situations when the available transmission power is insufficient for a channel-status report in addition to the acknowledgement.

17.3.1.4   Resource-block mapping for PUCCH

The signals described above, for both of the PUCCH formats, are, as already explained, transmitted on a resource-block pair with one resource block in each slot. The resource-block pair to use is determined from the PUCCH resource index. Multiple resource-block pairs can be used to increase the control-signaling capacity; when one resource-block pair is full the next PUCCH resource index is mapped to the next resource-block pair in sequence. The mapping is in principle done such that PUCCH format 2 (channel-status reports) is transmitted closest to the edges of the uplink cell bandwidth with PUCCH format 1 (hybrid-ARQ acknowledgements, scheduling requests) next as illustrated in Figure 17.18. A semi-static parameter, provided as part of the system information, controls on which resource-block pair the mapping of PUCCH format 1 starts. Furthermore, the semi-statically configured scheduling requests are located at the outermost parts of the format 1 resources. As the amount of resources necessary for hybrid-ARQ acknowledgements varies dynamically, this maximizes the amount of contiguous spectrum available for PUSCH.

In many scenarios, the configuration of the PUCCH resources can be done such that the two PUCCH formats are transmitted on separate sets of resource blocks. However, for the smallest cell bandwidths, this would result in a too high overhead. Therefore, it is possible to mix the two PUCCH formats in one of the resource-block pairs; for example in Figure 17.18 this is the case for the resource with index 2. Although this mixture is primarily motivated by the smaller cell bandwidths, it can equally well be used for the larger cell bandwidths.

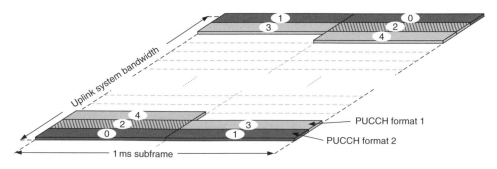

Figure 17.18   *Allocation of resource blocks for PUCCH.*

In the resource-block pair where the two PUCCH formats are mixed, the rotation angles are divided between the two formats. Furthermore, some of the phase rotations are reserved as 'guard'; hence the efficiency of such a mixed resource-block pair is slightly lower than a resource-block pair carrying only one of the PUCCH formats.

17.3.2   Uplink L1/L2 control signaling on PUSCH

If the terminal is transmitting data on PUSCH, that is, has a valid scheduling grant in the subframe, control signaling cannot be simultaneously sent on PUCCH as this would violate the single-carrier properties. Instead, control signaling is time multiplexed with data on the PUSCH. Only hybrid-ARQ acknowledgements and channel-status reports are transmitted on the PUSCH. Obviously there is no need to request a scheduling grant when the terminal is already scheduled; instead, in-band buffer-status reports are send as part of the MAC headers as described in Chapter 19.

Time multiplexing of channel-status reports and hybrid-ARQ acknowledgements is illustrated in Figure 17.19. However, although they both use time multiplexing there are some differences in the details for the two types of uplink L1/L2 control signaling motivated by their different properties.

The hybrid-ARQ acknowledgement is important for proper operation of the downlink. Robust QPSK modulation is used, regardless of the modulation scheme used for the data, and the hybrid-ARQ acknowledgement is transmitted next to the reference symbols as the channel estimates are of better quality close to the reference symbols. This is especially important at high Doppler frequencies where the channel may vary during a slot. Unlike the data part, the hybrid-ARQ acknowledgement cannot rely on retransmissions and strong channel coding to handle these variations.

In principle, the eNodeB knows when to expect a hybrid-ARQ acknowledgement from the terminal and can therefore perform the appropriate demultiplexing of the acknowledgement and the data part. However, there is a certain probability that the terminal has missed the scheduling assignment on the PDCCH in which case the eNodeB will expect a hybrid-ARQ acknowledgement while the terminal will not transmit one. If the rate matching pattern would depends on whether an acknowledgement is transmitted or not all the coded bits transmitted in the data part could be affected by a missed PDCCH assignment, which is likely to cause the UL-SCH decoding to fail. To avoid this error

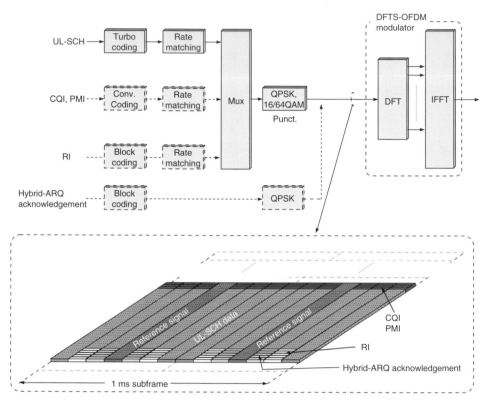

Figure 17.19 *Multiplexing of control and data onto PUSCH.*

case, the hybrid-ARQ acknowledgements are therefore punctured into the coded UL-SCH bit stream. Thus, the non-punctured bits are not affected by the presence/absence of hybrid-ARQ acknowledgements and the problem of a mismatch between the rate matching in the terminal and the eNodeB is avoided.

The contents of the channel-status reports is described in Chapter 19; at this stage it suffices to note that a channel-status report consists of *Channel-Quality Indicator* (CQI), *Precoding Matrix Indicator* (PMI), and *Rank Indicator* (RI). The channel-status reports are, similar to the hybrid-ARQ acknowledgements, time multiplexed with the data part, but unlike the acknowledgements the reports are not necessarily located next to the reference symbols. Channel-status reports are mainly useful for low-to-medium Doppler frequencies for which the radio channel is relatively constant, making the need for a special mapping less pronounced. The RI is mapped differently than the CQI and PMI; as illustrated in Figure 17.19, the RI is located close to the reference symbols using a similar mapping as the hybrid-ARQ acknowledgements. The more robust mapping of the RI is motivated by the fact that the RI is required in order to correctly interpret the CQI/PMI. The CQI/PMI, on the other-hand, is simply mapped across the full subframe duration.

The basis for channel-status reports on the PUSCH is *aperiodic reports*, where the eNodeB requests a channel-status report from the terminal by setting the CQI request bit in scheduling grant as mentioned in Chapter 16. The UL-SCH rate matching takes the presence of the channel-status reports into account; by using a higher code rate a suitable number of resource elements are made available for transmission of the channel-status report. Since the reports are explicitly requested by the eNodeB, their presence is known and the appropriate rate de-matching can be done at the receiver. If a periodic report is configured to be transmitted on PUCCH in a subframe when the terminal is scheduled to transmit on the PUSCH, then the periodic report is 'rerouted' and transmitted on the PUSCH resources. Also, in this case there is no risk of mismatch in rate matching; the transmission instants for periodic reports are configured by robust RRC signaling and the eNodeB knows in which subframes such reports will be transmitted.

The channel coding of the channel-status reports depends on the report size. For the smaller sizes such as a periodic report that otherwise would have been transmitted on the PUCCH, the same block coding as used for the PUCCH reports is used. For the larger reports, a tail-biting convolutional code is used for CQI/PMI, whereas the RI uses a (3, 2) block code.

Unlike the data part that relies on rate adaptation to handle different radio conditions, this cannot be used for the L1/L2 control-signaling part. Power control could, in principle, be used as an alternative, but this would imply rapid power variations in the time domain which negatively impacts the RF properties. Therefore, the transmission power is kept constant over the subframe and the amount of resource elements allocated to L1/L2 control signaling, that is, the code rate of the control signaling, is varied according to the scheduling decision for the data of the data part. High data rates are typically scheduled when the radio conditions are advantageous and hence a smaller amount of resource need to be used by the L1/L2 control signaling compared to the case of poor radio conditions. To account for different hybrid-ARQ operating points, an offset between the code rate for the control-signaling part and the modulation-and-coding scheme used for the data part can be configured via higher-layer signaling.

17.4 Uplink transport-channel processing

The LTE uplink transport-channel processing, more specifically the processing for the UL-SCH, can be outlined according to Figure 17.20. Most steps of the uplink transport-channel processing are similar to the corresponding steps of the downlink transport-channel processing as outlined in Section 16.5. However, as there is no spatial multiplexing or transmit diversity currently defined for the LTE uplink, there is no explicit multi-antenna-mapping function as part of the uplink transport-channel processing. As a consequence, there is also only a single transport block, of dynamic size, transmitted for each TTI.

In more details, the uplink transport-channel processing consists of the following steps:

- *CRC insertion per transport block*:
  A 24-bit CRC is calculated for and appended to each uplink transport block in the same way as for downlink transport channels.
- *Code-block segmentation and per-code-block CRC insertion*:
  In the same way as for the downlink, code-block segmentation is applied for transport blocks larger than 6144 bits. The code-block segmentation also includes per-code-block CRC (in case of more than one code block) and possible insertion of filler bits similar to the downlink.

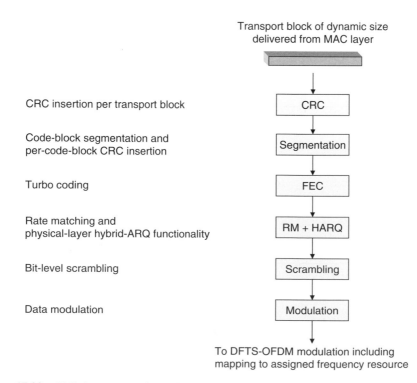

Figure 17.20   *Uplink transport-channel processing.*

- *Turbo coding*:
  The same rate 1/3 Turbo code with QPP-based internal interleaver as is used for the downlink is also used for the uplink.
- *Rate-matching and physical-layer hybrid-ARQ functionality*:
  The uplink physical-layer aspects of the LTE uplink hybrid ARQ are basically the same as the corresponding downlink functionality, that is sub-block interleaving and bit collection into a circular buffer, followed by bit selection. It should be noted that, in some aspects, there are some clear differences between the downlink and uplink hybrid-ARQ protocols, such as asynchronous versus synchronous operation (see Chapter 19). However, these differences are not really visible in the physical-layer aspects of the hybrid-ARQ functionality.
- *Bit-level scrambling*:
  Similar to the downlink, bit-level scrambling is also applied to the code bits on the LTE uplink. The aim of uplink scrambling is the same as for the downlink, that is to randomize the interference and thus ensure that the processing gain provided by the channel code can be fully utilized. To achieve this, the uplink scrambling is mobile-terminal specific, that is, different mobile terminals use different scrambling sequences.
- *Data modulation*:
  Similar to downlink DL-SCH transmission, QPSK, 16QAM, and 64QAM modulation can be used for UL-SCH transmission.

The block of modulation symbols is then applied to the DFTS-OFDM processing as outlined in Figure 17.1. The exact frequency-domain mapping is controlled by the scheduler.

17.5 PUSCH frequency hopping

In Chapter 16, it was described how the notion of VRBs in combination with the mapping from VRBs to physical resource blocks (PRBs) allowed for downlink distributed transmission, that is, the spreading of a downlink transmission in the frequency domain. As described, downlink distributed transmission consists of two separate steps: (1) a mapping from VRB pairs to PRB pairs such that consecutive VRB pairs are not mapped to frequency-consecutive PRB pairs and (2) a split of each resource-block pair such that the two resource blocks of the resource-block pair are transmitted with a certain frequency gap in between. This second step can be seen as frequency hopping on a slot basis.

Also for the LTE uplink the notion of VRBs can be used, allowing for frequency-domain-distributed transmission also for the uplink. However, in the uplink, where transmission from a terminal should always be over a set of consecutive subcarriers, distributing resource-block pairs in the frequency domain, as in the first step of downlink distributed transmission, is not possible. Rather, uplink distributed transmission is similar to the second step of downlink distributed transmission, that is, a frequency separation of the transmissions in the first and second slot of a subframe. Uplink distributed transmission for PUSCH can thus more directly be referred to as *uplink* (PUSCH) *frequency hopping*.

There are two types of uplink frequency hopping defined for PUSCH:

- Subband-based hopping according to cell-specific hopping/mirroring patterns.
- Hopping based on explicit hopping information in the scheduling grant.

17.5.1 *Hopping based on cell-specific hopping/mirroring patterns*

To support subband-based hopping according to cell-specific hopping/mirroring patterns, a set of consecutive subbands of a certain size are defined from the overall uplink frequency band as illustrated in Figure 17.21. It should be noted that the subbands do not cover the total uplink frequency band, mainly due to the fact that a number of resource blocks at the edges of the uplink frequency band are used for PUCCH transmission. For example in Figure 17.21, the overall uplink bandwidth corresponds to 50 resource blocks and there are a total of 4 subbands, each consisting of 11 resource blocks. Six resource blocks are not included in the hopping bandwidth and could, for example, be used for PUCCH transmission.

In case of subband-based hopping, the resource defined by a scheduling grant is not the actual set of resource blocks to use for transmission. Rather, what is provided in the scheduling grant is a *virtual resource* (a set of *VRBs*). The resource

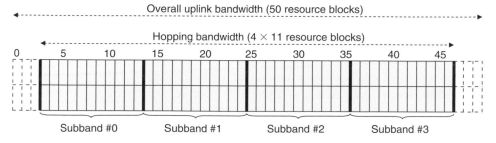

Figure 17.21 *Definition of subbands for PUSCH hopping. A total of four subbands, each consisting of eleven resource blocks.*

to use for transmission (the *PRBs*) is the resource provided in the scheduling grant shifted a number of subbands according to a cell-specific hopping pattern, where the hopping pattern can provide different shifts for each slot. As illustrated in Figure 17.22, a terminal is assigned VRBs 27, 28, and 29. In the first slot, the predefined hopping pattern takes the value 1, implying transmission using PRBs 38, 39, and 40. In the second slot, the predefined hopping pattern takes the value 3, implying transmission using resource blocks 16, 17, and 18. Note that the shifting 'wraps-around,' that is, in case of four subbands, a shift of three subbands is the same as a negative shift of one subband. As the hopping pattern is cell specific, that is the same for all terminals within a cell, different terminals will transmit on non-overlapping physical resources as long as they are assigned non-overlapping virtual resources.

In addition to the hopping pattern, there is also a cell-specific *mirroring pattern* defined in a cell. The mirroring pattern controls, on a slot basis, whether or not mirroring, within each subband, should be applied to the assigned resource. Mirroring (in combination with hopping) is illustrated in Figure 17.23. Here, the

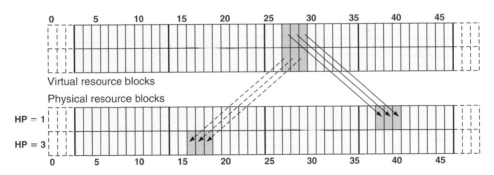

Figure 17.22 *Hopping according to predefined hopping pattern.*

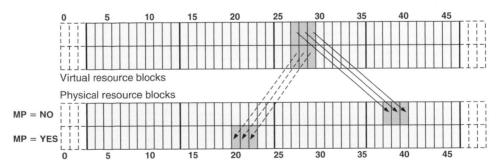

Figure 17.23 *Hopping/mirroring according to predefined hopping/mirroring patterns. Same hopping pattern as in Figure 17.22.*

mirroring pattern is such that mirroring is not applied to the first slot while mirroring is applied to the second slot.

Both the hopping pattern and the mirroring pattern depend on the physical-layer cell identity and are thus typically different in neighbor cells. Furthermore, the period of the hopping/mirroring patterns corresponds to one frame.

17.5.2   Hopping based on explicit hopping information

As an alternative to hopping/mirroring according to cell-specific hopping/mirroring patterns as described above, uplink slot-based frequency hopping for PUSCH can also be controlled by *explicit hopping information* provided in the scheduling grant. In such a case the scheduling grant includes the following:

- Information about the resource to use for uplink transmission in the first slot, exactly as in the non-hopping case.
- Additional information about the offset of the resource to use for uplink transmission is the second slot, relative to the resource of the first slot.

Selection between hopping according to cell-specific hopping/mirroring patterns as described above or hopping according to explicit information in the scheduling grant can be done dynamically. More specifically, in case of cell bandwidths less than 50 resource blocks, there is a single bit in the scheduling grant indicating if hopping should be according to the cell-specific hopping/mirroring patterns or should be according to information in the scheduling grant. In the later case, the hop is always half of the hopping bandwidth. In case of larger bandwidths (50 resource blocks and beyond), there are two bits in the scheduling grant. One of

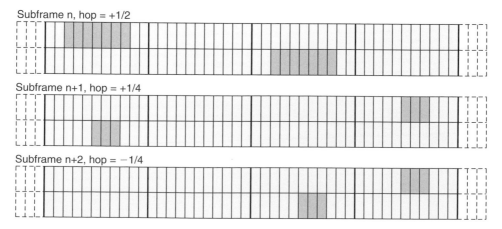

Figure 17.24   *Frequency hopping according to explicit hopping information.*

the combinations indicate that hopping should be according to the cell-specific hopping/mirroring patterns while the three remaining alternatives indicate hopping of 1/2, +1/4, and −1/4 of the hopping bandwidth. Hopping according to information in the scheduling grant for the case of a cell bandwidth corresponding to 50 resource blocks is illustrated in Figure 17.24. In the first subframe, the scheduling grant indicates a hop of one-half the hopping bandwidth. In the second subframe, the grant indicates a hop of one-quarter the hopping bandwidth (equivalent to a negative hop of three quarters the hopping bandwidth). Finally, in the third subframe, the grant indicates a negative hop of one-quarter the hopping bandwidth.

18

LTE access procedures

The previous chapters have described the LTE uplink and downlink transmission schemes. However, prior to transmission of data, the mobile terminal needs to connect to the network. This chapter describes the procedures necessary for a terminal to be able to access an LTE-based network.

18.1 Acquisition and cell search

Before an LTE terminal can communicate with an LTE network it has to do the following:

- Find and acquire synchronization to a cell within the network.
- Receive and decode the information, also referred to as the *cell system information*, needed to communicate with and operate properly within the cell.

Once the system information has been correctly decoded, the terminal can access the cell by means of the so-called *random-access procedure*.

The first of these steps, often simply referred to as *cell search*, is discussed in this section. The next section then discusses, in more details, the means by which the network provides the cell system information. Finally, the LTE random-access procedure is described in Section 18.3.

18.1.1 Overview of LTE cell search

A terminal does not only need to carry out cell search at power-up, that is, when initially accessing the system. Rather, to support mobility, terminals need to continuously search for, synchronize to, and estimate the reception quality of neighbor cells. The reception quality of the neighbor cells, in relation to the reception

quality of the current cell, is then evaluated to conclude if a *handover* (for terminals in connected mode) or *cell re-selection* (for terminals in idle mode) should
be carried out.

LTE cell search consists of the following basic parts:

- Acquire frequency and symbol synchronization to a cell.
- Acquire frame timing of the cell, that is, determine the start of the downlink
  frame.
- Determine the physical-layer cell identity of the cell.

As already mentioned (e.g., in Chapter 16), there are 504 different physical-
layer cell identities defined for LTE, where each cell identity corresponds to one
specific downlink reference-signal sequence. The set of physical-layer cell identities is further divided into 168 *cell-identity groups*, with three cell identities
within each group.

To assist the cell search, two special signals are transmitted on the LTE
downlink, the *Primary Synchronization Signal* (PSS) and the *Secondary
Synchronization Signal* (SSS). Although having the same detailed structure, the
time-domain positions of the synchronization signals within the frame differ
somewhat depending on if the cell is operating in FDD or TDD.

- In case of FDD (upper part of Figure 18.1), the PSS is transmitted within
  the last symbol of the first slot of subframes 0 and 5, while the SSS is
  transmitted within the second last symbol of the same slot (i.e., just prior to
  the PSS).

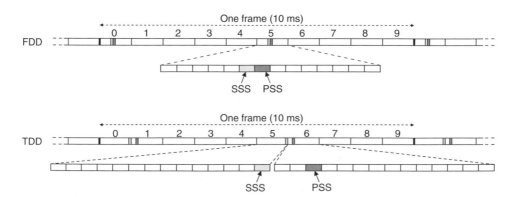

Figure 18.1   *Time-domain positions of PSSs in case of FDD and TDD.*

- In case of TDD (lower part of Figure 18.1), the PSS is transmitted within the third symbol of subframe 1 and 6 (i.e., within the DwPTS), while the SSS is transmitted in the last symbol of subframes 0 and 5 (i.e., three symbols ahead of the PSS).

There are several reasons for the difference in the positions of the synchronization signals in case of FDD and TDD. For example, the difference allows for the detection of the duplex scheme used on a carrier if this is not known in advance.

Within one cell, the two PSS within a frame are identical. Furthermore, the PSS of a cell can take three different values depending on the physical-layer cell identity of the cell. More specifically, the three cell identities within a cell-identity group always correspond to different PSS. Thus, once the terminal has detected and identified the PSS of the cell, it has found the following:

- 5 ms timing of the cell and thus also the position of the SSS which has a fixed offset relative to the PSS.
- the cell identity within the cell-identity group. However, the terminal has not yet determined the cell-identity group itself, that is, the number of possible cell identities has been reduced from 504 to 168.

Thus, from the SSS, the position of which is known once the PSS has been detected, the terminal should find the following:

- frame timing (two different alternatives given the found position of the PSS),
- the cell-identity group (168 alternatives).

Furthermore, it should be possible for a terminal to do this from the reception of one single SSS. The reason is that, for example, in the case when the terminal is searching for cells on other carriers, the search window may not be sufficiently large to cover more than one SSS.

To enable this, each SSS can take 168 different values corresponding to the 168 different cell-identity groups. Furthermore, the set of values valid for the two SSS within a frame (SSS_1 in subframe 0 and SSS_2 in subframe 5) are different, implying that, from the detection of a single SSS, the terminal can determine whether SSS_1 or SSS_2 has been detected and thus determine frame timing.

Once the terminal has acquired frame timing and the physical-layer cell identity, it has identified the cell-specific reference signal and can begin channel estimation. It can then decode the BCH transport channel which carries the most basic set of system information.

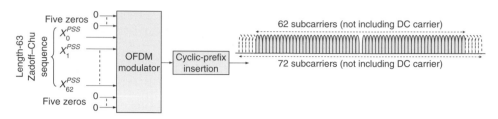

Figure 18.2   *Definition and structure of PSS.*

18.1.2   PSS structure

On a more detailed level, the three PSSs are three length-63 Zadoff–Chu (ZC) sequences extended with five zeros at the edges and mapped to the center 73 subcarriers, as illustrated in Figure 18.2. It should be noted though that the center subcarrier is actually not transmitted as it coincides with the DC subcarrier. Thus only 62 elements of the length-63 ZC sequences are actually transmitted (element X_{32}^{PSS} is not transmitted).

The PSS thus occupies 72 resource elements (not including the DC carrier) in subframes 0 and 5 (FDD) and subframes 1 and 6 (TDD). These resource elements are then not available for transmission of DL-SCH.

18.1.3   SSS structure

Similar to PSS, the SSS occupies the center 72 resource elements (not including the DC carrier) in subframes 0 and 5 (for both FDD and TDD). As described above, the SSS should be designed so that

- The two SSS (SSS_1 in subframe 0 and SSS_2 in subframe 5) should take their values from sets of 168 possible values corresponding to the 168 different cell-identity groups.
- The set of values applicable for SSS_2 should be different from the set of values applicable for SSS_1 to allow for frame-timing detection from the reception of a single SSS.

The structure of the two SSS is illustrated in Figure 18.3. SSS_1 is based on the frequency interleaving of two length-31 m-sequences X and Y, each of which can take 31 different values (actually 31 different shifts of the same *m*-sequence). Within a cell, SSS_2 is based on exactly the same two sequences as SSS_1. However, the two sequences have been swapped in the frequency domain as outlined in Figure 18.3. The set of valid combinations of X and Y for SSS_1 have then

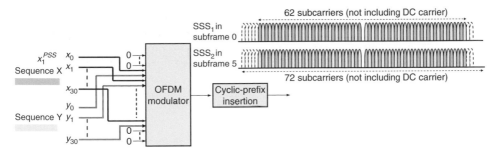

Figure 18.3 *Definition and structure of SSS.*

been selected so that a swapping of the two sequences in the frequency domain is not a valid combination for SSS_1. Thus the above requirements are fulfilled:

- The set of valid combinations of X and Y for SSS_1 (as well as for SSS_2) are 168, allowing for detection of the physical-layer cell identity.
- As the sequences X and Y are swapped between SSS_1 and SSS_2, frame timing can be found.

18.2 System information

By means of the basic cell-search procedure described in Section 18.1, a terminal synchronizes to a cell, acquires the physical-layer identity of the cell, and detects the cell frame timing. Once this has been achieved, the terminal has to acquire the cell *system information*. This is information that is repeatedly broadcast by the network and which needs to be acquired by terminals in order for them to be able access and, in general, operate properly within the network and within a specific cell. The system information includes, among other things, information about the downlink and uplink cell bandwidths, the uplink/downlink configuration in case of TDD, detailed parameters related to random-access transmission and uplink power control, etc.

In LTE, system information is delivered by two different mechanisms relying on two different transport channels:

- A limited amount of system information, corresponding to the so-called *Master Information Block* (MIB), is transmitted using the BCH.
- The main part of the system information, corresponding to different so-called *System Information Blocks* (SIBs), is transmitted using the downlink shared channel (DL-SCH).

It should be noted that system information in both the MIB and the SIBs correspond to the BCCH logical channel. Thus, as also illustrated in Figure 15.3 in Section 15.2, BCCH can be mapped to both BCH and DL-SCH depending on the exact BCCH information.

18.2.1 MIB and BCH transmission

As mentioned above, the MIB transmitted using BCH consists of a very limited amount of system information, mainly such information that is absolutely needed for a terminal to be able to read the remaining system information provided using DL-SCH. More specifically, the MIB includes the following information:

- *Information about the downlink cell bandwidth.* Four bits are available within the MIB to indicate the downlink bandwidth. Thus up to 16 different bandwidths, measured in number of resource blocks, can be defined for each frequency band.
- *Information about the PHICH configuration of the cell.* As mentioned in Section 16.4.2, the terminal must know the PHICH configuration to be able to receive the L1/L2 control signaling on PDCCH which, in turn, is needed to receive DL-SCH. Thus information about the PHICH configuration (three bits) is included in the MIB, that is, transmitted using BCH which can be received and decoded without first receiving any PDCCH.
- *The System Frame Number* (SFN) *or, more* exactly, all bits except the two least significant bits of the SFN are included in the MIB. As described below, the terminal can indirectly acquire the two least significant bits of the SFN from the BCH decoding.

BCH physical-layer processing, such as channel coding and resource mapping, differs quite substantially from the corresponding processing and mapping for DL-SCH as outlined in Chapter 16.

As can be seen in Figure 18.4, one BCH transport block, corresponding to the MIB, is transmitted once in every 40 ms. The BCH Transmissions Time Interval (TTI) thus equals 40 ms.

The BCH relies on a 16-bit CRC, in contrast to a 24-bit CRC used for all other downlink transport channels. The reason for the shorter BCH CRC is to reduce the relative CRC overhead, having the relatively small BCH transport-block size in mind.

BCH channel coding is based on the same rate-1/3 tail-biting convolutional code as is used for the PDCCH control channel. The reason for using convolutional

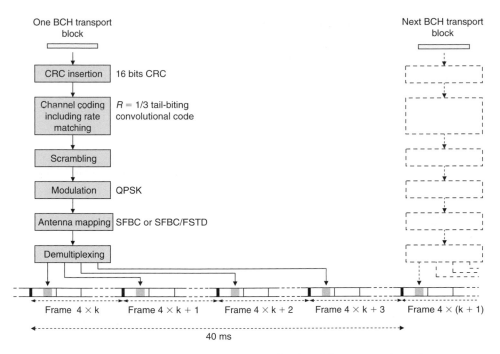

Figure 18.4 *Channel coding and subframe mapping for the BCH transport channel.*

coding for BCH, rather than the Turbo code used for all other transport channels, is the small size of the BCH transport block. With such small blocks, tail-biting convolutional coding actually outperforms Turbo coding. The channel coding is followed by rate matching, in practice repetition of the coded bits, and bit-level scrambling. *Quadrature Phase-Shift Keying* (QPSK) modulation is then applied to the coded and scrambled BCH transport block.

BCH multi-antenna transmission is limited to transmit diversity, that is, SFBC in case of two antenna ports and combined SFBC/FSTD in case of four antenna ports. Actually, if two antenna ports are available within the cell, SFBC *must* be used for BCH. Similarly, if four antenna ports are available, combined SFBC/FSTD *must* be used. Thus, by blindly detecting what transmit-diversity scheme is used for BCH, a terminal can, indirectly, determine the number of cell-specific antenna ports within the cell and also the transmit-diversity scheme used for the L1/L2 control signaling.

As can also be seen from Figure 18.4, the coded BCH transport block is mapped to the first subframe of each frame in four consecutive frames. However, as can be seen in Figure 18.5 and in contrast to other downlink transport channels, the BCH is not mapped on a resource-block basis. Instead, the BCH is transmitted

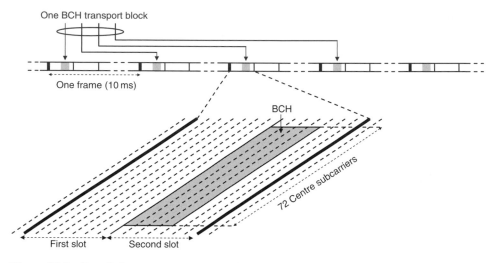

Figure 18.5   *Detailed resource mapping for the BCH transport channel.*

within the first four OFDM symbols of the second slot of subframe 0 and only over the 72 center subcarriers.[1] Thus, in case of FDD, BCH follows immediately after the PSS and SSS in subframe 0. The corresponding resource elements are then not available for DL-SCH transmission.

The reason for limiting the BCH transmission to the 72 center subcarriers, regardless of the cell bandwidth, is that a terminal may not know the downlink cell bandwidth when receiving BCH. Thus, when first receiving BCH of a cell, the terminal can assume a cell bandwidth equal to the minimum possible downlink bandwidth (i.e., six resource blocks corresponding to 72 subcarriers). From the decoded MIB, the terminal is then informed about the actual downlink cell bandwidth and can adjust the receiver bandwidth accordingly.

Clearly, the total number of resource elements to which the coded BCH is mapped is very large compared to the size of the BCH transport block, implying extensive repetition coding or, equivalently, massive processing gain for the BCH transmission. Such large processing gain is needed as it should be possible to receive and correctly decode the BCH also by terminals in neighbor cells, implying potentially very low receiver *Signal-to-Interference-and-Noise Ratio* (SINR) when decoding the BCH. At the same time, many terminals will receive BCH in much better channel conditions. Such terminals then do not need to receive the full set of four subframes over which a BCH transport block is transmitted

---

[1] Not including the DC carrier.

to acquire sufficient energy for correct decoding of the transport block. Instead, already by receiving only a few or perhaps only a single subframe, the BCH transport block may be decodable.

From the initial cell search, the terminal has found only the cell frame timing. Thus, when receiving BCH, the terminal does not know to what set of four subframes a certain BCH transport block is mapped to. Instead, a terminal must try to decode the BCH at four possible timing positions. Depending on which decoding is successful, indicated by a correct CRC check, the terminal can implicitly determine 40 ms timing or, equivalently, the two least significant bits of the SFN.[2] This is the reason these bits do not need to be explicitly included in the MIB.

18.2.2 System-Information Blocks

As already mentioned, the MIB on the BCH only includes a very limited part of the system information. The main part of the system information is instead included in different *System Information Blocks* (SIBs) that are transmitted using the DL-SCH. The presence of system information on DL-SCH in a subframe is indicated by the transmission of a corresponding PDCCH marked with a special *System Information RNTI* (SI-RNTI). Similar to the PDCCH providing the scheduling assignment for 'normal' DL-SCH transmission, this PDCCH also indicates the transport format and physical resource (set of resource blocks) used for the system-information transmission.

LTE defines at least eight different SIBs which are characterized by the type of information that is included within them:

- SIB1 includes information mainly related to whether a terminal is allowed to camp on the cell. This includes information about the operator/operators of the cell, if there are restrictions with regards to what users may access the cell, etc. SIB1 also includes information about the allocation of subframes to uplink/downlink and configuration of the special subframe in case of TDD. Finally, SIB1 includes information about the time-domain scheduling of the remaining SIBs (SIB2 and beyond).
- SIB2 includes information that terminals need in order to be able to access the cell. This includes information about the uplink cell bandwidth, random-access parameters, and parameters related to uplink power control.
- SIB3 mainly includes information related to cell-reselection.

---

[2] BCH scrambling is defined with 40 ms periodicity, hence even if the terminal successfully decodes the BCH after observing only a single transmission instant, it can determine the 40 ms timing.

- SIB4–SIB8 include neighbor-cell-related information, including information related to neighbor cells on the same carrier, neighbor cells on different carriers, and neighbor non-LTE cells, such as WCDMA/HSPA, GSM, and CDMA2000 cells.

Similar to the MIB, the SIBs are broadcasted repeatedly. How often a certain SIB needs to be transmitted depends on how quickly terminals need to acquire the corresponding system information when entering the cell. In general, a lower-order SIB is more time critical and is thus transmitted more often compared to a higher-order SIB. SIB1 is transmitted every 80 ms whereas the transmission period for the higher-order SIBs is flexible and can be different for different networks.

The SIBs represent the basic system information to be transmitted. The different SIBs are then mapped onto different *System Information messages* (SIs), which corresponds to the actual transport blocks to be transmitted on DL-SCH. SIB1 is always mapped, by itself, onto the first system-information message SI-1[3] whereas the remaining SIBs may be group-wise multiplexed onto the same SI subject to the following constraints:

- The SIBs mapped to the same SI must, obviously, have the same transmission period. Thus, as an example, two SIBs with a transmission period of 320 ms can be mapped to the same SI, whereas a SIB with a transmission period of 160 ms must be mapped to a different SI.
- The total number of information bits that is mapped to a single SI must not exceed what is possible to transmit within a transport block.

It should be noted that the transmission period for a given SIB might be different in different networks. For example, different operators may have different requirements on with what period different types of neighbor-cell information need to be transmitted. Furthermore, the amount of information that can fit into a single transport block very much depends on the exact deployment situation such as cell bandwidth, cell size, and so on.

Thus, in general, the SIB-to-SI mapping for SIBs beyond SIB1 is flexible and may be different for different networks or even within a network. An example of a SIB-to-SI mapping is illustrated in Figure 18.6. In this case, SIB2 is mapped to SI-2 with a transmission period of 160 ms. SIB3 and SIB4 are multiplexed into SI-3 with a transmission period of 320 ms whereas SIB5, that also requires a transmission period of 320 ms is mapped to a separate SI (SI-4). Finally, SIB6, SIB7, and SIB8 are multiplexed into SI-5 with a transmission period of 640 ms.

---

[3] Strictly speaking, as SIB1 is not multiplexed with any other SIBs, it is not even said to be mapped to an SI. Rather, SIB1 in itself directly corresponds to the transport block.

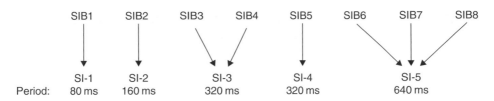

Figure 18.6 *Example of mapping of SIBs to SIs.*

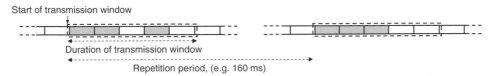

Figure 18.7 *Transmission window for the transmission of an SI.*

Information about the detailed SIB-to-SI mapping, as well as the transmission period of the different SIs, is provided in SIB1.

Regarding the more detailed transmission of the different system-information messages there is a difference between the transmission of SI-1, corresponding to SIB1, and the transmission of the remaining SIs.

The transmission of SI-1 has only a limited flexibility. More specifically, SI-1 is always transmitted within subframe #5. However, the bandwidth or, in general, the set of resource blocks, over which SI-1 is transmitted, as well as other aspects of the transport format may vary and is signaled on the associated PDCCH.

For the remaining SIs, the scheduling on DL-SCH is more flexible in the sense that each SI can, in principle, be transmitted in any subframe within *time windows* with well-defined starting points and durations. The starting point and duration of the time window of each SI is provided in SIB-1. It should be noted that an SI does not need to be transmitted on consecutive subframes within the time window, as is also illustrated in Figure 18.7. Within the time window, the presence of system information in a subframe is indicated by the SI-RNTI on PDCCH which also provides the frequency-domain scheduling as well as other parameters related to the system-information transmission.

Different SIs have different non-overlapping time windows. Thus, a terminal knows what SI is being received without the need for any specific identifier for each SI.

In case of a relatively small SI and a relatively large system bandwidth, a single subframe may be sufficient for the transmission of the SI. In other cases, multiple subframes may be needed for the transmission of a single SI. In the latter case, instead of segmenting each SI into sufficiently small blocks that are separately channel coded and transmitted in separate subframes, the complete SI is channel coded and mapped to multiple, not necessarily consecutive subframes.

Similar to the case of the BCH, terminals that are experiencing good channel conditions may then be able decode the complete SI after receiving only a subset of the subframes to which the coded SI is mapped, while terminals in bad positions need to receive more subframes for proper decoding of the SI. This approach has two benefits:

- Similar to BCH decoding, terminals in good positions need to receive fewer subframes, implying the possibility for reduced terminal power consumption.
- The use of larger code blocks in combination with Turbo coding leads to improved channel-coding gain.

18.3   Random access

A fundamental requirement for any cellular system is the possibility for the terminal to request a connection setup, commonly referred to as *random access*. In LTE, random access is used for several purposes, including:

- for initial access when establishing a radio link (moving from RRC_IDLE to RRC_CONNECTED; see Chapter 15 for a discussion on different terminal states);
- to re-establish a radio link after radio link failure;
- for handover when uplink synchronization needs to be established to the new cell;
- to establish uplink synchronization if uplink or downlink data arrives when the terminal is in RRC_CONNECTED and the uplink is not synchronized;
- as a scheduling request if no dedicated scheduling-request resources have been configured on PUCCH (see Chapter 19 for a discussion on uplink scheduling procedures).

Establishment of uplink synchronization is the main objective for all the cases above; when establishing an initial radio link (i.e., when moving from RRC_IDLE to RRC_CONNECTED), the random-access procedure also serves the purpose of assigning a unique identity, the C-RNTI, to the terminal.

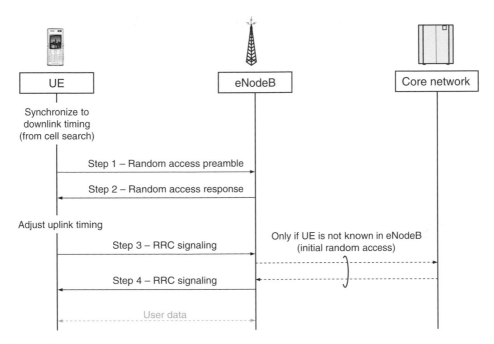

Figure 18.8 *Overview of the random-access procedure.*

The basis for random access is a contention-based procedure, illustrated in Figure 18.8, and consists of four steps:

1. The first step consists of transmission of a random-access preamble, allowing the eNodeB to estimate the transmission timing of the terminal. Uplink synchronization is necessary as the terminal otherwise cannot transmit any uplink data.
2. The second step consists of the network transmitting a timing advance command to adjust the terminal transmit timing, based on the timing estimate in the first step. In addition to establishing uplink synchronization, the second step also assigns uplink resources to the terminal to be used in the third step in the random-access procedure.
3. The third step consists of transmission of the mobile-terminal identity to the network using the UL-SCH similar to normal scheduled data. The exact content of this signaling depends on the state of the terminal, in particular whether it is previously known to the network or not.
4. The fourth and final step consists of transmission of a contention-resolution message from the network to the terminal on the DL-SCH. This step also resolves any contention due to multiple terminals trying to access the system using the same random-access resource.

Only the first step uses physical-layer processing specifically designed for random access. The last three steps utilize the same physical-layer processing as used for normal uplink and downlink data transmission. In the following, each of these steps is described in more detail.

Additionally, for handover purposes it is possible to use the random-access mechanism in a contention-free manner as described further below. In this case only the first two steps of the procedure are used as there is no need for contention resolution in a contention-free scheme.

18.3.1   Step 1: Random-access preamble transmission

The first step in the random-access procedure is the transmission of a random-access preamble. The main purpose of the preamble transmission is to indicate to the base station the presence of a random-access attempt and to allow the base station to estimate the delay between the eNodeB and the terminal. The delay estimate will be used in the second step to adjust the uplink timing.

The time-frequency resource on which the random-access preamble is transmitted upon is known as the *Physical Random Access Channel* (PRACH). The network broadcasts information to all terminals in which time-frequency resources random-access preamble transmission is allowed (i.e., the PRACH resources, in SIB-2). As part of the first step of the random-access procedure, the terminal selects one preamble to transmit on the PRACH.

In each cell, there are 64 preamble sequences available. Two subsets of the 64 sequences are defined as illustrated in Figure 18.9, where the set of sequences in each subset is signaled as part of the system information. When performing a (contention-based) random-access attempt, the terminal at random selects one sequence in one of the subsets. As long as no other terminal is performing a random-access attempt using the same sequence at the same time instant, no collisions will occur and the attempt will, with a high likelihood, be detected by the eNodeB.

The subset to select the preamble sequence from is given by the amount of data the terminal would like to transmit on the UL-SCH in the third random-access

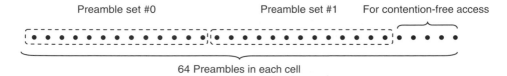

Figure 18.9   *Preamble subsets.*

step. Hence, from the preamble the terminal used, the eNodeB, will get some guidance on the amount of uplink resources to be granted to the terminal.

If the terminal has been requested to perform a contention-free random access, for example, for handover to a new cell, the preamble to use is explicitly indicated from the eNodeB. To avoid collisions, the eNodeB should preferably select the contention-free preamble from sequences outside the two subsets used for contention-based random access.

18.3.1.1 PRACH time-frequency resources

In the frequency domain, the PRACH resource, illustrated in Figure 18.10, has a bandwidth corresponding to six resource blocks (1.08 MHz). This nicely matches the smallest uplink cell bandwidth of six resource blocks in which LTE can operate. Hence, the same random-access preamble structure can be used, regardless of the transmission bandwidth in the cell.

In the time domain, the length of the preamble region depends on configured preamble as will be discussed further below. The basic random-access resource is 1 ms in duration, but there is also the possibility to configure longer preambles. Also, note that the eNodeB uplink scheduler in principle can reserve an arbitrary long random-access region by simply avoiding scheduling terminals in multiple subsequent subframes.

Typically, the eNodeB avoids scheduling any uplink transmissions in the time-frequency resources used for random access. This avoids interference between UL-SCH transmissions and random-access attempts from different terminals. The random-access preamble is said to be *orthogonal* to user data, unlike WCDMA where uplink data transmission and random-access attempts share the same resources. However, from a specification perspective, nothing prevents the uplink scheduler to schedule transmissions in the random-access region. Hybrid-ARQ retransmissions are an example of this; synchronous non-adaptive hybrid-ARQ retransmissions may overlap with the random-access region and it is up to the implementation to handle this, either by moving the retransmissions in the frequency domain as discussed in Chapter 19 or by handling the interference at the eNodeB receiver.

For FDD, there is at most one random-access region per subframe, that is, multiple random-access attempts are not multiplexed in the frequency domain. From a delay perspective, it is better to spread out the random-access opportunities in the time domain to minimize the average waiting time before a random-access attempt can be initialized. For TDD, multiple random-access regions can be

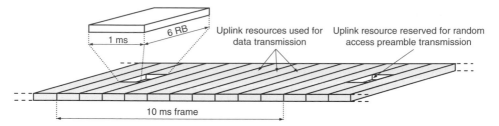

Figure 18.10 *Principal illustration of random-access-preamble transmission.*

configured in a single subframe. The reason is the smaller number of uplink subframes per radio frame in TDD and to maintain the same random-access capacity, frequency-domain multiplexing is sometimes necessary. The number of random-access regions is configurable and can vary from one per 20 to 1 ms for FDD; for TDD up to six attempts per 10 ms radio frame can be configured.

18.3.1.2 Preamble structure and sequence selection

The preamble consists of two parts:

- preamble sequence
- cyclic prefix.

Furthermore, the preamble transmission uses a guard period to handle the timing uncertainty. Prior to starting the random-access procedure, the terminal has obtained downlink synchronization from the cell-search procedure. However, as uplink synchronization has not yet been established prior to random access, there is, as already discussed, an uncertainty in the uplink timing[4] as the location of the terminal in the cell is not known. The larger the uplink timing uncertainty, the larger is the cell size and amounts to 6.7 μs/km. To account for the timing uncertainty and to avoid interference with subsequent subframes not used for random access, a guard time is used as part of the preamble transmission, that is the length of the actual preamble is shorter than 1 ms.

Including a cyclic prefix as part of the preamble is beneficial as it allows for frequency-domain processing at the base station (discussed further below), which can be advantageous from a complexity perspective. Preferably, the length of the cyclic prefix is approximately equal to the length of the guard period. With a preamble sequence length of approximately 0.8 ms, there is 0.1 ms cyclic prefix and 0.1 ms guard time. This allows for cell sizes up to 15 km and is the typical

---

[4]The start of an uplink frame at the terminal is defined relative to the start of a downlink frame received at the terminal.

random-access configuration, configuration 0 in Figure 18.11. To handle larger cells, where the timing uncertainty is larger, preamble configurations 1–3 can be used. Some of these configurations also support a longer preamble sequence to increase the preamble energy at the detector, which can be beneficial in larger cells. The preamble configuration used in a cell is signaled as part of the system information. Finally, note that guard times larger than those in Figure 18.11 can easily be created by not scheduling any uplink transmissions in the subframe following the random-access resource.

The preamble formats in Figure 18.11 are applicable to both FDD and TDD. However, for TDD, there is an additional fourth preamble configuration for random access. In this configuration, the random-access preamble is transmitted in the UpPTS field of the special subframe instead of in a normal subframe. Since this field is at most two DFTS-OFDM symbols long, the preamble and the possible guard time are substantially shorter than the preamble formats described above. Hence, format four is applicable to very small cells only. The location of the UpPTS, next to the downlink-to-uplink switch for TDD, also implies that the interference from distant base stations may interfere with this short random-access format and limits its use to smaller cell sizes.

18.3.1.3 PRACH power setting
The basis for setting the transmission power of the random-access preamble is a downlink pathloss estimate obtained from measuring on the cell-specific reference signals. From this pathloss estimate, the initial PRACH transmission power is obtained by adding a configurable offset.

The LTE random-access mechanism allows power ramping where the actual PRACH transmission power is increased for each unsuccessful random-access attempt. For the first attempt, the PRACH transmission power is set to the initial PRACH power. In most cases, this is sufficient for the random-access attempts to be successful. However, if the random-access attempt fails (random-access failures are detected at the second of fourth random-access step as described in the following sections), the PRACH transmission power for the next attempt is increased by a configurable step size to increase the likelihood of the next attempt to be successful.

A similar power-ramping procedure is defined for random access in WCDMA. However, the reasons are somewhat different. In WCDMA, random access is non-orthogonal to data transmissions from other terminals. Hence, to control the random-access-to-data interference in WCDMA, power ramping is used and several preamble transmission attempts are typically required. Hence, although

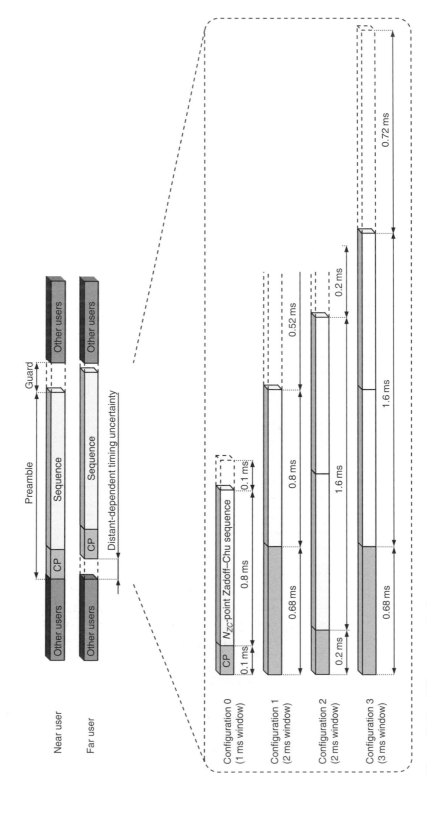

Figure 18.11   *Different preamble formats.*

power ramping efficiently handles the interference problem, it may also cause delays in the random-access procedure.

LTE however allows for orthogonality between preamble and data transmissions from different terminals. Hence, the need for power ramping to control intra-cell interference is significantly smaller and in many cases the transmission power is set such that the first random-access attempt with a high likelihood is successful. This is beneficial from a delay perspective.

18.3.1.4 Preamble sequence generation

The preamble sequences are generated from cyclic shifts of root Zadoff-Chu sequences [131]. Zadoff-Chu sequences are also used for creating the uplink reference signals as described in Chapter 17, where the structure of those sequences is described. From each root Zadoff-Chu sequence, $X_{ZC}^{(u)}(k)$, $\left\lfloor N_{ZC}/N_{CS} \right\rfloor$, cyclically shifted[5] sequences are obtained by cyclic shifts of N_{CS} each, where N_{ZC} is the length of the root Zadoff-Chu sequence. The generation of the random-access preamble is illustrated in Figure 18.12. Although the figure illustrates generation in the time domain, frequency-domain generation can equally well be used in an implementation.

Cyclically shifted Zadoff-Chu sequences possess several attractive properties. The amplitude of the sequences is constant, which ensures efficient power amplifier utilization and maintains the low PAR properties of the single-carrier uplink. The sequences also have ideal cyclic auto-correlation, which is important for obtaining an accurate timing estimation at the eNodeB. Finally, the cross-correlation between different preambles based on cyclic shifts of the same Zadoff-Chu root sequence is zero at the receiver as long as the cyclic shift N_{CS} used when generating the preambles is larger than the maximum round-trip propagation time in the cell plus the maximum delay spread of the channel. Therefore, due to the ideal cross-correlation property, there is no intra-cell interference from multiple random-access attempts using preambles derived from the same Zadoff-Chu root sequence.

To handle different cell sizes, the cyclic shift N_{CS} is signaled as part of the system information. Thus, in smaller cells, a small cyclic shift can be configured, resulting in a larger number of cyclically shifted sequences being generated from each root sequence. In cell sizes below 1.5 km, all 64 preambles can be generated from a single root sequence. In larger cells, a larger cyclic shift needs to be configured and to generate the 64 preamble sequences, multiple root Zadoff-Chu sequences must be used in the cell. Although the larger number of root sequence is not a problem in itself, the zero cross-correlation property only

---

[5] The cyclic shift is in the time domain. Similar to the uplink reference signals and control signaling, this can equivalently be described as a phase rotation.

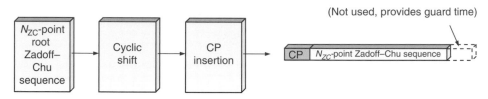

Figure 18.12 *Random-access preamble generation.*

holds between shifts of the *same* root sequence and from an interference perspective it is therefore beneficial to use as few root sequences as possible.

Reception of the random-access preamble is discussed further below, but in principle it is based on correlation of the received signal with the root Zadoff-Chu sequences. One disadvantage of Zadoff-Chu sequences is the difficulties in separating a frequency offset from the distance-dependent delay. A frequency offset results in an additional correlation peak in the time-domain; a correlation peak that corresponds to a spurious terminal-to-base-station distance. In addition, the true correlation peak is attenuated. At low frequency-offsets, this effect is small and the detection performance is hardly affected. However, at high Doppler frequencies, the spurious correlation peak can be larger than the true peak. This results in erroneous detection; the correct preamble may not be detected or the delay estimate may be incorrect.

To avoid the ambiguities from spurious correlation peaks, the set of preamble sequences generated from each root sequence can be restricted. Restrictions imply that only some of the sequences possible to generate from a root sequence are used to define random-access preambles. Whether restriction should be applied or not to the preamble generation is signaled as part of the system information. The location of the spurious correlation peak relative to the 'true' peak depends on the root sequence and hence different restrictions have to be applied to different root sequences. The restrictions to apply are broadcasted as part of the system information in the cell.

18.3.1.5 Preamble detection

The base-station processing is implementation specific, but due to the cyclic prefix included in the preamble, low-complexity frequency-domain processing is possible. An example hereof is shown in Figure 18.13. Samples taken in a time-domain window are collected and converted into the frequency-domain representation using an FFT. The window length is 0.8 ms, which is equal to the

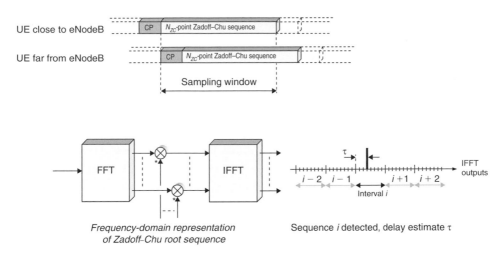

Figure 18.13 *Random-access preamble detection in the frequency domain.*

length of the ZC sequence without a cyclic prefix. This allows handling timing uncertainties up to 0.1 ms and matches the guard time defined for the basic preamble configuration.

The output of the FFT, representing the received signal in the frequency domain, is multiplied with the complex-conjugate frequency-domain representation of the root ZC sequence and the result is fed through an IFFT. By observing the IFFT outputs, it is possible to detect which of the shifts of the ZC root sequence has been transmitted and its delay. Basically, a peak of the IFFT output in interval *i* corresponds to the *i*th cyclically shifted sequence and the delay is given by the position of the peak within the interval. This frequency-domain implementation is computationally efficient and allows simultaneous detection of multiple random-access attempts using different cyclic shifted sequences generated from the same root ZC sequence; in case of multiple attempts there will simply be a peak in each of the corresponding intervals.

18.3.2 Step 2: Random-access response

In response to the detected random-access attempt, the eNodeB will, as the second step of the random-access procedure, transmit a message on the DL-SCH, containing:

- The index of the random-access preamble sequences the network detected and for which the response is valid.
- The timing correction calculated by the random-access preamble receiver.

- A scheduling grant, indicating resources the terminal shall use for the transmission of the message in the third step.
- A temporary identity, the TC-RNTI, used for further communication between the terminal and the network.

In case the network detected multiple random-access attempts (from different terminals), the individual response messages of multiple mobile terminals can be combined in a single transmission. Therefore, the response message is scheduled on the DL-SCH and indicated on a PDCCH using an identity reserved for random-access response, the RA-RNTI. All terminals, which have transmitted a preamble, monitor the L1/L2 control channels for random-access response within a configurable time window. The timing of the response message is not fixed in the specification in order to be able to respond to sufficiently many simultaneous accesses. It also provides some flexibility in the base-station implementation. If the terminal does not detect a random-access response within the time window, the attempt will be declared as failed and the procedure will repeat from the first step again, possibly with an increased preamble transmission power.

As long as the terminals that performed random access in the same resource used different preambles, no collision will occur and from the downlink signaling it is clear to which terminal(s) the information is related. However, there is a certain probability of contention, that is, multiple terminals using the same random-access preamble at the same time. In this case, multiple terminals will react upon the same downlink response message and a collision occurs. Resolving these collisions is part of the subsequent steps as discussed below. Contention is also one of the reasons why hybrid ARQ is not used for transmission of the random-access response. A terminal receiving a random-access response intended for another terminal will have incorrect uplink timing. If hybrid ARQ would be used, the timing of the hybrid-ARQ acknowledgement for such a terminal would be incorrect and may disturb uplink control signaling from other users.

Upon reception of the random-access response in the second step, the terminal will adjust its uplink transmission timing and continue to the third step. If a contention-free random access using a dedicated preamble is used, then this is the last step of the random-access procedure as there is no need to handle contention in this case. Furthermore, the terminal already has a unique identity allocated in the form of a C-RNTI.

18.3.3 *Step 3: Terminal identification*

After the second step, the uplink of the terminal is time synchronized. However, before user data can be transmitted to/from the terminal, a unique identity within the cell, the C-RNTI, must be assigned to the terminal. Depending on the terminal state, there may also be a need for additional message exchange for setting up the connection.

In the third step, the terminal transmits the necessary messages to the eNodeB using the UL-SCH resources assigned in the random-access response in the second step. Transmitting the uplink message in the same manner as scheduled uplink data instead of attaching it to the preamble in the first step is beneficial for several reasons. First, the amount of information transmitted in the absence of uplink synchronization should be minimized as the need for a large guard time makes such transmissions relatively costly. Secondly, the use of the 'normal' uplink transmission scheme for message transmission allows the grant size and modulation scheme to be adjusted to, for example, different radio conditions. Finally, it allows for hybrid ARQ with soft combining for the uplink message. The latter is an important aspect, especially in coverage-limited scenarios, as it allows for the use of one or several retransmissions to collect sufficient energy for the uplink signaling to ensure a sufficiently high probability of successful transmission. Note that RLC retransmissions are not used[6] for the uplink RRC signaling in step 3.

An important part of the uplink message is the inclusion of a terminal identity as this identity is used as part of the contention-resolution mechanism in the fourth step. In case the terminal is in the RRC_CONNECTED state, that is connected to a known cell and therefore has a C-RNTI assigned, this C-RNTI is used as the terminal identity in the uplink message.[7] Otherwise, a core-network terminal identifier is used and the eNodeB needs to involve the core network prior to responding to the uplink message in step 3.

UE-specific scrambling is, described in Chapter 17, used for transmission on UL-SCH. However, as the terminal may not yet have been allocated its final identity, the scrambling cannot be based on the C-RNTI. Instead, the temporary identity is used (TC-RNTI).

18.3.4 Step 4: Contention resolution

The last step in the random-access procedure consists of a downlink message for contention resolution. Note that, from the second step, multiple terminals

---

[6] This is known as RLC *transparent mode* as will be discussed in Chapter 19.
[7] The terminal identity is included as a MAC control element on the UL-SCH, see Chapter 19.

performing simultaneous random-access attempts using the same preamble sequence in the first step listen to the same response message in the second step and therefore have the same temporary identifier. Hence, in the fourth step, each terminal receiving the downlink message will compare the identity in the message with the identity they transmitted in the third step. Only a terminal which observes a match between the identity received in the fourth step and the identity transmitted as part of the third step will declare the random-access procedure successful. If the terminal has not yet been assigned a C-RNTI, the TC-RNTI from the second step is promoted to the C-RNTI; otherwise the terminal keeps its already assigned C-RNTI.

The contention-resolution message is transmitted on the DL-SCH, using the temporary identity from the second step for addressing the terminal on the L1/L2 control channel. Since uplink synchronization already has been established, hybrid ARQ is applied to the downlink signaling in this step. Terminals with a match between the identity they transmitted in the third step and the message received in the fourth step will also transmit a hybrid-ARQ acknowledgement in the uplink.

Terminals which do not find a match between the identity received in the fourth step and the respective identity transmitted as part of the third step are considered to have failed the random-access procedure and need to restart the procedure from the first step. Obviously, no hybrid-ARQ feedback is transmitted from these terminals. Furthermore, a terminal that has not received the downlink message in step four within a certain time from the transmission of the uplink message in step three will declare the random-access procedure as failed and need to restart from the first step.

18.4 Paging

Paging is used for network-initiated connection setup. An efficient paging procedure should allow the terminal to sleep with no receiver processing most of the time and to briefly wake up at predefined time intervals to monitor paging information from the network.

In WCDMA, a separate paging-indicator channel, monitored at predefined time instants, is used to indicate to the terminal that paging information is transmitted. As the paging indicator is significantly shorter than the duration of the paging information, this approach minimizes the time the terminal is awake.

In LTE, no separate paging-indicator channel is used as the potential power savings are very small due to the short duration of the L1/L2 control signaling, at

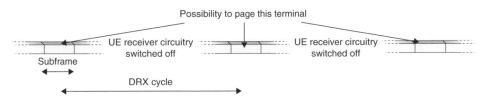

Figure 18.14 *DRX for paging.*

most three OFDM symbols as described in Chapter 16. Instead, the same mechanism as for 'normal' downlink data transmission on the DL-SCH is used and the mobile-terminal monitors the L1/L2 control signaling for downlink scheduling assignments. The DRX mechanism (see Chapter 19 for details) is used to allow the terminal to sleep most of the time and only briefly wake up to monitor the L1/L2 control signaling. If the terminal detects a group identity used for paging (the P-RNTI) when it wakes up, it will process the corresponding downlink paging message transmitted on the PCH. The paging message includes the identity of the terminal(s) being paged and a terminal not finding its identity will discard the received information and sleep according to the DRX cycle. Obviously, as the uplink timing is unknown during the DRX cycles, no hybrid-ARQ acknowledgements can be transmitted and consequently hybrid ARQ with soft combining is not used for paging messages.

The DRX cycle for paging is illustrated in Figure 18.14.

19

LTE transmission procedures

The LTE uplink and downlink transmission schemes have been described from a physical layer perspective in Chapters 16 and 17. In this chapter, some procedures and protocols required for data transmission, for example, retransmission protocols, scheduling, and uplink power control, will be described.

19.1 RLC and hybrid-ARQ protocol operation

In LTE, retransmissions of missing or erroneous data units are handled primarily by the hybrid-ARQ mechanism in the MAC layer, complemented by the retransmission functionality of the RLC protocol. The reasons for having a two-level retransmission structure can be found in the trade-off between fast and reliable feedback of the status reports. The hybrid-ARQ mechanism targets very fast retransmissions and, consequently, feedback on the decoding attempt is provided to the transmitter after each transmission. Although it is in principle possible to attain an arbitrarily low error probability of the hybrid-ARQ feedback, it comes at a cost in transmission power. Keeping the cost reasonable typically results in a feedback error rate of around 1% which results in a hybrid-ARQ residual error rate of a similar order. Such an error rate is in many cases far too high; high data rates with TCP may require virtually error-free delivery of packets to the TCP protocol layer. As an example, for sustainable data rates exceeding 100 Mbit/s, a packet-loss probability lower than 10^{-5} is required [76]. The reason is that TCP assumes packet errors to be due to congestion in the network. Any packet error therefore triggers the TCP congestion-avoidance mechanism with a corresponding decrease in data rate.

Compared to the hybrid-ARQ acknowledgments, the RLC status reports are transmitted relatively infrequently and thus the cost of obtaining a reliability of 10^{-5} or lower is relatively small. Hence, the combination of hybrid-ARQ and RLC attains a good combination of small roundtrip time and a modest feedback overhead where the two components complement each other. Thus, the two-level structure combines the best of two worlds – fast retransmissions due to the hybrid-ARQ mechanism and reliable packet delivery due to the RLC. Using two levels of retransmission protocols, RLC on top of hybrid-ARQ, is similar to HSPA described in Part III, partially for the same reasons. In HSPA, the two-layer structure is a consequence of the HSPA-extension to the WCDMA architecture with two types of radio-network nodes and the two protocols operate fairly independently. However, as the MAC and RLC protocol layers are located in the same network node for LTE, a much tighter interaction between the two protocols is possible in the LTE case. Hence, to some extent, the combination of the two can be viewed as *one* retransmission mechanism with *two* feedback channels.

In the following section, the principles behind the hybrid-ARQ and RLC protocols will be discussed in more detail.

19.1.1   *Hybrid-ARQ with soft combining*

The hybrid-ARQ functionality spans both the physical layer and the MAC layer; generation of different redundancy versions at the transmitter as well as the soft combining at the receiver are handled by the physical layer, while the hybrid-ARQ protocol is part of the MAC layer. Similar to HSPA, the LTE hybrid-ARQ protocol is based on a structure with multiple hybrid-ARQ processes. For each terminal, there is one hybrid-ARQ entity in the eNodeB and one in the terminal, where each hybrid-ARQ entity consists of a number of parallel hybrid-ARQ processes as illustrated in Figure 19.1. Upon receiving of a transport block for a certain hybrid-ARQ process, the receiver makes an attempt to decode the transport block and informs the transmitter about the outcome through a hybrid-ARQ acknowledgment, indicating whether the transport block was correctly decoded or not. In case of downlink spatial multiplexing, where two transport blocks are transmitted in parallel as described in Chapter 16, there are two parallel acknowledgments, one for each transport block. The details for the physical layer transmission of the downlink and uplink hybrid-ARQ acknowledgments were described in Chapters 16 and 17.

An important part of the hybrid-ARQ mechanism is the use of *soft combining*, which implies that the receiver combines the received signal from multiple transmission attempts. The principles of hybrid-ARQ with soft combining were outlined in Chapter 7. Clearly, the receiver needs to know when to

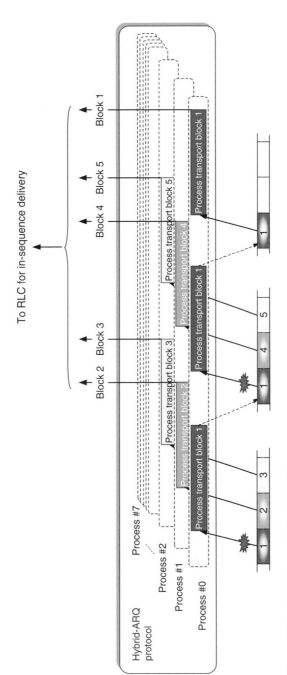

Figure 19.1 *Multiple parallel hybrid-ARQ processes.*

perform soft combining prior to decoding and when to clear the soft buffer, that is, the receiver needs to differentiate between the reception of an initial transmission (prior to which the soft buffer should be cleared) and the reception of a retransmission. Therefore, an explicit *new-data indicator* is included on the PDCCH.

In case of downlink data transmission, the new-data indicator is toggled for each new transport block, that is, it is in essence a single-bit sequence number. Upon reception of a downlink scheduling assignment, the terminal checks the new-data indicator to determine whether the current transmission should be soft combined with the received data currently in the soft buffer for the hybrid-ARQ process in question, or if the soft buffer should be cleared.

For uplink data transmission, there is also a new-data indicator transmitted on the downlink PDCCH. In this case it informs the terminal whether the scheduling grant relates to a retransmission or if an initial transmission of a new transport block is requested.

The use of multiple parallel hybrid-ARQ processes can result in data being delivered from the hybrid-ARQ mechanism out-of-sequence. For example, transport block 5 in Figure 19.1 was successfully decoded before transport block 1, which required two retransmissions. To resolve this, the RLC protocol includes an in-sequence-delivery mechanism as described in Section 18.1.2.

For LTE, as well as for HSPA, as discussed in earlier chapters, hybrid-ARQ protocols can be characterized as synchronous vs. asynchronous, related to the flexibility in the time domain, as well as adaptive vs. non-adaptive, related to the flexibility in the frequency domain.

- An *asynchronous* hybrid-ARQ protocol implies that retransmissions can occur at any time whereas a *synchronous* protocol implies that retransmissions occur at a fixed time after the previous transmission. The benefit of a synchronous protocol is that there is no need to explicitly signal the hybrid-ARQ process number as this information can be derived from the subframe number. On the other hand, an asynchronous protocol allows for more flexibility in the scheduling of retransmissions.
- An *adaptive* hybrid-ARQ protocol implies that the frequency location and possibly also the more detailed transmission format can be changed between retransmissions. A *non-adaptive protocol*, in contrast, implies that the retransmission must occur at the same frequency resources and with the same transmission format as the initial transmission.

In the case of LTE, asynchronous adaptive hybrid-ARQ is used for the downlink. For the uplink, synchronous hybrid-ARQ is used. Typically, the retransmissions are non-adaptive, but there is also a possibility to use adaptive retransmissions as a complement.

19.1.1.1 Downlink hybrid-ARQ

In the downlink, retransmissions are scheduled in the same way as new data, that is, they may occur at any time and at an arbitrary location within the downlink cell bandwidth. Hence, the downlink hybrid-ARQ protocol is *asynchronous* and *adaptive*. The support for asynchronous and adaptive hybrid-ARQ for the LTE downlink is motivated by the need to avoid collisions with, for example, transmission of system information and MBSFN subframes. Instead of dropping a retransmission that otherwise would collide with MBSFN subframes or transmission of system information, the eNodeB can move the retransmission in time and/or frequency to avoid the overlap in resources.

Support for soft combining is, as described in the introduction, provided through an explicit new-data indicator, toggled for each new transport block. In addition to the new-data indicator, hybrid-ARQ-related downlink control signaling consists of the hybrid-ARQ process number (three bits for FDD, four bits for TDD) and the redundancy version (two bits), both explicitly signaled in the scheduling assignment for each downlink transmission.

In case of downlink spatial multiplexing, there are, as already mentioned, two transport blocks transmitted in parallel to a terminal. To provide the possibility to retransmit only one of the transport blocks, which is beneficial as error events for the two transport blocks can be fairly independent, each transport block has its own separate new-data indicator and redundancy-version indication. However, there is no need to signal the process number separately as each process consists of two subprocesses in case of spatial multiplexing, or, expressed differently, once the process number for the first transport block is known, the process number for the second transport block is given implicitly.

Downlink transport blocks are acknowledged by transmitting one or two bits on the uplink as described in Chapter 17. In absence of spatial multiplexing, there is only a single transport block within a TTI and consequently only a single acknowledgment bit is required in response to this. However, the downlink transmission used spatial multiplexing, there are two transport blocks per TTI, each requiring its own hybrid-ARQ acknowledgment bit.

Obviously, the terminal should not transmit a hybrid-ARQ acknowledgment in response to reception of system information, paging message and other broadcast traffic. Hence, hybrid-ARQ acknowledgments are only sent in the uplink for 'normal' unicast transmission.

19.1.1.2   Uplink hybrid-ARQ

Shifting the focus to the uplink, a difference compared to the downlink case is the use of synchronous, non-adaptive operation as the basic principle of the hybrid-ARQ protocol. Hence, uplink retransmissions always occur at an *a priori* known subframe; in case of FDD operation uplink retransmissions occur eight subframes after the prior transmission attempt for the same hybrid-ARQ process. The set of resource blocks used for the retransmission is identical to the initial transmission. Thus, the only control signaling in the downlink is a single bit hybrid-ARQ acknowledgment, transmitted on the PHICH as described in Chapter 16. In case of a negative acknowledgment on the PHICH, the data is retransmitted.

Despite the fact that the basic mode of operation for the uplink is synchronous, non-adaptive hybrid-ARQ, there is also the possibility to operate the uplink hybrid-ARQ in a synchronous, *adaptive* manner, where the resource-block set and modulation-and-coding scheme for the retransmissions is changed. Although non-adaptive retransmissions are typically used due to the very low overhead in terms of downlink control signaling, adaptive retransmissions are sometimes needed to avoid fragmenting the uplink frequency resource or to avoid collisions with random-access resources. This is illustrated in Figure 19.2. A terminal is scheduled for an initial transmission in subframe n; a transmission that is not correctly received and consequently a retransmission is required in subframe $n + 8$ (assuming FDD, for TDD the timing obviously depends on the downlink–uplink allocation). With non-adaptive hybrid-ARQ, the retransmissions occupy the same part of the uplink spectrum as the initial transmission. Hence, in this example the spectrum is fragmented which limits the bandwidth available to another terminal. In subframe $n + 16$, an example of an adaptive retransmission is found; to make room for another terminal to be granted a large part of the uplink spectrum, the retransmission is moved in the frequency domain. It should be noted that the uplink hybrid-ARQ protocol is still synchronous, that is, a retransmission should always occur eight subframes after the previous transmission.

The support for both adaptive and non-adaptive hybrid-ARQ is realized by *not* flushing the transmission buffer when receiving a positive hybrid-ARQ acknowledgment on PHICH for a given hybrid-ARQ process. Instead, the actual control of whether data should be retransmitted or not is done by the new-data indicator included in the uplink scheduling grant sent on the PDCCH. The new-data indicator is toggled for each new transport block. If the new-data indicator is toggled, the terminal flushes the transmission buffer and transmits a new data packet. However, if the uplink grant does not request transmission of new data, the previous transport block is retransmitted. Hence, the actual retransmission request is not sent on the PHICH but on the PDCCH as part of the uplink grant. The negative hybrid-ARQ acknowledgment on the PHICH could instead be seen as a single-bit scheduling grant for retransmissions where the set of bits to transmit and all the resource information are known from the previous transmission attempt. A consequence of the above method of supporting both adaptive and non-adaptive hybrid-ARQ is that the PHICH and PDCCH related to the same uplink subframe have the same timing. If this were not the case, the complexity would increase as the terminal would not know whether to obey the PHICH or wait for a PDCCH overriding the PHICH.

As explained earlier, the new-data indicator is explicitly transmitted in the uplink grant. However, unlike the downlink case, the redundancy version is *not* explicitly signaled for each retransmission. With a single-bit acknowledgment on the PHICH, this is not possible. Instead, as the uplink hybrid-ARQ protocol is synchronous, the redundancy version follows a predefined pattern, starting with zero when the initial transmission is scheduled by the PDCCH. Whenever a retransmission is requested by a negative acknowledgment on the PHICH, the next redundancy version in the sequence is used. However, if a retransmission is explicitly scheduled by the PDCCH overriding the PHICH, there is a possibility to affect the redundancy version to use. Grants for retransmissions use the same format as ordinary grants (for initial transmissions). One of the information fields in an uplink grant is, as described in Chapter 16, the modulation-and-coding scheme. Of the 32 different combinations this five-bit field can take, three of them are reserved. Those three combinations represent different redundancy versions; hence, if one of these combinations is signaled as part of an uplink grant indicating a retransmission, the corresponding redundancy version is used for the transmission. The transport block size is already known from the initial transmission as it, by definition, cannot change between retransmission attempts. In Figure 19.2, the initial transmission in subframe n use the first redundancy version in sequence as the transport block size must be indicated for the initial transmission. The retransmission in subframe $n + 8$ uses the next redundancy version in the sequence, while the explicitly scheduled retransmission in subframe $n + 16$ can use any redundancy scheme as indicated on the PDCCH.

Whether to exploit the possibility to signal an arbitrary redundancy version in a retransmission scheduled by the PDCCH is a trade-off between

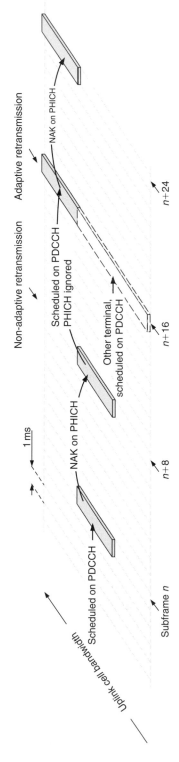

Figure 19.2 *Non-adaptive and adaptive hybrid-ARQ operation.*

incremental-redundancy gain and robustness. From an incremental-redundancy perspective, changing the redundancy value between retransmissions is typically beneficial to fully exploit the gain from incremental redundancy. However, as the modulation-and-coding scheme is normally not indicated for uplink retransmissions, as either the single-bit PHICH is used or the modulation-and-coding field is used to explicitly indicate a new redundancy version, it is implicitly assumed that the terminal did not miss the initial scheduling grant. If this is the case, it is necessary to explicitly indicate the modulation-and-coding scheme, which also implies that the first redundancy version in the sequence is used.

19.1.1.3 Hybrid-ARQ timing

Clearly, the receiver must know to which hybrid-ARQ process a received acknowledgment is associated. This is handled in the same way as for HSPA, that is, the timing of the hybrid-ARQ acknowledgment is used to associate the acknowledgment with a certain hybrid-ARQ process. Hence, there is a fixed timing relation between the reception of data in the downlink and transmission of the hybrid-ARQ acknowledgment in the uplink (and vice versa). From a latency perspective, the time between the reception of downlink data at the terminal and transmission of the hybrid-ARQ acknowledgment in the uplink should be as short as possible. At the same time, an unnecessarily short time would increase the demand on the terminal processing capacity and a trade-off between latency and implementation complexity is required. The situation is similar for uplink data transmissions. For LTE, this trade-off led to the decision to have eight hybrid-ARQ processes in both uplink and downlink for FDD. For TDD, the number of processes depends on the downlink–uplink allocation as discussed below.

Starting with the FDD case, illustrated in Figure 19.3, downlink data on the DL-SCH is transmitted to the terminal in subframe n and received by the terminal, after the propagation delay T_p, in subframe n. The terminal attempts to decode the received signal, possibly after soft combining with a previous transmission attempt, and transmits the hybrid-ARQ acknowledgment in uplink subframe $n + 4$ (note that the start of an uplink subframe at the terminal is offset by T_{TA} relative to the start of the corresponding downlink subframe at the terminal as a result of the timing-advance procedure described in Section 19.5). Upon reception of the hybrid-ARQ acknowledgment, the eNodeB can, if needed, retransmit the downlink data in subframe $n + 8$. Thus, eight hybrid-ARQ processes are used, that is, the hybrid-ARQ roundtrip time is 8 ms. Although the figure illustrates downlink data transmission, the same time relation applies for uplink data transmission, decoding in the eNodeB and transmission of the hybrid-ARQ.

In Figure 19.3 it is seen that the processing time available to the terminal, T_{UE}, depends on the value of the timing advance or, equivalently, on the

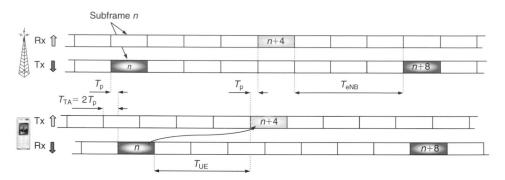

Figure 19.3 *Timing relation between downlink data in subframe n and uplink hybrid-ARQ acknowledgment in subframe n + 4 for FDD.*

terminal-to-base-station distance. As a terminal must operate at any distance up to the maximum size supported by the specifications, the terminal must be designed such that it can handle the worst-case scenario. With a timing advance of 0.67 ms, corresponding to the LTE requirement of 100 km as described in Chapter 13, there is approximately 2.3 ms left for the terminal processing, which is considered as reasonable trade-off between the processing requirements imposed on a terminal and the associated delays. As a comparison, the corresponding value for the HSPA downlink is 5 ms, despite the significantly lower data rates. The lower value for LTE is partially due to advancements in implementation technology and partially due to a more parallelization-friendly QPP interleaver in the turbo code as described in Chapter 16.

For the eNodeB, the processing time available, denoted as T_{eNB}, is 3 ms and thus in the same order as for the terminal. In case of downlink data transmission, the eNodeB performs scheduling of any retransmissions during this time and, for uplink data transmission, the time is used for decoding of the received signal. The timing budget is thus similar for the terminal and the eNodeB, which is motivated by the fact that, although the eNodeB typically have more processing power than a terminal, it also has to serve multiple terminals and to perform the scheduling operation.

Having the same number of hybrid-ARQ processes in uplink and downlink is beneficial for half-duplex FDD operation, discussed in Section 19.2.4. By proper scheduling, the uplink hybrid-ARQ transmission from the terminal will coincide with transmission of uplink data and the acknowledgments related to the reception of uplink data will be transmitted in the same subframe as the downlink data. Thus, using the same number of hybrid-ARQ processes in uplink and downlink results in a 50/50 split between transmission and reception for a half-duplex terminal.

Shifting to TDD operation, the time relation between the reception of data in a certain hybrid-ARQ process and the transmission of the hybrid-ARQ acknowledgment depends on the downlink–uplink allocation. Obviously, an uplink hybrid-ARQ acknowledgment can only be transmitted in an uplink subframe and a downlink acknowledgment only in a downlink subframe. The amount of processing time required in the terminal and the eNodeB remains the same though as the same turbo decoders are used and the scheduling decisions are similar. Therefore, for TDD the acknowledgment of a transport block in subframe n is transmitted in subframe $n + k$ where $k \geq 4$ and selected such that $n + k$ is an uplink subframe in case the acknowledgment is to be transmitted from the terminal and a downlink subframe in case the acknowledgment is transmitted from the eNodeB. The value of k depends on the downlink–uplink allocation as shown in Table 19.1 (see Section 16.1 for a description of the downlink–uplink allocations in different TDD configurations). The value of k is, as a consequence of the downlink–uplink allocation, in many cases larger than FDD, which uses $k = 4$. This also implies that the number of hybrid-ARQ processes used for TDD depends on the downlink–uplink allocation and can be larger compared to FDD. As seen in the table, for the downlink-heavy configurations 2, 3, 4, and 5, the number of downlink hybrid-ARQ processes is larger than for FDD. The reason is the limited number of uplink subframes resulting in a k in the relation above well beyond four for some subframes. A consequence is also the difference in roundtrip time between different hybrid-ARQ processes.

The downlink–uplink allocation for TDD has implications on the number of transport blocks to acknowledge in a single subframe. For FDD, there is always a one-to-one relation between uplink and downlink subframes and a subframe only need to carry acknowledgments for one subframe in the other direction. For TDD, in contrast, there is not necessarily a one-to-one relation between uplink and downlink subframes. This can be seen in the possible downlink–uplink allocations described in Chapter 16. Hence, for some configurations, there is, depending on the scheduling decisions, a need to acknowledge multiple downlink subframes in one uplink subframe as illustrated in Figure 19.4 (the illustration corresponds to the two entries shown in bold in Table 19.1). Therefore, the outcome of the decoding of downlink transport blocks from multiple downlink subframes can be combined into a single hybrid-ARQ acknowledgment transmitted in the uplink. Only if both of the downlink transmissions in subframes 0 and 3 in the example in Figure 19.4 are correctly decoded, a positive acknowledgment will be transmitted in uplink subframe 7. The figure also illustrates another property of TDD

Table 19.1 *Number of hybrid-ARQ processes and uplink acknowledgment timing k for different TDD configurations.*

| | | Processes | | Value of k upon reception of data in subframe | | | | | | | | | |
| | DL:UL | | | 0 | 1 | 2 | 3 | 4 | 5 | 6 | 7 | 8 | 9 |
Configuration	allocation	Uplink	Downlink										
0	2:3	7	4	4	6	–	–	–	4	6	–	–	–
1	3:2	4	7	7	6	–	–	4	7	6	–	–	4
2	4:1	2	10	**7**	6	–	**4**	8	7	6	–	4	8
3	7:3	3	9	4	11	–	–	–	7	6	6	5	5
4	8:2	2	12	12	11	–	–	8	7	7	6	5	4
5	9:1	1	15	12	11	–	9	8	7	6	5	4	13
6	5:5	6	6	7	7	–	–	–	7	7	–	–	5

mentioned earlier, namely the difference in hybrid-ARQ roundtrip time depending on the subframe in which the data was transmitted.

Combining acknowledgments related to multiple downlink transmissions into a single uplink message assumes that the terminal has not missed any of the scheduling assignments upon which the acknowledgment is based. Assume, as an example, that the eNodeB scheduled the terminal in two (subsequent) subframes, but the terminal missed the PDCCH transmission in the first of the two subframes and successfully decoded the data transmitted in the second subframe. Obviously, without any additional mechanism, the terminal will transmit an acknowledgment based on the assumption that it was scheduled in the second subframe only, while the eNodeB will interpret acknowledgment as the terminal successfully received both transmissions. To avoid such errors, the *downlink assignment index* (see Section 16.4.4) in the scheduling assignment on the PDCCH is used. The downlink assignment index in essence informs the terminal about the number of transmissions it should base the combined acknowledgment upon. If there is a mismatch between the assignment index and the number of transmissions the terminal received, the terminal concludes at least one assignment was missed and transmits no hybrid-ARQ acknowledgment, thereby avoiding acknowledging transmissions not received.

For uplink transmissions on UL-SCH, each uplink subframe is acknowledged individually by using the PHICH. Hence, in uplink-heavy asymmetries (downlink–uplink configuration 0), the terminal may need to receive two hybrid-ARQ acknowledgments in downlink subframes 0 and 5. There are also some subframes where no PHICH will be transmitted and therefore the amount of PHICH groups may vary between subframes in case of TDD.

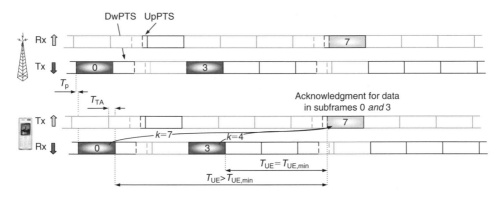

Figure 19.4 *Example of timing relation between downlink data and uplink hybrid-ARQ acknowledgment for TDD (configuration 2).*

19.1.2 Radio-link control

The *radio-link control* (RLC) protocol takes data in the form of RLC SDUs from PDCP and delivers them to the corresponding RLC entity in the receiver by using functionality in MAC and the physical layer. The relation between RLC and MAC, including multiplexing of multiple logical channels into a single transport channel, is illustrated in Figure 19.5. Multiplexing of multiple logical channels into a single transport channel is mainly used for priority handling as described in Section 19.2 in conjunction with downlink and uplink scheduling.

There is one RLC entity per logical channel configured for a terminal, where each RLC entity is responsible for:

- segmentation, concatenation, and reassembly of RLC SDUs;
- RLC retransmission;
- in-sequence delivery for the corresponding logical channel.

The basic structure of the RLC is similar to that of HSPA, although there are some differences of which two are of particular interest: (1) the handling of varying PDU sizes and (2) the possibility for close interaction between the hybrid-ARQ and RLC protocols

19.1.2.1 Segmentation, concatenation, and reassembly of RLC SDUs
The purpose of the segmentation and concatenation mechanism is to generate RLC PDUs of appropriate size from the incoming RLC SDUs. One possibility would be to define a fixed PDU size; a fixed size that would result in a compromise. If the size would be too large, it would not be possible to support

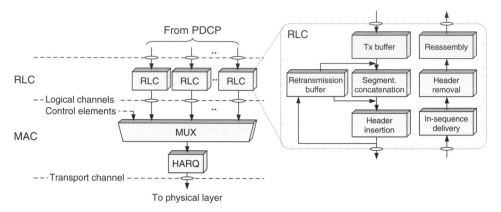

Figure 19.5 *MAC and RLC structure (single-terminal view).*

the lowest data rates. Also, excessive padding would be required in some scenarios. A single small PDU size, however, would result in a high overhead from the header included with each PDU. To avoid these drawbacks, the RLC PDU size varies dynamically in LTE. The use of rate adaptation and dynamic scheduling in LTE implies that the payload possible to transmit in a subframe, that is, the transport block size, varies dynamically and, as a consequence, the RLC PDU size varies dynamically. The use of a varying RLC PDU size is possible thanks to the RLC and the scheduler being located in the same node and is a major difference compared to the first releases of HSPA, which, as it is an extension to WCDMA with the RLC and scheduler being located in different nodes, uses fixed PDU sizes. Later releases of HSPA, as described in Chapter 12, has the possibility to segment PDUs to obtain similar benefits as the dynamically varying PDU size in LTE.

Segmentation and concatenation of RLC SDUs into RLC PDUs are illustrated in Figure 19.6. The header includes, among other fields, a sequence number, which is used by the reordering and retransmission mechanisms. The reassembly function at the receiver side performs the reverse operation to reassemble the SDUs from the received PDUs.

19.1.2.2   RLC retransmission
Retransmission of missing PDUs is the main functionality of the RLC. Although most of the errors can be handled by the hybrid-ARQ protocol, there are, as discussed at the beginning of Section 19.1, benefits of having a second-level retransmission mechanism as a complement. By inspecting the sequence numbers of the received PDUs, missing PDUs can be detected and a retransmission requested from the transmitting side.

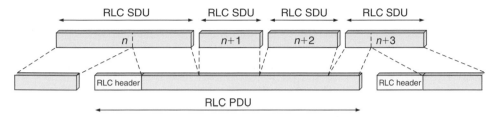

Figure 19.6 *Generation of RLC PDUs from RLC SDUs.*

Different services have different requirements; for some services, for example, transfer of a large file, error-free delivery of data is important, whereas for other applications, for example, streaming services, a small amount of missing packets is not a problem. The RLC can therefore be operated in three different modes, depending on the requirements from the application:

- *Transparent mode* (TM), where the RLC is completely transparent and is in essence bypassed. No retransmissions, no segmentation/reassembly, and no in-sequence delivery take place. This configuration is used for broadcast channels such as BCCH, CCCH, and PCCH where the information should reach multiple users. The size of these messages are selected such that all intended terminals are reached with a high probability and hence there is neither need for segmentation to handle varying channel conditions, nor retransmissions to provide error-free data transmission. Furthermore, retransmissions are not possible for these channels as there is no possibility for the terminal to feed back status reports as no uplink has been established.
- *Unacknowledged mode* (UM), supporting segmentation/reassembly and in-sequence delivery, but not retransmissions. This mode is used when error-free delivery is not required, for example for MCCH and MTCH using MBSFN and for VoIP.
- *Acknowledged mode* (AM) is the main mode of operation for TCP/IP packet data transmission on the DL-SCH. Segmentation/reassembly, in-sequence delivery and retransmissions of erroneous data are all supported.

In the following, the acknowledged mode is described. Unacknowledged mode is similar with the exception that no retransmissions are done and that each RLC entity is unidirectional.

In acknowledged mode, the RLC entity is bidirectional, that is, data may flow in both directions between the two peer entities. This is obviously needed as the reception of PDUs needs to be acknowledged back to the entity that transmitted those PDUs. Information about missing PDUs is provided by the receiving end to the transmitting end in the form of so-called *status reports*. Status reports can either be transmitted autonomously by the receiver or requested by the transmitter. To keep track of the PDUs in transit, the transmitter attaches a RLC header to each PDU, including a sequence number among other fields.

Both RLC entities maintain a window, the transmission and reception window, respectively. Only PDUs in the transmission window are eligible for transmission; PDUs with sequence number below the start of the window have already been acknowledged by the receiving RLC. Similarly, the receiver only accepts PDUs with sequence numbers within the reception window. The receiver also discards any duplicate PDUs as each PDU should be assembled into an SDU only once.

19.1.2.3   In-sequence delivery

In-sequence delivery implies that data blocks are delivered by the receiver in the same order as they were transmitted. This is an essential part of RLC; the hybrid-ARQ processes operate independently and transport blocks may therefore be delivered out-of-sequence as seen in Figure 19.1. In-sequence delivery implies that SDU n should be delivered prior to SDU $n + 1$. This is an important aspect as several applications require the data to be received in the same order as it was transmitted. TCP can, to some extent, handle IP packets arriving out-of-sequence, although with some performance impact, while for some streaming applications in-sequence delivery is essential. The basic idea behind in-sequence delivery is to store the received PDUs in a buffer until all PDUs with lower sequence number have been delivered. Only when all PDUs with lower sequence number have been used for assembling SDUs is the next PDU used. RLC retransmission, provided when operating in acknowledged mode only, operates on the same buffer.

19.1.2.4   RLC operation

The operation of the RLC is best understood by the simple example in Figure 19.7, where two RLC entities are illustrated, one in the transmitting node and one in the receiving node. When operating in acknowledged mode, as assumed below, each RLC entity has both transmitter and receiver functionality but in this example only one of the directions is discussed as the other direction is identical. In the example, PDUs numbered from n to $n + 4$ are awaiting transmission in the transmitter buffer. At time t_0, PDUs with sequence number up to and including n have been transmitted and correctly received, but only PDUs up to and including $n-1$ have been acknowledged by the receiver. As seen in the figure, the transmitter window starts from n, the first not-yet-acknowledged PDU while the receiver window starts from $n + 1$, the next PDU expected to be received. Upon reception of PDU n, the PDU is forwarded to the SDU reassembly functionality for further processing.

The transmission of PDUs continues and at time t_1, PDUs $n + 1$ and $n + 2$ have been transmitted, but, at the receiving end, only PDU $n + 2$ has arrived. One reason for this could be that the missing PDU, $n + 1$, is being retransmitted by the hybrid-ARQ protocol and therefore has not yet been delivered from the hybrid-ARQ to the RLC. The transmitter window remains unchanged compared to the previous figure as none of the PDUs n and higher have been acknowledged by the receiver. Hence, any of these PDUs may need to be retransmitted as the transmitter is not aware of whether they have been received correctly or not. The receiver window is not updated when PDU $n + 2$ arrives. The reason is that PDU $n + 2$ cannot be forwarded to SDU assembly as PDU $n + 1$ is missing. Instead, the receiver waits for the missing PDU $n + 1$. Clearly, waiting for the missing PDU for an infinite time would stall the queue. Hence, the receiver starts a timer, the *reordering timer*, for the missing PDU. If the PDU is not received before the timer expires, a retransmission is requested. Fortunately, in this example, the missing PDU arrives from the hybrid-ARQ protocol at time t_2, before the timer expires. The receiver window is advanced and the reordering timer is stopped as the missing PDU has arrived. PDUs $n + 1$ and $n + 2$ are delivered for reassembly into SDUs.

The example above illustrated the basic principle behind in-sequence delivery which is supported by both acknowledged and unacknowledged mode. However, the acknowledged mode-of-operation also provides retransmission functionality. To illustrate the principles behind this, consider Figure 19.8, which is a continuation of the example above. At time t_3, PDUs up to $n + 5$ have been transmitted. Only PDU $n + 5$ has arrived and PDUs $n + 3$ and $n + 4$ are missing. Similar to the case above, this causes the reordering timer to start. However, in this example no PDUs arrive prior to the expiration of the timer. The expiration of the timer at time t_4 triggers the receiver to send a control PDU containing a status report, indicating the missing PDUs, to its peer entity. Control PDUs have higher priority than data PDUs to avoid the status reports to be unnecessary delayed and negatively impact the retransmission delay. Upon reception of the status report at time t_5, the transmitter knows that PDUs up to $n + 2$ have been received correctly and the transmitter window is advanced. The missing PDUs $n + 3$ and $n + 4$ are retransmitted and, this time, correctly received.

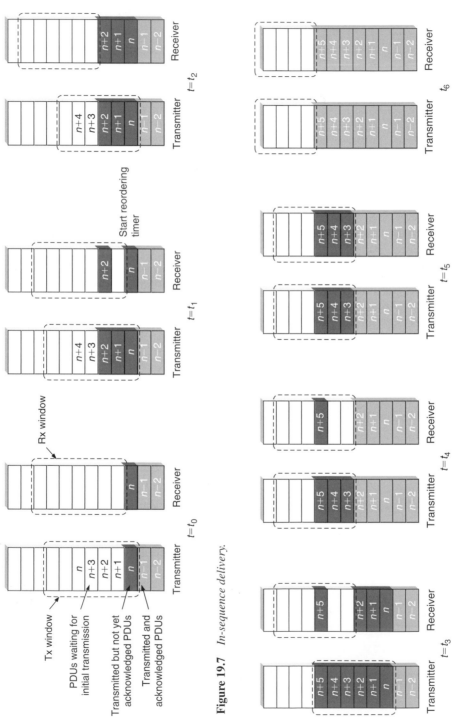

Figure 19.7 *In-sequence delivery.*

Figure 19.8 *Retransmission of missing PDUs.*

The retransmission was triggered by the reception of a status report in this example. However, as the hybrid-ARQ and RLC protocols are located in the same node, tight interaction between the two is possible. The hybrid-ARQ protocol at the transmitting end could therefore inform the RLC at the transmitting end in case the transport block(s) containing PDUs $n + 3$ and $n + 4$ have failed. The RLC can use this to trigger retransmission of missing PDUs without waiting for an explicit RLC status report. This could (significantly) reduce the delays associated with RLC retransmissions and is a benefit compared to HSPA, where the RLC and hybrid-ARQ protocols are in different nodes and such tight interaction is not possible.

Finally, at time t_6, all PDUs, including the retransmissions, have been delivered by the transmitter and successfully received. As $n + 5$ was the last PDU in the transmission buffer, the transmitter requests a status report from the receiver by setting a flag in the header of the last RLC data PDU. Upon reception of the PDU with the flag set, the receiver will respond by transmitting the requested status report, acknowledging all PDUs up to and including $n + 5$. Reception of the status report by the transmitter causes all the PDUs to be declared as correctly received and the transmitter window is advanced.

Status reports can, as mentioned earlier, be triggered for multiple reasons. However, to control the amount of status reports and to avoid flooding the return link with an excessive number of status reports, there is the possibility to use a status prohibit timer. With such a timer, status reports cannot be transmitted more often than once per time interval as determined by the timer.

For the initial transmission, it is relatively straightforward to rely on a dynamic PDU size as a means to handle the varying data rates. However, the channel conditions and the amount of resources may also change between RLC retransmissions. To handle these variations, already transmitted PDUs can be (re)segmented for retransmissions. The reordering and retransmission mechanisms described above still apply; a PDU is assumed to be received when all the segments have been received. Status reports and retransmissions operate on individual segments; only the missing segment of a PDU needs to be retransmitted.

19.2   Scheduling and rate adaptation

The purpose of the scheduler is to determine to/from which terminal(s) to transmit/receive data and on which set of resource blocks. The scheduler is a key element and to a large degree determines the overall behavior of the system. The basic operation is the so-called *dynamic* scheduling, where the eNodeB in each 1 ms TTI send scheduling information to the selected set of terminals, controlling the uplink and downlink transmission activity. The scheduling decisions are transmitted on the PDCCHs as described in Chapter 16. To reduce the control signaling overhead, there is also a possibility for *semi-persistent scheduling*. Semi-persistent scheduling will be described further in Section 19.2.3.

The terminal follows scheduling commands, for both uplink and downlink, from a single cell only, the *serving cell*. This is in contrast to enhanced uplink, described in Chapter 10, where the terminal may receive scheduling information also from non-serving cells. For LTE, inter-cell coordination between different eNodeBs relies on signaling over the X2 interface.

The *downlink scheduler* is responsible for dynamically controlling the terminal(s) to transmit to and, for each of these terminals, the set of resource blocks upon which the terminal's DL-SCH should be transmitted. Transport format selection (selection of transport-bock size, modulation scheme, and code rate) and logical channel multiplexing for downlink transmissions are controlled by the eNodeB as illustrated in the left part of Figure 19.9.

The *uplink scheduler* serves a similar purpose, namely to dynamically control which mobile terminals are to transmit on their UL-SCH and on which uplink

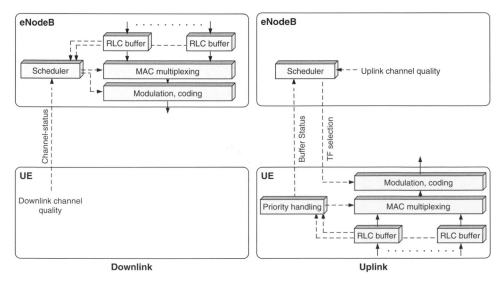

Figure 19.9   *Transport format selection in downlink (left) and uplink (right).*

resources. The uplink scheduler is in complete control of the transport format the mobile terminal shall use whereas the logical-channel multiplexing is controlled by the terminal according to a set of rules. Thus, uplink scheduling is *per mobile terminal* and not per radio bearer. This is illustrated in the right part of Figure 19.9, where the scheduler controls the transport format and the mobile terminal controls the logical channel multiplexing.

19.2.1 Downlink scheduling

The task of the downlink scheduler is to dynamically determine the terminal(s) to transmit to and, for each of these terminals, the set of resource blocks upon which the terminal's DL-SCH should be transmitted. In most cases, a single terminal cannot use the full capacity of the cell, for example, due to lack of data. Also, as the channel properties may vary in the frequency domain, it is useful to be able to transmit to different terminals on different parts of the spectrum. Therefore, multiple terminals can be scheduled in parallel in a subframe, in which case there is one DL-SCH per scheduled terminal, each dynamically mapped to a (unique) set of frequency resources. The scheduler is in control of the instantaneous data rate used and the RLC segmentation and MAC multiplexing will therefore be affected by the scheduling decision. Although formally part of the MAC layer but to some extent better viewed as a separate entity, the scheduler is thus controlling most of the functions in the eNodeB associated with downlink data transmission:

- *RLC*: Segmentation/concatenation of RLC SDUs is directly related to the instantaneous data rate. For low data rates, it may only be possible to deliver a part of an RLC SDU in a TTI in which case segmentation is needed. Similarly, for high data rates, multiple RLC SDUs may need to be concatenated to form a sufficiently large transport block.
- *MAC*: Multiplexing of logical channels depends on the priorities between different streams. For example, radio resource control signaling, such as handover commands, typically has a higher priority than streaming data, which in turn has higher priority than a background file transfer. Thus, depending on the data rate and the amount of traffic of different priorities, the multiplexing of different logical channels is affected. Hybrid-ARQ retransmissions also need to be accounted for.
- *L1*: Coding, modulation and, in case of spatial multiplexing, the number of transmission layers and the associated code book, are obviously affected by the scheduling decision. The choices of these parameters are mainly determined by the selected data rate, that is, the transport block size.

The scheduling decision is communicated to each of the scheduled terminals through the downlink L1/L2 control signaling as described in Chapter 16. The identity of the scheduled terminals, as well as the necessary information to demodulate the transmitted transport block, is encoded and sent on to one of the PDCCHs. Once the transport block is successfully decoded, the terminal will demultiplex the received data into the appropriate logical channels.

The scheduling strategy is implementation-specific and not part of the 3GPP specifications. In principle any of the schedulers described in Chapter 7 can be applied. However, the overall goal of most schedulers is to take advantage of the channel variations between terminals and preferably schedule transmissions to a terminal when the channel conditions are advantageous. This is common with the objectives of the downlink scheduler in HSPA, although LTE in addition to scheduling in the time domain also enables channel-dependent scheduling in the frequency domain. Most scheduling strategies need information about:

- channel conditions at the terminal;
- buffer status and priorities of the different data flows;
- interference situation in neighboring cells (if some form of interference coordination is implemented).

Information about the channel conditions at the terminal can be obtained in several ways. In principle, the eNodeB can use any information available, but typically the channel-status reports from the terminal, further described in Section 19.2.5, are used. However, additional sources of channel knowledge, for example exploiting channel reciprocity to estimate the downlink quality from uplink channel estimates in case of TDD, can also be exploited by a particular scheduler implementation.

In addition to the channel-quality, the scheduler should take buffer status and priority levels into account. Obviously it does not make sense to schedule a terminal with empty transmission buffers. Priorities of the different types of traffic may also vary; RRC signaling may be prioritized over user data. Furthermore, RLC and hybrid-ARQ retransmissions, which are in no way different from other types of data from a scheduler perspective, are typically also given priority over initial transmissions from the corresponding logical channel.

To support priority handling, multiple logical channels, where each logical channel has its own RLC entity, can be multiplexed into one transport channel by the MAC layer. At the terminal, the MAC layer handles the corresponding demultiplexing and forwards the RLC PDUs to their respective RLC entity for in-sequence delivery and the other functions handled by the RLC. To support the demultiplexing at the receiver, a MAC header, shown in Figure 19.10, is used. To each RLC PDU, there is an associated subheader in the MAC header. The subheader contains the identity of the logical channel (LCID) from which the RLC PDU originated and the length of the PDU in bytes. There is also a flag indicating whether this is the last subheader or not. One or several RLC PDUs, together with the MAC header and, if necessary, padding to meet the scheduled transport-block size, form one transport block which is encoded and processed as described in Chapter 16 prior to transmission over the radio interface.

In addition to multiplexing of different logical channels, the MAC layer can also insert the so-called *MAC control elements*. A MAC control element is used for inband control signaling, for example, timing-advance commands and random access response as described in Sections 19.5 and 18.3, respectively. Control elements are identified with reserved values in the LCID field, where the LCID value indicates the type of control information. Furthermore, the length field in the subheader is removed for control elements with a fixed length.

Downlink inter-cell interference coordination is also part of the implementation-specific scheduler strategy. A cell may signal to its neighboring cells the intention to transmit with a lower transmission power in the downlink on a set of resource blocks. This information can then be exploited by neighboring cells as a region of low interference where it is advantageous to schedule terminals at the cell edge; terminals that otherwise could not attain high data rates due to the interference level.

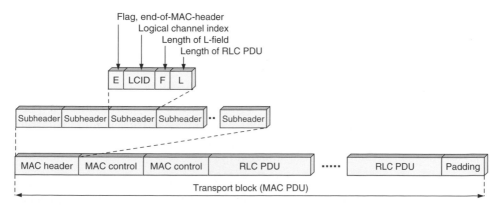

Figure 19.10 *MAC header and SDU multiplexing.*

19.2.2   Uplink scheduling

The basic function of the *uplink scheduler* is similar to the downlink counterpart, namely to dynamically determine, for each 1 ms interval, which mobile terminals are to transmit data on their UL-SCH and on which uplink resources. Uplink scheduling is used also for HSPA, but due to the different multiple-access schemes used, there are some significant differences between HSPA and LTE in this respect.

In HSPA, the shared uplink resource is primarily the acceptable interference at the base station as described in Chapter 10. The HSPA uplink scheduler only sets an upper limit of the amount of uplink interference the mobile terminal is allowed to generate. Based on this limit, the mobile terminal autonomously selects a suitable transport format. This strategy clearly makes sense for a non-orthogonal uplink as is the case for HSPA. A mobile terminal not utilizing all its granted resources will transmit at a lower power, thereby reducing the intra-cell interference. Hence, shared resources not utilized by one mobile terminal can be exploited by another mobile terminal through statistical multiplexing. Since the transport-format selection is located in the mobile terminal for the HSPA uplink, outband signaling is required to inform the NodeB about the selection made.

For LTE, the uplink is orthogonal and the shared resource controlled by the eNodeB scheduler is time–frequency resource units. An assigned resource not fully utilized by a mobile terminal cannot be partially utilized by another mobile terminal. Hence, due to the orthogonal uplink, there is significantly less gain in letting the mobile terminal select the transport format compared to HSPA. Consequently, in addition to assigning the time–frequency resources to the mobile terminal, the eNodeB scheduler is also responsible for controlling the transport format (payload size, modulation scheme) the mobile terminal shall use. As the scheduler knows the transport format the mobile terminal will use when it is transmitting, there is no need for outband control signaling from the mobile terminal to the eNodeB. This is beneficial from a coverage perspective taking into account that the cost per bit of transmitting outband control information can be significantly higher than the cost of data transmission as the control signaling needs to be received with a higher reliability. A consequence of the scheduler being responsible for selection of the transport format is that accurate and detailed knowledge about the terminal situation with respect to buffer status and power availability is more accentuated in LTE.

The basis for uplink scheduling are *scheduling grants*, containing the scheduling decision and providing the terminal information about the resources and the associated transport format (transport-block size, modulation scheme) to use for transmission of the UL-SCH. Only if the terminal has a valid grant is it allowed to transmit on the UL-SCH; there is no possibility for autonomous transmissions without a corresponding grant as in enhanced uplink. This is a direct consequence of the orthogonal uplink. Dynamic grants are valid for one subframe, that is, for each subframe in which the terminal is to transmit on the UL-SCH the scheduler transmits a grant on a downlink PDCCH.

The terminal monitors the set of PDCCHs as described in Chapter 16 for uplink scheduling grants. If a valid grant intended for the terminal is detected in subframe n, the actual transmission of the uplink data takes place in subframe $n + 4$ for FDD. For TDD the first uplink subframe in $n + 4$ or later is used. This is the same timing relation as used for the PHICH and is motivated by the possibility to override the PHICH by a dynamic scheduling grant as described in Section 19.1.1.2. The detailed timing between receiving a scheduling grant and the uplink transmission was described in Chapter 17.

Similarly to the downlink case, the uplink scheduler can exploit information about channel conditions, buffer status, and priorities of the different data flows, and, if some form of interference coordination is employed, the interference situation in neighboring cells. Channel-dependent scheduling, which typically is used for the downlink, can be used for the uplink as well. In the uplink, estimates of the channel quality can be obtained from the use of uplink channel-sounding as described in Chapter 17. For scenarios where the overhead from channel sounding is too costly, or when the variations in the channel are too rapid to be tracked, for example, at high terminal speeds, uplink diversity can be used instead. The use of frequency-hopping as discussed in Chapter 17 is one example of a way to obtain diversity in the uplink.

19.2.2.1 Priority handling

Multiple logical channels of different priorities can be multiplexed into the same transport block. The same MAC header structure as described in conjunction with downlink scheduling is used in the uplink as well. However, unlike the downlink case where the prioritization is under control of the scheduler and up to the implementation, the uplink multiplexing is done according to a set of well-defined rules in the terminal as a scheduling grant applies to a *terminal*, not to a specific radio bearer within the terminal. Using radio-bearer specific scheduling grants would increase the control signaling overhead in the downlink.

The simplest multiplexing rule would be to serve logical channels in strict priority order. However, this may result in starvation of lower-priority channels;

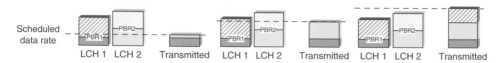

Figure 19.11   *Prioritization of two logical channels for three different uplink grants.*

all resources would be given to the high-priority channel until its transmission buffer is empty. Typically, an operator would instead like to provide at least some throughput also for low-priority services. Therefore, for each logical channel in an LTE terminal, a *prioritized data rate* is configured in addition to the priority value. The logical channels are then served in decreasing priority order up to their prioritized data rate, which avoids starvation as long as the scheduled data rate is at least as large as the sum of the prioritized data rates. Beyond the prioritized data rates, channels are served in strict priority order until the grant is fully exploited or the buffer is empty. This is illustrated in Figure 19.11.

19.2.2.2   *Scheduling requests and buffer status reports*

The scheduler needs knowledge about the amount of data awaiting transmission from the terminals to assign the proper amount of uplink resources. Obviously, there is no need in providing uplink resources to a terminal with no data to transmit as this would only result in the terminal performing padding to fill up the granted resources. Hence, as a minimum, the scheduler needs to know whether the terminal has data to transmit and should be given a grant. This is known as a *scheduling request*.

A scheduling request is a simple flag, raised by the terminal to request uplink resources from the uplink scheduler. Since the terminal requesting resources by definition has no PUSCH resource, the scheduling request is transmitted on the PUCCH. Each terminal is assigned a dedicated PUCCH scheduling request resource, occurring every nth subframe as described in Chapter 17. With a dedicated scheduling-request mechanism, there is no need to provide the identity of the terminal requesting to be scheduled as the identity of the terminal is implicitly known from the resources upon which the request is transmitted. When data with higher priority than already existing in the transmit buffers arrives to the terminal and the terminal has no grant and hence cannot transmit the data, the terminal transmits a scheduling request at the next possible instant as illustrated in Figure 19.12. Upon reception of the request, the scheduler can assign a grant to the terminal. If the terminal does not receive a scheduling grant until the next possible scheduling-request instant, than the scheduling request is repeated.

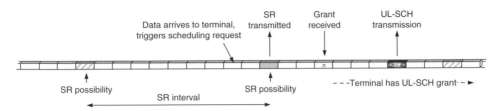

Figure 19.12 *Scheduling request transmission.*

The use of a single bit for the scheduling request is motivated by the desire to keep the uplink overhead small as a multi-bit scheduling request would come at a higher cost. A consequence of the single-bit scheduling request is the limited knowledge at the eNodeB about the buffer situation at the terminal when receiving such a request. Different scheduler implementations handle this differently. One possibility is to assign a small amount of resources to ensure that the terminal can exploit them efficiently without becoming power limited. Once the terminal has started to transmit on the UL-SCH, more detailed information about the buffer status and power headroom can be provided through the inband MAC control message as discussed later. Knowledge of the service type may also be used, for example, in case of voice the uplink resource to grant is preferably the size of a typical VoIP package. The scheduler may also exploit, for example, path-loss measurements used for mobility and handover decisions to estimate the amount of resources the terminal may efficiently utilize.

An alternative to a dedicated scheduling-request mechanism would be a contention-based design. In such a design, multiple terminals share a common resource and provide their identity as part of the request. This is similar to the design of the random access. The number of bits transmitted from a terminal as part of a request would in this case be larger, with the correspondingly larger need for resources. In contrast, the resources are shared by multiple users. Basically, contention-based designs are suitable for a situation where there are a large number of terminals in the cell and the traffic intensity, and hence the scheduling intensity, is low. In situations with higher intensities, the collision rate between different terminals simultaneously requesting resources would be too high and lend to an inefficient design.

Although the scheduling-request design for LTE relies on dedicated resources, a terminal that has not been allocated such resources obviously cannot transmit a scheduling request. Instead, terminals without scheduling-request resources configured rely on the random access mechanism, described in Chapter 18. In

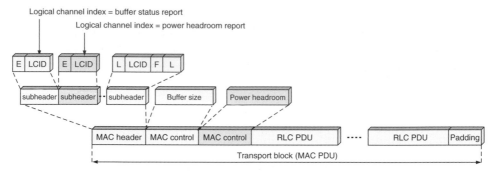

Figure 19.13   *Buffer status and power headroom reports.*

principle, an LTE terminal can therefore be configured to rely on a contention-based mechanism if this is advantageous in a specific deployment.

Terminals, which already have a valid grant, obviously do not need to request uplink resources. However, to allow the scheduler to determine the amount of resources to grant to each terminal in future subframes, information about the buffer situation and the power availability is useful as discussed above. This information is provided to the scheduler as part of the UL-SCH transmission through MAC control elements. The control elements are included in the MAC subheaders in a similar way as for the downlink. The LCID field in one of the MAC subheaders is set to a reserved value indicating the presence of a buffer status report as illustrated in Figure 19.13.

From a scheduling perspective, buffer information for each logical channel is beneficial, although this could result in a significant overhead. Logical channels are therefore grouped into logical-channel groups and the reporting is done per group. The buffer-size field in a buffer-status report indicates the amount of data awaiting transmission across all logical channels in logical-channel group. A buffer-status report represents one or all four logical-channel groups and can be triggered by the following reasons:

- arrival of data with higher priority than currently in the transmission buffer, that is, data in a logical-channel group with higher priority than the one currently being transmitted, as this may impact the scheduling decision;
- change of serving cell, in which case a buffer-status report is useful to provide the new serving cell with information about the situation in the terminal;
- periodically as controlled by a timer;

- instead of padding. If the amount of padding required to match the scheduled transport block size is larger than a buffer-status report, a buffer-status report is inserted. Clearly it is better to exploit the available payload for useful scheduling information instead of padding if possible.

In addition to buffer status, the amount of transmission power available in each mobile terminal is also relevant for the uplink scheduler. Obviously, there is little reason to schedule a higher data rate than the available transmission power can support. In the downlink, the available power is known to the scheduler as they are located in the same node and the total transmission power is typically constant. For the uplink, the power availability, or power headroom (see Section 19.3.2 for a discussion on power headroom), depends on the power-control mechanism and thereby indirectly on factors such as the interference in the system and the distance to the base stations. Information about the power headroom is fed back from the terminals to the eNodeB in a similar way as the buffer-status reports.

19.2.2.3 Uplink inter-cell interference coordination

Uplink interference coordination is a tool to improve the data rates experienced by terminals located at the cell edge by coordinating the interference due to uplink transmission in different cells. It can thus be seen as an example of multi-cell radio-resource management. Similarly to the downlink, the interference coordination scheme is not standardized but part of the implementation-specific scheduling strategy. This allows for different coordination strategies depending on the requirements in a particular scenario.

The basis for uplink inter-cell interference coordination is to restrict transmissions in the frequency domain. Hence, it is mainly applicable to services that, by nature, are narrow-band or to situations when the terminal cannot exploit the full cell bandwidth due to power limitations.

To aid the scheduler in inter-cell interference coordination, two indicators, exchanged between neighboring cells on the X2 interface, are defined: the *high-interference indicator* and the *overload indicator*. The indicators are exchanged on a relatively slow basis, that is, the coordination strategies operate on a time scale significantly longer that the subframe duration.

The high-interference indicator provides information to other cells about the resource blocks on which the current cell intends to schedule cell-edge terminals. The indicator consists of one bit per resource block, indicating the intention to use those resource blocks for cell-edge terminals transmitting close to their maximum output power. Hence, those resource blocks are susceptible to inter-cell interference as the terminals transmitting on these resources may have limited possibilities in increasing the transmission power to overcome interference from other cells. Furthermore, terminals on those resource blocks will also generate more interference in neighboring cells than terminals on the other resource blocks. Therefore, an eNodeB receiving a high-interference indicator for a certain resource block in a cell may want to avoid scheduling (cell-edge) terminals on this part of the spectrum if possible. These resources can instead be used to schedule cell-interior terminals as they are less susceptible to the interference. Downlink path-loss measurements, which are required for mobility, can be used to determine whether a terminal is close to the cell center or not.

The overload indicator provides information on the interference level: low, medium, or high, experienced in each resource block. An eNodeB receiving the overload indicator may reduce the interference generated on some of these resource blocks by adjusting its scheduling strategy, for example, by using a different set of resource blocks or lowering the data rate for terminals scheduled on these resource blocks.

Uplink interference coordination is illustrated in Figure 19.14.

19.2.3   Semi-persistent scheduling

The basis for uplink and downlinks scheduling is dynamic scheduling as described in Sections 19.2.1 and 19.2.2. Dynamic scheduling with a new scheduling decision taken in each subframe allows for full flexibility in terms of the resources used and can handle large variations in the amount of data to transmit

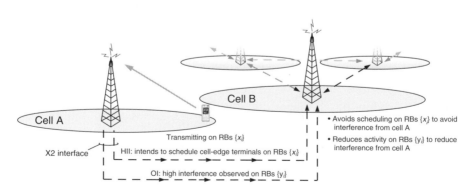

Figure 19.14   *Example of uplink inter-cell interference coordination.*

at the cost of the scheduling decision being sent on a PDCCH in each subframe. In many situations, the overhead in terms of control signaling on the PDCCH is well motivated and relatively small compared to the payload on DL-SCH/UL-SCH. However, some services, most notably VoIP, are characterized by regularly occurring transmission of relatively small payloads. To reduce the control signaling overhead for those services, LTE provides semi-persistent scheduling in addition to dynamic scheduling.

With semi-persistent scheduling, the terminal is provided with the scheduling decision on the PDCCH, together with an indication that this applies to every *n*th subframe until further notice. Hence, control signaling is only used once and the overhead is reduced as illustrated in Figure 19.15. The periodicity of semi-persistently scheduled transmissions, that is, the value of *n*, is configured by RRC signaling in advance, while activation (and deactivation) is done using the PDCCH. For example, for VoIP the scheduler can configure a periodicity of 20 ms for semi-persistent scheduling and once a talk spurt starts, the semi-persistent pattern is triggered by the PDCCH.

After enabling semi-persistent scheduling, the terminal continues to monitor the PDCCH for uplink and downlink scheduling commands. In case a dynamic scheduling command is detected, it takes precedence over the semi-persistent scheduling in that particular subframe, which is useful if the semi-persistently allocated resources occasionally need to be increased. For example, for VoIP in parallel with web browsing it may be useful to override the semi-persistent resource allocation with a larger transport block when downloading the web page.

For the downlink, only initial transmissions use semi-persistent scheduling. Retransmissions are explicitly scheduled using a PDCCH assignment. This follows directly from the use of an asynchronous hybrid-ARQ protocol in the downlink. Uplink retransmissions, in contrast, can either follow the semi-persistently allocated subframes or be dynamically scheduled.

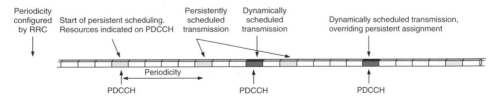

Figure 19.15 *Example of semi-persistent scheduling.*

19.2.4   Scheduling for half-duplex FDD

Half-duplex FDD implies that a single terminal cannot receive and transmit at the same time while the eNodeB still operates in full duplex. In LTE, half-duplex FDD is implemented as a scheduler constraint, implying it is up to the scheduler to ensure that a single terminal is not scheduled simultaneous in uplink and downlink. Hence, from a terminal perspective, subframes are dynamically used for uplink or downlink. Briefly, the basic principle for half-duplex FDD is that a terminal is receiving in the downlink unless it has been explicitly instructed to transmit in the uplink (UL-SCH transmission or hybrid-ARQ acknowledgments triggered by a downlink transmission). The timing and structure for control signaling is identical between half-and-full duplex FDD terminals.

An alternative approach would be to base half-duplex FDD on the TDD control signaling structure and timing, with a semi-static configuration of subframes to either downlink or uplink. However, this would complicate supporting a mixture of half-and-full duplex terminals in the same cell as the timing of the control signaling would differ. It would also imply a waste of uplink spectrum resources. All terminals need to be able to receive subframe zero and five as those subframes are used for system information and synchronization signals. Hence, if a fixed uplink–downlink allocation would be used, no uplink transmissions could take place in those two subframes, resulting in a loss in uplink spectral efficiency of 20%. Clearly this is not attractive and led to the choice of implementing half-duplex FDD as a scheduling strategy instead (Figure 19.15).

An example of half-duplex operation as seen from a terminal perspective is shown in Figure 19.16. In the leftmost part of the figure, the terminal is explicitly scheduled in the uplink and, consequently, cannot receive data in the downlink in the same subframe. The uplink transmission implies the reception of an acknowledgment on the PHICH four subframes later, and therefore the terminal cannot be scheduled in the uplink in this subframe. Similarly, when the terminal is scheduled to receive data in the downlink in subframe n, the corresponding hybrid-ARQ acknowledgment needs to be transmitted in the uplink subframe $n + 4$, preventing downlink reception in subframe $n + 4$. The scheduler can exploit this

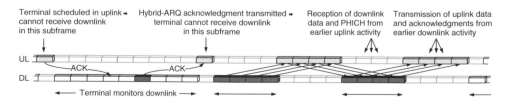

Figure 19.16   *Example of half-duplex FDD terminal operation.*

by scheduling downlink data in four consecutive subframes and uplink transmission in the four next subframes when the terminal anyway needs to transmit hybrid-ARQ acknowledgments in the uplink and so on. Hence, at most half of the time can be used in the downlink and half in the uplink or, in other words, the asymmetry in half-duplex FDD is 4:4. Efficient support of half-duplex FDD is one of the reasons why the same number of hybrid-ARQ processes was selected in uplink and downlink.

Note that, as the eNodeB is operating in full duplex, regardless of the duplex capability of the terminals, the cell capacity is hardly affected by the presence of half-duplex terminals as, given a sufficient number of terminals with data to transmit/receive, the scheduler can with a high likelihood find a set of terminal to schedule in the uplink and another set to schedule in the downlink in a given subframe.

Similar to TDD, a half-duplex terminal needs some guard time for switching between uplink and downlink. For half-duplex FDD, guard time for the downlink-to-uplink switch is created by allowing the terminal to skip reception of the last OFDM symbols in a downlink subframe immediately preceding an uplink subframe. Guard time for uplink-to-downlink switch is handled by setting the appropriate amount of timing advance in the terminals.

19.2.5 Channel-status reporting

As mentioned several times, the possibility for downlink channel-dependent scheduling, that is, selecting the downlink transmission configuration and related parameters depending on the instantaneous downlink channel conditions, is a key feature of LTE. An important part of the support for downlink channel-dependent scheduling is *channel-status reports* provided by terminals to the network, reports on which the latter can base its scheduling decisions.

Although referred to as channel-status reports, what a terminal delivers to the network are not explicit reports of the downlink channel status. Rather, what the terminal delivers are *recommendations* on what transmission configuration and related parameters the network should use if/when transmitting to the terminal on the downlink shared channel. The terminal has typically based these recommendations on estimates of the instantaneous downlink channel conditions, thus the term *channel-status report*.

The channel-status reports consist of one or several pieces of information:

- *Rank indication* (RI), providing information about the channel rank or, expressed differently, the number of layers that should, preferably, be used for downlink transmission to the terminal. RI only needs to be reported by terminals that are configured to be in one of the spatial-multiplexing transmission modes.

- *Precoder matrix indication* (PMI), providing a precoder matrix that should, preferably, be used for the downlink transmission. The reported precoder matrix should be determined assuming the number of layers indicated by the RI. PMI is only reported if the terminal is configured to be in closed-loop spatial-multiplexing mode. As discussed in Chapter 16, in case of open-loop spatial multiplexing, the network instead selects the precoder matrix to use for transmission according to a pre-defined rule. The precoder recommendation may be frequency-selective, implying that the terminal may recommend different precoders for different parts of the downlink spectrum.

- *Channel-quality indication* (CQI), representing the recommended modulation scheme and coding rate that should, preferably, be used for the downlink transmission. The CQI points into a table that consists of a set of pre-defined modulation-scheme/coding-rate combinations. The use of the term Channel-Quality Indication for the recommended modulation scheme and coding rate is for historical reasons. In the first releases of HSPA, when spatial-multiplexing was not supported, the modulation-scheme/coding-rate indication was the only information included as part of the channel status/quality reporting and it was then referred to as CQI. This term has continued to be used for the modulation-scheme/coding-rate indication, despite the fact that the channel-status reporting has since then, also for HSPA, been extended to include also other parameters.

Depending on the configuration, the RI, PMI, and CQI in different combinations form the channel-status report. Exactly what is included in a channel-status reporting obviously depends on the transmission mode the terminal is configured to be in. As mentioned earlier, RI and PMI do not need to be reported unless the terminal is in a spatial-multiplexing transmission mode. However, also given the transmission mode, there are different reporting modes that typically differ on what set of resource blocks the report is valid for.

Fundamentally, in the context of channel-status reports, the overall downlink cell bandwidth is divided into a number of subbands, where each subband consists of a set of consecutive resource blocks and where the size of each subband depends on the cell bandwidth. For example, for the most narrow bandwidths, each subband may consist of two resource blocks while, for the largest bandwidths, the subbands may be as large as eight resource blocks.

Channel-status reports can then be categorized as *wideband reports*, reflecting the status over the entire cell bandwidths, and *per-subband reports*, reflecting the status over each subband. The different reporting modes differ to what extent the different information being reported (RI, PMI, and CQI) are wideband reports or subband reports.

In general, what is delivered by the terminal is a *recommendation* on a suitable configuration for downlink transmission, a recommendation that the network may, but does not have to, follow. Information about the actual modulation scheme and coding rate used for DL-SCH transmission is always included in the downlink scheduling assignment and the terminal should always use this for demodulation and decoding of the actual DL-SCH transmission. With regards to the precoder-related recommendations, the network has two choices:

- The network may follow the terminal recommendation, in which case it only has to confirm (a one bit indicator in the downlink scheduling assignment) that the precoder configuration recommended by the terminal is used for the downlink transmission. On receiving such a confirmation, the terminal will use its recommended configuration when demodulating and decoding the corresponding DL-SCH transmission. Since the PMI can be frequency-selective, an eNodeB following the precoding matrix recommended by the terminal may have to apply different precoding matrices for different (sets of) resource blocks.
- The network may select a different precoder configuration, information about which then needs to be explicitly included in the downlink scheduling assignment. The terminal then uses this configuration when demodulating and decoding the DL-SCH. To reduce the amount of downlink signaling, only a single precoding matrix can be signaled in the scheduling assignment, implying that, if the network overrides the recommendation, than the precoding is frequency-non-selective.

Channel-status reporting in LTE, that is reporting of CQI, PMI, and RI, can either be *periodic* or *a-periodic*. The latter can also be referred to as *trigger-based reporting*.

An a-periodic or trigger-based channel-status report is delivered when explicitly requested by the network by means of a 'channel-status request' flag included in uplink scheduling grant (Section 16.5.4). An a-periodic channel-status report is always delivered using the PUSCH, that is, on a dynamically assigned resource. As the trigger for the a-periodic report is included in an uplink scheduling grant, the terminal will always have an uplink resource available for the reporting.

Periodic channel-status reports, in contrast, are configured by the network to be delivered with a certain periodicity, possible as often as once in every 2 ms.

Furthermore, the different types of information do not need to be reported with the same period. Typically, RI can be reported less often, compared to the reporting of PMI and CQI, reflecting the fact that the suitable number of layers typically varies on a slower basis, compared to the channel variations that impact the choice of precoder matrix and modulation rate, and coding scheme.

Normally, periodic channel-status reports are delivered using the PUCCH physical channel. However, similar to hybrid-ARQ acknowledgments normally delivered on PUCCH, channel-status reports are 're-routed' to the PUSCH if the terminal has a valid uplink grant and is anyway to transmit on the PUSCH.

19.3   Uplink power control

Uplink power control for LTE is the set of tools by which the transmit power for different uplink physical channels and signals are controlled to ensure that they are received at the cell site with an appropriate power. How to set the transmit power for the random-access preamble was briefly discussed in Chapter 18. What will be discussed in this section are the LTE power-control mechanisms for

- The physical uplink control channel (PUCCH)
- The physical uplink shared channel (PUSCH)
- Uplink sounding reference signals (SRS).

The uplink demodulation reference signals are always transmitted together with PUSCH or PUCCH and are then transmitted with the same power as the corresponding physical channel.

Fundamentally, LTE uplink power control is a combination of an *open-loop* mechanism, implying that the terminal transmit power depends on estimates of the downlink path-loss, and a *closed-loop* mechanism, implying that the network can, in addition, directly control the terminal transmit power by means of explicit *power-control commands* transmitted in the downlink.

19.3.1   Power control for PUCCH

For PUCCH, the appropriate received power is simply the power needed to achieve a sufficiently low error rate in the decoding of the L1/L2 control signaling transmitted on the PUCCH. However, it is then important to have the following in mind:

- In general, decoding performance is not determined by the *received signal strength* but rather by the *received signal-to-noise/interference ratio* (SINR).

What is an appropriate received power thus directly depends on the interference level at the receiver side, an interference level that may differ between different deployments and which may also vary in time as, for example, the load of the network varies.

- As described in Chapter 17, there are different PUCCH formats which are used to carry different types of uplink L1/L2 control signaling (hybrid-ARQ acknowledgments, scheduling requests, channel-status reports, or combinations thereof). The different PUCCH formats thus carry different number of information bits per subframe and the information they carry may also have different error-rate requirements. The required received SINR may therefore differ substantially between the different PUCCH formats, something that needs to be taken into account when setting the PUCCH transmit power in a given subframe.

Overall, power-control for PUCCH can be described according to:

$$P_{\mathrm{T}} = \min \left\{ P_{\max}, P_0 + PL_{\mathrm{DL}} + \Delta_{\mathrm{Format}} + \delta \right\} \tag{19.1}$$

In this expression, P_{T} is the transmit power to use in a given subframe, $P_{\max}$ is the maximum terminal transmit power, for example, 23 dBm, and PL_{DL} is the downlink path-loss as estimated by the terminal. The downlink path-loss can be estimated by measuring the received power of the downlink cell-specific reference signals, the transmit power of which are known to the mobile terminal.[1]

The '$\min \left\{ P_{\max}, ... \right\}$' part of the PUCCH power-control expression reflects the obvious fact that the terminal transmit power is upper limited by $P_{\max}$. It should be noted that, for PUCCH transmission, the terminal transmit power should preferably not reach this limit as this would imply that the transmit power cannot reach the level needed to achieve the required received SINR, with a too high error rate for the corresponding L1/L2 control signaling as a consequence.

The parameter P_0 in the expression above is a cell-specific parameter that is broadcast as part of the system information. Considering only the part $P_0 + PL_{\mathrm{DL}}$ in the PUCCH power-control expression and assuming that the uplink path-loss is the same as the (estimated) downlink path-loss, it is clear that P_0 can be seen as the *desired received power*. As discussed earlier, the required received power will depend on the uplink noise/interference level. Consequently, the value of P_0 should preferably take the interference level into account and may thus vary in time as the interference level varies. Alternatively, the value of P_0 should be

---

[1] The reference-signal power is broadcast as part of the LTE system information.

chosen to provide sufficient margin for the maximum interference level expected within the cell. This will, however, imply that terminals are typically transmitting with a too high power, causing additional interference to other cells.

For the transmit power to reflect the typically different SINR requirements for different PUCCH formats, the PUCCH power-control expression includes the term Δ_{Format} which adds a format-depending power offset to the transmit power. The power offsets are defined such that one format, more exactly the PUCCH format corresponding to the transmission of a single hybrid-ARQ acknowledgment (format 1 with BPSK modulation as described in Section 17.3.1.1), has an offset equal to 0 dB while the offsets for the remaining formats can be explicitly configured by the network. For example, PUCCH format 1 with QPSK modulation, carrying two simultaneous acknowledgments and used in case of downlink spatial multiplexing, should have a power offset of roughly 3 dB reflecting the fact that twice as high power is needed to communicate two acknowledgments, instead of just a single acknowledgment.

Finally there is a possibility for the network to directly adjust the PUCCH transmit power by providing the mobile terminal with explicit power-control commands that adjust the term δ in the power-control expression above. Similar to WCDMA/HSPA, these power-control commands are *accumulative*, that is, each received power-control command increases or decreases the term δ a certain amount. The power-control commands for PUCCH can be provided to the terminal by two different means:

- As mentioned in Chapter 16 (Section 16.4.4) a power-control command is included in each downlink scheduling assignment, that is, the mobile terminal receives a power-control command every time it is explicitly scheduled on the downlink. One reason for uplink PUCCH transmissions is the transmission of hybrid-ARQ acknowledgments as response to downlink DL-SCH transmissions. Such downlink transmissions are typically associated with downlink scheduling assignments on PDCCH and the corresponding power-control commands could thus be used to adjust the PUCCH transmit power prior to the transmission of the hybrid-ARQ acknowledgments.
- Power-control commands can also be provided on a special PDCCH that simultaneously provides power-control commands to multiple mobile terminals (PDCCH using DCI format 3/3 A, see Section 16.4.6). In practice, such power-control commands are then typically transmitted on a regular basis and can be used to adjust the PUCCH transmit power, for example, prior to (periodic) uplink channel-status reports.

In contrast to WCDMA/HSPA, power-control commands for PUCCH power control may be *multi-level*. The power-control command carried within the uplink scheduling grant consists of two bits, corresponding to update steps of $-1\,\mathrm{dB}$, $0\,\mathrm{dB}$, $+1\,\mathrm{dB}$, or $+3\,\mathrm{dB}$. The same is true for the power-control command carried on the special PDCCH assigned for power control when this is configured to Format 3 A. On the other hand, when the PDCCH is configured to Format 3, each power-control command consists of a single bit, corresponding to update steps of $+1\,\mathrm{dB}$ and $-1\,\mathrm{dB}$. In the latter case, twice as many terminals can be power controlled by a single PDCCH. The reason for including the possibility for $0\,\mathrm{dB}$ (no change of power) as one power-control step is that a power-control command is included in *every* downlink scheduling assignment and there may not be a need to update the PUCCH transmit power for each assignment.

One possible use for power adjustments by means of explicit power-control commands is to compensate for uplink multipath fading which is not reflected in the downlink path-loss and which can thus not be compensated for by the path-loss term in the PUCCH power-control expression.

Another possible use for power adjustments by means of explicit power-control commands is to compensate for uplink interference variations that are not captured in the P_0 parameter used by the mobile terminal. Typically, the mobile terminal does not read system information while in connected mode and it may thus not have acquired the most up-to-date value for P_0, reflecting the most up-to-date network estimate of the uplink interference level.

19.3.2 Power control for PUSCH

In case of WCDMA/HSPA, the DPCCH is the physical channel that is explicitly power controlled by the network and the transmit powers for the remaining uplink physical channels follow the DPCCH transmit power with different offsets. In contrast, power control for PUSCH in LTE is, although having many properties in common with power control for PUCCH, an independent process in the sense that it can be configured with completely different parameter values and can also be controlled by completely independent power-control commands.

Overall, power-control for PUSCH transmission can be described according to[2]:

$$P_{\mathrm{T}} = \min\left\{P_{\mathrm{max}}, P_0 + \alpha \cdot PL_{\mathrm{DL}} + 10 \cdot \log_{10}\left(M\right) + \Delta_{\mathrm{MCS}} + \delta\right\} \quad (19.2)$$

where M indicates the instantaneous PUSCH bandwidth measured in number of resource blocks and the term Δ_{MCS} is similar to the term Δ_{Format} in the expression

---

[2] Note that the parameters P_0 and δ may take different values compared to the corresponding parameters for PUCCH power control in expression (19.1).

for PUCCH power control, that is, it reflects the fact that different SINR is required for different modulation schemes and coding rates used for the PUSCH transmission. The above expression is clearly similar to the power-control expression for PUCCH transmission, with the main differences being the factor α in front of the downlink path-loss estimate and the term $10 \cdot \log_{10}(M)$. The latter term simply reflects the fact that what is fundamentally controlled is the power *per resource block*. Thus, for a larger resource assignment, a correspondingly higher received power and thus a correspondingly higher transmit power is needed.[3]

In case of PUSCH transmission, the explicit power-control commands controlling the term δ above are included in the uplink scheduling grants, rather than in the downlink scheduling assignments. Furthermore, in the same way as for PUCCH power control, explicit power-control commands for PUSCH can also be provided on a special PDCCH that simultaneously provides power-control commands to multiple terminals. Also similar to PUSCH, power-control commands for PUSCH can be multi-level.

Assuming α equal to one ($\alpha = 1$), also referred to as *full path-loss compensation*, the PUSCH power-control expression becomes very similar to the corresponding expression for PUCCH. Thus the network can select an MCS and the power-control mechanism, including the term Δ_{MCS}, will ensure that the received SINR will match the SINR required for that MCS, *assuming that the mobile-terminal transmit power does not reach its maximum value.*

To assist the network in the selection of a combination of MCS and resource size M that does not lead to the mobile terminal being power limited, the mobile terminal can be configured to provide regular reports on its power usage, the so-called *Power Headroom*, to the network (see also Section 19.2.2.2), with the power headroom defined as:

$$\text{Power headroom} = P_{\max} - \left(P_0 + \alpha \cdot PL_{\mathrm{DL}} + 10 \cdot \log_{10}(M) + \Delta_{\mathrm{MCS}} + \delta \right)$$

$$(19.3)$$

It should be noted that the power headroom is not a measure of the difference between the maximum terminal transmit power $P_{\max}$ and the actual uplink transmit power. Rather, the power headroom is a measure of the difference between the maximum terminal transmit power and the uplink transmit power that would have been used *assuming that the mobile terminal would not have been limited*

---

[3] One could have included a corresponding term also in the expression for PUCCH power control. However, as the PUCCH bandwidth always corresponds to one resource block the term would always equal zero.

by its maximum transmit power. Thus, the power headroom can very well be negative. More exactly, a negative power headroom indicates that the mobile terminal transmit power was limited by P_{max} at the time of the power headroom reporting. As the network knows what MCS and resource size the terminal used for transmission at the time of the power-headroom report it can determine what are the valid combinations of MCS and M, assuming that the downlink path loss PL_{DL} and the term δ have not changed substantially.

Furthermore, assuming that the network has a decent knowledge of δ, that is, of the accumulated power control, the power headroom report indirectly provides information about the downlink path-loss estimate.

In case of PUSCH transmission, it is also possible to 'turn off' the Δ_{MCS} function by setting all Δ_{MCS} to zero. In that case, the PUSCH received power will be matched to a certain MCS given by the selected value of P_0.

With the parameter α less than one ($\alpha < 1$), the PUSCH power control operates with so-called *partial path-loss compensation*, that is, an increased path loss is not fully compensated for by a corresponding increase in the uplink transmit power. In that case, the received power, and thus the received SINR per resource block, will vary with the path loss and, consequently, the scheduled MCS should vary accordingly. Clearly, in the case of fractional path-loss compensation, the Δ_{MCS} function should be disabled. Otherwise, the UE transmit power would be further reduced when the MCS is reduced to match the fractional path-loss compensation.

Figure 19.17 illustrates the differences between full path-loss compensation ($\alpha = 1$) and partial path-loss compensation ($\alpha < 1$). As can be seen, with fractional path-loss compensation, the terminal transmit power increases more slowly than the increase in path loss (left-hand figure) and, consequently, the received power, and thus also the received SINR, is reduced as the path-loss increases (right-hand figure). To compensate for this, the MCS, that is, the PUSCH data rate, should be reduced as the path loss increases.

The potential benefit of fractional path-loss compensation is a relatively lower transmit power for terminals closer to the cell border, implying less interference to other cells. At the same time, this also leads to a reduced data rate for these terminals. It should also be noted that a similar effect can be achieved with full path-loss compensation by having the scheduled MCS depend on the estimated downlink path-loss, which can be derived from the power headroom report, and rely on the Δ_{MCS} to reduce the relative terminal transmit power for terminals with higher path loss. However, an even better approach would then be to not

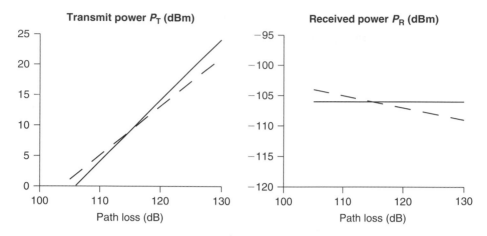

Figure 19.17 *Full vs. partial path-loss compensation. Solid curve. Full compensation ($\alpha = 1$); Dashed curve: Partial compensation ($\alpha = 0.8$).*

only base the MCS selection on the path loss to the current cell but also on the path loss to the neighbor interfered cells.

19.3.3 Power control for SRS

The SRS transmit power basically follows that of the PUSCH, compensating for the exact bandwidth of the SRS transmission. Thus, the power control for SRS transmission can be described according to:

$$P_T = \min\left\{P_{\max}, P_0 + \alpha \cdot PL_{DL} + 10 \cdot \log_{10}\left(M_{SRS}\right) + \delta + P_{SRS}\right\} \qquad (19.4)$$

where the parameters P_0, α, and δ are the same as for PUSCH power control, as discussed in Section 19.3.2. Furthermore, M_{SRS} is the bandwidth, in number of resource block, of the SRS transmission and P_{SRS} is a configurable offset.

19.4 Discontinuous reception (DRX)

Packet-data traffic is often highly bursty with occasional periods of transmission activity followed by longer periods of silence. Clearly, from a delay perspective, it is beneficial to monitor the downlink control signaling in each subframe to receive uplink grants or downlink data transmissions and instantaneously react on changes in the traffic behavior. At the same time this comes at cost in terms of power consumption at the terminal; the receiver circuitry in a typical terminal represents a non-negligible amount of the power consumption. To reduce the terminal power consumption, LTE includes mechanisms for *discontinuous*

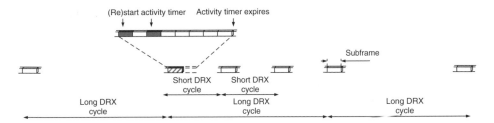

Figure 19.18   *Illustration of DRX operation.*

reception (DRX), which corresponds to the CPC feature for HSPA described in Chapter 12.

The basic mechanism for DRX is a configurable DRX cycle in the terminal. With a DRX cycle configured, the terminal monitors the downlink control signaling only in one subframe per DRX cycle, sleeping with the receiver circuitry switched off in the remaining subframes. This allows for a significant reduction in power consumption; the longer the cycle, the lower the power consumption. Naturally, this implies restrictions to the scheduler at the terminal can be addressed only in the active subframes.

In many situations, if the terminal has been scheduled and active with receiving or transmitting data in one subframe it is highly likely it will be scheduled again in the near future. One reason could be that it was not possible to transmit all the data in the transmission buffer in one subframe and additional subframes are required. Waiting until the next active subframe according to the DRX cycle, although possible, would result in additional delays. Hence, to reduce the delays, the terminal remains in the active state for a certain configurable time after being scheduled. This is implemented by the terminal (re)starting an inactivity timer every time it is scheduled and remain awake until the time expires as illustrated at the top of Figure 19.18.

Retransmissions take place, regardless of the DRX cycle. Thus, the terminal receives and transmits hybrid-ARQ acknowledgments as normal in response to data transmission. In the uplink, this also includes retransmissions in the subframes given by the synchronous hybrid-ARQ timing relation. In the downlink, where asynchronous hybrid-ARQ is used, there the retransmission instant is not fixed in the specifications. To handle this, the terminal monitors the downlink for retransmissions in a configurable time window after the previous transmission.

The above mechanism, a (long) DRX cycle in combination with the terminal remaining awake for some period after being scheduled, is sufficient for most scenarios. However, some services, most notably VoIP, are characterized by a periods of regular transmission instants, followed by periods of no or very little activity. To handle these services, a second short DRX cycle can optionally be used in addition to the long cycle described above. Normally, the terminal follows the long DRX cycle, but in case it has been recently scheduled, it follows a shorter DRX cycle for some time. Handling VoIP in this scenario can be done by setting the short DRX cycle to 20 ms as the voice codec typically delivers a VoIP packet per 20 ms. The long DRX cycle is then used to handle longer periods of silence between talk spurts.

19.5   Uplink timing alignment

The DFTS-OFDM-based LTE uplink transmission scheme allows for uplink intra-cell orthogonality, implying that uplink transmissions received from different mobile terminals within a cell do not cause interference to each other. A requirement for this *uplink orthogonality* to hold is that the signals transmitted from different mobile terminals within the same subframe but within different frequency resources arrive approximately time aligned at the base station. More specifically, any timing misalignment between signals received from different mobile terminals should fall within the cyclic prefix. Note that this should also include any time dispersion on the radio channel. For TDD, uplink timing alignment is also required to separate downlink and uplink transmissions; any misalignment in the timing between terminals at the base station would cause interference between uplink and downlink transmissions as discussed later. To ensure such receiver-side time alignment, LTE includes a mechanism known as *timing advance*. In principle, this is the same as the uplink transmission timing control discussed for uplink OFDM in Chapter 4.

In essence, timing advance is a negative offset, at the mobile terminal, between the start of a received downlink subframe and a transmitted uplink subframe. By controlling the offset appropriately for each mobile terminal, the network can control the timing of the signals received at the base station from the mobile terminals. In essence, mobile terminals far from the base station encounter a larger propagation delay and therefore need to start their uplink transmissions somewhat in advance, compared to mobile terminals closer to the base station, as illustrated in Figure 19.19. In this specific example, the first mobile terminal (UE1) is located close to the base station and experiences a small propagation delay, $T_{P,1}$. Thus, for this mobile terminal, a small value of the timing advance offset $T_{A,1}$ is sufficient to compensate for the propagation delay and to ensure the correct timing at the base station. However, a larger value of the timing advance is required for second mobile terminal (UE2), which is located at a larger distance from the base station and thus experiencing a larger propagation delay.

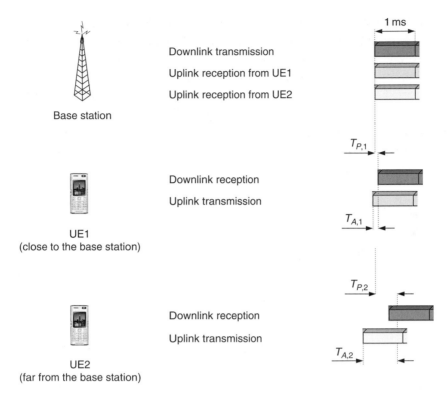

Figure 19.19 *Uplink timing advance.*

The timing-advance value for each mobile terminal is determined by the network based on measurements on the respective uplink transmissions. Hence, as long as a mobile terminal carries out uplink data transmission, this can be used by the receiving base station to estimate the uplink receive timing and thus be a source for the timing-advance commands. Uplink channel-sounding can be used as a regular signal to measure upon, but in principle the base station can use any signal transmitted from the terminals.

Based on the uplink measurements, the network determines the required timing correction for each terminal. If the timing of a specific terminal needs correction, the network issues a timing-advance command for this specific mobile terminal, instructing it to retard or advance its timing relative to the current uplink timing. The user-specific timing-advance command is transmitted as a MAC control element on the DL-SCH. The maximum value possible timing advance is 0.67 ms, corresponding to a terminal-to-base-station distance of slightly more than 100 km. This is also the value assumed when determining the processing time for decoding as discussed in Section 19.1.1. Typically, timing-advance commands to a mobile terminal are transmitted relatively infrequent, for example, one or few times per second.

If the terminal has not received a timing-advance command during a (configurable) period, the terminal assumes it has lost the uplink synchronization. Hence, prior to any PUSCH or PUCCH transmission in the uplink, an explicit timing-re-alignment phase using the random access procedure must be performed to restore the uplink time alignment.

For TDD, the timing alignment procedure is also closely related to the special subframes. As discussed at the beginning of Chapter 16, an essential aspect of any TDD system is to provide the possibility for a sufficiently large *guard period* (or guard time) where neither downlink nor uplink transmissions occur. This guard period is necessary for switching from downlink to uplink transmission and vice versa. In LTE, the special subframes (subframe one and, for some uplink–downlink configurations, subframe six) are split into three parts: a downlink part (DwPTS), a guard period (GP), and an uplink part (UpPTS).

The length of the guard period in the special subframe depends on several factors. First, it should be sufficiently large to provide the necessary time for the circuitry in the base station and the terminals to switch from downlink to uplink. Switching is typically relatively fast, in the order of a few tens of microseconds, and, in most deployments, does not significantly contribute to the required guard time.

Secondly, the guard time should also ensure that uplink and downlink transmissions do not interfere at the base station. This is handled by setting the timing-advance value such that, at the base station, the last uplink subframe before the uplink-to-downlink switch ends before the start of the first downlink subframe. Obviously, the guard period must be large enough to allow the terminal to receive the downlink transmission and switch to transmission before it should start the (timing-advanced) uplink transmission. Hence, in essence, some of the guard period of the special subframe is 'moved' to the uplink-to-downlink switch by the timing-advance mechanism. This is illustrated in Figure 19.20. As the timing advance, as discussed earlier, is proportional to the distance to the base station, a larger guard period is required when operating in large cells compared to small cells.

Finally, the selection of the guard period also needs to take interference between base stations into account. In a multi-cell network, inter-cell interference from downlink transmissions in neighboring cells must decay to a sufficiently low level before the base station can start to receive uplink transmissions. Hence, a larger guard period that motivated by the cell size itself may be required as

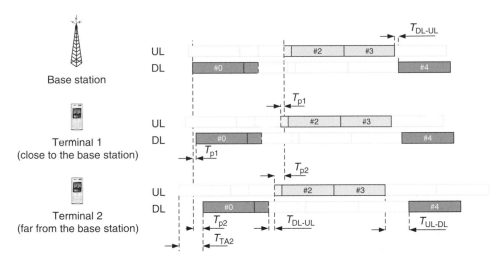

Figure 19.20 *Timing relation for TDD operation.*

the last part of the downlink transmissions form distant base stations otherwise may interfere with uplink reception. The amount of guard period depends on the propagation environments but in some cases the inter-base-station interference is the dominating factor when determining the guard period.

From the above discussion, it is clear that a sufficient amount of configurability of the guard period is needed to meet different deployment scenarios. Therefore, a set of DwPTS/GP/UpPTS configurations is supported, where each configuration corresponds to a given length of the three fields in the special subframes. The DwPTS/GP/UpPTS configuration used in the cell is signaled as part of the system information.

In addition to supporting wide range of different guard periods, an important aspect in the design of LTE was to simplify coexistence with and migration from systems based on the 3GPP TD-SCDMA standard (see Chapter 23 for details on TD-SCDMA). Basically, to handle inter-system interference from two different but co-sited TDD systems operating close in frequency, it is necessary to align the downlink/uplink switchpoints between the two systems. Since LTE supports configurable lengths of the DwPTS field, the switchpoints of LTE and TD-SCDMA can be aligned, despite the different subframe lengths used in the two systems. Aligning the switch-points between TD-SCDMA and LTE is the technical reasons for splitting the special subframe into the three fields DwPTS/GP/UpPTS instead of locating the switch-point at the subframe boundary. An example of LTE/TD-SCDMA coexistence is given in Figure 19.21.

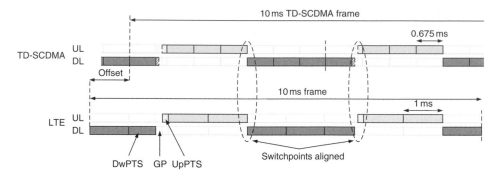

Figure 19.21   *Coexistence between TD-SCDMA and LTE.*

Table 19.2   *Resulting guard period for different DwPTS and UpPTS lengths (normal cyclic prefix).*

	UpPTS	
DwPTS	1	2
3	10	9
9	4	3
10	3	2
11	2	1
12	1	

The set of possible lengths of DwPTS/GP/UpPTS are selected to support common coexistence deployments, as well as to provide a high degree of guard-period flexibility for the reasons discussed earlier. The UpPTS length is one or two DFTS-OFDM symbols and the DwPTS length can vary from three[4] to twelve OFDM symbols, resulting in guard periods ranging from one to ten OFDM symbols. The resulting guard period for the different DwPTS and UpPTS configurations supported is summarized in Table 19.2 for the case of normal cyclic prefix. The DwPTS is in essence treated as a normal downlink subframe, although the amount of data possible to transmit is smaller due to the reduced length of the DwPTS compared to a normal subframe. The UpPTS, however, is due to the very short duration, not used for data transmission.

---

[4] The smallest DwPTS length is motivated by the location of the primary synchronization signal in the DwPTS.

Instead, it can be used for channel sounding or random access. It can also be left empty, in which case it in practice serves as guard period.

19.6 UE categories

The capabilities of the physical layer for a specific implementation depend on a set of parameters, for example, the modulation schemes a certain implementation supports. In principle, the different parameters could be specified separately, but to limit the number of combinations and avoid a parameter combination that does not make sense, a set of physical layer capabilities are lumped together to form a UE category. This is similar to the approach taken for HSPA.

In total five different UE categories have been specified for LTE, ranging from low-end category one not supporting spatial multiplexing and only a small amount of soft buffer to high-end category five supporting the full set of features supported by the physical layer specifications. The categories are summarized in Table 19.3. In the specifications, the peak data rates are expressed as the maximum number of transport block bits the terminal can receive in a TTI, whereas Table 19.3 directly states the peak rate to ease the reading. The size of the soft buffer has also been rounded; for the exact numbers see [129]. Note that, regardless of the category, a terminal is always capable of receiving transmissions from up to four antenna ports. This is necessary as the system information can be transmitted on up to four antenna ports.

In addition to the capabilities mentioned in the different UE categories, there are some capabilities specified outside the categories too. The supported duplexing schemes is one such example, the support of UE-specific reference signals for FDD being another. Whether the terminal supports other radio-access technologies, for example, GSM and WCDMA, are also declared separately.

Table 19.3 *UE categories.*

Category	1	2	3	4	5
Downlink peak rate (Mbit/s)	10	50	100	150	300
Uplink peak rate (Mbit/s)	5	25	50	50	75
Soft buffer size (Msoft bit)	0.25	1.2	1.2	1.8	3.7
Maximum downlink modulation			64QAM		
Maximum uplink modulation		16QAM			64QAM
Max layers for spatial multiplexing	1		2		4

20

Flexible bandwidth in LTE

Spectrum flexibility is a key feature of the LTE radio access and is set out in the LTE requirements [86]. It consists of several components, including deployment in different-sized spectrum allocations and deployment in diverse frequency ranges, both in paired and unpaired frequency bands.

There are a number of frequency bands identified for mobile use and for IMT-2000 today. Most of these bands are already defined for operation with WCDMA/HSPA, and LTE is the next step in the 3G evolution to be deployed in those bands. Both paired and unpaired bands are included in the LTE specifications. The additional challenge with LTE operation in some bands is the possibility of using channel bandwidths up to 20 MHz.

The use of OFDM in LTE gives flexibility both in terms of the size of the spectrum allocation needed and in the instantaneous transmission bandwidth used. The OFDM physical layer also enables frequency-domain scheduling. Beyond the physical layer implications described in Chapters 16 and 17, these properties also impact the RF implementation in terms of filters, amplifiers, and all other RF components that are used to transmit and receive the signal. This means that the RF requirements for the receiver and transmitter will have to be expressed with the flexibility in mind.

20.1   Spectrum for LTE

LTE can be deployed both in existing IMT bands and in future bands that may be identified. The possibility to operate a radio-access technology in different frequency bands is, in itself, nothing new. For example, quad-band GSM terminals are common, capable of operating in the 850, 900, 1800, and 1900 MHz bands. From a radio-access functionality perspective, this has no or limited impact and the LTE physical-layer specifications [106–109] do not assume any specific frequency band. What may differ, in terms of specification, between different bands

Table 20.1   *Paired frequency bands defined by 3GPP for LTE.*

Band	Uplink range (MHz)	Downlink range (MHz)	Main region(s)
1	1920–1980	2110–2170	Europe, Asia
2	1850–1910	1930–1990	Americas (Asia)
3	1710–1785	1805–1880	Europe, Asia (Americas)
4	1710–1755	2110–2155	Americas
5	824–849	869–894	Americas
6	830–840	875–885	Japan
7	2500–2570	2620–2690	Europe, Asia
8	880–915	925–960	Europe, Asia
9	1749.9–1784.9	1844.9–1879.9	Japan
10	1710–1770	2110–2170	Americas
11	1427.9–1452.9	1475.9–1500.9	Japan
12	698–716	728–746	Americas
13	777–787	746–756	Americas
14	788–798	758–768	Americas

are mainly the more specific RF requirements such as the allowed maximum transmit power, requirements/limits on out-of-band (OOB) emission, and so on. One reason for this is that external constraints, imposed by regulatory bodies, may differ between different frequency bands.

20.1.1   Frequency bands for LTE

The frequency bands where LTE will operate will be in both paired and unpaired spectrum, requiring flexibility in the duplex arrangement. For this reason, LTE supports both FDD and TDD as discussed in the previous chapters.

Release 8 of the 3GPP specifications for LTE includes fourteen frequency bands for FDD and eight for TDD. The paired bands for FDD operation are numbered from 1 to 14 [126] as shown in Table 20.1. The unpaired bands for TDD operation are numbered from 33 to 40 as shown in Table 20.2. Note that the frequency bands for UTRA FDD use the same numbers as the paired LTE bands, but are labeled with Roman numerals from I to XIV. All bands for LTE are summarized in Figure 20.1 and Figure 20.2, which also show the corresponding frequency allocation defined by the ITU.

Some of the frequency bands are partly or fully overlapping. This is in most cases explained by regional differences in how the bands defined by the ITU are implemented. At the same time, a high degree of commonality between the

Table 20.2 *Unpaired frequency bands defined by 3GPP for LTE.*

Band	Frequency range (MHz)	Main region(s)
33	1900–1920	Europe, Asia (not Japan)
34	2010–2025	Europe, Asia
35	1850–1910	–
36	1930–1990	–
37	1910–1930	–
38	2570–2620	Europe
39	1880–1920	China
40	2300–2400	Europe, Asia

bands is desired to enable global roaming. The set of bands have evolved over time as bands for UTRA, with each band originating in global, regional, and local spectrum developments. The complete set of UTRA bands was then transferred to the LTE specifications.

Bands 1, 33, and 34 are the same paired and unpaired bands that were defined first for UTRA in Release 99 of the 3GPPP specifications. *Band 2* was added later for operation in the US PCS1900 band and *Band 3* for 3G operation in the GSM1800 band. The unpaired *Bands 35, 36, and 37* are also defined for the PCS1900 frequency ranges, but are not deployed anywhere today.

Band 4 was introduced as a new band for the Americas following the addition of the 3G bands at WRC-2000. Its downlink overlaps completely with the downlink of Band 1, which facilitates roaming and eases the design of dual Band 1 + 4 terminals. *Band 10* is an extension of Band 4 from 2 × 45 to 2 × 60 MHz.

Band 9 overlaps with Band 3, but is also intended only for Japan. The specifications are drafted in such a way that implementation of roaming dual Band 3 + 9 terminals is possible. The 1500 MHz frequency band is also identified in 3GPP for Japan as *Band 11*. It is allocated globally to mobile service on a co-primary basis and was previously used for 2G in Japan.

With WRC-2000, the band 2500–2690 MHz was identified for IMT-2000 and it is identified as *Band 7* in 3GPP for FDD and *Band 38* for TDD operation in the 'center gap' of the FDD allocation. *Band 39* is an extension of the unpaired Band 33 from 20 to 40 MHz for use in China.

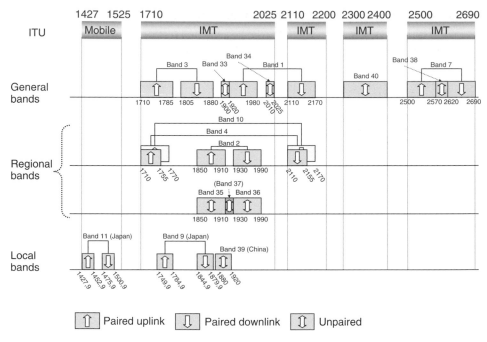

Figure 20.1   *Operating bands specified in 3GPP above 1 GHz and the corresponding ITU allocation.*

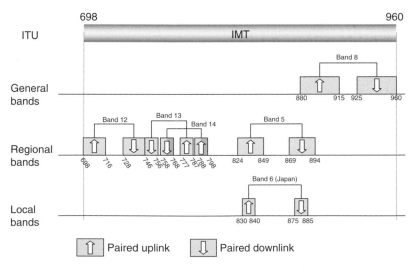

Figure 20.2   *Operating bands specified in 3GPP below 1 GHz and the corresponding ITU allocation.*

WRC-2000 also identified the frequency ranges 806–960 MHz for IMT-2000. As shown in Figure 20.2, *Bands 5, 6, and 8* are defined for FDD operation in this range. Band 8 uses the same band plan as GSM900. Bands 5 and 6 overlap, but are intended for different regions. Band 5 is based on the US cellular band, while Band 6 is restricted to Japan in the specifications. 2G systems in Japan had a very specific band plan and Band 6 is a way of aligning the Japanese spectrum plan in the 810–960 MHz range to that of other parts of the world.

Bands 12, 13, and 14 is the first set of bands defined for what is called the *digital dividend*, that is for spectrum previously used for broadcasting. This spectrum is partly migrated to be used by other wireless technologies, since TV broadcasting is migrated from analog to more spectrum efficient digital technologies.

20.1.2   New frequency bands

Additional frequency bands are continuously specified for UTRA and LTE. WRC-07 identified additional frequency bands for IMT, which encompasses both IMT-2000 and IMT-Advanced. Several bands were defined by WRC-07 that will be available partly or fully for deployment on a global basis:

- *450–470 MHz* was identified for IMT globally. It is already today allocated to mobile service globally, but it is a not a very large band.
- *698–806 MHz* was allocated to mobile service and identified to IMT to some extent in all regions. Together with the band 806–960 MHz identified at WRC-2000, it forms a wide frequency range from 698 to 960 MHz that is partly identified to IMT in all regions, with some variations.
- *2300–2400 MHz* was identified for IMT on a worldwide basis in all three regions.
- *3400–3600 MHz* was allocated to the mobile service on a primary basis in Europe and Asia and partly in some countries in the Americas. There is also satellite use in the bands today.

For the frequency ranges below 1 GHz identified at WRC-07, 3GPP has already specified several operating bands as shown in Figure 20.2. In addition to Bands 5, 6, and 8 described above, *Bands 12, 13, and 14* are defined for operation mainly for US allocations. Note that Band 14 has a special configuration, since the upper part of this band is intended for a public safety network that is to be operated in a private/public partnership by a commercial operator. Work is also ongoing in Europe within Electronic Communications Committee (ECC) Task Group 4 [138] on the technical feasibility for harmonized European spectrum allocations for fixed and mobile applications in the digital dividend (below 862 MHz).

Band 40 is an unpaired band specified for the new frequency range 2300–2400 MHz identified for IMT.

Work in 3GPP is initiated also for the frequency band 3.4–3.8 GHz [125]. In Europe, a majority of countries already license the band 3.4–3.6 GHz for both Fixed Wireless Access and mobile use. Licensing of 3.6–3.8 GHz for Wireless Access is more limited. There is a European spectrum decision for 3.4–3.8 GHz with 'flexible usage modes' for deployment of fixed, nomadic, and mobile networks. Frequency arrangements considered in the decision include FDD use with 100 MHz block offset between paired blocks and/or TDD use. In Japan, not only 3.4–3.6 GHz but also 3.6–4.2 GHz will be available to terrestrial mobile services such as IMT to use after 2010. The band 3.4–3.6 GHz has been licensed for wireless access also in Latin America.

20.2   Flexible spectrum use

Many of the frequency bands identified above for deployment of LTE are existing IMT-2000 bands and some also have other systems deployed in those bands, including WCDMA/HSPA and GSM. Bands are also in some regions defined in a 'technology neutral' manner, which means that coexistence between different technologies is a necessity.

The fundamental LTE requirement to operate in different frequency bands [85] does not, in itself, impose any specific requirements on the radio interface design. There are however implications for the RF requirements and how those are defined, in order to support the following:

- *Coexistence between operators in the same geographical area in the band*: These other operators may deploy LTE or other IMT-2000 technologies, such as UMTS/HSPA and GSM/EDGE. There may also be non-IMT-2000 technologies. Such coexistence requirements are to a large extent developed within 3GPP, but there may also be regional requirements defined by regulatory bodies in some frequency bands.
- *Co-location of BS equipment between operators*: There are in many cases limitations to where BS equipment can be deployed. Often sites must be shared between operators or an operator will deploy multiple technologies in one site. This puts additional requirement on both BS receivers and transmitters.
- *Coexistence with services in adjacent frequency bands and across country borders*: The use of the RF spectrum is regulated through complex international

agreements, involving many interests. There will therefore be requirements for coordination between operators in different countries and for coexistence with services in adjacent frequency bands. Most of these are defined in different regulatory bodies. Sometimes the regulators request that 3GPP includes such coexistence limits in the 3GPP specifications.

- *Release independent frequency band principles*: Frequency bands are defined regionally and new bands are added continuously. This means that every new release of 3GPP specifications will have new bands added. Through the 'release independence' principle, it is possible to design terminals based on an early release of 3GPP specifications that support a frequency band added in a later release.

20.3 Flexible channel bandwidth operation

The frequency allocations in Figures 20.1 and 20.2 are up to $2 \times 75\,\mathrm{MHz}$, but the spectrum available for a single operator may be from $2 \times 20\,\mathrm{MHz}$ down to $2 \times 5\,\mathrm{MHz}$ for FDD and down to $1 \times 5\,\mathrm{MHz}$ for TDD. Furthermore, the migration to LTE in frequency bands currently used for other radio-access technologies must often take place gradually to ensure that sufficient amount of spectrum remains to support the existing users. Thus the amount of spectrum that can initially be migrated to LTE may be relatively small, but may then gradually increase, as shown in Figure 20.3. The variation of possible spectrum scenarios

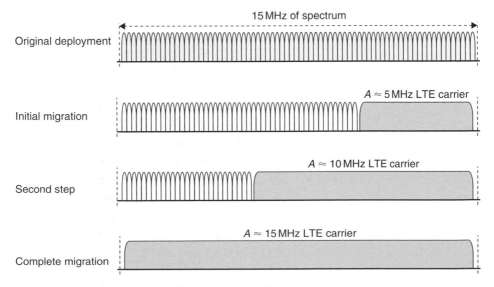

Figure 20.3 *Example of how LTE can be migrated step-by-step into a spectrum allocation with an original GSM deployment.*

will imply a requirement for spectrum flexibility for LTE in terms of the transmission bandwidths supported.

The spectrum flexibility requirement points out the need for LTE to be scalable in the frequency domain. This flexibility requirement is in [86] stated as a list of LTE spectrum allocations from 1.25 to 20 MHz. Note that the final channel bandwidths selected differ slightly from this initial assumption.

As shown in Chapters 16 and 17, the frequency-domain structure of LTE is based on resource blocks consisting of 12 subcarriers with a total bandwidth of 12×15 kHz $= 180$ kHz. The basic radio-access specification including the physical-layer and protocol specifications enable *transmission bandwidth configurations* from 6 up to 110 resource blocks on one LTE RF carrier. This allows for channel bandwidths ranging from 1.4 MHz up to beyond 20 MHz in steps of 180 kHz and is fundamental to providing the required spectrum flexibility.

In order to limit implementation complexity, only a limited set of bandwidths are defined in the RF specifications. Based on the frequency bands available for LTE deployment today and in the future as described above and considering the known migration and deployment scenarios in those bands, a limited set of six channel bandwidths are specified. The RF requirements for the BS and terminal are defined only for those six channel bandwidths. The channel bandwidths range from 1.4 to 20 MHz as shown in Table 20.3. The lower bandwidths 1.4 and 3 MHz are chosen specifically to ease migration to LTE in spectrum where CDMA2000 is operated, and also to facilitate migration of GSM and TD-SCDMA to LTE. The specified bandwidths target relevant scenarios in different frequency bands. For this reason, the set of bandwidths available for a specific band is not necessarily the same as in other bands. At a later stage, if new frequency bands are made available that have other spectrum scenarios requiring additional channel bandwidths, the corresponding RF parameters and requirements can be added in the RF specifications, without actually having to update the physical-layer specifications. The process of adding new channel bandwidths is in this way similar to adding new frequency bands.

Figure 20.4 illustrates in principle the relationship between the channel bandwidth and the number of resource blocks for one RF carrier. Note that for all channel bandwidths except 1.4 MHz, the resource blocks in the transmission bandwidth configuration fill up 90% of the channel bandwidth. The spectrum emissions shown in Figure 20.4 are for a pure OFDM signal, while the actual transmitted emissions will depend also on the transmitter RF chain and other components. The emissions outside the channel bandwidth are called *unwanted emissions* and the requirements for those are discussed further below.

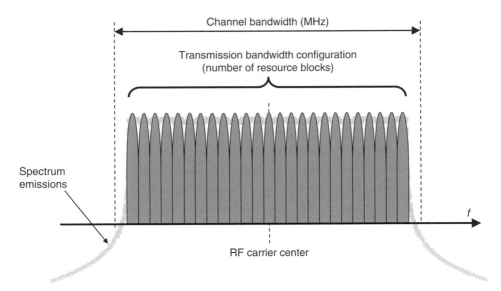

Figure 20.4 *The channel bandwidth for one RF carrier and the corresponding transmission bandwidth configuration.*

Table 20.3 *Channel bandwidths specified in LTE.*

Channel bandwidth (MHz)	Number of resource blocks
1.4	6
3	15
5	25
10	50
15	75
20	100

20.4 Requirements to support flexible bandwidth

20.4.1 RF requirements for LTE

The RF requirements define the receiver and transmitter RF characteristics of a BS or UE. The BS is the physical node that transmits and receives RF signals on one or more antenna connectors to cover one cell. UE is the 3GPP term for the terminal.

The set of RF requirements defined for LTE is fundamentally the same as those defined for UTRA or any other radio system. Some requirements are also based on regulatory requirements and are more related to the frequency band of operation and/or the place where the system is deployed, than it is related to the type of system.

What is particular to LTE is the flexible bandwidth and the related multiple channel bandwidths of the system, which makes some requirements more difficult to define. It has special implications for the transmitter requirements on unwanted emissions, where the definition of the limits in international regulation depends on the channel bandwidth, which becomes difficult for a system where the BS may operate with multiple channel bandwidths and the terminal may vary its channel bandwidth of operation. The properties of the flexible OFDM-based Layer 1 also have implications for specifying the transmitter modulation quality and how to define the receiver selectivity and blocking requirements.

There are also some differences in how the requirements for the terminal and BS requirements respectively are defined. For this reason, they are treated separately in this chapter. The detailed background of the RF requirements for LTE is described in [130] and [132]. The RF requirements for the BS are specified in [127] and for the terminal (UE) in [126]. The RF requirements are divided into transmitter and receiver characteristics. There are in addition 'performance characteristics' which define the receiver baseband performance and are thus not strictly RF requirements, though the performance will also depends on the RF to some extent.

Transmitter characteristics are maximum output power, output power dynamics, transmitted signal quality (mainly frequency error and Error Vector Magnitude, (EVM), unwanted emissions, and transmitter intermodulation.

Receiver characteristics are reference sensitivity level, receiver dynamic range, Adjacent Channel Selectivity (ACS), receiver blocking (including spurious response for the terminal), receiver intermodulation, and receiver spurious emissions.

Each RF requirement has a corresponding test defined in the LTE test specifications for the BS [128] and the terminal [131]. These specifications define the test setup, test procedure, test signals, tolerances, etc. needed to show compliance with the RF and performance requirements.

The discussion below will focus on requirements where the flexible bandwidth properties of LTE have particular implications.

20.4.2 Regional requirements

There are a number of regional variations to the RF requirements and their application. The variations originate in different regional and local regulation of spectrum and its use. The most obvious regional variation is the different frequency bands and their use as discussed above. Many of the regional RF requirements are also tied to specific frequency bands.

When there is a regional requirement on for example spurious emissions, this requirement should be reflected in the 3GPP specifications. For the BS it is entered as an optional requirement and is marked as 'regional.' For the terminal, the same procedure is not possible, since a terminal may roam between different regions and will therefore have to fulfill all regional requirements that are tied to an operating band in the regions where the band is used. For LTE, this becomes more complex than for UTRA, since there is an additional variation in the transmitter (and receiver) bandwidth used, making some regional requirements difficult to meet as a mandatory requirement. The concept of *network signaling* of RF requirements is therefore introduced for LTE, where a terminal can be informed at call setup of whether some specific RF requirements apply when the terminal is connected to a network.

Examples of regional requirements are:

- *Spurious emissions*: Different 'categories' of emission levels are defined by ITU-R [134] and applied in different regions. These are called categories A and B.
- *Coexistence with other systems in the same geographical area*: Since the type of system to coexist with varies between regions, this is often a regional requirement. In each region, the requirement is however usually mandatory. For terminals, it will normally be mandatory for any roaming device.
- *Co-location with other BS*: The type of BS to be potentially co-located with also varies between regions. Co-location requirements are however usually not mandatory from a regulatory point of view.

The way regional regulation is set also varies considerably. In Europe, most requirements are developed in cooperation between the standards body ETSI and the ECC, who work under mandate from the European Commission. The regulation for the US operating bands is developed by the FCC. Also Japan has a local radio regulation that is reflected in the 3GPP specifications.

20.4.3 BS transmitter requirements

Unwanted emissions from the transmitter are divided into *OOB emission* and *spurious emissions* in ITU-R recommendations [134]. OOB emissions are defined as emission on a frequency close to the RF carrier, which results from the modulation process. Spurious emissions are emissions outside the RF carrier that may be reduced without affecting the corresponding transmission of information, but excluding OOB emissions. Examples of spurious emissions are harmonic emissions, intermodulation products, and frequency conversion products. The frequency range where OOB emissions are normally defined is called the *OOB domain* whereas spurious emission limits are normally defined in the *spurious domain*.

ITU-R also defines the limit between the OOB and spurious domains at a frequency separation from the carrier center of 2.5 times the *necessary bandwidth*, which equals 2.5 times the Channel bandwidth for E-UTRA. This division of the requirements is applied for UTRA which has a fixed channel bandwidth, but becomes more difficult for LTE, which is a flexible bandwidth system implying that the frequency range where requirements apply would then vary with the channel bandwidth.

As shown in Chapter 4 the spectrum of an OFDM signal decays rather slowly outside of the transmission bandwidth configuration. Since the transmitted signal occupies 90% of the channel bandwidth, it is not possible to directly meet the unwanted emission limits with a 'pure' OFDM signal. The techniques used for achieving the transmitter requirements are however not specified or mandated in LTE specifications. Time-domain windowing is one method commonly used in OFDM-based transmission systems to control spectrum emissions. Filtering is always used, both time-domain digital filtering of the baseband signal and analog filtering of the RF signal. Since the RF signal in the downlink needs to be amplified with a power amplifier that has nonlinear characteristics, linearization schemes are also an essential part of controlling spectrum emissions.

The discussion below is related to the unwanted emission requirements. Those limits are to be fulfilled over a specified dynamic range of the transmitter, both in terms of variations of the total transmitted power and of the power per resource element in the OFDM signal. There are in addition requirements for the modulation quality in terms of *frequency error* and *EVM*, both of which define the difference between an ideal OFDM signal at the assigned channel frequency and the actual transmitted RF signal.

20.4.3.1 Operating band unwanted emissions

For the reasons above, a unified concept of *operating band unwanted emissions* is used for the LTE BS instead of the usual spectrum mask defined for OOB emissions. This requirement applies over the whole BS transmitter operating band, plus an additional 10 MHz on each side as shown in Figure 20.5. All requirements outside of that range are set by the 'regular' spurious emission

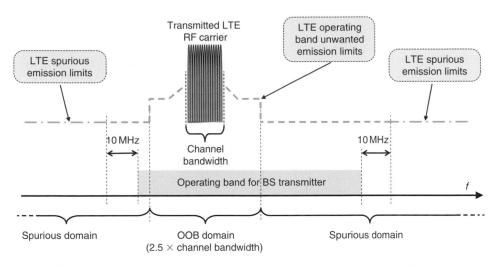

Figure 20.5 *Defined frequency ranges for spurious emissions and operating band unwanted emissions.*

limits, based on the regulatory limits. Since the operating band unwanted emissions are defined over a frequency range that for the wider channel bandwidths are completely in the OOB domain, while it for the smaller channel bandwidths can be both in spurious and OOB domain, the limits are for all cases set in a way that complies with the ITU-R recommendations for spurious emissions [134]. The operating band unwanted emissions are defined with a 100 kHz measurement bandwidth.

There are special limits defined by FCC regulation [143] for the operating bands used in the US. Those are specified as separate limits in addition to the operating band unwanted emission limits.

20.4.3.2 *Adjacent Channel Leakage Ratio*

In addition, the OOB emissions are defined by an *Adjacent Channel Leakage Ratio* (ACLR) requirement. The ACLR concept is very useful for analysis of coexistence between two systems that operate on adjacent frequencies. ACLR defines the ratio of the power transmitted within the assigned channel bandwidth, to the power of the unwanted emissions transmitted on an adjacent channel. There is a corresponding receiver requirement called *Adjacent Channel Selectivity* (ACS), which defines a receiver's ability to suppress a signal on an adjacent channel.

The definitions of ACLR and ACS are illustrated in Figure 20.6 for a wanted and an interfering signal received in adjacent channels. The interfering signal's leakage of unwanted emissions at the wanted signal receiver is given by the ACLR and the ability of the receiver of the wanted signal to suppress the interfering

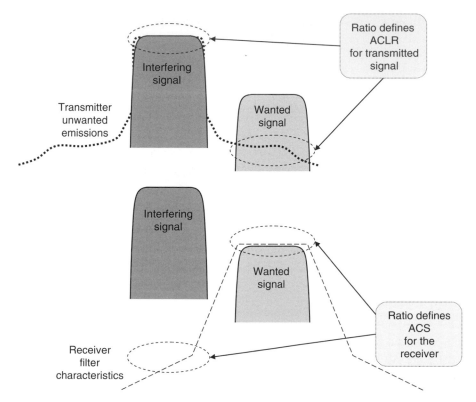

Figure 20.6 *Definitions of ACLR and ACS, using example characteristics of an 'aggressor' interfering and a 'victim' wanted signal.*

signal in the adjacent channel is defined by the ACS. The two parameters when combined define the total leakage between two transmissions on adjacent channels. That ratio is called *Adjacent Channel Interference Ratio* (ACIR) and is defined as the ratio of the power transmitted on one channel to the total interference received by a receiver on the adjacent channel, due to both transmitter (ACLR) and receiver (ACS) imperfections.

This relation between the adjacent channel parameters is [135]

$$\text{ACIR} = \frac{1}{\dfrac{1}{\text{ACLR}} + \dfrac{1}{\text{ACS}}} \tag{20.1}$$

Both ACLR and ACS can be defined with different channel bandwidths for the two adjacent channels, which is the case for some requirements set for LTE due to the bandwidth flexibility. The equation above will also apply for different channel bandwidths, but only if the same two channel bandwidths are used for defining all three parameters ACIR, ACLR, and ACS used in the equation.

The ACLR limits for LTE are derived based on extensive coexistence analysis [133] between LTE and potential LTE or other systems on adjacent carriers. Requirements on ACLR and operating band unwanted emissions both cover the OOB domain, but the operating band unwanted emission limits are set slightly more relaxed compared to the ACLR, since they are defined in a much narrower measurement bandwidth of 100 kHz. This allows for some variations in the unwanted emissions due to intermodulation products from varying power allocation between resource blocks within the channel.

For an LTE BS, there are ACLR requirements both for an adjacent channel with a UTRA receiver and with an LTE receiver of the same channel bandwidth.

20.4.3.3 *Spurious emissions*
The limits for spurious emissions are taken from international recommendations [134], but are only defined in the region outside the frequency range of operating band unwanted emissions limits as described above, that is at frequencies that are separated from the BS transmitter operating band with at least 10 MHz. There are also the additional regional or optional limits for protection of other systems that LTE may coexist with or even be co-located with. Examples of other systems considered in those additional spurious emissions requirements are GSM, UTRA FDD/TDD, CDMA, and PHS.

20.4.4 *BS receiver requirements*

The set of receiver requirements for LTE is quite similar to what is defined for UTRA, but many of them need to be defined differently, due to the flexible bandwidth properties. The receiver characteristics are fundamentally specified in three parts:

- Requirements for receiving the wanted signal in itself, including reference sensitivity and dynamic range.
- Requirements for the receiver's susceptibility to different types of interfering signals.
- Requirements on unwanted emissions from the receiver.

20.4.4.1 *Reference sensitivity and receiver dynamic range*
The primary purpose of the *reference sensitivity requirement* is to verify the receiver *Noise Figure*, which is a measure of how much the receiver's RF

signal chain degrades the SNR of the received signal. For this reason, a low SNR transmission scheme using QPSK is chosen as reference channel for the reference sensitivity test. The reference sensitivity is defined at a receiver input level where the throughput is 95% of the maximum throughput for the reference channel.

A terminal in LTE may be assigned only a small part of the uplink channel bandwidth, implying that the sensitivity should be defined for smaller bandwidths, ideally per resource block. For complexity reasons, a maximum granularity of 25 resource block has been chosen, which means that for channel bandwidths larger than 5 MHz, sensitivity is verified over multiple adjacent 5 MHz blocks, while it is only defined over the full channel for smaller channel bandwidths.

The intention of the *dynamic range requirement* is to ensure that the BS can receive with high throughput also in the presence of increased interference and corresponding higher wanted signal levels, thereby testing the effects of different receiver impairments. In order to stress the receiver a higher SNR transmission scheme using 16QAM is applied for the test. In order to further stress the receiver to higher signal levels, an interfering AWGN signal at a level 20 dB above the assumed noise floor is added to the received signal.

20.4.4.2   *Receiver susceptibility to interfering signals*
There is a set of requirements for defining the BS ability to receive a wanted signal in the presence of an interfering signal. The reason for the multiple requirements is that depending on the frequency offset of the interferer from the wanted signal, the interference scenario may look very different and different types of receiver impairments will impact the performance. The intention of the different combinations of interfering signals is to model as far as possible the range of possible scenarios with interfering signals of different bandwidths that may be encountered inside and outside the BS receiver operating band.

The following requirements are defined, starting from interferers with large frequency separation and going close-in (see also Figure 20.7). In all cases where the interfering signal is an LTE signal, it has the same bandwidth as the wanted signal, but at the most 5 MHz.

- *Blocking*: Corresponds to the scenario with strong interfering signals received outside the operating band (out-of-band) or inside the operating band, but not adjacent to the wanted signal (in-band, including the first 20 MHz outside the band). The scenarios are modeled with a *Continuous Wave* (CW) signal for the out-of-band case and an LTE signal for the in-band case. There are additional

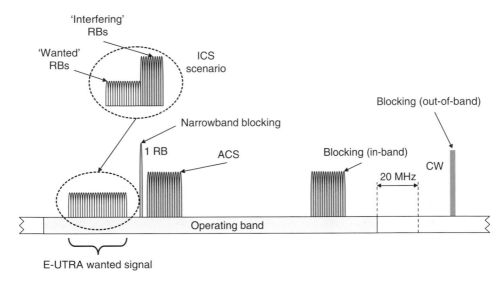

Figure 20.7 *Requirements for receiver susceptibility to interfering signals in terms of blocking, ACS, narrowband blocking, and in-channel selectivity (ICS).*

(optional) blocking requirements for the scenario when the BS is co-sited with another BS in a different operating band.

- *Adjacent Channel Selectivity*: The ACS scenario is a strong signal in the channel adjacent to the wanted signal and is closely related to the corresponding ACLR requirement for the terminal (see also the discussion in Section 20.4.3.2). The adjacent interferer is an LTE signal.
- *Narrowband blocking*: The scenario is an adjacent strong narrowband interferer, which in the requirement is modeled as a single resource block LTE signal.
- *In-channel selectivity* (ICS): The scenario is multiple received signals of different received power levels inside the channel bandwidth, where the performance of the weaker 'wanted' signal is verified in presence of the stronger 'interfering' signal.
- *Receiver intermodulation*: The scenario is *two* interfering signals near adjacent to the wanted signal, where the interferers are one CW and one LTE signal (not shown in Figure 20.7). The interferers are placed in frequency in such a way that the main intermodulation product falls inside the wanted signal's channel bandwidth. There is also a *narrowband intermodulation* requirement where the CW signal is very close to the wanted signal and the LTE interferer is a single RB signal.

For all requirements except in-channel selectivity, the wanted signal uses the same reference channel as in the reference sensitivity requirement. With the interference added, the same 95% relative throughput is met for the reference channel, but at a 'de-sensitized' higher wanted signal level, for most requirements at 6 dB above the reference sensitivity level.

20.4.5   Terminal transmitter requirements

The type of transmitter requirements defined for the terminal is very similar to what is defined for the BS as explained above and the definitions of the requirements are often similar. The output power levels are however considerably lower for a terminal, while the restrictions on the terminal implementation are much higher. There is a tight pressure on cost and complexity for all telecommunications equipment, but this is much more pronounced for terminals, due to the scale of the total market which is approximately one *billion* devices per year.

In the following, the focus will be on requirements where the terminal requirements differ in definition from the corresponding ones for the BS and where there are particular implications from the flexible bandwidth in LTE.

20.4.5.1   Channel bandwidths supported

For the terminal, the channel bandwidths supported are not only a function of the E-UTRA band, but also have a relation to the transmitter and receiver RF requirements. For some of the higher channel bandwidths supported in paired frequency bands where the duplex band gap between uplink and downlink is small, there are certain relaxations of the terminal performance. These relaxations and limitations may consist of a reduced lower power level or an allowed receiver sensitivity reduction (for the highest transmission bandwidths) due to duplex filter constrains.

20.4.5.2   Terminal power level

The terminal output power level is defined in three steps:

- *Terminal power class* defines a *nominal* maximum output power for QPSK modulation. It may be different in different operating bands, but the main terminal power class is today set at 23 dBm.
- *Maximum Power Reduction (MPR)* defines an allowed reduction of maximum power level for certain combinations of modulation used (QPSK or 16QAM) and number of resource blocks that are assigned.
- *Additional Maximum Power Reduction (A-MPR)* may be applied in some regions and is usually connected to specific transmitter requirements such as regional emission limits. For each such set of requirement, there is an associated network signaling value that identifies the allowed A-MPR and the associated conditions.

20.4.5.3   Unwanted emission limits

The unwanted emissions are defined in a slightly different way for the LTE terminal than for the BS. The limits are divided into three parts:

- *In-band emissions* are emissions within the occupied bandwidth (channel bandwidth). The requirement limits how much a terminal can transmit into other resource blocks within the channel bandwidth. Unlike the OOB emissions, the in-band emissions are measured after cyclic prefix removal and FFT, since this is how a terminal transmitter affects a real eNodeB receiver. In-band emissions are in the specification defined as a part of the transmit signal quality requirements.
- *OOB emissions* are defined in terms of a *Spectrum Emissions Mask* (SEM) and an ACLR requirement.
  - The *SEM* is defined as a general mask and a set of additional masks that can be applied to reflect different regional requirements. Each additional mask has an associated network signaling value.
  - *ACLR* limits are set both with assumed UTRA and LTE receivers on the adjacent channel. As for the BS, the limit is also set stricter than the corresponding SEM, thereby accounting for variations in the spectrum emissions resulting from variations in resource block allocations. The ACLR limits are set based on extensive coexistence analysis [133].
- *Spurious emission* limits are defined for all frequency ranges outside the frequency range covered by the SEM. The limits are in general based on international regulations [134], but there are also additional requirements for coexistence with other bands when the mobile is roaming. The additional spurious emission limits can have an associated network signaling value.

The limit between the frequency ranges for OOB limits and spurious limits do not follow the same principle as for the BS. For 5 MHz channel bandwidth, it is set at 250% of the necessary bandwidth as recommended by ITU-R, but for higher channel bandwidths it is set closer than 250%.

20.4.6   Terminal receiver requirements

Also the set of terminal transmitter requirements is similar to what is defined for the BS. The requirements are defined for the full channel bandwidth signals and with all resource blocks allocated for the wanted signal. All receiver requirements assume that the receiver is equipped with two Rx ports using antenna diversity, which does not preclude that a single port terminal can meet a specific requirement.

20.4.6.1 Reference sensitivity

The reference sensitivity is defined using a low SNR reference channel with QPSK modulation in order to verify the terminal noise figure. For the higher channel bandwidths (≥ 5 MHz) in some operating bands, the nominal reference sensitivity needs to be met with a minimum number of allocated resource blocks. For larger allocation, a certain relaxation is allowed.

20.4.6.2 Receiver susceptibility to interfering signals

The set of requirements that defines the terminal's ability to receive a wanted signal in the presence of an interfering signal is very similar to the corresponding BS requirements, as illustrated in Figure 20.7. The requirement levels are different for the terminal, since the interference scenarios for the BS and terminal are very different. There is also no terminal requirement corresponding to the BS in-channel selectivity requirement.

The following requirements are defined:

- *Blocking*: There are both out-of-band blocking and in-band blocking requirements, where in-band includes the first 15 MHz outside the operating band and excludes frequencies adjacent to the carriers. Limits are defined with CW interferers for out-of-band and LTE signals for in-band blocking. A fixed number of *exceptions* are allowed from the terminal out-of-band blocking requirement, for each assigned frequency channel and at the respective *spurious response frequencies*. At those frequencies, the terminal must comply with the more relaxed spurious response requirement.
- *Adjacent Channel Selectivity*: The ACS is specified for two cases with a lower and a higher signal level. The adjacent signal is an LTE signal.
- *Narrowband blocking*: The narrowband blocking requirement is defined with an adjacent CW interfering signal.
- *Receiver intermodulation*: The requirement is defined with one CW and one LTE signal, placed in such a way that the main intermodulation product falls inside the wanted signal. There is also in addition a narrowband intermodulation requirement.

21

System Architecture Evolution

In this chapter an overview of the *System Architecture Evolution* (SAE) work in 3GPP is given. Furthermore, in order to understand from where the SAE is coming from, the core network used by WCDMA/HSPA is discussed. Thus, the system architecture of WCDMA/HSPA and LTE, their connections, similarities, and differences are briefly described. The term system architecture describes the allocation of necessary functions to logical nodes and the required interfaces between the nodes. In the case of a mobile system, such as WCDMA/HSPA and LTE/SAE, most of the necessary functions for the radio interface have been described in the previous chapters. Those functions are normally called radio access network functions. However, in a mobile network several additional functions are needed to be able to provide the services: charging is needed for the operator to charge a user; authentication is needed to ensure that the user is a valid user; service setup is needed to ensure that there is an end-to-end connection; etc. Thus there are functions not directly related to the radio access technology itself, but needed for any radio access technology (and in fact there are functions that are needed also for fixed accesses). Those functions are normally called core network functions. The fact that there are different types of functions in a cellular system have lead to that the system architecture is divided into a radio-access network part and a core-network part (Figure 21.1).

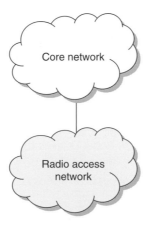

Figure 21.1 *Radio access network and core network.*

21.1 Functional split between radio access network and core network

In the process of specifying the WCDMA/HSPA and the LTE/SAE systems, the first task in both cases was to distribute functions to the *Radio Access Network* (RAN) and *Core Network* (CN), respectively. Although this may initially appear to be a simple task, it can often turn out to be relatively complicated. The vast majority of the functions can easily be located in either RAN or the core network, there are some functions requiring careful attention.

21.1.1 Functional split between WCDMA/HSPA radio access network and core network

For WCDMA/HSPA, the philosophy behind the functional split is to keep the core network unaware of the radio access technology and its layout. This means that the RAN should be in control of all functionality optimizing the radio interface and that the cells should be hidden from the core network. As a consequence, the core network can be used for any radio access technology that adopts the same functional split.

To find the origin of the philosophy behind the WCDMA/HSPA functional split, it is necessary to go back to the architecture of the GSM system, designed during the 1980s. One of the problems with the GSM architecture was that the core network nodes have full visibility of the cells in the system. Thus, when adding a cell to the system, the core network nodes need to be updated. For WCDMA/HSPA, the core network does not know the cells. Instead, the core network knows about service areas and the RAN translates service areas into cells. Thus, when adding a new cell in a service area, the core network does not need to be updated.

The second major difference compared to GSM is the location of retransmission protocols and data buffers in the core network for GSM. Since the retransmission protocols were optimized for the GSM radio interface, those protocols were radio interface specific and hence were not suitable for the WCDMA/HSPA radio interface. This was considered as a weakness of the core network and hence all the buffers and the retransmission protocols were moved to the RAN for WCDMA. Thus, as long as the radio access network uses the same interface to the core network, the *Iu* interface, the core network can be connected to radio access networks based on different radio access technologies.

Still, there are functional splits in WCDMA/HSPA that cannot solely be explained with the philosophy of making the core network radio-access-technology independent. The security functions are a particularly good example. Again, the background can be traced back to GSM, which has the security functions located at different positions for circuit-switched connections and packet-switched connections. For circuit-switched connections, the security functions are located in the GSM RAN, whereas for packet-switched connections, the security functions are located in the GSM core network. For WCDMA/HSPA, this was considered too complicated and a common security location was desired. The location was decided to be in the RAN as the radio resource management signaling and control needed to be secure.

Thus the RAN functions of WCDMA/HSPA are:

- coding, interleaving, modulation, and other typical physical layer functions;
- ARQ, header compression, and other typical link layer functions;
- *Radio Resource Management* (RRM), handover and other typical radio resource control functions; and
- security functions (that is, ciphering and integrity protection).

Functions necessary for any mobile system, but not specific to a radio access network and that do not boost performance, was placed in the core network. Such functions are:

- charging;
- subscriber management;
- mobility management (that is, keeping track of users roaming around in the network and in other networks);
- bearer management and quality-of-service handling;
- policy control of user data flows; and
- interconnection to external networks.

The reader interested in more details about the functional split is referred to the relevant 3GPP documents [89, 90].

21.1.2   Functional split between LTE RAN and core network

The functional split of the LTE RAN and core network is similar to the WCDMA/HSPA functional split. However, a key design philosophy of the LTE RAN was to minimize the number of nodes and find a solution where the RAN consists of only one type of node. At the same time, the philosophy behind the LTE core network is, to the extent possible, to be independent of the radio access technology. The resulting functional split is that most of the functions that for WCDMA/HSPA were classified as RAN functions remain RAN functions. Thus LTE RAN functions are:

- coding, interleaving, modulation and other typical physical layer functions;
- ARQ, header compression, and other typical link layer functions;
- user plane security functions (that is, ciphering) and RAN signaling security (that is, ciphering and integrity protection of RAN originated signaling to the UE); and
- RRM, handover, and other typical radio resource control functions.

Consequently CN functions are:

- charging;
- *Non-Access Stratum* (NAS) security functions (that is, ciphering and integrity protection of core network signaling to the UE);
- subscriber management;
- mobility management (that is, keeping track of users roaming around in the network and in other networks);
- bearer management and quality-of-service handling;
- policy control of user data flows; and
- interconnection to external networks.

The interested reader is referred to [79] and [91] for more information on the functional split between the LTE RAN and the SAE core network.

21.2   HSPA/WCDMA and LTE radio access network

In addition to the functional split between RAN and CN, the RAN-internal architecture also needs to be specified. While any RAN of any radio-access technology at least need a node that connects the antenna of one cell, different radio-access technologies have found different solutions to how many types of nodes and interfaces the RAN shall have.

The RAN architectures of HSPA/WCDMA and LTE are different. Fundamentally, the reason is not only the difference in design philosophy of the RAN/CN split, but also the difference of the radio access technologies and their adopted functions. The following sections will describe the HSPA/WCDMA RAN and the LTE RAN, highlighting their differences and similarities, and providing additional details compared to the previous chapters.

21.2.1 WCDMA/HSPA radio access network

In essence, one important driver for the WCDMA/HSPA RAN architecture is the macro-diversity functionality used by the DCH transport channels. As discussed in Chapter 8, macro-diversity requires an anchor point in the RAN[1] that splits and combines data flows to and from cells that the terminals are currently using. Those cells are called the active set of the terminal.

While it is perfectly possible to have the anchor in the node that connects to the antenna of one cell and have the data flow of other cells go through that node, it is not desirable from a transport-network point of view. Most radio-access networks have transport-network limitations, mainly in the last mile, that is the last hop to the antenna site. Furthermore, the antenna sites are normally leafs in a tree branch and hence an anchor in a leaf often implies that the last mile have to be traversed several times as illustrated in Figure 21.2. Due to this fact, the anchor point was specified to be in a separate node from the node connecting the antenna.

As a consequence of locating the macro-diversity combining above the node connecting to the antenna, the link layer needs to terminate in the same node as the macro-diversity or in a node higher up in the RAN hierarchy. Since the only reason for terminating the link layer in another node than the macro-diversity combining node would be to save transport resources, and having them separated would cause significant complexity, it was decided to have them in the same node. With the same reasoning also the control plane signaling of the RAN was located in the node doing the macro-diversity. The node was named *Radio Network Controller* (RNC), since it basically controls the RAN.

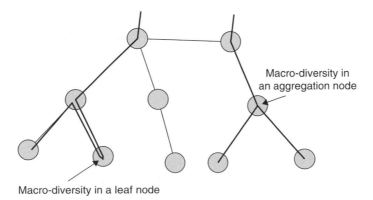

Macro-diversity in an aggregation node

Macro-diversity in a leaf node

Figure 21.2 *Transport network topology influencing functional allocation.*

---

[1] It is fundamentally possible to locate the macro-diversity function in the core network, but since that was against the RAN/CN functional split design philosophy it was never considered.

Although macro-diversity is not used for HSPA in the downlink, it is used in the uplink. This fact and the principle that the architecture also shall support WCDMA Release 99 with minimum changes leads to the RNC being present also in the WCDMA/HSPA architecture.

Figure 21.3 shows an overview of the WCDMA/HSPA radio access network. As can be seen in the figure, the RAN consists of two fundamental logical nodes: the RNC and the node connecting to the antenna of the cells, the *NodeB*.

The RNC is the node connecting the RAN to the core network via the Iu interface. The principle of the Iu interface is that it should be possible to use it toward different RANs, not only WCDMA/HSPA RAN.

Each RNC in the network can connect to every other RNC in the same network using the *Iur* interface. Thus, the Iur interface is a network wide interface making it possible to keep one RNC as an anchor point for a terminal and hide mobility from the core network. Furthermore, the Iur interface is necessary to be able to perform macro-diversity between cells belonging to different RNCs.

As can be seen from Figure 21.3, one RNC connects to one or more NodeBs using the *Iub* interface. However, in contrast to the fact that one RNC can connect to any other RNC in the network, one NodeB can only connect to one RNC. Thus only one RNC is controlling the NodeB. This means that the RNC owns the radio resources of the NodeB. In case of a macro-diversity connection across RNCs, the two RNCs agree between themselves about the use of the radio resources.

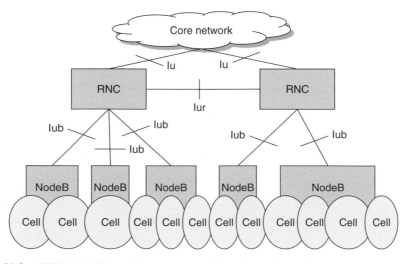

Figure 21.3   *WCDMA/HSPA radio access network: nodes and interfaces.*

The NodeB is a logical node handling the transmission and reception of a set of cells. Logically, the antennas of the cells belong to the NodeB but they are not necessarily located at the same antenna site. For example, in an indoor environment many small cells can be handled by one NodeB in the basement with the antennas in different corridors on different floors. It is the ability of serving cells not transmitted from the same antenna site that makes a NodeB different compared to a *Base Transceiver Station* (BTS), *Base Station* (BS), or *Radio Base Station* (RBS)[2] and therefore a new name was needed – the NodeB was born.

The NodeB owns its hardware but not the radio resources of its cells. Thus, the NodeB can reject a connection due to hardware limitations, but not due to radio resource shortage. With its hardware, the NodeB performs the physical layer functions except for macro-diversity. For HSPA the NodeB also performs the scheduling and hybrid ARQ protocols in the MAC-hs and MAC-e protocol layers as explained in Chapters 9 and 10, respectively.

21.2.1.1 Serving and drift RNC

When specifying where the RAN functionalities should reside, the property of the WCDMA radio interface made it necessary to have a centralized node handling the macro-diversity combining and splitting, as well as being in control of the radio resources in multiple cells. Albeit the NodeB controls a set of cells, the RNC controls several NodeBs and thus a greater area. Furthermore, the Iur interface makes it possible to have a coordinated approach in the whole coverage area of the network.

It is only one RNC, the *controlling RNC*, which is the master of one NodeB. The controlling RNC sets the frequencies the NodeB shall use in its cells; it allocates power and schedules the common channels of the cells of the NodeB; and it configures what codes that shall be used for HS-DSCH and the maximum power used. Furthermore, the controlling RNC is the RNC deciding whether a user is allowed to use the radio resources in a cell belonging to one of its NodeBs and in that case which radio resource. These are tasks not directly related to any user in particular, but to the configurations of the cells.

When a user makes access to the WCDMA/HSPA RAN, it accesses one cell controlled by one NodeB. The NodeB in its turn are controlled by one RNC, the controlling RNC of that NodeB and cell. This controlling RNC will be the RNC terminating the RAN-related control and user planes for that specific terminal. The RNC will become the *serving RNC* for the user. The serving RNC

---

[2] An RBS may be a physical implementation of a NodeB.

is the RNC evaluating measurement reports from the terminal and, based on those reports, deciding which cell(s) should be part of the terminals active set.[3] Furthermore, the serving RNC sets the quality targets of the terminal and it is the RNC that connects the user to the core network. It is also the serving RNC that configures the terminal with radio-bearer configurations enabling the different services that the user wishes to use.

During the connection, the terminal may move and at some point may need to connect to another cell that belongs to another RNC. In such case, the serving RNC of the terminal needs to contact the RNC owning the cell the terminal intends to use, asking for permission to add the new cell to the active set. If the controlling RNC owning the (target) cell accepts, the serving RNC instructs the terminal that it shall add the cell to its active set. The controlling RNC owning the target cell will then become a *drift RNC*. It should be noted that a drift RNC can be a serving RNC for another terminal at the same time. Thus, serving and drift are two different roles an RNC can take in a connection to a terminal. The serving and drift roles are illustrated in Figure 21.4.

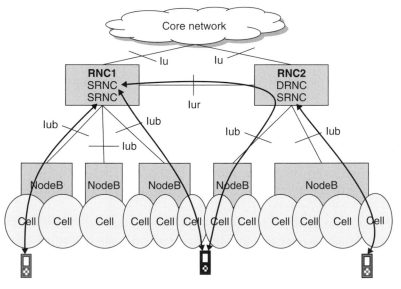

Figure 21.4 *Roles of the RNC. RNC1 is controlling RNC of the NodeBs and cells that are connected to it via Iub interfaces. RNC1 is Serving RNC for the leftmost terminal and the middle terminal. RNC2 is controlling RNC for the NodeBs and cells that are connected to it via Iub interfaces. RNC2 is Serving RNC for the rightmost terminal and at the same time Drift RNC for the middle terminal.*

---

[3] The active set is a set of cells that the terminal is active in. When using dedicated channels all the cells are sending and receiving the DL and the UL transmissions, respectively.

When a terminal has a long-lived connection it is possible that the serving RNC does not control any of the cells that the terminal is currently using. In such case, it is possible to change serving RNC. This is done by means of the *SRNS relocation* procedure.[4]

For the MBMS service, the RNC takes a special role. It is the RNC that decides whether to use broadcast channels in a cell or to use unicast channels. When using unicast channels, the operation is as for normal unicast traffic whereas when using the broadcast channel the RNC has the option to ensure that the same data is transmitted in the surrounding cells owned by the same RNC. By doing that, the terminal can perform macro-diversity combining of the streams from the different cells and the system throughput can be increased.

The basis whether to use unicast or broadcast for MBMS in a cell is typically based on the number of mobile terminals supposed to receive the same content at the same time in the same cell. If there are few users in the cell a unicast approach is more efficient whereas if there are many users in the cell (or in the surroundings of the cell), it is more efficient to use the broadcast channel. The techniques used for the MBMS broadcast channel were discussed in Chapter 11.

21.2.1.2 Architecture of HSPA evolution

3GPP has introduced WCDMA/HSPA RAN architecture migration steps toward a flatter architecture. One simple way is to move the complete RNC to the NodeB. This is in principle already possible with the Release 99 architecture, but with a few issues:

- The number of RNCs is limited to 4096 on the Iu interface. This has been extended and is thus not a problem.
- A more significant issue is the location of the security functions at the NodeB site. The NodeB site is normally considered as an unsecured and remote site. Having the security functions in the NodeB means that important and confidential cryptographic keys need to be transported to the NodeB. This is done over the last mile and it can easily be eavesdropped. Thus also the last mile needs to be secured by some security mechanism, for example *IPsec*. However, this is not sufficient to make the connection secure as also the equipment itself needs to be tamper resistant. This can make the solution complex and expensive. Thus if the operator know that the NodeB is at a secure site, then the operator can deploy a network with NodeBs and RNCs co-located (or with products that have them implemented in the same physical equipment). For sites not secure enough, the operator may use the conventional solution

---

[4] The SRNS relocation procedure is not further elaborated upon in this book. Interested readers are referred to 3GPP TS 25.413 [104] and TS 25.832 [83].

3G Evolution: HSPA and LTE for Mobile Broadband

with an RNC at a secure site higher up in the network and a NodeB at the vulnerable site.

- A third issue with the RNC functionality at the NodeB site is the macro-diversity functionality needed for the HSPA uplink to reach good capacity and quality. As discussed in Section 21.2.1, the macro-diversity location is often better located 'higher' up in the network.

There is also an alternative supported by the 3GPP specifications where the RNC functionality is placed in the NodeB, but the only user plane supported is the interface toward the packet domain, the *Iu_ps* interface. Thus the interface toward the circuit domain, the *Iu_cs* user plane, is not supported. In this case, the Iu_ps functions as normal (HSPA access, etc. as briefly explained in Section 21.3), whereas if the UE wishes to use the CS services, a relocation to a RNC supporting the Iu_cs user plane is necessary, possibly needing to change NodeB and cell too.

In any case, the most important requirement on the RAN architecture for HSPA Evolution is to be able to serve legacy traffic and cooperate with legacy nodes (RNCs and NodeBs). Thus, the Release 99 architecture is a valid architecture also for HSPA Evolution. Furthermore, as discussed above, the possibility of a flatter deployment exist also in this architecture.

21.2.2   LTE radio access network

At the time of adopting the single-node architecture for LTE, the function of macro-diversity was heavily discussed in 3GPP. Although it is technically possible to place the macro-diversity functionality in the corresponding LTE node to a WCDMA/HSPA NodeB, the *eNodeB*, and have one of those nodes as an anchor, the fundamental need for macro-diversity for LTE was questioned. Quite quickly it was decided that downlink macro-diversity is not needed for unicast traffic but the uplink was heavily debated. In the end it was decided that uplink macro-diversity does not give the gains for LTE that motivates the complexity increase. Thus, macro-diversity between eNodeBs is not supported in LTE.

For broadcast and multicast traffic it was decided very early that the eNodeBs need to be capable of transmitting the same data in a synchronized manner in order to support MBSFN operation. The needed synchronization is within microseconds as discussed in Chapter 16.

At first, it may seem obvious to move all the RAN functionality to the eNodeB when not supporting macro-diversity. However, terminal mobility needs to be considered as well. There are basically two issues with mobility that need

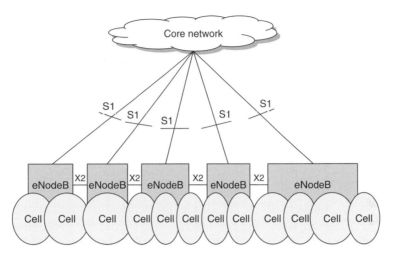

Figure 21.5 *LTE radio access network: nodes and interfaces.*

attention: guarantee of no loss of data when changing cell and minimizing the core network impact when changing cell. For LTE, the latter was not considered as a major problem; proper design of the core network will solve the issue. The former was, however, a more difficult problem to solve. It was in fact agreed that having a centralized anchor with a retransmission layer outside the eNodeB would make it easier for mobility. However, 3GPP decided that the added complexity with not having the anchor was better than requiring a node with RAN functionality outside the eNodeB.[5]

Figure 21.5 shows an overview of the LTE radio access network with its nodes and interfaces. Contrary to the WCDMA/HSPA RAN, the LTE RAN only has one node type: the eNodeB. Thus, there is no equivalent node to an RNC for LTE. The main reason for this is that there is no support for uplink or downlink macro-diversity for dedicated user traffic and the design philosophy of minimizing the number of nodes.

The eNodeB is in charge of a set of cells. Similar to the NodeB in the WCDMA/HSPA architecture, the cells of an eNodeB do not need to be using the same antenna site. Since the eNodeB has inherited most of the RNC functionality, the eNodeB is a more complicated node than the NodeB. The eNodeB is in charge of single cell RRM decisions, handover decisions, scheduling of users in both uplink and downlink in its cells, etc.

---

[5] There was a proposal where the retransmission functionality was located in the core network and not in the RAN to minimize the number of nodes. This was rejected due to the principle of keeping radio functions in the RAN.

The eNodeB is connected to the core network using the *S1* interface. The S1 interface is a similar interface as the Iu interface. There is also an interface similar to the Iur interface of WCDMA/HSPA, the *X2* interface. The X2 interface connects any eNodeB in the network with any other eNodeB. However, since the mobility mechanism for LTE is somewhat different compared to WCDMA/HSPA as there is no anchor point in the LTE RAN, the X2 interface will only be used between eNodeBs that has neighboring cells.

The X2 interface is mainly used to support active-mode mobility. This interface may also be used for multi-cell RRM functions. The X2 control-plane interface is similar to its counter part of WCDMA/HSPA, the Iur interface, but it lacks the RNC drift-functionality support. Instead, it provides the eNodeB relocation functionality support. The X2 user plane interface is used to support loss-less mobility (packet forwarding).

21.2.2.1   eNodeB roles and functionality

The eNodeB has the same functionality as the NodeB and in addition, it has most of the WCDMA/HSPA RNC functionality. Thus, the eNodeB is in charge of the radio resources in its cells, it decides about handover, and makes the scheduling decisions for both uplink and downlink. Obviously, it also performs the classical physical layer functions of coding, decoding, modulation, demodulation, interleaving, de-interleaving, etc. Furthermore, the eNodeB hosts the two layer retransmission mechanisms; the hybrid ARQ and an outer ARQ as described in Chapter 15.

Since the handover mechanism of LTE is different from WCDMA/HSPA, there is no other role of the eNodeB than serving eNodeB. The serving eNodeB is the eNodeB that serves the terminal. The concepts of controlling and drift do not exist. Instead the handover is done by means of eNodeB relocations.

For MBMS type of traffic the LTE RAN decides whether to use unicast or broadcast channels as is the case for WCDMA/HSPA. In case of broadcast channels, the coverage and capacity of those increases significantly if MBSFN operation can be used as described in Chapter 4. In order for the eNodeBs to be able to send the data streams simultaneously, an *MBMS Coordination Entity* (MCE) synchronizes the eNodeB transmissions and data streams. The synchronization is done via a global clock, for example GPS.

21.3   Core network architecture

As discussed previously in this chapter, the mobile system needs a core network to perform the core network functionality. At the same time as the RAN-internal architecture was discussed in 3GPP, the core network architecture was also discussed. Contrary to starting from scratch with both the WCDMA/HSPA RAN and the LTE RAN, the core network used for WCDMA/HSPA and LTE is based on an evolution from the GSM/GPRS core network. The core network used for WCDMA/HSPA is very close to the original core network of GSM/GPRS except for the difference in the functional split with the GSM RAN. The core network used to connect to the LTE RAN is however a more radical evolution of the GSM/GPRS core network. It has therefore got its own name: the *Evolved Packet Core* (EPC).

21.3.1   GSM core network used for WCDMA/HSPA

For WCDMA/HSPA, the core network is based on the GSM core network with the same nodes as for GSM. As discussed in Section 21.1.1, the functional split of GSM and WCDMA/HSPA is different. This led to the use of a different interface between the core network and the WCDMA/HSPA RAN compared to between the core network and the GSM RAN. For WCDMA/HSPA the Iu interface is used, whereas for GSM the *A* and *Gb* interfaces are used.

Figure 21.6 shows an overview of the core network architecture[6] used for WCDMA/HSPA. The figure shows a logical view and, as always when it comes

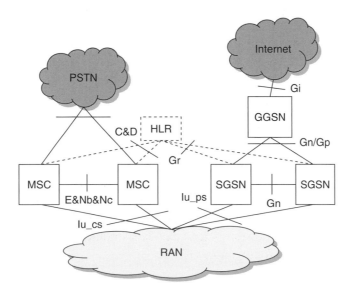

Figure 21.6   *Overview of GSM and WCDMA/HSPA core network – somewhat simplified figure.*

---

[6] The figure shows a simplified view of the core network. The interested reader is referred to [89] and [90] for more detailed information.

to architecture, does not necessarily translates into physical entities. The core network consists of two distinct domains:

1. the *Circuit-Switched* (CS) domain with the *Mobile Switching Center* (MSC);
2. the *Packet-Switched* (PS) domain with the *Serving GPRS Support Node* (SGSN) and *Gateway GPRS Support Node* (GGSN).

Common for the two domains is the *Home Location Register* (HLR), a data base in the home operator's network keeping track of the subscribers of that operator. As can be seen in the figure, the Iu interface is connecting the WCDMA/HSPA RAN to the MSC via the Iu_cs interface and to the SGSN via the Iu_ps interface. Not visible in the figure is that the A interface connects the MSC and the Gb connects the SGSN to the GSM RAN.

The Iu_cs is used to connect the RNC of WCDMA/HSPA to the circuit switch domain of the core network, that is to the MSC. The MSC is used for connecting phone calls to *Public Switched Telecommunications Networks* (PSTN). The MSC and the circuit-switched domain use the functions from the *Integrated Service Digital Network* (ISDN) as switching mechanism. Thus, the signaling to the MSC is based on ISDN.

The Iu_PS interface is used to connect the RNC to the packet-switch domain of the core network, that is the SGSN. The SGSN is in turn connected to a GGSN via a *Gn* or *Gp* interface. The GGSN is then having a *Gi* interface out to external packet networks (for example the Internet), to the operator's service domain or the *IP Multimedia Subsystem* (IMS). The packet-switch domain is using IP routing.[7]

Common for both CS and PS domain is the HLR, which is a database in the home operator's network keeping track of the subscribers of that operator. The HLR contains information about subscribed services, current location of the subscriber's *Subscriber Identity Module* (SIM)/UMTS SIM (USIM) card (that is in which location and routing area the terminal that the SIM/USIM is attached to currently is registered to be in), etc. The HLR is connected to the MSC via the *C* and *D* interfaces, and to the SGSN via the *Gr* interface.

21.3.1.1   Iu flex
The Iu interface supports a function called *Iu flex*. This function allows one RNC to connect to more than one SGSN or MSC and vice versa. This function is

---

[7] The direct connection between the GGSN and the RNC is not shown in the figure. It has been made possible to bypass the SGSN user plane. It is called the 'One Tunnel solution for Optimization of Packet Data Traffic.'

useful to reduce the effects if one of the core network nodes is unavailable, that is an SGSN or MSC is not working properly. The Iu flex mechanism is used to distribute the terminal connections over several SGSNs and MSCs. If one SGSN or MSC is unavailable, the other SGSNs and MSCs keep their allocated traffic and can take all the incoming calls or packet session setup requests (many incoming calls are expected when a core network node becomes unavailable, since most terminals will try to reconnect when they are disconnected without warning).

21.3.1.2 MBMS

For MBMS, the packet-switched domain of the core network is used. Consequently the Iu_ps interface is used to connect to the WCDMA/HSPA RAN. For MBMS, it is the core network that decides whether to use broadcast bearer or a multicast bearer. In case of broadcast bearers, the core network does not know the identity of mobile terminals receiving the information, whereas for the multicast bearers this is known to the core network. Thus the terminals do not need to inform the core network of their intentions when receiving a service that uses broadcast bearer, whereas when receiving a service that is using a multicast bearer the terminals need to inform the core network of its intention to use the service.

For both multicast and broadcast MBMS bearers, the RAN can decide whether to use unicast transport channels or a broadcast transport channel in a cell. This is done by means of the counting procedure, briefly mentioned in Chapter 11. Essentially, the RAN asks the users in a cell to inform it if the user is interested in a specific service. Then if sufficient amount of users is interested the broadcast transport channel is selected otherwise unicast transport channels arc used.

21.3.1.3 Roaming

It is the roaming functionalities of the core network that makes it possible for a user to use another operator's network. Roaming is supported both for the circuit-switched domain and the packet-switched domain. For both domains, different possibilities exist, but in practice traffic is routed via the home operators GGSN for the PS domain. For the CS domain, the common case for terminal-originated calls (outgoing calls) is to do the switching in the visited network. For terminal-terminated calls (incoming calls), the call is always routed through the home network. This is illustrated in Figure 21.7. In the figure, two terminals belonging to two different operators (A and B) are shown. The terminals are roaming in the network of the other operator (bright and dark networks in Figure 21.7) and both terminals have packet-switched connections. Furthermore, the A terminal is calling the B terminal via the circuit-switched domain. As can be seen in Figure 21.7 the packet-switched connections are routed from the SGSN in the visited network to the GGSN in the home network using the Gp interface (the Gn interface are used between SGSN and GGSN in the same network whereas the Gp interface is used

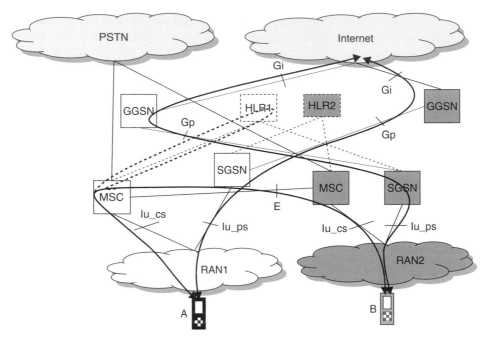

Figure 21.7   *Roaming in GSM/and WCDMA/HSPA.*

between SGSN and GGSN in different networks). For the circuit-switched call, the A terminal, which is originating the call, connects to the (bright) MSC of the visited network. The MSC realize that the called terminal belongs to its network and therefore contacts the bright HLR. The bright HLR responds with information that the B terminal is served by the dark MSC in the dark network. The bright MSC then contacts the dark MSC which sets up a connection to the B terminal.

21.3.1.4   *Charging and policy control*

Charging, a for the operator important function, is located in the core network. For the circuit-switched domain it is in the MSC, whereas for the packet-switched domain it is handled either in the SGSN or in the GGSN. Traditionally, it has been possible to do charging by minutes used and charging by volume. The former commonly used for the circuit-switched domain whereas the latter is more commonly used in the packet-switched domain. However, other charging principles are also possible, for example flat rate with or without opening tariffs. Different tariffs are used depending on the subscriber's subscription and whether the user is roaming or not. With GGSN handling the charging for packet-switched services, more advanced charging schemes, for example content- or event-based charging is supported. This allows the operator to charge the end users depending on the service used.

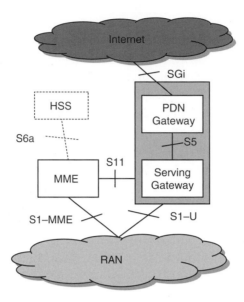

Figure 21.8 *Overview of SAE core network – simplified figure.*

Policy control is a function in the core network used to control the usage of packet-switched services, that is to ensure that the user does not use more bandwidth than allowed or that the user only is accessing 'approved' services or web sites. The policy control is effectuated in the GGSN and it exists only in the packet-switched domain.

21.3.2 The 'SAE' core network: The Evolved Packet Core

When the LTE RAN standardization was started, the corresponding work was started for the core network. This work was called the System Architecture Evolution (SAE). The core network defined in the SAE work is a radical evolution from the GSM/GPRS core network and therefore it has got a new name, Evolved Packet Core (EPC). The SAE scope only covers the packet-switched domain, not the circuit-switched domain. Looking back at the discussions in standardization, the philosophy of minimizing the number of nodes also reigns in the core network standardization. As a consequence, the EPC network started off as single-node architecture with all the functions in one node, except the *Home Subscriber Server* (HSS) which were kept outside the node. HSS is a node/database corresponding to the HLR in GSM/WCDMA core network. Figure 21.8 illustrates how the EPC fits into the total SAE architecture. The nodes of the EPC are:

- the *Mobility Management Entity* (MME), which is the control plane node of the EPC;

- the *Serving Gateway*, which is the user plane node connecting the EPC to the LTE RAN; and
- the *Packet Data Network Gateway* (PDN Gateway), which is the user plane node connecting the EPC to the internet using the *SGi* interface.

It should be noted that the Serving Gateway and the PDN Gateway can be configured as a single entity, thereby reducing the number of user plane nodes.

S1–U is the interface between eNodeBs and Serving Gateways. The S1–U interface is very similar to the Iu_ps user plane interface. The S1–U and Iu_ps user planes are transport tunnels based on IP, agnostic to the content of the packet sent. The IP packets of the end user are put into the S1–U IP tunnel by the Serving Gateway or the eNodeB and retrieved at the other end (eNodeB or Serving Gateway).

It is the *S1–MME* interface that is the interface between the MME and the LTE RAN. The difference between S1–MME and Iu control plane is not large. In fact, it is only in the details of the bearer establishment that it is visible. The difference is in how to indicate the assigned quality of service of a specific flow of a user. For WCDMA/HSPA it is done by means of *Radio Access Bearer* (RAB) parameters whereas for LTE it is done by means of pointing to a specific priority class.

There is also an interface between the MME and the Serving Gateway. It is the *S11* interface which is used by the MME to control the Serving Gateway.

The *S5* interface between the Serving Gateway and the PDN Gateway may be omitted if the single gateway option is configured, that is the Serving Gateway and the PDN Gateway is implemented as one entity. The *S8* interface is used between Serving Gateways in visited networks and PDN Gateways in home networks.

Perhaps the biggest difference between WCDMA/HSPA and LTE is the handling of mobility. In LTE, the PDN Gateway acts as an anchor in the core network for mobility, that is the PDN Gateway handling the user plane of the terminal is not changed during a connection. The PDN Gateway here takes the role of a GGSN for GSM/GPRS and WCDMA/HSPA. However, it is the Serving Gateway that connects to the eNodeBs, thus it is the Serving Gateway that need to be updated regarding which eNodeB it shall rout the packets of the user. This is a large difference compared to WCDMA/HSPA RAN where the RNC hides this kind of mobility from the core network.

The *S6a* interface shown in Figure 21.8 is the interface connecting the MME to the HSS. It is an evolution of the Gr interface used by the WCDMA/LTE core network to connect to the HLR. Thus, a combined HLR/HSS can be the same for EPC as for the legacy GSM and WCDMA core network.

21.3.2.1 S1 flex and EPC nodes in pools

Similar to Iu flex, *S1 flex* enables a more robust core network. If one of the EPC nodes becomes unavailable another EPC node of the same type can take over the lost traffic. Furthermore, the scaling of the network is easier due to that EPC nodes can be added when needed due to traffic demands and not due to increase in coverage.

21.3.2.2 Roaming, policy control, and charging in SAE

Roaming is of course supported by the EPC. Depending on operator policy the IP address can be allocated by the home EPC (traditional roaming) or visited EPC (local break out). Figure 21.9 shows some examples on roaming. In the figure, two terminals belonging to two different operators (A and B) are shown. The terminals are roaming in the network of the other operator (bright and dark networks in Figure 21.9) and both terminals have connections to the internet. As can be seen in Figure 21.9 the dark terminals connection is routed via the visited Serving Gateway to the home PDN Gateway using the S8 interface. This is the traditional way of doing roaming. However, EPC also supports local breakout, illustrated with the bright terminal. The bright terminal has two connections, one doing the traditional roaming with routing via the visited Serving Gateway

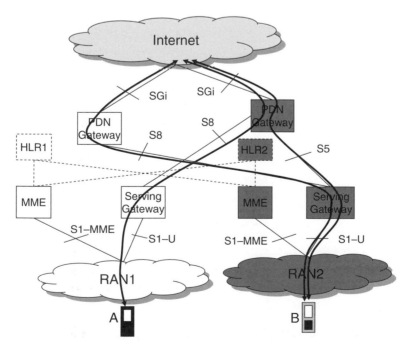

Figure 21.9 *Roaming in LTE/EPC.*

to the home PDN Gateway. The second connection is routed via the visited Serving Gateway to a visited PDN Gateway using the S5 interface, thereby enabling connection to the internet in the visited network.[8]

21.3.3 WCDMA/HSPA connected to Evolved Packet Core

When LTE/SAE technology is being introduced in the network, handover to WCDMA/HSPA will be needed by many operators. The way it is solved is to allow WCDMA/HSPA to connect to the EPC network. In fact, it is the SGSN of the GSM core network used for WCDMA/HSPA that is connected to the EPC, both the Serving Gateway and the PDN Gateway. The PDN Gateway acts as a GGSN when the traffic is routed through the WCDMA/HSPA RAN using the *S4* interface (which is based on the Gn/Gp interface used between GGSN and SGSN) and as a normal PDN Gateway when the traffic is routed through the LTE RAN. This is possible since the user plane termination in the PDN Gateway is kept and thus the IP address of the terminal. The control plane parts of the EPC (in the MME) are not used when the terminal is connected to the WCDMA/HSPA RAN. Instead the core network protocols of the SGSN are used. With this approach, minimal changes are needed to the packet core network used for WCDMA/HSPA while still being able to provide a fast and seamless handover to and from LTE (Figure 21.10).[9]

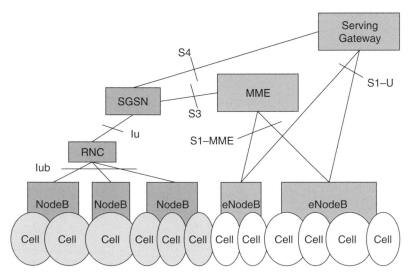

Figure 21.10 *WCDMA/HSPA connected to LTE/SAE.*

---

[8] The visited Serving Gateway and PDN Gateway can be a singe gateway configuration. In that case the S8 interface is used toward the home PDN Gateway and the S5 interface is not used.

[9] In the figure, the *S12* interface between the PDN Gateway and the WCDMA RAN is not shown. It is possible to bypass the SGSN for the user plane. It is called the One Tunnel solution for Optimization of Packet Data Traffic.'

When a handover from WCDMA/HSPA to LTE is needed, the connection is taken over by a MME and a Serving Gateway from the SGSN. This is done through the *S3* interface between the MME and the SGSN. The S3 interface is based on the Gn interface used between SGSNs for SGSN relocations. Thus, the handover is close to the SGSN relocations with a user plane switch in the PDN Gateway instead of in the GGSN.

21.3.4 Non-3GPP access connected to Evolved Packet Core

The EPC may not only connect to 3GPP defined radio access networks. There is also work done on connecting EPC to non-3GPP accesses. In particular WiFi, WiMAX, and CDMA2000/HRPD accesses are being worked upon.[10] The main approach is that non-3GPP accesses shall have a generic interworking with EPC, that is no special optimizations for a given access. However, for CDMA2000/ HRPD there exist special optimizations for handover between LTE and CDMA2000/HRPD and vice versa. Work is also going on for WiMAX.

The general approach is that, when a UE detects another access (either a non-3GPP access when connected to a 3GPP access or a 3GPP access when connected to a non-3GPP access), the UE need to establish a connection to the found access from scratch. This means that the quality of the handover depends on the capabilities of the UE. If the UE can do the necessary signaling to the new access in parallel to the communication to the old access, the performance of the handover will be good. However, if the UE can only communicate with one access at a time, the performance of the handover will be bad.

21.3.4.1 CDMA2000 and HRPD connected to Evolved Packet Core

In order to make it easier to ensure a handover performance between CDMA2000/ HRPD and LTE (and vice versa) that is equally good as between LTE and WCDMA/HSPA, there are special optimizations standardized (both in 3GPP and in 3GPP2). With these special optimizations, the performance of the handover will be equally good as the performance of the handover between LTE and WCDMA/ HSPA. Figure 21.11 shows the setup. As can be seen from the figure, there are a couple of new interfaces introduced between the EPC and the CDMA2000/HRPD network. It is the *S101* between the MME and the *CDMA2000/HRPD access network* (HRPD AN), the *S103* between the Serving Gateway and the PDSN of the CDMA2000/HRPD core network, and the *S2a* interface between the PDN Gateway and the PDSN.[11] In the figure, the *IOS* interface between the PDSN and the CDMA2000/HRPD AN is also shown. The CDMA2000/HRPD AN consists

---

[10] Further details on WiMAX, CDMA2000 and HRPD are given in Chapter 24.

[11] The S2a interface is also used in the generic case of non-3GPP accesses connected to EPC. S101 and S103 are only used by the CDMA2000/HRPD connection to EPC.

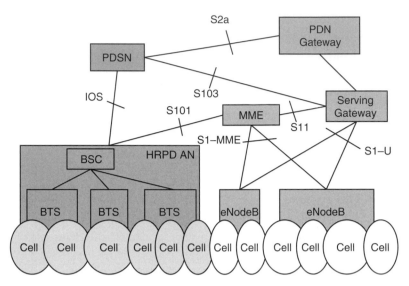

Figure 21.11   *CDMA/HRPD connected to LTE/SAE.*

of *Base Station Controllers* (BSCs) and *Base Transceiver Stations* (BTSs). The BSC controls one or several BTS, which in turn handles one or several cells.

The S101 interface is used to do pre-registration and handover signaling, whereas the S103 interface is used to forward downlink data in order to minimize packet losses. Pre-registration is initiated in the source system and all the necessary signaling to be registered in the target system is conducted whilst still in the source system. The UE is using transparent containers through the source system routed via the *S101* interface. If a handover then becomes necessary, the handover signaling is the done according to the custom of the target system using the same transparent container technique through the source system. When the UE is ready to switch to the target system, the target system is ready to handle the UE and the packet loss is minimized.

22

LTE-Advanced

22.1 IMT-2000 development

As discussed in Chapter 1, early discussions on a global standard for 3G mobile communication started within ITU already in the late 1980s, initially under the working name *FPLMTS* (Future Public Land Mobile Telephone System) and later renamed *IMT-2000*. For several years the work on IMT-2000 progressed as a relatively isolated activity within ITU with discussions mainly focusing on high-level scenarios and requirements.

In the end, the actual technical specification of 3G radio-access technologies was to take place elsewhere, mainly within 3GPP but also within 3GPP2. However, the 3GPP and 3GPP2 radio-access technologies were then submitted to ITU and received ITU approval as IMT-2000 technologies. Thus, IMT-2000 should not be seen as a radio-access technology in itself, but as a family of radio-access technologies, all fulfilling the ITU requirements on IMT-2000 and all being approved by ITU as IMT-2000 technologies.

Already in the late 1990s, that is when the first phase of the 3G radio-access technologies were being finalized, ITU started early considerations on the step beyond IMT-2000. Initially, this work was simply referred to as *Systems Beyond IMT-2000* but it was later renamed *IMT-Advanced*. Similar to IMT-2000, during several years, the discussions on IMT-Advanced were going on within ITU with focus on the application process and the high-level scenarios and requirements. However, during 2007, the work on IMT-Advanced intensified with an increased focus on the more detailed technical requirements. Finally, in March 2008, although with the detailed requirement discussions yet to be finalized, ITU issued a *Circular Letter* [139] inviting submissions for candidate radio-interface technologies for IMT-Advanced.

As outlined in the ITU time schedule in Figure 22.1, such submissions are expected to take place during 2009. After submission, ITU will evaluate the different candidate

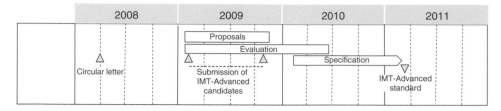

Figure 22.1   *Current time schedule for IMT-Advanced within ITU.*

technologies and assess if they fulfill the requirements defined for IMT-Advanced. Based on the outcome of the evaluation, the formal ITU-R recommendations for the IMT-Advanced radio-interface specifications will be drafted.

It is important to understand though, that similar to IMT-2000, the task of producing the detailed technical specifications for the different IMT-Advanced radio-access technologies will, in practice, be the task of the organizations submitting the different technologies to ITU. One such organization will be 3GPP and the candidate radio-access technology will be *LTE-Advanced*.

22.2   LTE-Advanced – The 3GPP candidate for IMT-Advanced

Partly triggered by the ITU activities and anticipating the issuing of the ITU circular letter, 3GPP already in March 2008 initiated a Study Item on *LTE-Advanced*. The task of the Study Item is to define requirements and investigate and propose technology components to be part of LTE-Advanced. Within 3GPP, LTE-Advanced is seen as the next major step in the evolution of LTE, which is very similar to HSPA being the first major step in the evolution of the WCDMA radio access. It is generally anticipated that LTE-Advanced will coincide with LTE Release 10 with the intermediate Release 9 mainly implying minor updates to the current LTE specifications.[1]

Furthermore, LTE-Advanced is anticipated to be the radio-access technology submitted to ITU as the 3GPP candidate for IMT-Advanced radio access. It should be noted that this is very much aligned with what was already from the start stated for LTE, namely that *LTE should provide the starting point for a smooth transition to 4G (= IMT-Advanced) radio access*. With the initiation of the LTE-Advanced Study Item and the work on defining LTE-Advanced ramping up, this smooth transition to '4G' radio access is now ongoing.

---

[1] Similar to HSPA being Release 5 of WCDMA with Release 4 implying only minor updates to the original WCDMA Release 99.

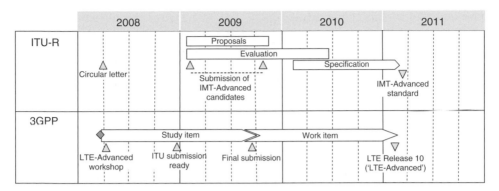

Figure 22.2 *3GPP time schedule for LTE-Advanced in relation to ITU time schedule on IMT-Advanced.*

The time schedule for the 3GPP work on LTE-Advanced, in relation to the ITU time schedule on IMT-Advanced, is outlined in Figure 22.2. As a first step of the LTE-Advanced Study Item, a workshop was held in April 2008, where different requirements and basic technology components for LTE-Advanced were discussed. According to the time schedule, the Study Item is to proceed until mid-2009. At that time, a Work Item is expected to be initiated for the detailed specification of LTE-Advanced. This work, making LTE-Advanced ready for initial commercial deployment, is assumed to be finalized in early 2011. Note that this timing is well aligned with the planned finalization of the IMT-Advanced recommendation in ITU-R. It should also be noted that the initial 3GPP submission to ITU is expected approximately halfway into the Study Item (early 2009). The initial submission will then be followed by complementary submissions, filling in details of LTE-Advanced as these emerge as part of the 3GPP work. The final submission, in the fall of 2009, will then correspond to the finalization of the 3GPP Study Item, when all main components of LTE-Advanced should have been agreed upon.

22.2.1 Fundamental requirements for LTE-Advanced

As LTE-Advanced is anticipated to be the 3GPP candidate radio-access technology for IMT-Advanced radio access, an obvious requirement for LTE-Advanced is the complete fulfillment of all the requirements for IMT-Advanced defined by ITU. Another basic prerequisite for the work on LTE-Advanced is that LTE-Advanced *is an evolution of LTE*. The implication of this is that LTE-Advanced has to fulfill a set of basic backward compatibility requirements.

LTE-Advanced should provide backward compatibility in terms of spectrum coexistence, implying that it should be possible to deploy LTE-Advanced in spectrum already occupied by LTE with no impact on existing LTE terminals. A direct consequence of this requirement is that, for an LTE release terminal, an LTE-Advanced cell should appear as an LTE Release 8 cell. This is similar to HSPA, where an early WCDMA terminal can access a cell supporting HSPA, although from the point-of-view of this terminal, the cell will appear as a WCDMA Release 99 cell. Such spectrum compatibility is of critical importance for a smooth, low-cost transition to LTE-Advanced capabilities within the network.

LTE-Advanced should also be 'backward compatible' in terms of infrastructure, in practice implying that it should be possible to upgrade already installed LTE infrastructure equipment to LTE-Advanced capability with a reasonable cost. Also this is a critical prerequisite for a smooth and low-cost transition to LTE-Advanced network capability.

Finally, LTE-Advanced should be 'backward compatible' in terms of terminal implementation, implying that it should be possible to introduce LTE-Advanced functionality in mobile terminals with a reasonable incremental complexity and associated cost, compared to current LTE capability. This is clearly vital to ensure a fast adoption of LTE-Advanced terminal capability.

22.2.2 Extended requirements beyond ITU requirements

It is a common understanding within 3GPP that LTE-Advanced should not be limited to the fulfillment of the ITU requirements on IMT-Advanced. Rather, LTE-Advanced should go beyond the IMT-Advanced requirements and hence the targets for LTE-Advanced are substantially more ambitious, including:

- Support for peak-data up to 1 Gbps in the downlink and 500 Mbps in the uplink.
- Substantial improvements in system performance such as cell and user throughput with target values significantly exceeding those of IMT-Advanced.
- Possibility for low-cost infrastructure deployment and terminals.
- High power efficiency, that is, low power consumption for both terminals and infrastructure.
- Efficient spectrum utilization, including efficient utilization of fragmented spectrum. This includes the possibility for spectrum aggregation as further discussed in Section 22.3.1.

22.3 Technical components of LTE-Advanced

In the following, a brief outline of the potential technology components of LTE-Advanced will be discussed. It is important to understand though that at the time of writing, the discussions on LTE-Advanced in 3GPP is still in its very initial phase. Exactly what technology components eventually will be part of LTE-Advanced is therefore far from decision and the discussion below should be seen as examples illustrating the potential of evolving LTE.

22.3.1 Wider bandwidth and carrier aggregation

The peak-data rate targets for LTE-Advanced can only be fulfilled in a reasonable way with a further increase of the transmission bandwidth, compared to what is supported with the first release of LTE. Thus, it can be expected that one component of the LTE evolution toward LTE-Advanced will be an increase of the maximum transmission bandwidth beyond 20 MHz, perhaps up to as high as 100 MHz or even beyond. This is true for both the downlink and the uplink.

To align such an extension of the transmission bandwidth with the requirement on spectrum compatibility as discussed in Section 22.2.1, some care must be taken. One promising approach to wider bandwidth for LTE-Advanced is a *carrier aggregation* as outlined in Figure 22.3. In case of carrier aggregation, the extension to wider bandwidth is accomplished by the aggregation of basic component carriers of a more narrow bandwidth. Each component carrier would then, to an LTE terminal, appear as, and provide all the capabilities of, an LTE carrier. At the same time, an LTE-Advanced-capable terminal would be able to access the entire aggregation of multiple carriers, thus experiencing an overall wider bandwidth with corresponding possibilities for higher data rates.

The carrier aggregation outlined in Figure 22.3 assumes the component carriers being adjacent to each other. However, from a basic core-specification point-of-view, there is no need for such a constraint. Rather, a general carrier aggregation should allow for non-adjacent component carriers, including carriers in different frequency bands as outlined in Figure 22.4. Thus, the introduction of carrier aggregation as part of LTE-Advanced allows for *spectrum aggregation*, that

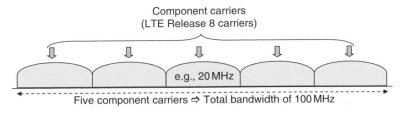

Figure 22.3 *LTE carrier aggregation for extension to wider overall transmission bandwidth.*

Two aggregated carriers ⇨ Total bandwidth of 40 MHz

Figure 22.4 *Carrier aggregation as a tool for spectrum aggregation and efficient utilization of fragmented spectrum.*

is the simultaneous usage of different non-contiguous spectrum fragments for communication to/from a single mobile terminal.

It is a general understanding though that, although straightforward from a basic core-specification point-of-view, aggregation of non-adjacent carriers in different bands has major impacts on actual terminal implementation. Thus, although spectrum aggregation would be supported by the basic specifications, the actual implementation will be strongly constrained, including specification of only a limited number of aggregation scenarios and aggregation over dispersed spectrum only being supported by the most advanced terminals.

22.3.2 *Extended multi-antenna solutions*

As discussed extensively in Chapter 16, LTE already supports a wide range of multi-antenna transmission technologies. As part of the extension of LTE toward LTE-Advanced, further enhancement of this support is expected.

As a minimum, support for spatial multiplexing on the uplink is anticipated to be part of LTE-Advanced. The reason for this is that even by just considering the ITU requirements, uplink spatial multiplexing is, in practice, needed to fulfill the peak spectral-efficiency requirements.

Other multi-antenna technologies to be considered for the evolution of LTE toward LTE-Advanced include the extension of downlink spatial multiplexing to more than four layers. It should be noted though that the application of this is probably limited in the sense that the benefits of eight-layer spatial multiplexing are only present in special scenarios where high SINR can be achieved.

Already in current networks, multiple, geographically dispersed antennas connected to a central baseband processing unit are used as a cost-efficient way of building networks. Such deployments open up for the use of *coordinated multi-point transmission/reception* beyond the traditional three-sector sites. This is illustrated in Figure 22.5. Coordinating the transmission from the multiple antennas can be used to increase the signal-to-noise ratio for users far from the antenna

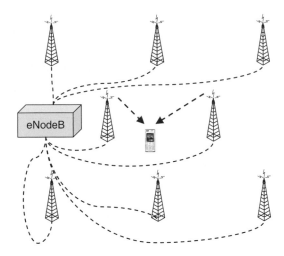

Figure 22.5   *Coordinated multi-point transmission.*

sites, for example by transmitting the same signal from multiple sites. Such strategies can also improve the power-amplifier utilization in the network, especially in a lightly loaded network where otherwise some power amplifiers would be idle.

22.3.3   Advanced repeaters and relaying functionality

The link performance of LTE is already quite close to the Shannon limit and from a pure link-budget perspective, the very high data rates targeted by LTE-Advanced require a higher signal-to-noise ratio than what is typically experienced in wide-area cellular networks. To achieve such high signal-to-noise ratios, and thus to be able to deliver the required data rates over a wide area, denser infrastructure deployment and/or various beam-forming techniques are required. Coordinated multipoint transmission, described above, is one possibility for deploying a denser infrastructure, although not the only possibility.

Difference relaying solutions (Figure 22.6) can also be used to provide very high data rates over large geographical areas. A wide range of relay types can be envisioned, ranging from simple repeaters, which are already used today as a low-cost tool for coverage improvement in troublesome spots, to more advanced solutions where the relay can be seen as a (small) base station. The latter is sometimes also referred to as *self-backhauling* as the transport network from the relay uses the same air-interface as the terminals. In contrast, current network deployments often use dedicated radio-link equipment for the transport network.

To what extent standardization support is needed depends on the relaying strategy. Repeaters are transparent to both the eNodeB and the terminal and thus no

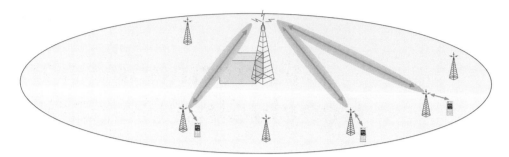

Figure 22.6 *Relaying as a tool to improve the coverage of high data rates in a cell.*

changes to the Rel-8 specifications are needed, whereas for some of the more advanced strategies additions to the specifications might be needed, for example to be able to activate the repeater only when users are present in the area covered by the repeater. Self-backhauling, if introduced, would also need some standardization support, mainly related to transport-network aspects.

22.4 Conclusion

In this chapter, the evolution of LTE to LTE-Advanced in order to fulfill the IMT-Advanced requirements as well as additional 3GPP requirements has been discussed. As seen from the brief discussion above, although LTE already in its first release provides very high performance, it can also serve as a solid framework for evolving into even higher performance. Most likely, the evolution will also continue beyond LTE-Advanced to meet future requirements emerging with raising user expectations.

Part V

Performance and Concluding Remarks

23

Performance of 3G evolution

23.1 Performance assessment

Computer simulations of mobile systems is a very powerful tool to assess the system performance. The 'real life' performance can, of course, be measured and evaluated in the field for an already deployed system and such values represent a valid example of performance for a certain system configuration. But there are several advantages with computer simulations:

- Evaluation can be made of system concepts that are not deployed or still under development such as LTE.
- There is full control of the environment, including propagation parameters, traffic, system lay-out, etc., and full traceability of all parameters affecting the result.
- Well-controlled 'experiments' comparing similar system concepts or parts of concepts can be done under repeatable conditions.

In spite of the advantages, simulation results do obviously not give a full picture of the performance of a system. It is impossible to model all aspects of the mobile environment and to properly model the behavior of all components in a system. Still, a very good picture of system performance can be obtained and it can often be used to find the potential limits for performance. Because of the difficulty in modeling all relevant aspects, relative capacity measures for introducing features will be more accurate than absolute capacity numbers, if a good model is introduced of the feature in question.

The capacity of a system is difficult to assess without comparisons, since any system performance number in itself does not provide very much information. It is when set in relation to how other systems can perform that the number becomes interesting. But since making comparisons is an important component in assessing performance, it also makes performance numbers a highly contentious issue. One always has to watch out for 'apples and oranges' comparisons, since the system performance depends on so many parameters. If parameters are not properly selected to give comparable conditions for two systems, the performance numbers will not at all be comparable either.

In this context, it is also essential to take into account that system performance and capacity will be feature dependent. Many features such as MIMO and advanced antenna techniques that are introduced in evolved 3G systems are very similar between systems. If a certain feature is a viable option for several systems that are evaluated in parallel, the feature should be included in the evaluation for all systems.

Any simulated performance number should be viewed in the context that real radio network performance will depend on many parameters that are difficult to control, including:

- The mobile environment, including channel conditions, angular spreads, clutter type, mobile speeds, indoor/outdoor usage, and coverage holes.
- User-related behavior, such as voice activity, traffic distribution, and service distribution.
- System tuning of service quality and network quality.
- Deployment aspects such as site types, antenna heights and types, and frequency reuse plan.
- A number of additional parameters that are usually not modeled, such as signaling capacity and performance, and measurement quality.

There is no single universal measure of performance for a telecommunications system. Indeed, end users (subscribers) and system operators define good performance quite differently. On the one hand, end-users want to experience the highest possible level of quality. On the other hand, operators want to derive maximum revenue, for example, by squeezing as many users as possible into the system. Performance-enhancing features can improve perceived quality-of-service (end-user viewpoint) or system performance (operator viewpoint). The good news, however, is that the 3G evolution through HSPA and LTE have the potential to do both. Compared to earlier releases of WCDMA, these evolution steps yield better data rates and shorter delay. That is, they can greatly improve both the service experience (end-user viewpoint) and the system capacity (operator viewpoint).

23.1.1 End-user perspective of performance

Users of circuit-switched services are assured of a fixed data rate. The quality of service in the context of speech or video telephony services is defined by perceived speech or video quality. Superior-quality services have fewer bit errors in the received signal.

By contrast, users who download a web page or movie clip via packet data describe quality of service in terms of the delay they experience from the time they start the download until the web page or movie clip is displayed. Best-effort services do not guarantee a fixed data rate. Instead, users are allocated whatever data rate is available under present conditions. This is a general property of packet-switched networks, that is, network resources are not reserved for each user. Given that delay increases with the size of the object to be downloaded, absolute delay is not a fair measure of quality of service.

A lone user in a radio network experiencing good radio conditions may enjoy the *peak data rate* of the radio interface, see Figure 23.1. A user will, however, normally share radio resources with other users. If radio conditions are less than optimum or there is interference from other users, the radio interface data rate will be less than the peak data rate. In addition, some data packets might be lost, in which case the missing data must be retransmitted, further reducing the effective data rate as seen from higher protocol layers. Furthermore, the effective data rate diminishes even further as the distance from the cell increases (due to poorer radio conditions at cell edges). The data rate experienced above the MAC layer, after sharing the channel with other users is denoted *user throughput*.

The *Transmission Control Protocol* (TCP) – the protocol at the transport layer – is commonly used together with IP traffic. However due to its slow-start algorithm, which is sensitive to latency in the network, it is especially prone to cause

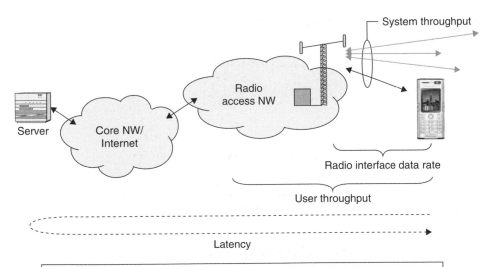

- *Radio interface data rate*: Data rate of the physical layer achieved under certain radio conditions with specific coding and modulation.
- *Peak data rate*: Peak data rate of the radio interface under ideal radio conditions.
- *User throughput*: Data rate experienced above the MAC layer, under real channel conditions and after sharing the channel with other users.
- *Latency*: End-to-end round-trip time of a small packet.
- *System throughput*: Total number of bits per second transmitted over the radio interface (per sector).

Figure 23.1 *Definitions of data rates for performance.*

delay for small files. The slow-start algorithm is meant to ensure that the packet transmission rate from the source does not exceed the capability of network nodes and interfaces.

Network *latency*, which in principle is a measure of the time it takes for a packet to travel from a client to server and back again, has a direct impact on performance with TCP. Therefore, an important design objective in both HSPA and LTE has been to reduce network latency. One other quality-related criterion (end-user viewpoint) relates to the setup time for initiating, for example, a web-browsing session.

23.1.2 Operator perspective

Radio resources need to be shared when multiple users are in the network. As a result, all data must be queued before it can be transmitted, which restricts the effective data rate to each user. Notwithstanding, by scheduling radio resources, operators can improve system throughput or the total number of bits per second transmitted over the radio interface. A common measure of system performance is 'spectrum efficiency' which is the system throughput per MHz of spectrum in each sector of the system.

HSPA and LTE both employ intelligent scheduling methods to optimize performance (end-user and operator viewpoint). One example evaluation of HSDPA and Enhanced Uplink can be found in [22].

An important performance measure for operators is the number of active users who can be connected simultaneously. Given that system resources are limited, there will thus be a trade-off between number of active users and perceived quality of service in terms of user throughput.

23.2 Performance in terms of peak data rates

The design targets for LTE are documented in 3GPP TR 25.913 [86] as outlined in Chapter 13. The target capability when operating in a 20 MHz spectrum allocation is a peak data rate of 100 Mbit/s in the downlink and 50 Mbit/s in the uplink. The numbers assume two receive antennas in the terminal for the downlink capability and one transmit antenna for the uplink capability.

These target numbers are exceeded with a good margin by the peak rate capability of the specified LTE standard. LTE supports up to 150 Mbit/s in the downlink and 75 Mbit/s in the uplink for a 20 MHz allocation. These peak numbers assume 2×2 MIMO for the downlink and 64QAM modulation for both uplink and downlink.

To reach even higher data rates in the downlink, 4×4 MIMO can be used requiring four receive antennas in the terminal. The peak data rate with 64QAM modulation will then be 300 Mbit/s.

23.3   Performance evaluation of 3G evolution

As explained above, it is difficult to assess how well a system performs without comparisons. An evaluation that is of interest is one including both advanced WCDMA (HSPA) as described in Part III of this book and the *Long-Term Evolution* (LTE) as described in Part IV. In addition, the existing WCDMA (HSPA) standard Release 6 is a suitable reference for comparison. An evaluation comparing these three systems is presented in [122] and is also outlined below.

The evaluation is based on static simulations. Performance numbers are included for a baseline Release 6 system, an evolved HSPA system and for LTE. The evolved HSPA system has advanced receivers, a downlink with 2×2 MIMO and an uplink with 16QAM. The combined results makes possible also relative assessments of the gains associated with OFDM and MIMO, since the other features included for the OFDM-based LTE-system and the advanced HSPA system are very similar.

With the level of details applied to model the protocols for LTE and HSPA Evolution and the simulation assumptions used for the evaluation described below, the technology potential of LTE and HSPA Evolution is demonstrated. As a part of the 3GPP work on the LTE physical layer, the system performance of LTE was evaluated with a similar set of assumptions, but using dynamic simulations. The results and conclusion of the 3GPP evaluation is presented in Section 23.4.

23.3.1   Models and assumptions

This section presents the models and assumptions used for the evaluation in [122]. A summary of models and assumptions grouped into traffic, radio network, and system models is provided in Table 23.1. Three different systems are studied:

1. *A reference Release 6 WCDMA system*, as described in the 3GPP requirements [86] using single-stream transmission and a baseline Rake-based receiver.
2. *An evolved HSPA system* with more advanced receivers, 2×2 MIMO for downlink, and 16QAM for uplink.
3. *An LTE system* configured according to the 3GPP requirements [86], with 2×2 MIMO for the downlink.

It should also be noted that many control-plane and user-plane protocol aspects above the physical layer are omitted in the simulations, making absolute values

Table 23.1   *Models and assumptions for the evaluations (from [122]).*

Traffic Models	
User distribution	Uniform, in average 10 users per sector
Terminal speed	0 km/h
Data generation	On–off with activity factor 5%, 10%, 20%, 40%, 60%, 80%, 100%
Radio network models	
Distance attenuation	$L = 35.3 + 37.6 \times \log(d)$, d = distance in meters
Shadow fading	Log-normal, 8 dB standard deviation
Multipath fading	3GPP Typical Urban and Pedestrian A
Cell layout	Hexagonal grid, 3-sector sites, 57 sectors in total
Cell radius	167 m (500 m inter-site distance)
System models	
Spectrum allocation	5 MHz for DL and 5 MHz for UL (FDD)
Base station and UE output power	20 W and 125 mW into antenna
Max antenna gain	15 dBi
Modulation and coding schemes	QPSK and 16QAM, Turbo coding according to WCDMA Release 6. Only QPSK for basic WCDMA uplink
Scheduling	Round robin in time domain
Basic WCDMA characteristics	
Transmission scheme	Single stream in DL and UL
Receiver	Two-branch antenna diversity with rake receiver, maximum ratio combining of all channel taps. 9 dB noise figure in UE, 5 dB in NodeB
Advanced WCDMA characteristics	
Transmission scheme	DL : 2 stream PARC UL : Single stream
Receiver	DL : GRAKE [29] with Successive Interference Cancellation UL : GRAKE with 2-branch receive diversity, soft handover with selection combining between sites
LTE characteristics	
Transmission scheme	DL : 2 stream PARC UL : Single stream
Receiver	DL : MMSE with Successive Interference Cancellation UL : MMSE with 2-branch receive diversity, soft handover with selection combining between sites

optimistic. For LTE, frequency-domain adaptation and other higher-layer improvements are not included.

The simulation methodology is 'static,' where mobiles are randomly positioned over a model of a radio network and the radio channel between each base station and UE antenna pair is calculated according to the propagation and fading models. Statistics are collected and then new mobiles are randomly positioned for a new iteration step. Different system load levels are simulated setting a random activity factor for each base station from 5% to 100%. Channel-independent time-domain scheduling is used, corresponding to round-robin scheduling.

Based on the channel realizations and the active interferers, a *signal-to-interference and noise ratio* (SINR) is calculated for each UE (or base station) receive antenna. The SINR values are then mapped to active radio link data rates R_u, for each active user u, when it is scheduled. In the case of MIMO, R_u is modeled as the sum of the rates achieved per MIMO stream. The data rate experienced above the MAC layer, after sharing the channel with other users, is denoted user throughput S_u, and is calculated based on the activity factor. Active base stations and users differ between iterations, and statistics are collected over a large number of iterations.

The served traffic per sector T is calculated as the sum of the active radio-link data rates for the active users in the sector, assuming that the users are scheduled an equal amount of time. Statistics for end-user quality are taken from the distribution of user throughput as the mean (average quality) and the 5th percentile (cell-edge quality). With increasing activity factor, the served traffic per sector will increase, while the individual data rates (user quality) will decrease because of the decreasing SINR and the less frequent access to the shared channel for each user.

23.3.2   *Performance numbers for LTE with 5 MHz FDD carriers*

Figure 23.2 shows the mean and 5th percentile (cell-edge) downlink user throughput (S_u) vs. served traffic (T) for Typical Urban propagation conditions. The simulation results show that compared to the basic WCDMA system,[1] LTE has a significantly improved user throughput for both average and cell-edge users. The evolved HSPA system, however, gives almost the same performance as the LTE system.

The relative improvements in mean and cell-edge user throughput can be estimated by comparing the throughputs achieved by the different systems at the same traffic load. A user throughput gain exceeding a factor of 3× is achieved over the range of loads in Figure 23.2. As an example, at a served traffic of 2 Mbps per sector, basic WCDMA achieves a mean user throughput of about

---

[1] Note that the 'basic WCDMA system' used as a reference in the comparison is not identical to the Release 6 WCDMA reference used in the evaluation of LTE in 3GPP described.

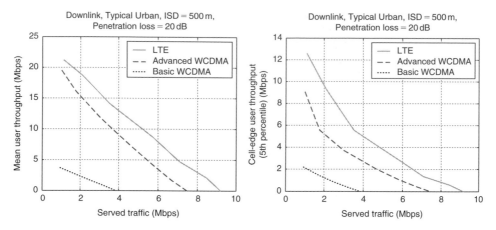

Figure 23.2 *Mean and cell-edge downlink user throughput vs. served traffic, Typical Urban propagation.*

2.5 Mbps, as compared to about 18 Mbps for LTE. Note that the advanced WCDMA system reaches 15 Mbps at this load.

The faster links of both the LTE and advanced HSPA systems will impact the throughput in two ways:

 (i) For a given SINR a higher throughput is achieved.
(ii) For the same served traffic, this results in lower link utilization, and thereby less interference and a higher SINR.

Also the cell-edge throughput gain exceeds a factor $2\times$ for the range of traffic loads in Figure 23.2. Spectrum efficiency gains can be estimated by comparing the served traffic for a given requirement on cell-edge throughput. For example, for a cell-edge throughput requirement of 1 Mbps, basic WCDMA can serve some 2.5 Mbps per sector. The corresponding number for LTE is 8 Mbps, that is a gain with more than a factor $3\times$. However for lower cell-edge throughput requirements the 3GPP LTE requirement is not met. In a fully loaded network, the served traffic is about 4 and 9 Mbps for basic WCDMA and LTE, respectively, that is a gain of a factor $2.25\times$.

Similar results are shown for the less time dispersive Pedestrian A channel in Figure 23.3. In lower time dispersion, WCDMA performs better, especially the basic WCDMA system without advanced receivers. High performance gains for LTE relative to basic WCDMA are still achieved, with $3\times$ in mean user throughput and $2\times$ in cell-edge user throughput. Spectrum efficiency gains of more than $3\times$ are also reached for cell-edge user throughput requirements exceeding 2 Mbps.

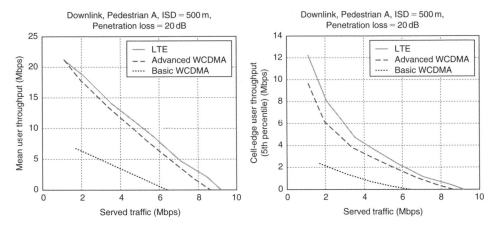

Figure 23.3 *Mean and cell-edge downlink user throughput vs. served traffic, Pedestrian A propagation.*

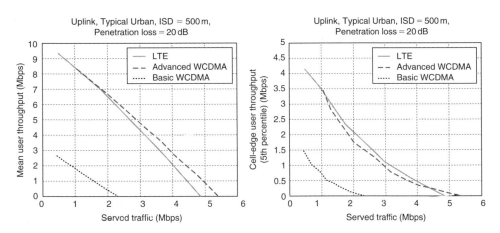

Figure 23.4 *Mean and cell-edge uplink user throughput vs. served traffic, Typical Urban propagation.*

The corresponding uplink results for Typical Urban propagation are presented in Figure 23.4 and for Pedestrian A propagation in Figure 23.5. The results show a gain of more than 2× in mean user throughput and cell-edge throughput compared to the WCDMA reference. For the cell-edge throughput the gain is smaller than for the mean user throughput.

23.4 Evaluation of LTE in 3GPP

23.4.1 LTE performance requirements

The system performance targets for LTE were defined by 3GPP in 2005 and documented in 3GPP TR25.913 [86] together with the goals for capabilities,

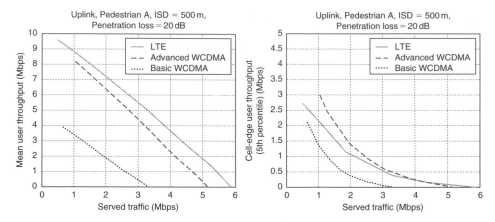

Figure 23.5   *Mean and cell-edge uplink user throughput vs. served traffic, Pedestrian A propagation.*

complexity, deployment, and architecture as discussed in Chapter 13. The target measures for system performance are:

- *Average user throughput*: Measured per MHz as the average over all users.
- *'Cell-edge' user throughput*: Measured per MHz at the 5th percentile of the user distribution (95% of the users have better performance).
- *Spectrum efficiency*: The system throughput per sector in bps/MHz/site.
- *Coverage*: Performance in larger cells.

While the detailed technology solutions for LTE should not be a basis for the performance requirements, the number of Tx and Rx antennas configured for the BS and UE must be agreed as prerequisites for the performance targets. The reason is that an increase in number of antennas can also be seen as a limitation for the selected solution because of the complexity increase. It is in theory possible to get unrealistic large gains assuming an unrealistic number of antennas. The following downlink and uplink configurations are therefore chosen for the LTE targets in [86]:

- An LTE downlink with a maximum of 2 Tx antennas at the NodeB and 2 Rx antennas at the UE.
- An LTE uplink with a maximum of a single Tx antenna at the UE and 2 Rx antennas at the NodeB.

The performance is evaluated in uplink and downlink separately and the targets are set relative to the reference performance of a baseline Release 6 system. This **baseline** system consists of:

- A reference Release 6 downlink based on HSDPA with a single Tx antenna at the NodeB with enhanced performance Type 1 receiver at the UE (requirements based on dual antenna diversity).
- A reference Release 6 uplink based on Enhanced Uplink with a single Tx antenna at the UE and 2 Rx antennas at the NodeB.

Table 23.2 *LTE performance targets in [86, 93].*

Performance measure	Downlink target relative to baseline Rel-6 HSDPA	Uplink target relative to baseline Rel-6 Enhanced UL
Average user throughput (per MHz)	3–4×	2–3×
Cell-edge user throughput (pcr MHz, 5th percentile)	2–3×	2–3×
Spectrum efficiency (bps/Hz/site)	3–4×	2–3×
Coverage	Meet above targets up to 5 km cell range	Meet above targets up to 5 km cell range

The agreed LTE performance targets are shown in Table 23.2. Since it is expected that both average user throughput and spectrum efficiency will benefit from the increase from 1 to 2 Tx antennas in the downlink, it is the target increase of cell-edge user throughput that would be the challenge in the downlink.

23.4.2 LTE performance evaluation

During the 3GPP feasibility study for LTE, several downlink and uplink physical layer concepts were studied and evaluated. The evaluation results and the chosen physical layer concept are reported in TR25.814 [81]. The chosen concept was also evaluated in terms of user throughput, spectrum efficiency, coverage, and the other performance targets defined in [86].

The assumptions and methodology established for the 3GPP evaluation are reported in [81], together with summaries of the results. More detailed description of simulations that the 3GPP evaluation of LTE is based on can be found in [57] for the downlink and [58] for the uplink. This LTE evaluation for 3GPP consists of dynamic simulations where users are randomly positioned over a model radio network and then move according to a mobility model. Propagation and fading is modeled and statistics collected for each simulation step of one TTI.

All performance numbers in the 3GPP evaluation are for LTE with 5 MHz carriers, set in relation to the baseline Release 6 3G system as defined above. Results in [57, 58] are presented for a variation of cell sizes, different scheduling (round robin and proportional fair), multiple channel models, and different bandwidths. The traffic and data generation and scheduling used differ from the results used in [122] and Section 23.3, which are based on static simulations and only uses round-robin scheduling. There are also some other differences such as cell sizes and propagation models. Both sets of simulations are based on ideal channel and

CQI estimation. The results presented in [57, 58] are somewhat different from the LTE results in [122] due to the difference in assumptions and they also contain results verifying the coverage performance.

The 2 × 2 antenna configuration agreed for LTE evaluation can also be used for downlink single-stream beam-forming. It is shown in [59] that the mean user throughput for high traffic load and the cell-edge throughput for all loads are improved compared to 2 × 2 MIMO. This comes at the expense of mean user throughput at a lower load. When combining the beam-forming results in [59] with the 2 × 2 MIMO results in [57, 58], all performance targets are met using the agreed antenna configuration from Table 23.2. It is also shown in [59] that a 4 × 2 multi-stream beam-forming concept will outperform the 2 × 2 MIMO and the single-stream beam-forming concept at all loads.

As a conclusion of the feasibility study, the performance of LTE was reported and compared to the WCDMA baseline in TR25.912 [85] and it was concluded that the targets are met.

23.4.3 Performance of LTE with 20 MHz FDD carrier

The evaluations presented above based on static simulations [122] and the 3GPP evaluation in [57, 58] are all based on LTE with 5 MHz carriers. LTE supports a range of different carrier bandwidths up to 20 MHz. A higher bandwidth gives a potential for higher peak data rates for the users and also higher mean user throughput. Simulation results are shown for a 20 MHz LTE downlink carrier in Figure 23.6 for 2 × 2 MIMO and 4 × 4 MIMO, and with 5 MHz results as reference. The numbers are based on a dynamic simulation with the same assumptions as in the 3GPP downlink evaluation [57], except for the parameters shown in Table 23.3.

The performance numbers in Figure 23.6 show a mean user throughput for a 20 MHz carrier that is on the order of 4 times that of a 5 MHz carrier for both low and high traffic loads. Though the 20 MHz carrier gives this fourfold increase in data rate, the maximum spectral efficiency (Mbps/MHz) at high loads is virtually the same. The numbers for 4 × 4 MIMO indicate that using twice as many antennas has a potential to almost double the spectral efficiency compared to 2 × 2 MIMO.

23.5 Conclusion

Although based on simplified models and excluding higher layer protocol improvements, the simulation results presented in this chapter demonstrate the

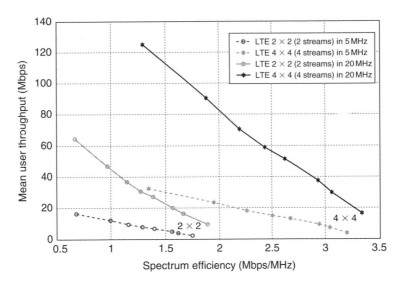

Figure 23.6 *Mean downlink user throughput vs. spectral efficiency for 5 and 20 MHz LTE carriers.*

Table 23.3 *Assumptions for the results in Figure 23.6, in addition to the ones in [57].*

Base-station power	40 W (20 MHz) and 20 W (5 MHz)
Propagation model	Suburban Macro (spatial channel model)
Inter-site distance	500 m
Scheduling	Proportionally fair in time and frequency

high potential of both LTE and HSPA evolution to improve user quality, capacity, and coverage, thereby reducing overall infrastructure cost in both coverage and capacity limited scenarios.

The performance numbers presented here together with the results used in the 3GPP evaluation indicate that LTE indeed fulfills the 3GPP targets on user throughput and spectrum efficiency. These requirements are partly formulated as relative comparisons to a rather basic WCDMA system. A more advanced WCDMA system, employing MIMO and GRAKE receivers, reaches performance similar to that of the LTE concept. In summary, it should be noted that many of the improvements for LTE also will be applied in the WCDMA/HSPA evolution.

24

Other wireless communications systems

Most mobile and cellular systems go through a continuous development of the underlying technology and of the features, services, and performance supported. The direction of development is from the baseline 'narrowband' systems of the 1990s supporting mainly voice services to today's 'wideband' systems that target a much wider set of services, including broadband wireless data.

The fundamental technologies introduced in the systems to support new and better performing services all fall within the scope of Part II of this book. When the problems to solve are fundamentally the same, the solutions engineered tend to be quite similar. Wideband transmission, higher-order modulation, fast scheduling, advanced receivers, multi-carrier, OFDM, MIMO, etc. are added as continuous developments and, at other times, as more revolutionary steps for the different technologies.

The application of these different technical solutions to the evolution of 3G radio access in terms of HSPA and LTE was discussed in Parts III and IV of this book. A similar development is occurring for the other technologies in the IMT-2000 family, and also for non-IMT-2000 technologies, including the MBWA standard developed within IEEE (*Institute of Electrical and Electronics Engineers*). This chapter gives an overview of those technologies and a high-level description of some of the solutions chosen in their ongoing evolution (Figure 24.1).

24.1 UTRA TDD

The evolution of UTRA TDD is part of the same effort in 3GPP of evolving the 3G radio access as described in Part III of this book. UTRA TDD is a part of the IMT-2000 family of technologies as 'IMT-2000 CDMA TDD' [46]. Very similar

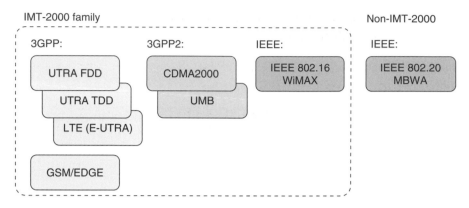

Figure 24.1 *The wireless technologies discussed in this book. The evolution tracks for UTRA are described in Parts III and IV of this book, while this chapter gives an overview for the other technologies.*

ideas have been applied as for WCDMA/HSDPA (UTRA FDD), but they were completed at a later stage in the 3GPP standardization process:

- *HSDPA*: Development of high-speed downlink packet access for UTRA TDD was performed as a common work item for WCDMA and UTRA TDD, but was performed as a separate task in practice [84]. HSDPA for TDD has many similarities with the FDD solution, such as the use of 16QAM higher-order modulation and the use of hybrid ARQ. Some main differences are that due to the TDD frame structure, timing relations are very different and a much larger TTI of 10 ms is used. Also, ACK/NACK information is not sent on a dedicated channel, but rather on a shared uplink resource.
- *Enhanced Uplink*: There is a corresponding Enhanced Uplink specified for UTRA TDD [82], also with some similarities to the FDD Enhanced Uplink. Some main differences are the 10 ms TTI due to the TDD frame structure and the use of a common scrambling sequence for uplink resources in the same time slot. This implies that the number of codes available in a time slot is limited requiring that the NodeB scheduler dynamically reassigns the available uplink code resources between users. A second implication is that the TDD uplink can be code limited and therefore both QPSK and 16QAM modulation schemes are supported.
- *MBMS*: As described in Chapter 11, the main impact of MBMS is not on the radio access network [102]. There are however some additional transport channels. For MBMS in TDD, an additional feature added is simulcast combining. Unlike for FDD, selective or soft combining in the terminal for macro-diversity reception is not mandatory for TDD prior to the introduction of MBMS. Since macro-diversity is essential for efficient broadcast service, a combination of simulcast reception with time slot reuse between cells is introduced. Time slots exclusively used for MBMS are assigned to groups

of cells in a reuse pattern and the terminal attempts to receive the time slots from each set and then combine them to enhance reception.

There are also some steps taken in UTRA TDD standardization that have no corresponding feature in WCDMA/HSPA:

- *1.28 Mcps TDD (low chip rate TDD)*: Described in Section 24.2.
- *7.68 Mcps rate TDD*: There is also a 7.68 Mcps option for UTRA TDD, with standardization work just completed, including a high-speed HSDPA enhancement. It is based on the same time slot structure as the 3.84 Mcps mode and is otherwise also very similar.

24.2 TD-SCDMA (low chip rate UTRA TDD)

A UTRA TDD mode with a 1.28 Mcps chip rate was introduced in Release 4 of the 3GPP specifications. This *Low chip rate* UTRA TDD is often referred to as *Time Division-Synchronous CDMA* (TD-SCDMA). The main differences from WCDMA/HSPA are, apart from the use of TDD instead of FDD, the lower chip rate and resulting (approximate) 1.6 MHz carrier bandwidth, the optional 8PSK higher-order modulation and a different 5 ms time slot structure.

The high-speed downlink HSDPA enhancement to the low chip rate TDD mode is very similar to the high chip rate, with the main difference being the time slot structure that gives a shorter 5 ms TTI. There is also an Enhanced Uplink mode for 1.28 Mcps UTRA TDD.

TD-SCDMA is mainly developed in China as an industry standard within CCSA. Some features reflected in the 3GPP low chip rate TDD standard origin from the work within CCSA [137]. These features include:

- *Multi-frequency operation*: In this mode, multiple 1.28 Mcps carriers are supported in one cell, but BCH is only sent on one carrier called *primary frequency*, in order to decrease inter-cell interference. The carrier on the primary frequency contains all common channels while traffic channels can be on both primary and secondary carriers. Each terminal still operates on a single 1.6 MHz carrier.
- *Beam forming improvement*: Angle-of-Arrival parameters for beam forming can be signaled in the RACH and FACH data frames.
- *Multi-carrier HSDPA*: In a cell using multi-carrier HSDPA, the HS-DSCH can be transmitted to a terminal on more than one carrier. There is a UE capability defined for receiving up to 6 carriers.

As discussed in Chapter 19, the LTE TDD mode has been designed to ease coexistence with and migration from TD-SCDMA to LTE.

24.3 CDMA2000

CDMA2000 evolved as a cellular standard under the name IS-95 and later became a part of the IMT-2000 family of technologies as described in Chapter 1. When it became a more global IMT-2000 technology, the name was changed to CDMA2000 and the specification work was moved from the *US Telecommunications Industry Association* (TIA) to 3GPP2. Being a sister organization to 3GPP, 3GPP2 is responsible for CDMA2000 specifications.

The CDMA2000 standard is going through an evolution similar to that of WCDMA/HSPA. In the different evolution steps, the same gradual shift of focus from voice and circuit-switched data to best-effort data and broadband data can be seen as for WCDMA/HSPA. It turns out that also for CDMA2000, basically the same means and technologies are used to incorporate the new-data capabilities and improve performance, as described in Part III of this book.

The evolution steps of CDMA2000 are shown in Figure 24.2. After the CDMA2000 1x standard was formed as an input to ITU for IMT-2000, two parallel evolution tracks were initiated for better support of data services. The first one was EV-DO (*Evolution–Data Only*)[1] which has continued to be the main track as further described below. It is also called HRPD (*High Rate Packet Data*). A parallel track was EV-DV (*Evolution for integrated Data and Voice*)

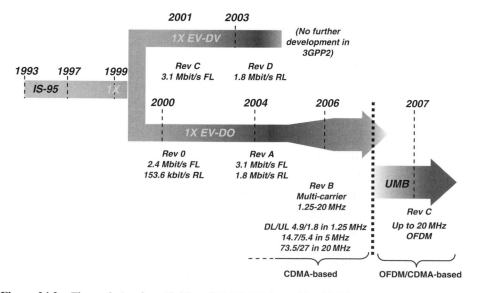

Figure 24.2 *The evolution from IS-95 to CDMA2000 1x and 1x EV-DO.*

---

[1] The acronym 'DO' is also interpreted as 'Data Optimized,' since the EV-DO carrier targets data services.

was developed to give parallel support of data and circuit-switched services on the same carrier. It is at the moment not developed further within 3GPP2.

24.3.1 CDMA2000 1x

The CDMA2000 standard, that originally supported both a single-carrier (1x) and a multi-carrier (3x) mode, was adopted by ITU-R under the name IMT-2000 CDMA *Multi-carrier* (MC) [46]. It offers several improvements over the earlier versions of IS-95 that give better spectral efficiency and higher data rates. The most important aspect from a 3G evolution perspective however is that 1x has been a platform for further evolution of packet-data modes as shown in Figure 24.2. While 1x EV-DV is not undergoing any new development, 1x EV-DO is deployed today and is going through several new evolution steps in Rev 0, Rev A, Rev B, and Rev C that are further described below. Rev C is also called *Ultra Mobile Broadband* (UMB). While all steps up to Rev A are fundamentally based on direct sequence spread spectrum and a 1.25 MHz carrier bandwidth, Rev B and Rev C diverge from this by including wider carrier bandwidths and for Rev C also OFDM operation.

The 3x mode of CDMA2000 was never deployed in its original form, but was still an essential component of the submission of CDMA2000 to the ITU-R. However, multi-carrier CDMA is today once again included in the CDMA2000 evolution through CDMA2000 Rev B.

24.3.2 1x EV-DO Rev 0

EV-DO Rev 0 defines a new uplink and downlink structure for CDMA2000 1x, where DO originally implied '*Data Only.*' The reason is that an EV-DO carrier has a structure optimized for data that does not support the voice and circuit-switched data services of a CDMA2000 1x carrier. In this way, the whole carrier works as a shared downlink resource for data transmission. An operator would deploy an additional carrier for EV-DO, thereby separating circuit-switched and packet-switched connections on different carriers. A drawback is a reduced flexibility in that there cannot be a simultaneous packet data and legacy circuit-switched service (such as voice) to the same user on one carrier 1x EV-DO has later also been named HRPD (*High Rate Packet Data*).

With EV-DO Rev 0, a peak data rate of 2.4 Mbps is supported in the downlink on a 1.25 MHz carrier. There are several components in EV-DO Rev 0, some of which have similarities with the HSPA evolution described in Chapter 9:

- *Shared-channel transmission*: EV-DO Rev 0 has a *Time Division Multi-plexing* (TDM) downlink, with transmission to only one user at the time with the full power of the Base Station (BS). This makes the downlink a resource that is shared between the users in the time domain only. This is similar to the 'shared-channel transmission' for HSDPA, with the difference that HSDPA can also share the downlink in the code domain, primarily with non-HSDPA users.

- *Channel-dependent scheduling*: The adaptive data scheduler of EV-DO Rev 0 takes into account fairness, queue sizes, and measures of the channel state. It thereby exploits multi-user diversity in fading channels in a way similar to the channel-dependent scheduling used in HSPA

- *Short TTI*: The transmission time interval is reduced from 20 ms in CDMA2000 to 1.6 ms in EV-DO Rev 0. This is important to enable fast channel-dependent scheduling and rapid retransmissions, resulting also in lower latency. The TTI for EV-DO Rev 0 is thereby of the same order as the 2 ms TTI used for HSPA.

- *Rate control*: EV-DO Rev 0 employs rate control through adaptive modulation and coding, thereby maximizing the throughput for a given channel condition. This is similar to HSPA, but the system in CDMA2000 follows the rate request from the mobile, while in HSPA the feedback is a recommendation and the NodeB takes the final decision.

- *Higher-order modulation*: EV-DO Rev 0 supports 16QAM modulation in the downlink, which is the same as for the HSPA downlink.

- *Hybrid ARQ*: The EV-DO Hybrid ARQ scheme is similar to the scheme used for HSPA.

- *Virtual SOHO*: EV-DO Rev 0 does not use soft handover in the downlink like CDMA2000 does. Instead the terminal supports 'virtual soft handover' via adaptive server selection initiated by the terminal, which can be seen as a fast cell selection within the 'active set' of base stations. These server changes may result in some packet transmission delays [12].

- *Receive diversity in the mobile*: EV-DO Rev 0 has terminal performance numbers specified assuming receive diversity, similar to the 'Type 1' performance requirements specified for HSPA advanced receivers as discussed in Chapter 12.

24.3.3   1x EV-DO Rev A

The next step in the evolution of CDMA2000 is 1x EV-DO Rev A. The focus is on an uplink improvement similar to the Enhanced Uplink of HSPA, but it also includes an updated downlink, a more advanced quality-of-service handling, and an add-on multicast mode [12].

The downlink of EV-DO Rev A is based on the EV-DO Rev 0 downlink, with the following differences:

- *Higher peak rates*: EV-DO Rev A downlink supports 3.1 Mbps as compared to the 2.4 Mbps of EV-DO Rev 0. Rev A also offers a finer quantization of data rates.
- *Shorter packets*: New transmission formats for EV-DO Rev A enable 128-, 256-, and 512-bit packets. This together with new multi-user packets, where data to multiple terminals share the same packet in the downlink, improves support for lower rate, delay-sensitive services.

The major enhancements in EV-DO Rev A compared to Rev 0 are in the uplink. This results in a more packet-oriented uplink with higher capacity and data rates. Peak uplink data rates of up to 1.8 Mbps are supported:

- *Higher-order modulation*: In addition to BPSK modulation in EV-DO Rev 0, the uplink physical layer of Rev A supports QPSK and optionally 8PSK modulation.
- *Hybrid ARQ*: Improved performance is achieved through an uplink hybrid ARQ scheme, similar to HSPA.
- *Reduced latency*: The use of smaller packet sizes and a shorter TTI enables a reduced latency of up to 50% compared to EV-DO Rev 0 [8].
- *Capacity/latency trade-off*: Each packet can be transmitted in one of two possible transmission modes; *LoLat* gives low latency through higher power level ensuring that the packet is received within the latency target, while *HiCap* gives higher total capacity by allowing for more retransmissions and lower transmit power levels. This is similar to the use of hybrid ARQ profiles for Enhanced Uplink in HSPA as described in Chapter 10.

24.3.4   1x EV-DO Rev B

The next step for the 1x EV-DO series of standards is Rev B, which enables higher data rate by aggregation of multiple carriers. This fundamental way of increasing the data rate is also discussed in Chapter 3. Rev B permits up to sixteen 1.25 MHz carriers to be aggregated, forming a 20 MHz wide system [13], and giving a theoretical peak data rate of up to 46.5 Mbps in the downlink. For reasons of cost, size, and battery life, Rev B devices will most likely support up to 3 carriers [8], giving a peak downlink data rate of 9.3 Mbps.

The lower layers of the radio interface are similar to and compatible with those in Rev A, making it possible for both single-carrier Rev 0 and Rev A terminals

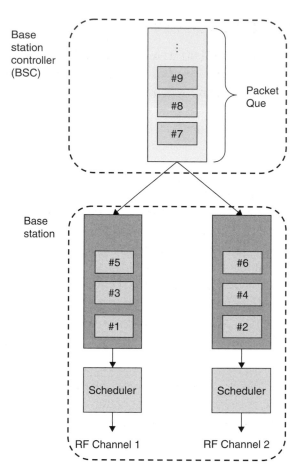

Figure 24.3 *In 1x EV-DO Rev B, multi-carrier operation can occur on multiple independent BS channel cards to allow a simple upgrade of existing base stations.*

to function on a Rev B network that supports multi-carrier operation. This is implemented in Rev B networks through a *multilink radio link protocol* (ML-RLP) that enables operation of a Rev B terminal assigned to multiple carriers on multiple legacy EV-DO Rev A base-station channel cards as shown in Figure 24.3. The channel cards do not have to communicate with each other and can have independent scheduling and physical layer processes on each card.

Carriers do not have to be symmetrically allocated in the uplink and the downlink. For asymmetric applications such as file download, a larger number of carriers can be set up for downlink than for uplink. This reduces the amount of overhead for the uplink transmission. One uplink channel can carry feedback information for multiple forward link channels operating on multiple downlink carriers to a single terminal in asymmetric mode.

24.3.5 UMB (1x EV-DO Rev C)

A further step of the CDMA2000 development in 3GPP2 is Revision C of the 1x EV-DO standard, also called *Ultra Mobile Broadband* (UMB). It includes OFDM, work on smart-antenna technologies and fundamental channel bandwidths up to 5, 10, or 20 MHz [8]. This evolution step is thus not backward compatible with the previous revisions of the CDMA2000 standard.

The UMB standard is based on a joint framework proposal made in 3GPP2 in 2006 [1], detailing a physical layer and some MAC layer aspects for FDD operation. Objectives included in the framework were introduction of higher peak date rates, better spectral efficiency, lower latency, improved terminal battery life and higher capacity and enhanced user experience for delay-sensitive applications. It is very similar to LTE in terms of technologies and features used, with one main difference that UMB uses OFDM in the uplink while LTE uses a single-carrier modulation. UMB also supports a CDMA mode in the uplink. The theoretical peak data rates for a 20 MHz downlink is 288 Mbps and for the uplink 75 Mbps [142].

The downlink is OFDM based while for low-rate traffic there is also a CDMA mode in the uplink. Several varieties of multi-antenna techniques including multi-layer transmission are supported. Some important features of the downlink are:

- OFDM data transmission with 9.6 kHz subcarrier spacing and assuming an FFT size of 128, 256, 512, 1024, or 2048, depending on the sampling rate and bandwidth. There is a variable length cyclic prefix with a length of between 6% and 23% of the OFDM symbol duration.
- The downlink dedicated channel supports QPSK, 8PSK, 16QAM, and 64QAM modulations.
- Downlink scheduling and rate adaptation are based on CQI reports from the terminal. The CQI report also indicates the desired downlink for handover and can be used for power control purposes.
- MIMO (spatial multiplexing) with *Single Codeword* (SCW) and *Multi-Codeword* (MCW) design is supported.
- *Space Time Transmit Diversity* (STTD) with up to four transmit antennas are used in place of MIMO when the propagation environment is not suitable for MIMO.
- *Spaced Division Multiple Access* (SDMA) is included as an advanced antenna technique with uplink user feedback as part of the scheme in FDD deployments.

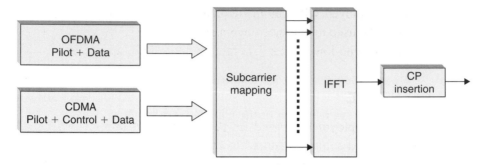

Figure 24.4 *UMB enables multiplexing of OFDMA and CDMA traffic on the uplink.*

The uplink is OFDMA based with optional CDMA transmission:

- OFDMA and CDMA uplink traffic are frequency multiplexed as shown in Figure 24.4.
- CDMA is mainly targeting bursty low-rate traffic and delay-sensitive applications. The lower *Peak-to-Average Ratio* (PAR) of CDMA is also beneficial for power-limited users.
- Uplink power control is used for both OFDMA and CDMA with a scheme called *Fast Distributed Reverse Link Power Control*. In OFDMA mode the control channels are closed loop controlled, while the traffic has a delta power setting relative to the control channels. A fast *Other Cell Interference Channel* (OSICH) from base stations in the active set is used to further control the uplink power.
- The OFDMA mode includes frequency hopping which can be used universally across the band for full frequency diversity. It can also be done on a subband level to achieve multiplexing diversity in combination with the subband scheduling.
- There is also a 'quasi-orthogonal' uplink scheme that uses spatial processing (with multiple receive antennas) to demodulate the individual uplink signals. Frequency hopping is done with hopping sequences randomized between sectors to achieve intra-sector interference diversity.

Other prominent features of UMB are:

- A half-duplex mode is supported for use in fragmented spectrum. This may be scenarios where the mobile receive and transmit bands are too close, making the implementation of a duplexer in the mobile too complex. In half-duplex mode, non-overlapping frames for uplink and downlink transmission are interlaced in two 'half-duplex interlaces,' with control information replicated on both interlaces.

- UMB can be deployed with single frequency reuse, but frequency planning can optionally be used to enhance coverage and QoS. The trade-off is a bandwidth reduction with lower peak data rates and usually lower capacity.

24.4 GSM/EDGE

Worldwide GSM deployment started in 1992 and GSM quickly became the most widely deployed cellular standard in the world with more than 2.5 billion subscribers today. A major evolution step for GSM was initiated in late 1996, at the same time as UTRA development started in ETSI. This step sprung out of the study of higher-order modulation for a 3G TDMA system that was performed within the European FRAMES project (see also Chapter 1).

The enhancement of GSM is called EDGE (*Enhanced Data Rates for GSM Evolution*). The EDGE development was initially focused on higher end-user data rates by introducing higher-order modulation in GSM, both for circuit-switched services and the GPRS packet-switched services. During the continued work, focus moved to the enhancement of GPRS (EGPRS), where other advanced radio interface components were added, including link adaptation, hybrid ARQ with soft combining, and advanced scheduling techniques as described in Part II of this book. In this way GSM became the first cellular standard to add such enhancements,[2] later followed by WCDMA/HSPA, CDMA2000, and other technologies. GSM/EDGE is however a more narrowband technology than WCDMA/HSDPA and CDMA2000, implying that the peak data rates achievable are not as high. HSDPA and CDMA2000 have however added a time division structure to make full use of advanced scheduling techniques for high-rate data services. Being a TDMA system, EDGE already has a time division structure.

24.4.1 Objectives for the GSM/EDGE evolution

An evolved GSM/EDGE standard is developed in 3GPP based on a feasibility study for an 'Evolved GSM/EDGE Radio Access Network (GERAN)' [87]. The objective is to improve service performance for GERAN for a range of services, including interactive best-effort services as well as conversational services including *Voice-over-IP* (VoIP).

The evolved GERAN is compatible with the existing GSM/EDGE in terms of frequency planning and coexistence with legacy terminals. It also reuses the

---

[2] The first cellular technology to deploy higher-order modulation was iDEN that uses 'M-16QAM.' It is however a proprietary technology and is as such not standardized.

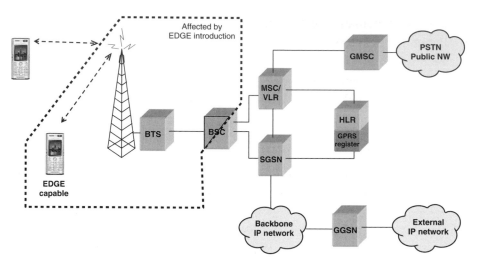

Figure 24.5   *GSM/EDGE network structure.*

existing network architecture (Figure 24.5) and has minimum impact on the BTS, BSC, and core network hardware. For the standardization of Evolved GERAN, a number of performance target were set [87]:

- *Improved spectrum efficiency*: The target is a 50% increase in an interference-limited scenario (measured in kbps/MHz/cell for data or Erlang/MHz/cell for voice).
- *Increased peak data rates*: The target is a 100% increase in both downlink and uplink.
- *Improved coverage for voice and data*: The target is a sensitivity increase of 3 dB in downlink (noise-limited scenario).
- *Improved service availability*: The target is a 50% increase in mean data rate for uplink and downlink at cell edges (when cells are planned for voice).
- *Reduced latency*: The target is a *roundtrip time* (RTT) of less than 450 ms for initial access and less than 100 ms after initial access (in non-ideal radio conditions, counting from the terminal to the GGSN and back, as shown in Figure 24.5).

A number of alternative solutions were studied in 3GPP to achieve the performance and compatibility targets listed above. These are not fundamentally different from the solutions selected for LTE or other technologies and are all described in a more general context in Part II of this book. The technologies chosen for standardization in 3GPP for GERAN Evolution were:

- Dual-antenna terminals.
- Multi-carrier EDGE.
- Reduced TTI and fast feedback.

- Improved modulation and coding.
- Higher symbol rates.

The specific application of these technologies to GERAN is discussed below. The solutions are included in Release 7 of the GSM/EDGE specifications.

24.4.2 Dual-antenna terminals

As was shown in Chapter 6, multiple receiver antennas are an effective means against multipath fading and to provide an improved signal-to-noise ratio through 'power gain' when combining the antenna signals. There will thus be improvements for both interference-limited and noise-limited scenarios. There are also possibilities for interference cancellation through multiple antennas. While multiple antennas have an implementation impact for the terminal, there is no impact for the base station hardware or software.

For GSM/EDGE, a dual-antenna solution called *Mobile Station Receive Diversity* (MSRD) is standardized. Analysis shows that a dual-antenna solution in GSM terminals can give a substantial coverage improvement of up to 6 dB. In addition, the dual-antenna terminals could potentially handle almost 10 dB more interference [20, 85].

24.4.3 Multi-carrier EDGE

The GSM radio interface is based on a TDMA structure with 8 time slots and 200 kHz carrier bandwidth. To increase data rates, today's GPRS and EDGE terminals can use multiple time slots for transmission, reception, or both. Using 8PSK modulation (Figure 24.6) and assigning the maximum of eight time slots, the GSM/EDGE standard gives a theoretical peak data rate of close to 480 kbps. However, from a design and complexity point of view, it is best to avoid simultaneous transmission and reception. Today's terminals typically receive on a maximum of five time slots because they must also transmit (on at least one time slot) as well as measure the signal strength of neighboring cells.

To increase data rates further, multiple carriers for the downlink and the uplink can be introduced, similar to what is discussed in Chapter 3. This straightforward enhancement increases peak and mean data rates in proportion to the number of carriers employed. For example, given two carriers with eight time slots each, the peak data rate will be close to 1 Mbps. Using dual carriers can be seen as a straightforward extension of the multi-slot principle, allowing a multi-slot configuration to span more than one carrier. The limiting factor in this case is the

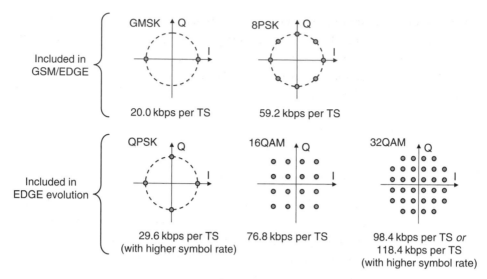

Figure 24.6 *Existing and new modulation schemes for GSM/EDGE. The highest specified radio interface data rate using GPRS is shown for each scheme [20]. Note that the view of the non-linear binary GMSK scheme is simplified in the figure.*

complexity and cost of the terminal, which must have either multiple transmitters and receivers or a wideband transmitter and receiver. The use of multiple carriers has only a minor impact on base transceiver stations. A dual-carrier downlink solution is standardized in Release-7 of the GSM/EDGE standard.

24.4.4   Reduced TTI and fast feedback

Latency is usually defined as the roundtrip time over the radio access network. It has a major influence on user experience and especially conversational services such as voice and video telephony require low latency, but also services such as web browsing and e-mail download are improved considerably. A major parameter in the radio interface that has impact on latency is the *Transmission Time Interval* (TTI) as discussed for HSPA and Enhanced Uplink in Chapters 9 and 10. It is difficult to substantially improve latency without reducing the TTI.

The *roundtrip time* (RTT) in GSM/EDGE networks can be 150 ms [20], including network delays but not retransmissions over the radio interface. Radio blocks are transmitted interleaved over four consecutive bursts on one assigned time slot over 20 ms. In GSM/EDGE evolution, a reduced TTI with radio blocks interleaved over four consecutive burst assigned on *two time slots* in 10 ms is introduced. This reduction of the TTI from 20 to 10 ms can reduce the roundtrip time from 150 to less than 100 ms [20]. Multiplexing of users with the two different interleaving schemes is also possible on the same RF carrier.

Faster responses to incorrectly received radio blocks speed up the retransmission of radio blocks and can also help in reducing the latency. In GSM/EDGE evolution, a fast ACK/NACK reporting mechanism is introduced to reduce time for the network to realize that a block is lost. ACK/NACK reports can also 'piggyback' on user data, which reduces the overhead. Combined with the shorter TTI, the total time to retransmit lost radio blocks is in this way reduced significantly and the throughout is increased.

24.4.5 Improved modulation and coding

The main evolution step taken for GSM when EDGE was introduced was the higher-order modulation to enhance the data rates over the radio interface – Enhanced Data Rates for GSM Evolution. 16QAM was considered for EDGE, but 8PSK was finally chosen as a smooth evolution step that gives 3 bits per modulated symbol instead of one. This increases the peak data rate from approximately 20 to 60 kbps.

Figure 24.6 shows the further steps that are standardized for Evolved EDGE, giving 4 bits/symbol for 16QAM and 5 bits/symbol for 32QAM. Since the signal points in the new schemes move more closely together, they are more susceptible to interference. With more bits per symbol, however, the higher data rates allow for more robust channel coding which can more than compensate for the increased susceptibility to interference. Also QPSK modulation with 2 bits/symbol is included in Evolved GSM/EDGE, but used only with the higher symbol rate described below.

Both GSM and EDGE use convolutional codes, while also Turbo codes are standardized for Evolved EDGE. Decoding Turbo codes is more complex than regular convolutional codes. However, since Turbo codes are already used today for WCDMA and HSPA, and many terminals support both GSM/EDGE and WCDMA/HSPA, the decoding circuitry for Turbo codes already exists in the terminals and can be reused for EDGE.

Turbo codes perform well for large code block sizes, making them better suited for higher data rate EDGE channels using 8PSK or higher-order modulation. It is estimated that compared to the existing EDGE scheme with 8PSK modulation, the combination of Turbo codes and 16QAM improves the user data rates 30–40% for the median user in a system [87]. Turbo codes are standardized for the downlink in Release 7 of GSM/EDGE.

24.4.6 Higher symbol rates

The modulation can also be improved by simply increasing the modulation symbol rate. As a part of GERAN Evolution in 3GPP, the combination of higher-order

Table 24.1 *Combinations of modulation schemes and symbol rates in GSM/EDGE evolution.*

Terminal capability	Symbol rate	Modulation schemes
Uplink Level A	271 ksymbols/s	GMSK, 8PSK, 16QAM
Uplink Level B	271 ksymbols/s	GMSK
	325 ksymbols/s	QPSK, 16QAM, 32QAM
Downlink Level A	271 ksymbols/s	GMSK, 8PSK, 16QAM, 32QAM
Downlink Level B	271 ksymbols/s	GMSK
	325 ksymbols/s	QPSK, 16QAM, 32QAM

modulation and a 20% increase in symbol rate has been standardized [65]. For both uplink and downlink, two 'levels' of terminal capabilities are defined, called Level A and Level B. Each level defines a set of modulation schemes, where Level B also applies the higher symbol rate for some modulations schemes. Table 24.1 shows the combinations that are defined.

The higher symbol rate of 325 ksymbols/s will operate with the same nominal carrier bandwidth and carrier raster as legacy GERAN having 271 ksymbols/s, which puts some requirements on the transmitter filter. The amount of transmitter filtering has to be weighed against the receiver complexity and performance, and the amount of interference in adjacent channels. An alternative wider Tx-filter has been standardised for the uplink. A wider Tx-filter is also studied for the downlink.

24.5 WiMAX (IEEE 802.16)

The IEEE 802 LAN/MAN Standards Committee develops both wired and wireless network standards. The wireless standards currently developed within IEEE 802 range from the short-range *Personal Area Network* (PAN) Standards in 802.15 to the Wide Area Network Standard 802.20 (see also Section 20.5).

WiMAX[3] is the commonly used name for broadband wireless access based on the IEEE 802.16 family of standards. The WiMAX forum is an industry-led, non-profit corporation formed to promote and certify compatibility and interoperability of 802.16 broadband wireless products.

IEEE 802.16 is an IEEE Standard for *Wireless Metropolitan Area Networks* (WMAN). The first version was developed for fixed wireless broadband access

---

[3] Worldwide Interoperability for Microwave Access.

in the 10–66 GHz bands and line-of-sight communication in 2001. The next step called IEEE 802.16a that was completed in 2003 supports also non-line-of-sight communication, focusing on the 2–11 GHz bands. The first products based on WiMAX will be based on the revised and consolidated version of the standard published in 2004 called 802.16–2004. The most recent addition to the WiMAX family of standards is 802.16e, which is also called 'Mobile WiMAX.'

The IEEE standards specify the *physical layer* (PHY) and the *Medium Access Layer* (MAC), with no definition of higher layers. For IEEE 802.16, those are addressed in the WiMAX Forum Network Working Group. There is a range of options specified in IEEE 802.16, making the standards much more fragmented than what is seen in 3GPP and 3GPP2 standards. The 802.16 standard defines four different physical layers, of which two are certified by the WiMAX forum:

- OFDM-PHY: based on an FFT size of 256 and aimed at fixed networks.
- OFDMA-PHY (scalable): based on an FFT size from 128 to 2048 for 802.16e.

In addition to the multiple physical layers, the 802.16 standards support a range of options, including:

- TDD, FDD, and *half-duplex FDD* (H-FDD) operation.
- TDM access with variable frame size (2–20 ms).
- OFDM with a configurable cyclic prefix length.
- A wide range of bandwidths supported (1.25–28 MHz).
- Multiple modulation and coding schemes: QPSK, 16QAM, and 64QAM combined with convolutional codes, convolutional Turbo codes, block Turbo codes, and LDPC (*Low-Density Parity Check*) codes.
- Hybrid ARQ
- *Adaptive antenna system* (AAS) and MIMO.

There is also a range of *Radio Resource Management* (RRM) options, MAC features and enhancements in the standards. The WiMAX forum defines system profiles that reduce all the optional features to a smaller set to allow interoperability among different vendors. This is done through an industry selection of features for MAC, PHY, and RF from 802.16 specifications and forms the basis for testing conformance and interoperability. Products certified by the WiMAX forum adhere to a Certification Profile that is based on a combination of band of operation, duplexing option and bandwidth.

The intended applications with the original 802.16 standard were fixed access and backhaul, mainly for line-of-sight operation. The addition of a physical layer for non-line-of-sight applications in IEEE 802.16-2004 and support for mobility in IEEE 802.16e opens up the standard for nomadic and mobile use. In addition, provisions for *multicast and broadcast services* (MBS) are also included. This makes the standard more similar to the evolved 3G standards, but coming from a completely different direction. The IEEE standards such as 802.16 are driven by the datacom industry as Layer 1 and 2 standards, starting with line-of-sight use for limited mobility, targeting best-effort data applications and now moving to higher mobility and encompassing also other applications such as conversational services. The evolved 3G standards are driven by the telecom industry, targeting non-line-of-sight use and mobility from the beginning, optimized end-to-end standards for voice and later also data services, now moving to broader data applications including best-effort services.

There are several versions of the IEEE 802.16 standards published from 2001 to 2005, with multiple physical layers and features as described above. The description of the WiMAX technology and its features below is based on the amended IEEE 802.16e-2005 version of the standard [2] and what is referred to as the Release 1 Mobile WiMAX system profiles [116] unless otherwise stated.

24.5.1   Spectrum, bandwidth options and duplexing arrangement

The Release 1 WiMAX profiles cover operation in licensed spectrum allocations in the 2.3, 2.5, 3.3, and 3.5 GHz bands. The channel bandwidths supported are 5, 7, 8.75, and 10 MHz.

While WiMAX supports TDD, FDD, and half-duplex FDD, the first release only supports TDD operation. TDD enables adjustment of the downlink/uplink ratio for asymmetric traffic, does not require paired spectrum, and has a less complex transceiver design. To counter interference issues, TDD does however require system-wide synchronization and use of the same uplink/downlink ratio in neighboring cells [116]. The reason is the potential for mobile-to-mobile and base station-to-base station interference if uplink and downlink allocations overlap, which becomes an issue in multi-cell deployments. Because of adjacent channel interference, system-wide synchronization may also be required for TDD operators deployed on adjacent or near-adjacent channels.

While the initial profiles and deployment of WiMAX use TDD as the preferred mode, FDD may be introduced in the longer term. The reason may be local regulatory requirements or to address need for more extended multi-cell coverage where FDD may become more suitable [115].

24.5.2 Scalable OFDMA

The physical layer is based on OFDM and supports FFT sizes of 128, 512, 1024, or 2048 for operation with various channel bandwidths. The 5 and 10 MHz bandwidths in the WiMAX profile are supported with FFT sizes 512 and 1024. Each subcarrier is 10.94 kHz wide and a 'subchannel' consists of 48 data subcarriers for data transmission and pilot subcarriers to facilitate coherent detection. A subchannel can consist of *contiguous* subcarriers or a pseudo-random *diversity* permutation of subcarriers for better frequency diversity and inter-cell interference averaging. Diversity permutations supported include DL-FUSC (*Fully Used Subcarriers*) and DL-PUSC (*Partially Used Subcarriers*) for the downlink and UL-PUSC for the uplink. The number of pilot subcarriers per subchannel is different for the adjacent subcarrier, FUSC and PUSC permutations.

The frame can be divided into permutation *zones* with different subchannelization scheme that enable deployment with fractional frequency reuse as described below. There are also optional allocation schemes for later WiMAX profiles, including *Adaptive Modulation and Coding* (band AMC), where contiguous subcarriers form a subchannel, relying more on the time coherence of the channel. This is also better suited for the channel quality feedback in helping an adaptive scheduler that can adapt to channel quality variations in the frequency domain.

24.5.3 TDD frame structure

The OFDMA frame structure for WiMAX TDD is divided into a downlink and an uplink subframe with a transmission gap in-between as shown in Figure 24.7. The downlink subframe starts with a preamble for synchronization and base-station identification, followed by a *Frame Control Header* (FCH) with decoding information for the *MAP* messages. The attached downlink and uplink MAP messages provide downlink and uplink allocations of the user data bursts, plus other control information.

The uplink subframe contains a *ranging* subchannel to gain access, for bandwidth requests and for closed-loop time, frequency and power adjustments for the users. In addition, there are uplink allocations for fast channel quality feedback (CQICH) and for *Mobile Station* (MS) feedback of hybrid ARQ acknowledgments (ACK-CH).

24.5.4 Modulation, coding and Hybrid ARQ

Support for QPSK, 16QAM, and 64QAM modulations are mandatory in the downlink, while QPSK and 16QAM are mandatory in the uplink. There are

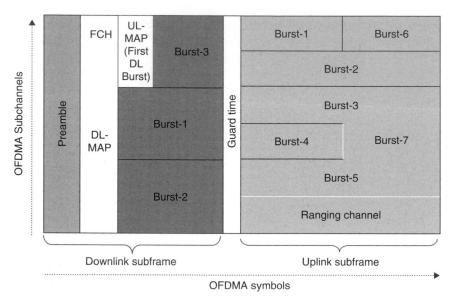

Figure 24.7   *Example OFDMA frame structure for WiMAX (TDD).*

multiple coding schemes defined, but only the first two in the list below are mandatory. In addition repetition coding is supported to achieve low effective coding rates:

- Tail-biting *Convolutional Code* (CC)
- *Convolutional Turbo Code* (CTC)
- *Block Turbo Code* (BTC) – Optional
- *Low-Density Parity Check Code* (LDPC) – Optional
- *Zero Tailed Convolutional Code* (ZTCC) – Optional

The appropriate combination of modulation and coding scheme is determined by the base-station scheduler, based on channel state, buffer size, and other parameters. The downlink channel quality information is reported by the mobile station through the CQICH channel mentioned above. It is also possible to take channel reciprocity into account in TDD operation.

Hybrid ARQ is supported and allows for fully asynchronous operation to allow variable delay between retransmissions. Chase combining is mandatory in the WiMAX profiles, while incremental redundancy is optional.

24.5.5   *Quality-of-service handling*

A connection-oriented *Quality-of-Service* (QoS) mechanism is implemented, enabling end-to-end QoS control. The QoS parameters are set per service flow, with multiple service flows possible to/from a mobile station. The parameters define transmission ordering and scheduling on the air interface and can be negotiated statically or dynamically through MAC messages.

Applications supported through the WiMAX QoS mechanism are:

- *Unsolicited Grant Service* (UGS): VoIP
- *Real-Time Polling Service* (rtPS): Streaming Audio or Video
- *Extended Real-Time Polling Service* (ErtPS): Voice with Activity
- *Non-Real-Time Polling Service* (nrtPS): File Transfer Protocol (FTP)
- *Best-Effort Service* (BE): Data Transfer, Web Browsing, etc.

24.5.6   Mobility

Mobile WiMAX supports both sleep mode and idle mode for more efficient power management. In sleep mode, there is a pre-negotiated period of absence from the radio interface to the serving base station, where the mobile station may power down or scan other neighboring base stations. There are different 'power saving classes' suitable for applications with different QoS types, each having different sleep mode parameters. There is also an idle mode, where the terminal is not registered to any base station and instead periodically scans the network at discrete intervals.

There are three handover methods supported, with *Hard Handover* (HHO) being mandatory and *Fast Base-Station Switching* (FBSS) and *Macro-Diversity handover* (MDHO) being optional [2, 116]. Soft handover is not supported:

- *Hard Handover* (HHO): The mobile station scans neighboring Base Stations (BS) and maintains a list of candidate BS to make potential decisions for cell re-selection. The decision to handover to a new BS is made at the MS or at the serving BS. After decision, the MS immediately starts synchronization to the downlink of the target BS and obtains uplink and downlink transmission parameters and establishes a connection. Finally, the context of all connections to the previous serving BS are terminated.
- *Fast Base-Station Switching* (FBSS): The MS can maintain a diversity set (same as 'active set' for WCDMA) of multiple suitable Base Stations. One BS in the active set is selected as Anchor BS through which all uplink and downlink communication is done, both traffic and control messages. The MS selects the best BS in the diversity set as its Anchor and can change Anchor BS without any explicit handover signaling by reporting the selected BS on the CQICH channel.

- *Macro-Diversity Handover* (MDHO): An Anchor BS and a diversity set of the most suitable BSs are maintained also in MDHO. However, the MS communicates with all BSs in the diversity set in case of MDHO. In the downlink, multiple BS provides synchronized transmission of the same messages and in the uplink, multiple BS receive messages from the MS and provide selection diversity.

A requirement for FBSS and MDHO is that all BS in the active set need to operate on the same frequency, be time synchronized and also share and transfer all MAC contexts between BSs.

24.5.7 Multi-antenna technologies

In addition to conventional receive diversity, WiMAX supports additional options using smart-antenna technologies:

- *Beam-forming or Adaptive/Advanced Antenna System* (AAS): Provides increased coverage and capacity through uplink and downlink beam-forming. AAS and non-AAS mobiles are time division multiplexed using a different time zone for the different options. Other MIMO schemes may be used in the non-AAS zones.
- *Space–time coding* (STC): With two transmit antennas, the use of transmit diversity through Alamouti code is supported to achieve spatial diversity. It is also possible to have transmit diversity using Linear Dispersion Codes (LDC) for more than two transmit antennas.
- *Spatial multiplexing through MIMO*: Higher peak rates and throughput in favorable propagation environments are provided through a downlink 2×2 MIMO scheme. The uplink supports 'Uplink collaborative MIMO,' where two users transmit with one antenna in the same slot and the BS receives the two multiplexed streams as if it were a MIMO transmission from a single user.

24.5.8 Fractional frequency reuse

WiMAX can operate with a frequency reuse of one, but co-channel interference may in this case degrade the quality for users at the cell edge. However, a flexible subchannel reuse is made possible by dividing the frame into permutation zones as described above. In this way, it is possible to have a subchannel reuse by proper configuration of the subchannel usage for the users. For users at the cell edges, the Base Station operates on a zone with a fraction of the subchannels, while users close to the Base Station can operate on a zone with all subchannels. As shown in the example in Figure 24.8, there can be an effective reuse of

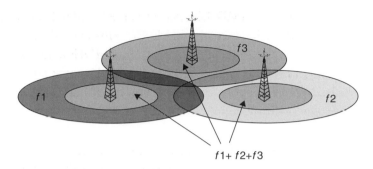

f1+ f2+f3

Figure 24.8 *Fractional frequency reuse. In this example there is a reuse of 3 for users at the cell edge, while users closer to the BSs have a single frequency reuse.*

frequencies for users at the cell edge, while still maintaining a reuse of one for the OFDMA carrier as a whole.

24.5.9 Advanced Air Interface (IEEE 802.16m)

Within the IEEE 802.16 committee, there is also a Task Group working on an advanced air interface to be specified as an amendment to 802.16. The intention is to offer support for legacy 802.16 equipment, while also meeting the requirements of IMT-Advanced as defined by ITU-R. The amended standard would then become a candidate for consideration in the ITU-R evaluation process for IMT-Advanced.

A set of system requirements is defined for 802.16m [140] and include the following:

- *Legacy support* in terms of compatibility with legacy equipment, co-existence and sharing on the same RF carrier and handover to and from legacy BS.
- *Complexity* of architecture and protocols should be minimized and low cost devices should be supported.
- *Services* should be supported as for the legacy system, plus the additional services identified for IMT-Advanced.
- *Spectrum flexibility* shall be supported in terms of operation at different RF frequencies (below 6 GHz), different Duplex modes and scalable operating bandwidths from 5 to 20 MHz.
- *Advanced antenna techniques* including MIMO and beam forming shall be supported with a minimum of 2 Tx and 2 Rx antennas at the BS and 1 Tx and 2 Rx antennas at the MS.
- *Functional requirements* including requirements on peak data rates, latency, QoS, RRM, Security, Handover, Multicast/Broadcast, Location-Based Services

and multi-RAT (*Radio Access Technology*) operation, for example, 3GPP technologies.

- *Performance requirements* including targets for data throughput, capacity, and mobility.
- *Operational requirements* including support for multi-hop relay, co-existence with other networks (for example, 3GPP technologies) and support of self-organizing mechanisms.

The development of the standard is still in its initial phase (May 2008), but higher level properties of the radio interface are determined in order to support the above requirements [141]. While most PHY and MAC layer solutions are not detailed yet, the following preliminary properties can be noted:

- Procedures to request that an MS report measurements for Self-configuration and Self-optimization.
- Multi-carrier support, where a single MAC entity can control multiple frequency channels, each with a bandwidth of up to 20 MHz bandwidth.
- Interference management for inter-cell interference including special measurement reports, flexible frequency re-use, power control, interference randomization, interference cancellation, and transmitter beam forming.
- Physical layer support of TDD, FDD and half-duplex FDD operation.
- A downlink multiple access scheme based on OFDMA with 10.94 kHz sub-carriers as for legacy 802.16e. The frame structure is defined with a fixed offset to the legacy frame structure to accommodate new synchronization and other common channels.

24.6 Mobile Broadband Wireless Access (IEEE 802.20)

Within the IEEE 802 LAN/MAN Standards Committee, the 802.20 group specifies a *Mobile Broadband Wireless Access* (MBWA) standard. The group is targeting a system optimized for IP-data transport, operating in licensed bands below 3.5 GHz with peak data rates per user in excess of 1 Mbps. High mobility with mobile speeds up to 250 km/h and spectral efficiency on par with mobile systems is part of the MBWA system's scope, making it potentially more than a MAN system.

The work on the 802.20 standard was initiated in 2002, and has since then moved to become a high data rate, flexible bandwidth system that is quite similar to the other technologies discussed in this book, such as LTE, CDMA2000 Rev C, and IEEE 802.16. The IEEE 802.20 standard was approved in June 2008. The standard consists of a *wideband* mode and a *625 kHz-spaced Multi-Carrier* (625k-MC) mode. The two modes have separate PHY and MAC defined that are not compatible.

The Wideband mode was originally proposed in 2005 [3] based to a large extent on the same concept and solutions used for UMB in 3GPP2 and there are for this reason many similarities. Some key features of the Wideband mode are listed below.

- The radio interface has OFDM data transmission with 9.6 kHz subcarrier spacing and an FFT size of 512, 1024, and 2048, supporting operation bandwidths of 5, 10, and 20 MHz, respectively. The cyclic prefix is configurable from 6% to 23% of the OFDM symbol duration. Most of the uplink control channels are transmitted with CDMA on a contiguous set of OFDM subcarriers.
- Hopping of subcarriers is possible at symbol level, occurring every two symbol intervals (to allow for Alamouti coding), or at a block level. Blocks are about the same size as a resource block in LTE.
- There is one frame structure for FDD and one for TDD where guard times are added between up-and downlink parts.
- There is support of QPSK, 8PSK, 16QAM, and 64QAM modulations.
- Convolutional coding is used for small packets (mainly signaling) and Turbo codes for larger packets. Hybrid ARQ is used with a modulation step down for retransmissions.
- Fast closed-loop uplink power control is used for control channel power. Traffic channel power is set relative to control channels. Uplink interference from neighboring *Access Points* (AP) is controlled through an *Other Sector Indication Channel* (F-OSICH) that signals a load indication to the mobiles.
- Handovers are mobile station initiated. The mobile station makes SINR measurements on candidate AP pilots and keeps an active set of up to 8 APs. All have allocated MAC IDs and control resources, but only one is the serving AP. Handover can be disjoint (independent) between uplink and downlink.
- The uplink can be quasi-orthogonal where multiple mobiles are assigned the same bandwidth resources.
- A fractional frequency reuse scheme enables mobiles in different channel conditions to have different frequency reuse.
- *Space Time Transmit Diversity* (STTD) is supported.
- There is support for downlink MIMO with single codeword and multiple codeword designs.
- Eigen-beam-forming is supported using feedback from the mobile station. A special beam-forming mode is supported for TDD operation.
- There is possibility of embedding other physical layers such as single frequency network technologies for broadcast services.
- Scalable bandwidth options are supported so that mobiles capable of only receiving on a lower carrier bandwidth (say 5 MHz) can still be operated on a 20 MHz carrier being transmitted by the base station.

The 625 k-MC mode is based on an ATIS-standard called HC-SDMA (*High Capacity-Space Division Multiple Access*) or *iBurst*. It supports TDD only and has special features for smart antenna solutions (SDMA).

24.7 Summary

The IMT-2000 technologies and the other technologies introduced above are developed in different standardization bodies, but all show a lot of commonalities. The reason is that they target the same type of application and operate under similar conditions and scenarios. The fundamental constraints for achieving high data rates, good quality of service, and system performance will require that a set of efficient tools are applied to reach those targets.

The technologies described in Part II of this book form a set of such tools and it turns out that many of those are applied across most of the technologies discussed here. To cater for high data rates, different ways to transmit over wider bandwidth is employed, such as single-and multi-carrier transmission including OFDM, often with the addition of higher-order modulation. Multi-antenna techniques are employed to exploit the 'spatial domain,' using techniques such as receive and transmit diversity, beam-forming and multi-layer transmission. Most of the schemes also employ dynamic link adaptation and scheduling to exploit variations in channel conditions. In addition, coding schemes such as Turbo codes are combined with advanced retransmission schemes such as hybrid ARQ.

As mentioned above, one reason that solutions become similar between systems is that they target similar problems for the systems. It is also to some extent true that some technologies and their corresponding acronyms go in and out of 'fashion.' Most 2G systems were developed using TDMA, while many 3G systems are based on CDMA and the 3G evolution steps taken now are based on OFDM. Another reason for this step-wise shift of technologies is of course that as technology develops, more complex implementations are made possible. A closer look at many of the evolved wireless communication systems of today also show that they often combine multiple techniques from previous steps, and are built on a mix of TDMA, OFDM, and spread spectrum components.

25

Future evolution

This book has described the evolution of 3G mobile systems from WCDMA to HSPA and its continued evolution, and finally the 3G Long-Term Evolution. The overview of the technologies used for 3G evolution in Part II of the book served as a foundation for the detailed discussion and explanation of the evolution steps taken for both HSPA and LTE, and for the future evolution steps being planned. The enhancements introduced enable higher peak data rates, improved system performance, and other enhanced capabilities. The resulting performance was also presented in Chapter 23 and it was shown that the targets set up by 3GPP for LTE are met.

It was also demonstrated that many of the enhancing technologies applied are the same for HSPA and LTE and also give the same types of improvements. They are also similar to the enhancements applied for other wireless technologies as shown in the overview made in Chapter 24.

Naturally, the technology evolution does not stop with HSPA Evolution and LTE. The same driving forces to further enhance performance and capabilities are still there, albeit the targets are always moved further ahead when state-of-the-art improves. The next step of wireless evolution is sometimes called 4G. But since the evolution is not stepwise, but more of a continuous process, also being a set of parallel evolution processes for similar and often related systems, it may be difficult to identify specific technology steps as being a new next generation. Some may want to put a 4G label on technologies that other consider not meeting even 3G requirements, while other try to label intermediate steps as 3.5G or 3.9G.

The technology race for the next generation of mobile communication has already started with regulatory bodies, standards organizations, market forums, research bodies and other bodies taking various initiatives. These bodies all work on concepts that are candidates for the forthcoming process in ITU-R defining what IMT-Advanced should be.

25.1 IMT-Advanced

Within the ITU, Working Party 5D works on IMT-2000 and systems beyond IMT-2000. The capabilities of IMT-2000, its enhancements, and systems that include new radio interfaces beyond IMT-2000 are shown in Figure 25.1. *ITU-R* anticipated in [47] that there will be a need for new mobile radio-access technologies for capabilities beyond enhanced IMT-2000, but the exact point where this may be needed was not identified. The term IMT-Advanced is used for systems that include new radio interfaces supporting the new capabilities of systems beyond IMT-2000. Note that the 3G evolution into enhanced IMT-2000 as described in this book covers a large part of the step toward IMT-Advanced in Figure 25.1 and that IMT-Advanced will also encompass the capabilities of previous systems.

The process for defining IMT-Advanced is also defined by WP5D [43] and will be quite similar to the process used in developing the IMT-2000 recommendations. It is based on a set of minimum technical requirements and evaluation criteria. All ITU members and other organizations are invited to the process through a circular letter [139] and proposed technologies will be evaluated according to the agreed criteria. The target is harmonization through consensus

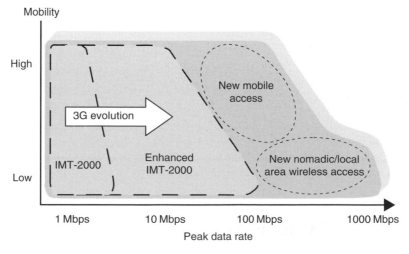

Figure 25.1 *Illustration of capabilities of IMT-2000 and systems beyond IMT-2000, based on the framework described in ITU -R Recommendation M.1645 [47].*

building, resulting in a recommendation for IMT-Advanced similar to the ITU-R Recommendation M.1457 for IMT-2000 [46]. The evaluation will be done in cooperation with external bodies such as standards-developing organizations. Since the process will be based on consensus, the number of technologies that will finally be encompassed by IMT-Advanced cannot be determined in advance. It is a trade-off between economies of scale, support of different user environments, and the capabilities of different technologies. In addition, the possibility for global circulation of terminals will be an important aspect.

Another major activity within ITU-R concerning IMT-Advanced has been to identify spectrum that is suitable for IMT-Advanced and that is available globally. Adequate spectrum availability and globally harmonized spectrum are identified as essential for IMT-Advanced, and a number of new frequency bands were also identified for IMT systems at WRC'07 as discussed in Chapters 1 and 20. This 'new' spectrum will in the near future gradually become available on a regional basis for both IMT-2000 and IMT-Advanced.

25.2 The research community

In the research community, several research projects are run in the area of IMT-Advanced and the next generation of radio access. One example is the Winner project, which was partly funded by the European Union. The Winner concept has many components that are very close to LTE. However, Winner is targeting higher data rates than LTE and is therefore designed for a wider bandwidth than 20 MHz. Another key difference is that the Winner concept will work with relaying and multi-hop modes. For further details, see the Winner homepage [118].

Other regions are running research projects similar to the European ones, such as the Future project in China, all with the goal of making an IMT-Advanced radio interface proposal. In the end however, these research communities will not make the final inputs of IMT-Advanced concepts. It is expected that the established standards developing organizations (ETSI, ARIB, CWTS, etc.) will do down selections of proposals from their respective regions. A global standard body such as 3GPP will most likely also have a role to play in harmonizing proposals across the different regions and standards bodies.

25.3 Standardization bodies

3GPP has initiated activity on an LTE-Advanced concept as discussed in Chapter 21, with the target of providing a candidate proposal to ITU-R.

IEEE802.16 is also targeting a proposal for IMT-Advanced in its work on 802.16m. Discussions are also ongoing in the community around CDMA2000, where 3GPP2 has formed a New Technology Ad Hoc Group that is planning for an evolved UMB standard.

25.4   Concluding remarks

Even before the standardization for LTE was completed, the race for the next radio interface was already ongoing. Many activities are now initiated in the research community and some standards developing organizations have just started their work toward IMT-Advanced. However, IMT-Advanced is still several years away whereas deployment of HSPA Evolution and LTE are just around the corner. HSPA Evolution is built as a continuing evolution of the existing WCDMA/HSPA technology whereas LTE is a new radio access optimized purely for IP based traffic. These two technologies promise to give more services, capabilities, and performance to the end users than any other radio interface technology has been able to do to date.

References

[1] China Unicom et al., 'Joint Proposal for 3GPP2 Physical Layer for FDD Spectra', 3GPP2 TSG-C WG3 Contribution C30-20060731-040R2, July 2006.

[2] IEEE, 'IEEE Standard for Local and Metropolitan Area Networks. Part 16: Air Interface for Fixed and Mobile Broadband Wireless Access Systems; Amendment 2: Physical and Medium Access Control Layers for Combined Fixed and Mobile Operation in Licensed Bands', *IEEE Std 802.16-2005*.

[3] IEEE, *MBFDD and MBTDD Wideband Mode: Technology Overview*, IEEE C802.20-05/68r1.

[4] H. Holma and A. Toskala, *WCDMA for UMTS: Radio Access for Third Generation Mobile Communications*, John Wiley & Sons, Chichester, UK, 2000.

[5] B. Hagerman et al., 'WCDMA Uplink Interference Cancellation Performance – Field Measurements and System Simulations', *Nordic Radio Symposium NRS'04*, Oulo, Finland, August 16–18, 2004.

[6] A. Huebner, F. Schuehlein, M. Bossert, E. Costa and H. Haas, 'A Simple Space-frequency Coding Scheme with Cyclic Delay Diversity for OFDM', *Proceedings of the 5th European Personal Mobile Communications Conference*, Glasgow, Scotland, April 2000, pp. 106–110.

[7] A. Milewski, 'Periodic Sequences with Optimal Properties for Channel Estimation and Fast Start-up Equalization', *IBM Journal of Research and Development*, Vol. 27, No. 5, September 1983, pp. 426–431.

[8] M.W. Thelander, 'The 3G Evolution: Taking CDMA2000 into the Next Decade', Signal Research Group, LLC, White Paper developed for the CDMA Development Group, October 2005.

[9] D. Chase, 'Code Combining – A Maximum-likelihood Decoding Approach for Combining and Arbitrary Number of Noisy Packets', *IEEE Transactions on Communications*, Vol. 33, May 1985, pp. 385–393.

[10] J.-F. Cheng, 'Coding Performance of Hybrid ARQ Schemes', *IEEE Transactions on Communications*, Vol. 54, June 2006, pp. 1017–1029.

[11] S.T. Chung and A.J. Goldsmith, 'Degrees of Freedom in Adaptive Modulation: A Unified View', *IEEE Transactions on Communications*, Vol. 49, No. 9, September 2001, pp. 1561–1571.

[12] N. Bhushan et al., 'CDMA2000 1xEV-DO Revision A: A Physical Layer and MAC Layer Overview', *IEEE Communications Magazine*, February 2006, pp. 75–87.

[13] R. Attar et al., 'Evolution of CDMA2000 Cellular Networks: Multicarrier EV-DO', *IEEE Communications Magazine*, March 2006, pp. 46–53.

[14] E. Dahlman, B. Gudmundsson, M. Nilsson and J. Sköld, 'UMTS/IMT-2000 Based on Wideband CDMA', *IEEE Communication Magazine*, 1998, pp. 70–80.

[15] 'Views on OFDM Parameter Set for Evolved UTRA Downlink', NTT DoCoMo et al., Tdoc R1-050386, *3GPP TSG-RAN WG 1*, Athens, Greece, May 9–13, 2005.

[16] J. Pautler, M. Ahmed and K. Rohani, 'On Application of Multiple-input Multiple-output Antennas to CDMA Cellular Systems', *IEEE Vehicular Technology Conference*, Atlantic City, NJ, USA, October 7–11, 2001.

[17] 'Digital Video Broadcasting (DVB): Framing Structure, Channel Coding and Modulation for Digital Terrestrial Television', ETSI, ETS EN 300 744 v 1.1.2.

[18] H. Ekström, A. Furuskär, J. Karlsson, M. Meyer, S. Parkvall, J. Torsner and M. Wahlqvist, 'Technical Solutions for the 3G Long-term Evolution', *IEEE Communications Magazine*, March 2006, pp. 38–45.

[19] A. Furuskär, J. Näslund and H. Olofsson, 'Edge – Enhanced Data Rates for GSM and TDMA/136 Evolution', *Ericsson Review*, No. 01, 1999, pp. 28–37. Telefonaktiebolaget LM Ericsson, Stockholm, Sweden.

[20] H. Axelsson, P. Björkén, P. de Bruin, S. Eriksson and H. Persson, 'GSM/EDGE Continued Evolution', *Ericsson Review*, No. 01, 2006, pp. 20–29. Telefonaktiebolaget LM Ericsson, Stockholm, Sweden.

[21] S. Parkvall, E. Englund, P. Malm, T. Hedberg, M. Persson and J. Peisa, 'WCDMA evolved-high-speed packet-data services', *Ericsson Review*, No. 02, 2003, pp. 56–65. Telefonaktiebolaget LM Ericsson, Stockholm, Sweden.

[22] J. Sköld, M. Lundevall, S. Parkvall and M. Sundelin, 'Broadband Data Performance of Third-generation Mobile Systems', *Ericsson Review*, No. 01, 2005 Telefonaktiebolaget LM Ericsson, Stockholm, Sweden.

[23] P.V. Etvalt, 'Peak to Average Power Reduction for OFDM Schemes by Selective Scrambling', *Electronics Letters*, Vol. 32, No. 21, October 1996, pp. 1963–1964.

[24] D. Falconer et al., 'Frequency Domain Equalization for Single-carrier Broadband Wireless Systems', *IEEE Communications Magazine*, Vol. 40, No. 4, April 2002, pp. 58–66.

[25] G. Forney, 'Maximum Likelihood Sequence Estimation of Digital Sequences in the Presence of Intersymbol Interference', *IEEE Transactions on Information Theory*, Vol. IT-18, May 1972, pp. 363–378.

[26] G. Forney, 'The Viterbi Algorithm', *Proceedings of the IEEE*, Piscataway, NJ, USA, Vol. 61, No. 3, March 1973, pp. 268–278.

[27] P. Frenger, S. Parkvall and E. Dahlman, 'Performance Comparison of HARQ with Chase Combining and Incremental Redundancy for HSDPA',

Proceedings of the IEEE Vehicular Technology Conference, Atlantic City, NJ, USA, October 2001, pp. 1829–1833.

[28] A.J. Goldsmith and P. Varaiya, 'Capacity of Fading Channels with Channel Side Information', *IEEE Transactions on Information Theory*, Vol. 43, November 1997, pp. 1986–1992.

[29] G. Bottomley, T. Ottosson and Y.-P. Eric Wang, 'A Generalized RAKE Receiver for Interference Suppression', *IEEE Journal on Selected Areas in Communications*, Vol. 18, No. 8, August 2000, pp. 1536–1545.

[30] J. Hagenauer, 'Rate-compatible Punctured Convolutional Codes (RCPC Codes) and Their Applications', *IEEE Transactions on Communications*, Vol. 36, April 1988, pp. 389–400.

[31] L. Hanzo, M. Munster, B.J. Choi and T. Keller, '*OFDM and MC-CDMA for Broadband Multi-user Communications, WLANs, and Broadcasting*', Wiley-IEEE Press, Chichester, UK, ISBN 0-48085879.

[32] S. Haykin, *Adaptive Filter Theory*, Prentice-Hall International, NJ, USA, 1986, ISBN 0-13-004052-5 025.

[33] J.M. Holtzman, 'CDMA Forward Link Waterfilling Power Control', *Proceedings of the IEEE Vehicular Technology Conference*, Tokyo, Japan, Vol. 3, May 2000, pp. 1663–1667.

[34] J.M. Holtzman, 'Asymptotic Analysis of Proportional Fair Algorithm', *Proceedings of the IEEE Conference on Personal Indoor and Mobile Radio Communications*, San Diego, CA, USA, Vol. 2, 2001, pp. 33–37.

[35] M.L. Honig and U. Madhow, 'Hybrid Intra-cell TDMA/inter-cell CDMA with Inter-cell Interference Suppression for Wireless Networks', *Proceedings of the IEEE Vehicular Technology Conference*, Secaucus, NJ, USA, 1993, pp. 309–312.

[36] R. Horn and C. Johnson, 'Matrix Analysis', *Proceedings of the 36th Asilomar Conference on Signals, Systems and Computers*, Pacific Grove, CA, USA, November 2002.

[37] A. Hottinen, O. Tirkkonen and R. Wichman, *Multi-antenna Transceiver Techniques for 3G and Beyond*, John Wiley & Sons, Chichester, UK, ISBN 0470 84542 2.

[38] J. Karlsson and J. Heinegard, 'Interference Rejection Combining for GSM', *Proceedings of the 5th IEEE International Conference on Universal Personal Communications*, Cambridge, MA, USA, 1996, pp. 433–437.

[39] K.J. KIM, 'Channel Estimation and Data Detection Algorithms for MIMO–OFDM Systems', *Proceedings of the 36th Asilomar Conference on Signals, Systems and Computers*, Pacific Grove, CA, USA, November 2002.

[40] R. Knopp and P.A. Humblet, 'Information Capacity and Power Control in Single-cell Multi-user Communications', *Proceedings of the IEEE International Conference on Communications*, Seattle, WA, USA, Vol. 1, 1995, pp. 331–335.

[41] W. Lee, *Mobile Communications Engineering*, McGraw-Hill, New York, NY, USA, ISBN 0-07-037039-7.

[42] S. Lin and D. Costello, *Error Control Coding*, Prentice-Hall, Upper Saddle River, NJ, USA.

[43] 'Principles for the process of development of IMT-Advanced', ITU-R, Resolution ITU-R 57, October 2007.

[44] 'Frequency Arrangements for Implementation of the Terrestrial Component of International Mobile Telecommunications-2000 (IMT-2000) in the Bands 806–960 MHz, 1710–2025 MHz, 2110–2200 MHz and 2500–2690 MHz', ITU-R, Recommendation ITU-R M.1036-3, July 2007.

[45] 'Guidelines for Evaluation of Radio Transmission Technologies for IMT-2000', ITU-R, Recommendation ITU-R M.1225, February 1997.

[46] 'Detailed Specifications of the Radio Interfaces of International Mobile Telecommunications-2000 (IMT-2000)', ITU-R, Recommendation ITU-R M. 1457-7, October 2007.

[47] 'Framework and Overall Objectives of the Future Development of IMT-2000 and Systems Beyond IMT-2000', ITU-R, Recommendation ITU-R M.1645, June 2003.

[48] 'International Mobile Telecommunications-2000 (IMT-2000)', ITU-R, Recommendation ITU-R M.687-2, February 1997.

[49] 'Comparison of PAR and Cubic Metric for Power De-rating', Motorola, Tdoc R1-040642, 3GPP TSG-RAN WG1, May 2004.

[50] J.G. Proakis, *Digital Communications*, McGraw-Hill, New York, 2001.

[51] S.-J. Oh and K.M. Wasserman, 'Optimality of Greedy Power Control and Variable Spreading Gain in Multi-class CDMA Mobile Networks', *Proceedings of the AMC/IEEE MobiComp,* Seattle, Washington, USA, 1999, pp. 102–112.

[52] A. Oppenheim and R.W. Schafer, *Digital Signal Processing*, Prentice-Hall International, ISBN 0-13-214107-8 01.

[53] S. Grant et al., 'Per-Antenna-Rate-Control (PARC) in Frequency Selective Fading with SIC-GRAKE Receiver', 60th IEEE *Vehicular Technology Conference*, Los Angeles, CA, USA, September 2004, Vol. 2, pp. 1458–1462.

[54] G.F. Pedersen and J.B. Andersen, 'Handset Antennas for Mobile Communications: Integration, Diversity, and Performance', in R.W. Stone (ed.), *Review of Radio Science 1996–1999*, Wiley-IEEE Press, Chichester, UK, September 1999.

[55] J. Peisa, S. Wager, M. Sågfors, J. Torsner, B. Göransson, T. Fulghum, C. Cozzo and S. Grant, 'High Speed Packet Access Evolution – Concept and Technologies', *IEEE Vehicular Technology Conference*, Maryland, MD, USA, September 30–October 3, 2007.

[56] M.B. Pursley and S.D. Sandberg, 'Incremental-redundancy Transmission for Meteor-burst Communications', *IEEE Transactions on Communications*, Vol. 39, May 1991, pp. 689–702.

[57] 'E-UTRA Downlink User Throughput and Spectrum Efficiency', Ericsson, Tdoc R1-061381, 3GPP TSG-RAN WG1, Shanghai, China, May 8–12, 2006.

[58] 'E-UTRA Uplink User Throughput and Spectrum Efficiency', Ericsson, Tdoc R1-061382, 3GPP TSG-RAN WG1, Shanghai, China, May 8–12, 2006.

[59] 'E-UTRA Downlink User Throughput and Spectrum Efficiency', Ericsson, Tdoc R1-061685, 3GPP TSG-RAN WG1 LTE Ad Hoc, Cannes, France, June 27–30, 2006.

[60] 'Implementation of International Mobile Telecommunications in the bands 1 885-2 025 MHz and 2 110-2 200 MHz', ITU-R, Resolution 212 (Rev. WRC-07).

[61] 'Minutes of HSDPA Simulation Ad-hoc', Document R4-040770, 3GPP TSG-RAN WG4 meeting 33, Shin-Yokohama, Japan, November 2004.

[62] S. Ramakrishna and J.M. Holtzman, 'A Scheme for Throughput Maximization in a Dual-class CDMA System', *IEEE Journal on Selected Areas in Communications*, Vol. 16, No. 6, 1998, pp. 830–844.

[63] A. Shokrollahi, 'Raptor Codes', *IEEE Transactions on Information Theory*, Vol. 52, No. 6, 2006, pp. 2551–2567.

[64] 'Robust Header Compression (ROHC): Framework and Four Profiles: RTP, UDP, ESP, and Uncompressed', IETF, RFC 3095.

[65] 'Updated New WID on Higher Uplink Performance for GERAN Evolution (HUGE)', GERAN, Tdoc GP-061901, 3GPP TSG GERAN #31, Denver, USA, September 4–8, 2006.

[66] C. Schlegel, *Trellis and Turbo Coding*, Wiley – IEEE Press, Chichester, UK, March 2004.

[67] M. Schnell et al., 'IFDMA – A New Spread-spectrum Multiple-access Scheme', *Proceedings of the ICC'98*, Atlanta, GA, USA, June 1998, pp. 1267–1272.

[68] Lal C. Godara, 'Applications of Antenna Arrays to Mobile Communications, Part I: Beam-forming and Direction-of-Arrival Considerations', *Proceedings of the IEEE*, Vol. 85, No. 7, July 1997, pp. 1029–1030.

[69] Lal C. Godara, 'Applications of Antenna Arrays to Mobile Communications, Part II: Beam-forming and Direction-of- Arrival Considerations', *Proceedings of the IEEE*, Vol. 85, No. 7, July 1997, pp. 1031–1060.

[70] C.E. Shannon, 'A Mathematical Theory of Communication', *Bell System Technical Journal*, Vol. 27, July and October 1948, pp. 379–423, 623–656..

[71] M.K. Varanasi and T. Guess, 'Optimum Decision Feedback Multi-user Equalization with Successive Decoding Achieves the Total Capacity of the Gaussian Multiple-access Channel', *Proceedings of the Asilomar conference on Signals, Systems, and Computers*, Monterey, CA, November 1997.

[72] J. Sun and O.Y. Takeshita, 'Interleavers for Turbo Codes Using Permutation Polynomials Over Integer Rings', *IEEE Transactions on Information Theory*, Vol. 51, No. 1, January 2005, pp. 101–119.

[73] O.Y. Takeshita, 'On Maximum Contention-free Interleavers and Permutation Polynomials Over Integer Rings', *IEEE Transactions on Information Theory*, Vol. 52, No. 3, March 2006, pp. 1249–1253.

[74] V. Tarokh, N. Seshadri and A. Calderbank, 'Space–time Block Codes from Orthogonal Design', *IEEE Transactions on Information Theory*, Vol. 45, No. 5, July 1999, pp. 1456–1467.

[75] X. Zhaoji and B. Sébire, 'Impact of ACK/NACK Signalling Errors on High Speed Uplink Packet Access (HSUPA)', *IEEE International Conference on Communications*, Vol. 4, May 2005, pp. 2223–2227, 16–20.

[76] J. Padhye, V. Firoiu, D.F. Towsley and J.F. Kurose, 'Modelling TCP Reno Performance: A Simple Model and Its Empirical Validation', *ACM/IEEE Transactions on Networking*, Vol. 8, No. 2, 2000, pp. 133–145.

[77] C. Kambiz and L. Krasny, 'Capacity-achieving Transmitter and Receiver Pairs for Dispersive MISO Channels', *IEEE Transactions on Communications*, Vol. 42, April 1994, pp. 1431–1440.

[78] J. Tellado and J.M. Cioffi, 'PAR Reduction in Multi-carrier Transmission Systems', ANSI T1E1.4/97-367.

[79] '3rd Generation Partnership Project; Technical Specification Group Services and System Aspects; 3GPP System Architecture Evolution: Report on Technical Options and Conclusions (Release 7)', 3GPP, 3GPP TR 23.882.

[80] '3rd Generation Partnership Project; Technical Specification Group Radio Access Network; S-CCPCH Performance for MBMS (Release 6)', 3GPP, 3GPP TR 25.803.

[81] '3rd Generation Partnership Project; Technical Specification Group Radio Access Network; Physical Layer Aspects for Evolved Universal Terrestrial Radio Access (UTRA) (Release 7)', 3GPP, 3GPP TR 25.814.

[82] '3rd Generation Partnership Project; Technical Specification Group Radio Access Network; 3.84 Mcps TDD Enhanced Uplink; Physical Layer Aspects (Release 7)', 3GPP, 3GPP TR 25.826.

[83] '3rd Generation Partnership Project; Technical Specification Group Radio Access Network; Manifestations of Handover and SRNS Relocation (Release 4)', 3GPP, 3GPP TR 25.832.

[84] '3rd Generation Partnership Project; Technical Specification Group Radio Access Network; High Speed Downlink Packet Access: Physical Layer Aspects (Release 5)', 3GPP, 3GPP TR 25.858.

[85] '3rd Generation Partnership Project; Technical Specification Group Radio Access Network; Feasibility Study for Evolved Universal Terrestrial Radio Access (UTRA) and Universal Terrestrial Radio Access Network (UTRAN) (Release 7)', 3GPP, 3GPP TR 25.912.

[86] '3rd Generation Partnership Project; Technical Specification Group Radio Access Network; Requirements for Evolved UTRA (E-UTRA) and Evolved UTRAN (E-UTRAN) (Release 7)', 3GPP, 3GPP TR 25.913.

[87] '3rd Generation Partnership Project; Technical Specification Group GSM/EDGE Radio Access Network; Feasibility Study for Evolved GSM /EDGE Radio Access Network (GERAN) (Release 7)', 3GPP, 3GPP TR 45.912.

[88] '3rd Generation Partnership Project; Technical Specification Group Services and System Aspects; Service Requirements for Evolution of the 3GPP System (Release 8)', 3GPP, 3GPP TS 22.278.

[89] '3rd Generation Partnership Project; Technical Specification Group Services and System Aspects; Network Architecture', 3GPP, 3GPP TS 23.002.

[90] '3rd Generation Partnership Project; Technical Specification Group Services and System Aspects; General Packet Radio Service (GPRS); Service Description; Stage 2', 3GPP, 3GPP TS 23.060.

[91] '3rd Generation Partnership Project; Technical Specification Group Services and System Aspects; 3GPP System Architecture Evolution: GPRS Enhancements for LTE Access; Release 8', 3GPP, 3GPP TS 23.401.

[92] '3rd Generation Partnership Project; Technical Specification Group Radio Access Network; User Equipment (UE) Radio Transmission and Reception (FDD)', 3GPP, 3GPP TS 25.101.

[93] '3rd Generation Partnership Project; Technical Specification Group Radio Access Network; User Equipment (UE) Radio Transmission and Reception (TDD)', 3GPP, 3GPP TS 25.102.

[94] '3rd Generation Partnership Project; Technical Specification Group Radio Access Network; Physical Channels and Mapping of Transport Channels onto Physical Channels (FDD)', 3GPP, 3GP TS 25.211.

[95] '3rd Generation Partnership Project; Technical Specification Group Radio Access Network; Multiplexing and Channel Coding (FDD)', 3GPP, 3GP TS 25.212.

[96] '3rd Generation Partnership Project; Technical Specification Group Radio Access Network; Spreading and Modulation (FDD)', 3GPP, 3GP TS 25.213.

[97] '3rd Generation Partnership Project; Technical Specification Group Radio Access Network; Physical Layer Procedures (FDD)', 3GPP, 3GP TS 25.214.

[98] '3rd Generation Partnership Project; Technical Specification Group Radio Access Network; Radio Interface Protocol Architecture', 3GPP, 3GPP TS 25.301.

[99] '3rd Generation Partnership Project; Technical Specification Group Radio Access Network; UE Radio Access Capabilities', 3GPP, 3GPP TS 25.306.

[100] '3rd Generation Partnership Project; Technical Specification Group Radio Access Network; High Speed Downlink Packet Access (HSDPA); Overall Description; Stage 2', 3GPP, 3GPP TS 25.308.

[101] '3rd Generation Partnership Project; Technical Specification Group Radio Access Network; FDD Enhanced Uplink; Overall Description; Stage 2', 3GPP, 3GPP TS 25.309.

[102] '3rd Generation Partnership Project; Technical Specification Group Radio Access Network; Introduction of the Multimedia Broadcast Multicast Service (MBMS) in the Radio Access Network (RAN); Stage 2 (Release 6)', 3GPP, 3GPP TS 25.346.

[103] '3rd Generation Partnership Project; Technical Specification Group Radio Access Network; UTRAN Overall Description', 3GPP, 3GPP TS 25.401.

[104] '3rd Generation Partnership Project; Technical Specification Group Radio Access Network; UTRAN Iu Interface RANAP Signalling (Release 7)', 3GPP, 3GPP TS 25.413.

[105] '3rd Generation Partnership Project; Technical Specification Group Services and System Aspects; Multimedia Broadcast/multicast Service (MBMS); Protocols and Codecs', 3GPP, 3GPP TS 26.346.

[106] '3rd Generation Partnership Project; Technical Specification Group Radio Access Network; Physical Channels and Modulation (Release 8)', 3GPP, 3GPP TS 36.211.

[107] '3rd Generation Partnership Project; Technical Specification Group Radio Access Network; Multiplexing and Channel Coding (Release 8)', 3GPP, 3GPP TS 36.212.

[108] '3rd Generation Partnership Project; Technical Specification Group Radio Access Network; Physical Layer Procedures (Release 8)', 3GPP, 3GPP TS 36.213.

[109] '3rd Generation Partnership Project; Technical Specification Group Radio Access Network; Physical Layer – Measurements (Release 8)', 3GPP, 3GPP TS 36.214.

[110] '3rd Generation Partnership Project; Technical Specification Group Radio Access Network; Evolved Universal Terrestrial Radio Access (E-UTRA) and Evolved Universal Terrestrial Radio Access Network (E-UTRAN); Overall Description; Stage 2 (Release 8)', 3GPP, 3GPP TS 36.300.

[111] D. Tse, 'Optimal Power Allocation Over Parallel Gaussian Broadcast Channels', *Proceedings of the International Symposium on Information Theory*, Ulm, Germany, June 1997, p. 7.

[112] Y.-P. Eric Wang, A.S. Khayrallah and G.E. Bottomley, 'Dual Branch Receivers for Enhanced Voice and Data Communications in WCDMA', *Proceedings of the Globecom*, San Francisco, CA, USA, November 2006.

[113] S. Verdu, *Multiuser Detection*, Cambridge University Press, New York, 1998.

[114] S.B. Wicker and M. Bartz, 'Type-I Hybrid ARQ Protocols Using Punctured MDS Codes', *IEEE Transactions on Communications*, Vol. 42, April 1994, pp. 1431–1440.

[115] 'WiMAX: From Fixed Wireless Access to Internet in the Pocket – Technology White Paper', Alcatel, *Alcatel Telecommunications Review*, 2nd Quarter, 2005.

[116] 'Mobile WiMAX – Part I: A Technical Overview and Performance Evaluation', WiMAX Forum, White Paper, August 2006.

[117] R. van Nee and R. Prasad, *OFDM for Wireless Multimedia Communications*, Artech House Publishers, London, January 2000.

[118] '*Wireless World Initiative New Radio*', Eurescom GmbH 2006. https://www.ist-winner.org.

[119] P. Viswanath, D. Tse and R. Laroia, 'Opportunistic Beamforming Using Dumb Antennas', *IEEE Transactions on Information Theory*, Vol. 48, No. 6, 2002, pp. 1277–1294.

[120] J.M. Wozencraft and M. Horstein, 'Digitalised Communication Over Two-way Channels', *Fourth London Symposium on Information Theory*, London, UK, September 1960.

[121] E. Dahlman, P. Beming, J. Knutsson, F. Ovesjö, M. Persson and C. Roobol, 'WCDMA – The Radio Interface for Future Mobile Multimedia Communications', *IEEE Transactions on Vehicular Technology*, Vol. 47, November 1998, pp. 1105–1118.

[122] E. Dahlman, H. Ekström, A. Furuskär, Y. Jading, J. Karlsson, M. Lundevall and S. Parkvall, 'The 3G Long-term Evolution – Radio Interface Concepts and Performance Evaluation', *63rd Vehicular Technology Conference*, VTC 2006-Spring, Vol. 1, pp. 137–141, IEEE, 2006.

[123] D.C. Chu, 'Polyphase Codes with Good Periodic Correlation Properties', *IEEE Transactions on Information Theory*, Vol. 18, No. 4, July 1972, pp. 531–532.

[124] W. Zirwas, *Single Frequency Network Concepts for Cellular OFDM Radio Systems*, International OFDM Workshop, Hamburg, Germany, September 2000.

[125] 'Regional 3500 MHz band arrangements and use', Ericsson, Document RP-080133, 3GPP TSG-RAN meeting #39, Puerto Vallarta, Mexico, March 2008.

[126] '3rd Generation Partnership Project; Technical Specification Group Radio Access Network; Evolved Universal Terrestrial Radio Access (E-UTRA); User Equipment (UE) radio transmission and reception', 3GPP, 3GPP TS 36.101.

[127] '3rd Generation Partnership Project; Technical Specification Group Radio Access Network; Evolved Universal Terrestrial Radio Access (E-UTRA); Base Station (BS) radio transmission and reception', 3GPP, 3GPP TS 36.104.

[128] '3rd Generation Partnership Project; Technical Specification Group Radio Access Network; Evolved Universal Terrestrial Radio Access (E-UTRA); Base Station (BS) conformance testing', 3GPP, 3GPP TS 36.141.

[129] '3rd Generation Partnership Project; Technical Specification Group Radio Access Network; Evolved Universal Terrestrial Radio Access (E-UTRA); User Equipment (UE) radio access capabilities', 3GPP, 3GPP TS 36.306.

[130] '3rd Generation Partnership Project; Technical Specification Group Radio Access Network; Evolved Universal Terrestrial Radio Access (E-UTRA); User Equipment (UE) radio transmission and reception', 3GPP, 3GPP TR 36.803.

[131] '3rd Generation Partnership Project; Technical Specification Group Radio Access Network; Evolved Universal Terrestrial Radio Access (E-UTRA) and Evolved Universal Terrestrial Radio Access Network (E-UTRAN); User Equipment (UE) conformance specification; Radio transmission and reception (Part 1, 2 and 3)', 3GPP, 3GPP TS 36.521.

[132] '3rd Generation Partnership Project; Technical Specification Group Radio Access Network; Evolved Universal Terrestrial Radio Access (E-UTRA); Base Station (BS) radio transmission and reception', 3GPP, 3GPP TR 36.804.

[133] '3rd Generation Partnership Project; Technical Specification Group Radio Access Network; Evolved Universal Terrestrial Radio Access (E-UTRA); Radio Frequency (RF) system scenarios', 3GPP, 3GPP TR 36.942.

[134] 'Unwanted emissions in the spurious domain', ITU-R, Recommendation ITU-R SM.329-10, February, 2003.

[135] '3rd Generation Partnership Project; Technical Specification Group Radio Access Network; Radio Frequency (RF) system scenarios', 3GPP, 3GPP TR 25.942.

[136] '3rd Generation Partnership Project; Technical Specification Group Radio Access Network; Feasibility study on interference cancellation for UTRA FDD User Equipment (UE) (Release 7)', 3GPP, 3GPP TR 25.963.

[137] 'Introduce TD-SCDMA industry standard in CCSA to 3GPP', Alcatel Shanghai Bell et.al., Document R4-071394, 3GPP TSG-RAN WG4 meeting #44, Athens, Greece, August 2007.

[138] 'Technical Feasibility of Harmonising a Sub-Band of Bands IV and V for Fixed/Mobile Applications (Including Uplinks), Minimising the Impact on GE06', ECC TG4, Report B to ECC, November 2007.

[139] 'Invitation for submission of proposals for candidate radio interface technologies for the terrestrial components of the radio interface(s) for IMT-Advanced and invitation to participate in their subsequent evaluation', ITU-R SG5, Circular Letter 5/LCCE/2, March 2008.

[140] 'IEEE 802.16m System Requirements', Motorola, IEEE 802.16m-07/002r4, IEEE 802.16 Task Group m, October 2007.

[141] 'The Draft IEEE 802.16m System Description Document', IEEE 802.16m-08/003r1, IEEE 802.16 Task Group m, April 2008.

[142] 'Ultra Mobile Broadband Technology Overview and Competitive Advantages', Qalcomm Incorporated, January 2008.

[143] 'Title 47 of the Code of Federal Regulations (CFR)', Federal Communications Commission.

Index